HADRON PHYSICS 2002

Topics on the Structure and Interaction of Hadronic Systems

Proceedings of the VIII International Workshop on

HADRON PHYSICS 2002

Topics on the Structure and Interaction of Hadronic Systems

Bento Gonçalves, Rio Grande do Sul, Brazil
14–19 April 2002

Editors

César A. Z. Vasconcellos
Universidade Federal do Rio Grande do Sul

Victoria E. Herscovitz
Universidade Federal do Rio Grande do Sul

Dimiter Hadjimichef
Universidade Federal de Pelotas

Bardo E. J. Bodmann
Universidade do Vale do Rio dos Sinos
Universidade Federal de Pelotas

World Scientific
New Jersey • London • Singapore • Hong Kong

Published by

World Scientific Publishing Co. Pte. Ltd.

5 Toh Tuck Link, Singapore 596224

USA office: Suite 202, 1060 Main Street, River Edge, NJ 07661

UK office: 57 Shelton Street, Covent Garden, London WC2H 9HE

British Library Cataloguing-in-Publication Data
A catalogue record for this book is available from the British Library.

Cover: The Editors gladly acknowledge the celebrated artist Mr. Federico Reilly, for kindly authorize the reproduction of one of his gaucho paintings on the cover of this Proceedings.

STRUCTURE AND INTERACTION OF HADRONIC SYSTEMS
Proceedings of the VIII International Workshop on Hadron Physics 2002

ISBN 981-238-192-9

Printed by FuIsland Offset Printing (S) Pte Ltd, Singapore

PREFACE

Hadron Physics is a biennial International Workshop held in Brazil. Hadron Physics 2002 has been held at Bento Gonçalves, a pleasant town in the wine producing region of the southernmost part of Brazil, the state of Rio Grande do Sul.

The activities of the Workshop included four sets of lectures, several talks and the presentation of numerous posters contributions, an expressive part of this material being published in these Proceedings. Invited Lecturers were Frieder Lenz, Fridolin Weber, Horst Stöcker and Peter Lepage. Talks were presented by Theodor August Johannes Maris, Attilio Cucchieri, Carlos Alberto Garcia Canal, Helio Takai, Ignacio Bediaga, Laurent Favart, Lev Leinson, Manuel Fiolhais and Marzia Rosati. The Organizing Committee is thankful to the invited speakers for the excellent presentations, responsible for the success of the event. The stimulating atmosphere of the Workshop favored discussions and interaction among the participants. Concerning the set of lectures of Professor Peter Lepage and the conferences by Drs. Marzia Rosati and Helio Takai, in the impossibility of offering the related papers, one has opted to include a reproduction of the slides presented during the Meeting.

The Organization of the 2002 Workshop has prepared two Homage events. The first one was held during the Opening Session of the Meeting, in honor of Professor Theodor August Johannes Maris, who had introduced the research at the *Instituto de Física da Universidade Federal do Rio Grande do Sul* in the earlies 60's and precisely in the field of Hadron Physics. The second Homage was a tribute to the memory of Dr. Juan Alberto Mignaco, an enthusiastic member of the Hadron Physics community and also of the Organizing Committee of this Workshop, that unfortunately has prematurely passed away. An award with his name was attributed to the poster contributions of three students selected by the invited speakers, namely Daniel Tavares da Silva, Guilherme Frederico Marranghello and Karen Silva Torres.

The organization of the Workshop was possible thanks to the financial support of the agencies *Coordenação de Aperfeiçoamento de Pessoal de Nível Superior (CAPES)*, *Centro Latino-Americano de Física (CLAF)*, *Conselho Nacional de Desenvolvimento Científico e Tecnológico (CNPq)*, *Fundação de Amparo à Pesquisa do Estado do Rio Grande do Sul (FAPERGS)*, *Fundação de Amparo à Pesquisa do Estado do Rio de Janeiro (FAPERJ)*, *Fundação de Amparo à Pesquisa do Estado de São Paulo (FAPESP)*. The Organizing Committee thanks also the support of the following institutions: *Centro Brasileiro de Pesquisas Físicas (CBPF)*, *Centro de Ciências Matemáticas da Natureza*

- *Universidade Federal do Rio de Janeiro (CCMN-UFRJ), Instituto de Física Teórica da Universidade Estadual Paulista (IFT-UNESP), Instituto de Física da Universidade Federal do Rio Grande do Sul (IFUFRGS), Observatório Nacional (ON), Universidade Federal do Rio Grande do Sul (UFRGS), Universidade do Estado de São Paulo (USP).*

We would also like to extend our thankfulness to the financial support of foreign countries that allowed in part the participation of Helio Takai, Laurent Favart, Lev Leinson, Manfred Dillig, Manuel Fiolhais and Marzia Rosati. Finally we wish to thank Ana Trotto, Arthur Boos Jr., Luisa Superina de Ferrero, Mauro Pedro Wyrzykowski and Waldomiro da Silva Olivo for their support during the organizing activities of the Workshop.

The Editors

CONTENTS

CONTRIBUTIONS

1. Lattice, Light-Front and Effective Theories

2. Hadron Models

3. High-Energy Collisions

4. Decays and Medium-Energy Reactions

5. Nuclear Matter, Astrophysics, Gravitation

HOMAGES

THE ROLE OF NUCLEAR PHYSICS

THEODOR AUGUST JOHANNES MARIS
Instituto de Física, Universidade Federal do Rio Grande do Sul
Porto Alegre, Brazil
maris@if.ufrgs.br

Address of Professor Theodor A. J. Maris at the Opening Homage Ceremony of the VIII International Workshop on Hadron Physics

Magnífica Reitora, Professora Wrana Maria Panizzi, demais autoridades, prezados colegas e estudantes!

Thank you for your kindness! I am not anymore accustomed to give long talks, but let me briefly explain to you how I came to this beautiful country.

First a small general remark: it is interesting to observe how attitudes, which a human being learned in his early youth, often become dominant for the rest of his life. If he learns to be obedient, he may forever depend on the opinion of the others. If one is taught to think mainly about oneself, one may remain always egotistic. If one is taught to like mainly important people, one may tend to become a politician.

My parents taught me to try to solve difficult problems, which are basically important, but not yet recognized as such. Therefore, as I started to study, I did not feel attracted to fields that most friends wanted to study, like Law or Medicine. And later, doing research in Physics, I did not enter the fields of research that were in fashion, but tried to find different subjects.

I worked at several University Research Institutes in Europe, the United States of America and Canada. During my stay at the Florida State University in Tallahassee, a colleague of mine participated at a conference in Mexico City and afterwards told me that two Brazilian physicists, Darcy Dillenburg and Gerhard Jacob, whom he had met there, asked him whether he could find somebody who would be helpful in building up a Physics Institute in Porto Alegre, Brazil. It would be the first Physics Institute in a region where millions of people lived.

That really seemed to be a non-trivial problem that was worthwhile! My dear wife and children, already used to change and travel, agreed and our family moved to this country. Here we were warmly received by Brazilian colleagues and students, in first place Gerhard Jacob and Darcy Dillenburg, who had called me, but also Paulo Petry and Victoria Herscovitz, among others.

With the support of the University, we managed to bring the theoretical and experimental research of our Institute on an international level and to introduce master and doctoral programs of a good quality.

It is not very easy to believe, how kind and helpful Brazilian physicists and their families have been to us! A decade ago I retired and I still have my office at our Institute. This, in spite of the fact that, since I have always been encouraged to communicate in English with my graduate students and my colleagues, my knowledge of Portuguese is still somewhat limited.

Thinking of the future, allow me to make a remark: it seems to me that basic Physics, and in particular theoretical Physics, should get a somewhat modified task in the human society. Physics has and will always have a direct value for human beings, like the art of music has in a different way. We want to understand the background of the world in which we live. Theoretical Physics is basically the art of finding mathematical models that connect experimental results and has a somewhat deeper value, besides the practical one.

But this understanding has now in most fields been accomplished to a considerable degree. And in practice it is getting almost as difficult to live from basic Physics, as to live from Music! And how expensive experimental investigations can be! This could become a especially serious problem for our students, who after all are the future of our science!

To save on the long run the possibilities to study our field, I believe that we should not only emphasize the philosophical side of our science. I see the main future of Physics less in producing or explaining new experimental results, than in finding ways to make the theoretical and experimental results that we have found as useful as possible for other sciences and for techniques of daily life. This is the kind of possibility the musicians hardly have!!

Once again, many thanks for your kindness and in particular to Cesar Vasconcellos and Victoria Herscovitz, the organizers of this Conference.

THEODOR AUGUST JOHANNES MARIS: THE MAN, THE SCHOLAR, THE PROFESSOR

DARCY DILLENBURG AND GERHARD JACOB
Instituto de Física, Universidade Federal do Rio Grande do Sul
Porto Alegre, Brazil
dillen@if.ufrgs.br

The Early Academic Years

Theodor August Johannes Maris was born in 1920. After going through the havoc of the Second World War, he started studying Physics in Munich immediately after its end. It is pertinent to recall that Theo, at his first enrollment, took as one of the subjects Ordinary Differential Equations (not yet knowing Differential and Integral Calculus!), reasoning that "ordinary" must be "simple". This may be viewed as a significant (although somewhat harsh) example of the environment of academic freedom in which students and professors pursued knowledge at those times. With the current re-shuffling of the German University system such freedom, for better or for worse, is getting lost for good!

His first area of research was Thermodynamics, in which he got his Ph. D. (*summa cum laude*) in 1954, under the supervision of F. Bopp, with a thesis on Entropy[1].

His interest being theoretical Nuclear Physics, he went to Sweden, first to the Nobel Institute in Stockholm and afterwards to the Gustaf Werner Institute in Uppsala, where a (at those times) high energy ($185 MeV$) brand new proton accelerator existed. After some theoretical work on inelastic "high energy" proton scattering, close to ongoing experimental work, Theo's first fundamental contribution to Nuclear Physics originated as early as 1957, namely the proposal to investigate the Nuclear Shell Model (specifically inner shells of light and possibly medium light nuclei) through the quasi-free proton-proton scattering: (p,2p) reactions. In these processes, a high energy incident proton knocks out a nuclear one; the two outgoing particles have their energy and angle measured, thus allowing the determination of both the separation energy and the momentum of the knocked-out nuclear proton. Altough the

[1] D. Dillenburg, G. Jacob, P. Kitching and C.A.Z. Vasconcellos, *in Current Topics in Nuclear Physics and Quantum Field Theory, Festschrift for Th. A. J. Maris*, Editora da Universidade (UFRGS), Porto Alegre, 1994, p12.

majority of the published papers are co-authored by experimental physicists (at that time there were theoretical as well as experimental papers on the subject), it is recognized by all that it was Theo's original idea to start these investigations of the nuclear shell model.

From Uppsala Maris moved to the Florida State University in Tallahassee, as a Research Associate. There he got involved in optical model calculations, especially those related to his previous work in inelastic and quasi-free scattering of protons.

At Florida State the construction of a (again at those times) large Tandem Van de Graaff accelerator was being started, with the prospect of large turmoil at the Physics Department, a fact that made Theo look for a more quiet place to work.

The Many Years in Porto Alegre

And in this way he landed in 1959 here in Porto Alegre, where he came for one year and is still around!

Maris' work at the Institute of Physics of the Federal University of Rio Grande do Sul started in theoretical Nuclear Physics, following closely what he had done in Uppsala. There was very little research being conducted at those times at the Institute in Porto Alegre, so that he had to start from the very beginning to build up an institute that deserved this name. And he did it with great competence. As early as 1960 the first paper was published in an international refereed journal. And very soon some characteristics of Theo's personality, as a man as well as a scholar, appeared very clearly. He had a deep understanding of Physics, rebuilding in his mind and in his own way everything he learned. And he managed to give away the knowledge thus acquired in a clear, understandable manner. He was extremely generous in what authorship was concerned, giving high value to any contribution of his students and collaborators; being in most cases the senior author of a paper, he insisted on using the alphabetical order in the publications. To give formal lectures was not his strong point, but he enjoyed teaching and discussing in small groups; any question, no matter how stupid, deserved serious attention. Working in a new and under-developed institution, he was very much concerned with some administrative matters which could impair research, but he definitely and categorically refused to do any sort of bureaucracy himself or to have any administrative power; that he left to his Brazilian colleagues.

Having himself worked close to experimentalists, Theo believed strongly that a good, well-rounded Institute had to do research in experimental Physics as well. So he insisted that experimental work should be initiated as soon as possible. Taking into account the fact that research in experimental Nuclear Physics had been going on in Brazil for two decades, and that in Solid State Physics work was just being initiated, it seemed like a good starting point to investigate the applications of experimental methods of Nuclear Physics to Solid State problems. The chosen area was that of Perturbed Angular Correlations, which studies the effects of the internal electromagnetic fields of matter on nuclear angular correlations. And, as his nature "forced" him to do, he also started himself to work theoretically on the same subject, of course always in collaboration with colleagues and on problems directly related to experiments planned or going on at the Institute. So, whereas the first paper published under the heading of the Institute was in quasi-free scattering (1960), already two years later a paper on perturbed angular correlations appeared.

But he had already since a long time a much more fundamental question in his mind. Could the mass of leptons possibly be of a purely electromagnetic origin? In other words, the electromagnetic interaction being switched off, electrons and muons would have no mass at all? In that case, in the absence of a mass (length) unit (dilatational invariance, a broken symmetry in Quantum Electrodynamics), the only magnitudes that make sense are mass ratios such as the electron-muon one. And in 1963 a first paper on the subject was published. Many other followed, and the problem is still an open one today.

Maris was certainly not willing to commit "scientific suicide" and to have his (former) students to do so. Therefore, taking advantage of his scientific prestige, he took visiting positions in first world scientific centres and arranged for other members of the Institute to have opportunities to stay abroad. In Porto Alegre, he went on developing work in his two main areas of research: quasi-free scattering and spontaneously broken symmetries in Quantum Electrodynamics. A proposal to investigate the nuclear structure using electrons instead of protons, $(e, e'p)$ reactions, was put forward, stimulating experimental work with high energy electrons in several centres abroad, and further work on Field Theories was published.

With a deep understanding for the needs of a developing country, he insisted that work in Basic Physics performed in the Institute should include areas more closely related to possible applications. Therefore he suggested and strongly supported the creation of a group in Laser Physics, which developed in experimental as well as in theoretical aspects, and together with which a larger set of problems in Solid State Physics were dealt with. Not totally happy, later on he conducted the creation of a group in theoretical Plasma Physics. Being very busy with his original projects mentioned above, he did not do, as he usually would, his own research in such fields. Instead, he extended his quasi-free reactions to the use of polarized protons as incoming particles, and to neutrons as outgoing as well as to pions as incoming particles.

From the start, and all the time, Theo insisted that more attention be given to applied subjects, so as to show that an Institute of Physics doing good research can qualify people needed for the development of technology and innovation in industries. The idea turned out to be much more difficult to implement. But in recent years contracts of co-operation with some companies have been successfully undertaken; and a graduate programme in Materials Science has been developed jointly with other Institutes of this University.

Not enough emphasis can be given to the fact that very seldom a paper came out under his name alone; he published together with the many students and collaborators he educated along all these years. The paper of the contributing authors[2] in the "Festschrift for Th. A. J. Maris", and the list of his own publications[3] therein, give a good insight into this fact and on his latest interests. Nowadays, at age 82, Theo is still active in research, working daily at the Institute, an unfailing presence at the weekly colloquia.

To pay the tribute to Theo on the occasion the Institute of Physics is hosting this workshop seems to be very appropriate, since research in Hadron Physics here was certainly initiated and stimulated by him.

Summing Up

To close, we believe it is very appropriate to quote some personal reflections written on Maris by one of his former students, Cylon E. T. Gonçalves da Silva, in the introduction to a paper[4] entitled *The Development of Microtechnologies and the Brazilian National Laboratory for Synchrotron Light (LNLS)*:

"As a former student of Professor Maris, I admire in him the scholar, deeply committed to his discipline. However, I believe that his main contribution to Physics in Brazil goes deeper that what can be guessed from an appraisal of his work as a physicist: he was a role model in a society which lacked such models in Science. He showed us what is a professional scientist. By his example, more than by words, he told us: this is a scientist. If you want to grow up to be one, this is how you are expected to act in your chosen profession. In the rather provincial (provinces can be nice too!) Porto Alegre of diletante intellectuals in the late fifties and sixties, he set an example of professionalism, which I am not sure many of us, at the time, had a clear perception

[2] Ibidem, pp. 19-352.
[3] Ibidem, pp. 353-360.
[4] Ibidem, pp. 344-345.

of what it really meant. However, the nice thing about role models is that you do not need to be conscious of what is going on: the information worms itself into your brain before you can rationalize it. It is not a logic process, it is an intuitive one. That is why it takes time for us to perceive the importance of such persons. Professor Maris has contributed significantly for the development of theoretical Physics and for the national and international reputation of the *Instituto de Física da Universidade Federal do Rio Grande do Sul.* He has also contributed decisively for the development of experimental Physics in Porto Alegre. He realized from the very beginning that a strong effort in Theory, unaccompanied by a parallel endeavor in Experimental Science, has scant chance of success. Later on, I remember Professor Maris expounding on the need to bring the academic world nearer to the realities of economic life, especially in a developing country. Excellence in research is essential, but society must get some return on the investments made to achieve excellence. This is the reason I have chosen to describe in this paper, dedicated to Professor Theodor August Johannes Maris, an experiment which I am sure will be dear to his heart; the building of a scientific institution, begun almost a quarter of century after he came to Brazil to help build science in our country. Photons and electrons played an important part in Professor Maris' research. My first real work in Physics amounted to a small contribution to this research ("On a New Formulation of Quantum Electrodynamics", Th. A. J. Maris, D. Dillenburg, G. Jacob, C.E.T. Gonçalves da Silva and B. Liberman, Il Nuovo Cimento 55 A, 590, 1968). To find myself, after so many years, again involved with electrons and photons is a tribute to the training received from many teachers, not the least of whom was Th. A. J. Maris."

Thank You Very Much.

JUAN ALBERTO MIGNACO
(1940-2001)

Instituto de Física, UFRJ, Brazil

Juan Alberto Mignaco was born in Rosario, Argentina, in 03/January/1940. Graduated in Instituto de Física J.A. Balseiro, San Carlos de Bariloche, in 1962, immediately started teaching activities in the Universidad de Buenos Aires. At the same time he was doing research with Carlos G. Bollini, who supervised the work for his PhD thesis, On the Electrodynamics of Charged Vector Mesons, which was presented at the Universidad Nacional de Cuyo, San Luis, in September 1966.

In July 1966 troops of the military government invaded the Universidad de Buenos Aires, acting with physical violence against teachers and students. Most of faculty members resigned collectively, in protest, Mignaco among them.

Escaping the extreme difficulties of the moment, Mignaco had a fortunate chance to go to the Theory Division in CERN, with a fellowship for two years, followed by another fellowship at the University of Geneva covering the period 1968-1970. During this time Mignaco was doing research in fundamental problems of Quantum Electrodynamics.

In 1970 Mignaco returned to Argentina, competing for a position at Universidad Nacional de Rosario, where he stayed as long as he could: another military government took power in March 1976, and soon after established an intervention in the University, and arbitrarily dismissed large part of the academic body, Mignaco there included.

During these years in Rosario, Mignaco played important role in the Asociación Física Argentina, organizing many of its activities, and being elected President of its Directive Commission in 1975.

Fortunately, during the years starting in 1972, Mignaco established contacts of scientific collaboration with Brazilian groups, in particular at Universidade Federal do Rio de Janeiro, where he helped the Instituto de Física in the beginning of its life.

As a result of the cup against the Argentinian universities in 1976, Mignaco emigrated permanently to Brazil, taking a position at Centro Brasileiro de Pesquisas Físicas, where he stayed until 1993. He then moved for a position as full professor at UFRJ, where he could follow his inherent will of contacts with students and of a wider spectrum of educational activities.

The scientific career of Mignaco covered strong and electromagnetic interactions, quantum field theory, mathematical methods. His calculations in QED published in 1972 (e.g. with R. Barbieri and E. Remidi in Nuovo Cimento A11, 824-864, 1972) are until today important references in the area, permanently receiving many citations. Mignaco worked in the use of Pade Approximants in strong interactions, potential scattering, three body problems, solitons in the Skyrme Model, algebraic and geometric structures in field theories, among other subjects. Lately, Mignaco was mainly concerned with the study of mathematical methods and structures.

During the years in Brazil, until his death in 2001, Mignaco contributed in all ways for the development of our institutions, of our academic standards and of our scientific organization. Extremely dedicated to students, highly competent and ethical as a professional, as a scientist and as a citizen, Mignaco left with us valuable heritage in all these aspects.

In particular, in the area of Hadron Physics Mignaco was a fundamental leader for both Brazilian and Argentinian physics, in the creation and development of our communities, contributing to the effort to organize meetings, to estimulate students, to establish connections among groups.

HADRON PHYSICS, the workshop for whose tradition JUAN ALBERTO MIGNACO much contributed, created a prize with his name, to be given to students that present during the meeting the best posters describing their reasearch work. It is hoped that the prize becomes a tradition, being granted in the future workshops of the series. The enthusiasm that the prize created in the students renders justice to the enthusiasm that Mignaco has always put in the effort to estimulate younger generations.

The organizers of the workshop have also prepared a plate, to be passed to Mignaco's widow Mirian and two sons, Leonardo and Julio, manifesting the recognition of the Hadron Physics community for the great contributions received from him.

S. Wulck and E. Ferreira

LECTURES

HADRON PRODUCTION AT RHIC - CHEMICAL FREEZE-OUT, EFFECTIVE MASSES AND EXPANDING MATTER

D. ZSCHIESCHE

Institut für Theoretische Physik, D-60054 Frankfurt am Main, Germany

S. SCHRAMM

Argonne National Laboratory, Argonne IL 60439, USA

J. SCHAFFNER-BIELICH

Department of Physics, Columbia University, New York, NY 10027, USA

H. STÖCKER*

Institut für Theoretische Physik, D-60054 Frankfurt am Main, Germany
stoecker@th.physik.uni-frankfurt.de

Laboratoire de Physique Subatomique et des Technologies Associées, University of Nantes - IN2P3/CNRS - École des Mines de Nantes, F-44072 Nantes Cedex 03, France

W. GREINER

Institut für Theoretische Physik, D-60054 Frankfurt am Main, Germany

The properties of the strongly interacting matter produced at RHIC are investigated within a chemical and thermal equilibrium chiral $SU(3)$ $\sigma - \omega$ approach. The focus is on the freeze-out conditions, the effective masses of the produced hadrons and the expansion of the system. For the freeze-out conditions the commonly adopted noninteracting gas calculations yield temperatures and chemical potentials close to or above the critical temperature for the chiral phase transition, but without taking into account any interactions. Contrary, the chiral $SU(3)$ model predicts temperature and density dependent effective hadron masses and effective chemical potentials in the medium and a transition to a chirally restored phase at high temperatures or chemical potentials. Three different parameterizations of the model, which show different types of phase transition behaviour, are investigated. We show that if a chiral phase transition occurred in those collisions, "freezing" of the relative hadron abundances in the symmetric phase is excluded by the data. Therefore, either very rapid chemical equilibration must occur in the broken phase, or the measured hadron ratios are the outcome of the dynamical symmetry breaking.

---

*LECTURES DELIVERED BY HORST STÖCKER.

Furthermore, the extracted chemical freeze-out parameters differ considerably from those obtained in simple noninteracting gas calculations. In particular, the three models yield up to 35 MeV lower temperatures than the free gas approximation. The in-medium masses turn out to differ up to 150 MeV from their vacuum values.

1 Introduction

Thermodynamical equilibrium calculations of particle production in high energy particle and nuclear collisions have been carried out for a long time[1−11]. Recently hadron abundances and particle ratios have been measured in heavy-ion collisions from SIS, AGS, SPS to RHIC energies. These data have revived the interest in the extraction of temperatures and chemical potentials from thermal equilibrium "chemical" model analysis. The experimentally determined hadron ratios can be fitted well with straightforward noninteracting gas model calculations [7−13], if a sudden breakup of a thermalized source is assumed and once the subsequent feeding of the various channels by the strongly decaying resonances is taken into account. From the χ^2 freeze-out fits one has constructed a quite narrow band of freeze-out values in the $T - \mu_B$ plane (see e.g.[12,13]). The extracted freeze-out parameters are fairly close to the phase transition curve for SPS and RHIC energies. However, when we are indeed so close to the phase transition or to a cross-over as suggested by the data for T and μ_B, we cannot afford to neglect the very in-medium effects we are after - and which, after all, do produce the phase transition. Thus, since noninteracting gas models neglect any kind of possible in-medium modifications they cannot yield information about the phase transition.

Therefore, we will employ below a relativistic self-consistent chiral model of hadrons and hadron matter developed in [14,15,16]. This model can be used as a thermodynamically consistent effective theory or as a toy model, which embodies the restoration of chiral symmetry at high temperatures or densities. Therefore the model predicts temperature and density dependent hadronic masses and effective chemical potentials, which have already been proposed and considered in [4,16−20]. Thus, using the chiral $SU(3)$ model we can investigate, whether the freeze-out in fact takes place close to the phase transition boundary (if it exists) and if the extracted T, μ_B parameters are strongly model dependent. Furthermore, not only the values of the effective masses at freeze-out can be predicted but also their behaviour during the expansion of the hot and dense matter produced in the collision. Depending on the chosen parameters and degrees of freedom different scenarios for the chiral phase change are predicted by the model: strong or weak first order phase transition or a cross-over. The transitions take place around $T_c = 155$ MeV [21,16], which is in qualitative agreement with lattice predictions [22] for the critical temperature for the onset of a deconfined phase which coincides with that of a chirally restored phase [23].

2 Model description

The chiral $SU(3)$ model is presented in detail in [15,16]. We will briefly introduce the model here: we consider a relativistic field theoretical model of baryons and mesons built on chiral symmetry and broken scale invariance. The general form of the Lagrangean looks as follows:

$$\mathcal{L} = \mathcal{L}_{\text{kin}} + \sum_{W=X,Y,V,\mathcal{A},u} \mathcal{L}_{\text{BW}} + \mathcal{L}_{\text{VP}} + \mathcal{L}_{\text{vec}} + \mathcal{L}_0 + \mathcal{L}_{\text{SB}}. \tag{1}$$

$\mathcal{L}_{\text{kin}}$ is the kinetic energy term, $\mathcal{L}_{\text{BW}}$ includes the interaction terms of the different baryons with the various spin-0 and spin-1 mesons (see [15] for details). The baryon masses are generated by both, the nonstrange σ ($< q\bar{q} >$) and the strange ζ ($< s\bar{s} >$) scalar condensate. $\mathcal{L}_{\text{VP}}$ contains the interaction terms of vector mesons with pseudoscalar mesons. $\mathcal{L}_{\text{vec}}$ generates the masses of the spin-1 mesons through interactions with spin-0 fields, and $\mathcal{L}_0$ gives the meson-meson interaction terms which induce the spontaneous breaking of chiral symmetry: it also includes a scale-invariance breaking logarithmic potential. Finally, $\mathcal{L}_{\text{SB}}$ introduces an explicit symmetry breaking of the $U(1)_A$, the $SU(3)_V$, and the chiral symmetry. All these terms have been discussed in detail in [16,15].

The hadronic matter properties at finite density and temperature are studied in the mean-field approximation [24]. Then the Lagrangean (1) becomes

$$\mathcal{L}_{BX} + \mathcal{L}_{BV} = -\sum_i \overline{\psi_i}[g_{i\omega}\gamma_0\omega^0 + g_{i\phi}\gamma_0\phi^0 + m_i^*]\psi_i \,, \tag{2}$$

$$\mathcal{L}_{vec} = \frac{1}{2}m_\omega^2\frac{\chi^2}{\chi_0^2}\omega^2 + \frac{1}{2}m_\phi^2\frac{\chi^2}{\chi_0^2}\phi^2 + g_4^4(\omega^4 + 2\phi^4)\,, \tag{3}$$

$$\begin{aligned}\mathcal{V}_0 = {} & \frac{1}{2}k_0\chi^2(\sigma^2+\zeta^2) - k_1(\sigma^2+\zeta^2)^2 - k_2(\frac{\sigma^4}{2}+\zeta^4) - k_3\chi\sigma^2\zeta \\ & + k_4\chi^4 + \frac{1}{4}\chi^4\ln\frac{\chi^4}{\chi_0^4} - \frac{\delta}{3}\chi^4\ln\frac{\sigma^2\zeta}{\sigma_0^2\zeta_0}\,,\end{aligned} \tag{4}$$

$$\mathcal{V}_{SB} = \left(\frac{\chi}{\chi_0}\right)^2\left[m_\pi^2 f_\pi\sigma + (\sqrt{2}m_K^2 f_K - \frac{1}{\sqrt{2}}m_\pi^2 f_\pi)\zeta\right], \tag{5}$$

where m_i is the effective mass of the hadron species i. σ and ζ correspond to the scalar condensates, ω and ϕ represent the non-strange and the strange vector field respectively, and χ is the scalar-isoscalar dilaton field, which mimics the effects of the gluon condensate [25]. Only the scalar ($\mathcal{L}_{BX}$) and the vector meson terms ($\mathcal{L}_{BV}$) contribute to the baryon-meson interaction, since for all other mesons the expectation value vanishes in the mean-field approximation. The grand canonical potential Ω per volume V as a function of chemical potential μ and temperature T can be written as:

$$\frac{\Omega}{V} = -\mathcal{L}_{vec} - \mathcal{L}_0 - \mathcal{L}_{SB} - \mathcal{V}_{vac} \mp T \sum_i \frac{\gamma_i}{(2\pi)^3} \int d^3k \left[\ln\left(1 \pm e^{-\frac{1}{T}[E_i^*(k) - \mu_i^*]}\right)\right], \quad (6)$$

with the baryons (top sign) and mesons (bottom sign). The vacuum energy $\mathcal{V}_{vac}$ (the potential at $\rho_B = 0, T = 0$) has been subtracted in order to get a vanishing vacuum energy. γ_i denote the hadronic spin-isospin degeneracy factors. The single particle energies are $E_i^*(k) = \sqrt{k_i^2 + m_i^{*2}}$ and the effective chemical potentials read $\mu_i^* = \mu_i - g_{i\omega}\omega - g_{i\phi}\phi$.

The mesonic fields are determined by extremizing $\frac{\Omega}{V}(\mu, T = 0)$. The density of particle i can be calculated by differentiating Ω with respect to the corresponding chemical potential μ_i. This yields:

$$\rho_i = \gamma_i \int \frac{d^3k}{(2\pi)^3} \left[\frac{1}{\exp\left[(E_i^* - \mu_i^*)/T\right] \pm 1} \right]. \quad (7)$$

All other thermodynamic quantities can also be obtained from the grand canonical potential. In the present calculation the lowest lying baryonic octet and decuplet and the lowest lying mesonic nonets are coupled to the relativistic mean fields. Depending on the coupling of the baryon resonances (the decuplet) to the field equations, the model shows a first order phase transition or a cross-over (for details see [21]). We will use three different parameter sets: parameter set CI treats the members of the baryon decuplet as free particles, which yields a cross-over behaviour. Parameter sets CII and CIII include also the (anti)-baryon decuplet as sources for the meson field equations. They differ by an additional explicit symmetry breaking for the baryon resonances along the hypercharge direction, as described in [15] for the baryon octet. This is included in CII and not used in CIII. This leads to a weak first order phase transition at $\mu = 0$ for CII and two first order phase transitions for CIII, which can be viewed as one strong first order phase transition. Heavier resonances up to $m = 2$ GeV are always included as free particles. The resulting baryon masses for CI and CIII are shown in figure 1. We observe a continous

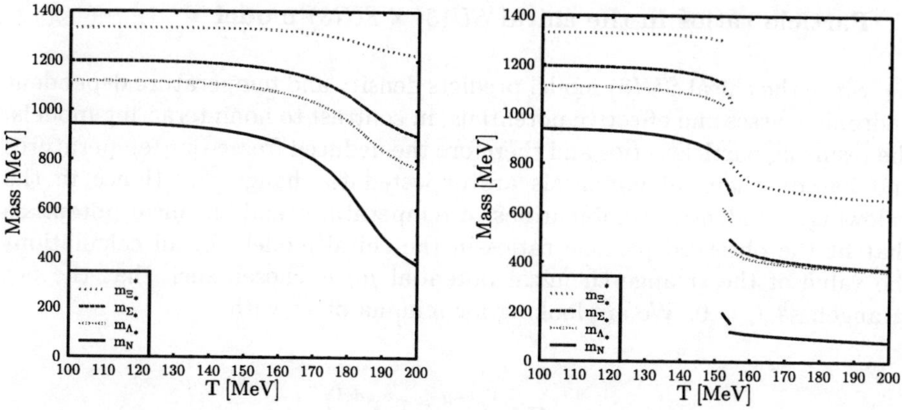

Figure 1. Baryon octet masses as function of temperature for vanishing chemical potential. Left CI, right CIII. Note the continuous change of the masses starting around $T = 150$ MeV for CI. In contrast for CIII two phase-transitions occur around $T_c \approx 155$ MeV. These result from the separate jumps in the nonstrange (σ) and the strange (ζ) condensate.

decrease of the baryon masses for CI starting at $T \approx 150$ MeV. In contrast, CIII shows two jumps around $T = 155$ MeV. These result from the jumps in the nonstrange (σ) and the strange (ζ) condensate, which occur at different temperatures due to the considerably smaller coupling of the ζ-field to the (anti-)nucleons [15]. The critical energy densities, the entropy densities and the transition temperatures for $\mu_q = \mu_s = 0$ ($\mu_q = \mu_B/3, \mu_s = \mu_B/3 - \mu_S$) are specified in table 1.

	ϵ^-/ϵ_0	ϵ^+/ϵ_0	$s^-[fm^{-3}]$	$s^+[fm^{-3}]$	T_c[MeV]
CII	2.8	7.2	2.8	6.7	156.3
CIII - 1st PT	2.3	8.3	2.4	7.9	153.4
CIII - 2nd PT	10.5	17.1	9.8	15.7	155.5

Table 1. Energy density, entropy density and phase transition temperatures for CII,CIII, $\mu_q = \mu_s = 0$. The $(-), (+)$ signs refer to an approach to the phase transition from below and above, respectively. T_c denotes the phase transition temperature. $\epsilon_0 = 138.45$ MeV/fm^3 denotes the energy density of nuclear matter in the ground state.

3 Particle ratios in the chiral $SU(3) \times SU(3)$ model

Since the chiral $SU(3)$ model predicts density and temperature dependent hadronic masses and effective potentials, in contrast to noninteracting models, the resulting particle ratios and therefore the deduced freeze-out temperatures and baryon chemical potentials are expected to change [26]. Hence in the following, we identify combinations of temperatures and chemical potentials that fit the observed particle ratios in the chiral model. In all calculations the value of the strange chemical potential μ_S is chosen such that the net strangeness $f_s = 0$. We are looking for minima of χ^2 with

$$\chi^2 = \sum_i \frac{\left(r_i^{exp} - r_i^{model}\right)^2}{\sigma_i^2}. \tag{8}$$

Here r_i^{exp} is the experimental ratio, r_i^{model} is the ratio calculated in the model and σ_i represents the error in the experimental data points. We use the same ratios as in [11]: $\bar{p}/p$, $\bar{\Lambda}/\Lambda$, $\bar{\Xi}/\Xi$, π^-/π^+, K^-/K^+, K^-/π^-, K_0^*/h^-, $\bar{K}_0^*/h^-$.

Even though the only parameters in a thermal and chemical equilibrium approach on first sight are the temperature and the baryon chemical potential, there exist further unknowns: on the one hand, some decays of high mass resonances are not well known and on the other hand the effect of weak decays in the experiments strongly depends on the detector geometry and on the reconstruction efficiency of the experiments. The feeding correction from the strong and electromagnetic decays of the hadronic resonances used here employs the procedure used in the UrQMD model [27,28]. Weak decays are not considered here. We rather focus on the principal question whether an interacting chiral $SU(3)$ approach with $m^* \neq m_{vac}$ can at all describe the particle yields at RHIC. Fine tuning of the χ^2 by adjustment of the weak decay scheme is not our intention. Even though it has been shown [29] that χ^2 values may be improved by including weak decays.

To compare the quality of the fits obtained in the chiral model with those obtained from the noninteracting gas approach, we set all masses and chemical potentials contained in the chiral model to their vacuum values and again use the same UrQMD feeding procedure as for the interacting model. This yields the ideal gas denoted ig_{FFM}. We find that the resulting ideal gas ratios are not identical but comparable to those obtained in the literature [26,29,11,30]. The differences should only result from a different treatment of weak interactions and from the uncertainty in the decay scheme of high mass resonances.

4 Results for Au + Au collisions at RHIC

First, we find that a reasonable fit of the measured particle ratios at RHIC is possible in all three phase transition scenarios of the chiral model and the ideal gas case with comparable quality.

Second, the resulting freeze-out values depend on the model employed, i.e. cross-over, weak first order, strong first order or free thermal gas.

Third, a reasonable description of the data is impossible above T_c in the models showing a first order phase transition. This shows that no direct freeze-out from the restored phase is observed.

Fourth, the effective masses at freeze-out change up to 15% from their vacuum values but are nearly identical for a given hadron in the different models. However, during the adiabatic expansion the masses depend strongly on the applied model.

Fifth, thermodynamic quantities like energy density and pressure significantly depend on the phase transition behaviour. This is due to the different constraints on temperature and possible energy density values in the investigated scenarios.

Figure 2 shows the value of χ^2 in the $T - \mu_B$ plane for the cross-over case and for the strong first order phase transition. We see that the best fit $T - \mu_B$ values differ in both models. Furthermore, in the cross-over case χ^2 is well behaved as a function of T and μ_B. In contrast, the model with a strong first order phase transition shows a very steep increase of χ^2 at the phase transition boundary: the quality of the fit decreases drastically due to the jump of the effective masses at the phase transition boundary. Above T_c the χ^2 values are unacceptable, $\chi^2 > 500$.

The resulting best-fit particle ratios, χ^2-values and thermodynamic quantities are shown in table 2 and figure 4. The χ^2 values for the chiral model are: $\chi^2_{CI} = 5.50$, $\chi^2_{CII} = 5.73$ and $\chi^2_{CIII} = 5.40$. Thus, all three parameter sets describe the data equally well. Furthermore, the agreement is as good as in the noninteracting gas calculation ($\chi^2_{BMRS} = 5.72$ [11], $\chi^2_{igFFM} = 5.66$). The best fit $T - \mu_B$ parameters vary quite considerably between the different models. The noninteracting gas calculation (ig_{FFM}) yields $T = 187.6$ MeV and $\mu_B = 44.1$ MeV. These freeze-out values can be compared to those obtained in other ideal gas calculations: $T = 174$ MeV, $\mu_B = 46$ MeV in[11], $T = 165$ MeV, $\mu_B = 41$ MeV in [30] and $T = 190$ MeV, $\mu_B = 45$ MeV in [31]. The cross-over case in the interacting chiral model (CI) yields $T = 170.8$ MeV, $\mu_B = 48.3$ MeV. Very strong deviations are found for the models with a first order phase transition (CII,CIII): the freeze-out temperatures are $T = 155$ MeV (CII) and $T = 153.3$ MeV (CIII), more than 30 MeV lower

Au+Au	Experiment	CI	CII	CIII	ig_{FFM}	BMRS
T_{chem} $[MeV]$		**170.8**	**155.0**	**153.3**	**187.6**	**174.0**
μ^B_{chem} $[MeV]$		**48.3**	**54.6**	**51.0**	**44.1**	**46.0**
μ^S_{chem} $[MeV]$		**11.1**	**9.8**	**9.4**	**13.5**	**13.6**
χ^2		**5.5**	**5.7**	**5.4**	**5.7**	**5.7**
ρ_{had} $[fm^{-3}]$		**0.66**	**0.38**	**0.35**	**1.12**	
$\rho_B + \rho_{\bar{B}}$ $[fm^{-3}]$		**0.15**	**0.08**	**0.07**	**0.28**	
p $\frac{MeV}{fm^3}$		**108**	**55**	**51**	**207**	
ϵ $\frac{MeV}{fm^3}$		**695**	**356**	**326**	**1324**	
E/A $[MeV]$		**1053**	**937**	**931**	**1182**	**≈ 1100**
S/A		**157**	**164**	**177**	**142**	
$\bar{p}/p$	0.65(7) [STAR] 0.64(8) [PHENIX] 0.60(7) [PHOBOS] 0.61(6) [BRAHMS]	0.640	0.648	0.652	0.629	0.629
$\overline{\Lambda}/\Lambda$	0.77(7) [STAR]	0.714	0.695	0.702	0.721	0.753
$\overline{\Xi}/\Xi$	0.82(8) [STAR]	0.787	0.731	0.743	0.834	0.894
π^-/π^+	1.00(2) [PHOBOS] 0.95(6) [BRAHMS]	1.000	1.000	1.000	1.000	1.007
K^-/K^+	0.88(5) [STAR] 0.78(13) [PHENIX] 0.91(9) [PHOBOS] 0.89(7) [BRAHMS]	0.919	0.914	0.915	0.916	0.894
K^-/π^-	0.15(2) [STAR]	0.183	0.168	0.168	0.179	0.145
$\bar{p}/\pi^-$	0.08(1) [STAR]	0.082	0.084	0.078	0.083	0.078
$\overline{K^*_0}/h^-$	0.058(17) [STAR]	0.055	0.049	0.049	0.046	0.032
K^*_0/h^-	0.060(17) [STAR]	0.049	0.044	0.044	0.041	0.037

Table 2. Chiral fit of the particle ratios measured at RHIC at $\sqrt{s} = 130$ GeV. Data and BMRS-fit taken from reference[11].

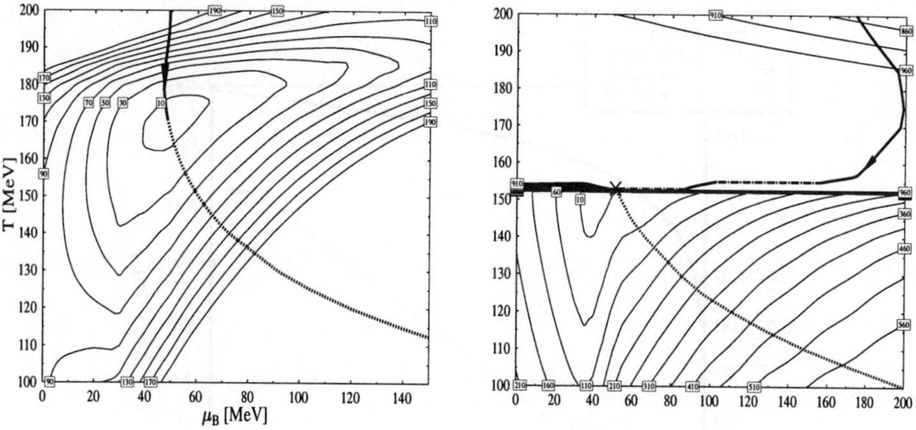

Figure 2. χ^2 contours in the $T - \mu_B$ plane for CI (left) and CIII (right). Data are taken from[11]. The adiabatic path (constant entropy per net baryon S/A), corresponding to expansive cooling of an ideal fluid, is also shown for the two cases. The dash-dotted lines in the right picture correspond to the path of the system along the phase transition boundary, i.e. the expansion in the mixed phases. μ_S is chosen such that $f_s = 0$.

than for ig_{FFM}. The fitted baryon chemical potentials μ_B increase by about $7-10$ MeV. These $T-\mu_B$ pairs are very close to the phase boundary (CII) or even right on it (CIII) and are about 10 MeV higher than the values obtained at SPS-energies[26,32]. This can also be seen in figure3, where temperature and pressure of the expanding matter as a function of energy density are depicted. For CIII the freeze-out takes place right at the phase transition border. Furthermore, figure 3 shows that mainly due to the different constraints by the phase transition behaviour on temperature and possible energy densities, the values of the corresponding thermodynamic quantities vary between the different approaches. However, the energy per particle E/A is approximately 1 GeV in all cases. This "unified freeze-out condition" has already been proposed in [33]. The particle densities of the expanding matter are plotted in figure 5. The crosses and stars mark the densities at the assumed global freeze-out. Also the hadron densities show a quite different behaviour during expansion and also at freeze-out. This will lead to a different volume if total yields are considered.

However, one should be aware that even though the particle ratios can satisfactorily be described by a global freeze-out scenario, this does not verify the assumption. As has been shown in [34], measured particle ratios at SPS can

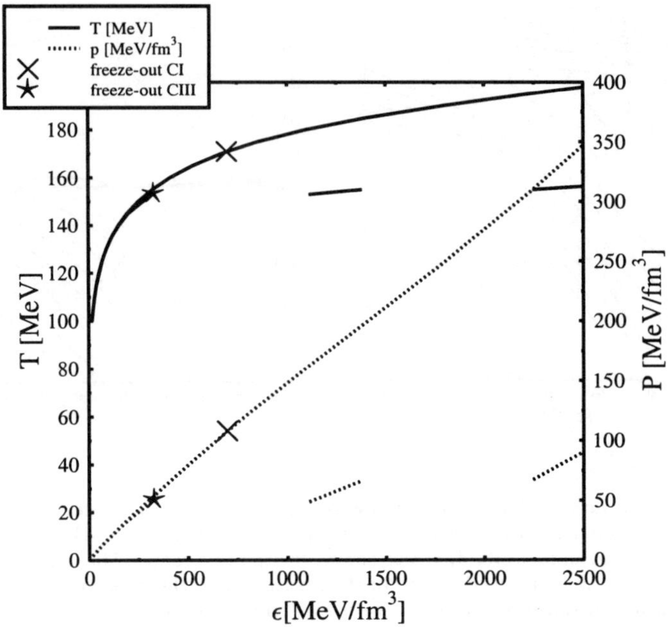

Figure 3. Temperature and pressure as a function of energy density for CI (continuous curves) and CIII (curves showing two jumps) along the adiabatic path, corresponding to expansive cooling of an ideal fluid.

also be described by continuous evaporation of hadrons from a hadronizing quark gluon plasma. That means, the measured hadrons origin from different T, μ states of the expanding system but integrating over the whole evaporation process still yields the correct values for the ratios.

The fact that the freeze-out appears right at the phase boundary or at cross-over implies that there are large in-medium corrections, in particular for the effective masses, a phenomenon observed already in [17]. The effective masses shown in figure 6 are shifted up to 15% from their vacuum values. However, all the interacting models show similar values for the effective mass of a given hadron. The strongest in-medium modifications are observed for the non-strange baryons ($\Delta m_i^*/m_i \approx 15\%$). Mesons and strange baryons show smaller changes of the effective masses, e.g. about 10% for Λ, π, K^*, about 5% for the kaons and nearly no change for the Ξ's.

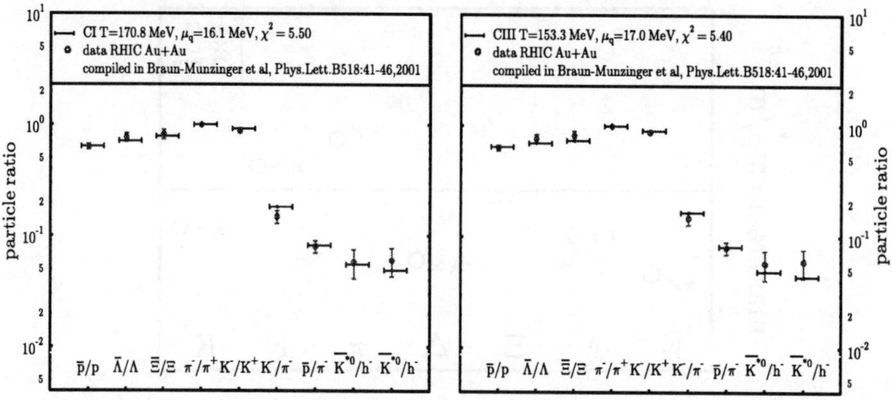

Figure 4. Particle ratios calculated with CI (left) and CIII (right) compared to RHIC data as compiled in[11].

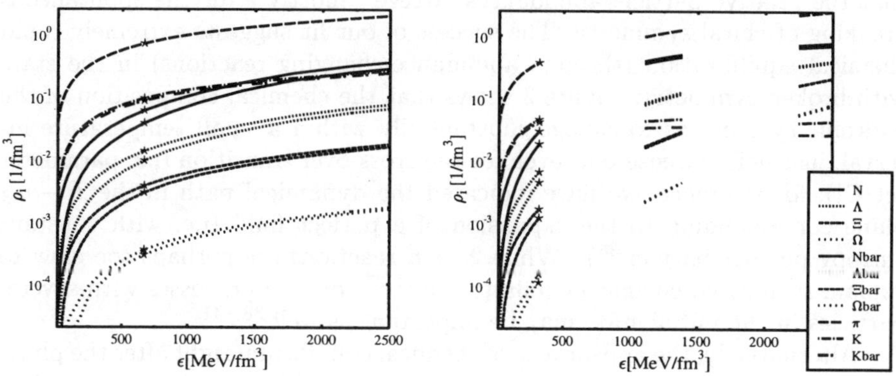

Figure 5. Hadron densities as a function of energy density for CI (left) and CIII (right) along the adiabatic path, corresponding to expansive cooling of an ideal fluid.

In figure 7 the effective hadron masses for the expanding system are plotted as a function of energy density. While for CI the hadron masses approach their freeze-out values continously, for CIII two separate jumps occur and the freeze-out takes place right after the system reaches the symmetry broken phase.

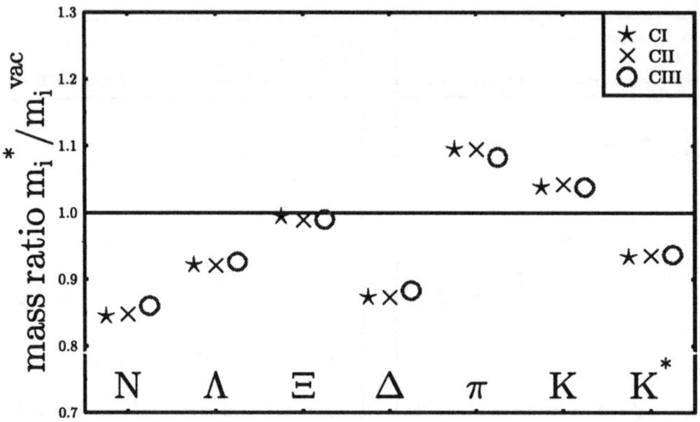

Figure 6. Effective masses for the different interacting chiral models and the ideal gas (vacuum values) case. The differences among the interacting approaches for a given hadron are less than 2%.

These results, together with the steep χ^2 contours from figure 2, suggest that the relative particle abundances "freeze" shortly after the spontaneous breaking of chiral symmetry. The success of our fit suggests extremely rapid chemical equilibration (through abundance-changing reactions) in the state with broken symmetry. Figure 2 shows that the chemical composition of the hadronic system has to change substantially within a small temperature interval, just before freeze-out, even for the cross-over transition (i.e. parameter set CI); for reference, we have indicated the dynamical path in the $T - \mu_B$ plane corresponding to the expansion of a perfect fluid (i.e. with constant entropy per net baryon [35]). While $2 \to n$ reactions are perhaps too slow to explain such rapid chemical equilibration [36,37], $m \to n$ processes with several particles in the initial state may be important as well [38–41].

Alternatively, the appearance of chemical equilibrium right after the phase transition (or the cross-over) to the state of broken chiral symmetry might just be the outcome of the dynamical symmetry breaking process itself [42], with statistical occupation of the various hadronic channels according to phase space [43–46]. If so, number-changing reactions in the broken phase need not proceed at a high rate. To test this picture experimentally, it might be useful to consider central collisions of small ions like protons or deuterons, at similar energy and particle densities in the central region as for central Au+Au. For systems of transverse extent comparable to the correlation lengths of the chiral condensates, the dynamical symmetry breaking process should be different from that in large systems (for example, the mean field approximation should

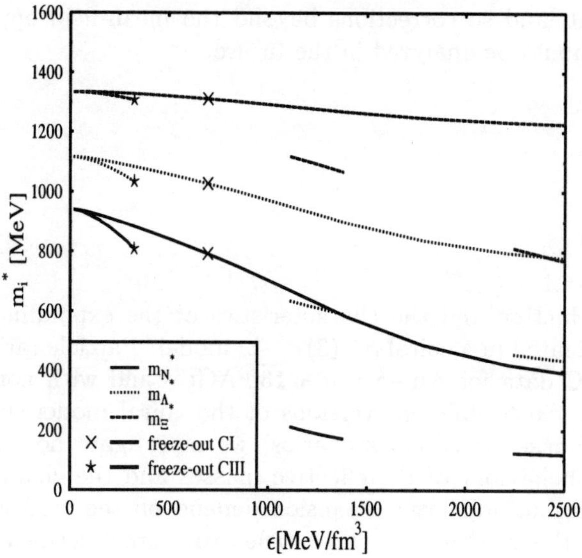

Figure 7. Baryon masses as a function of energy density for CI (continuous curves) and CIII (curves showing two jumps) along the adiabatic path, corresponding to expansive cooling of an ideal fluid.

not apply). The correlation lengths $\xi_{\sigma,\zeta}$ are given by

$$\xi_\sigma^{-2} = \frac{\partial^2 (\Omega/V)}{\partial \sigma^2} \,, \tag{9}$$

and accordingly for ξ_ζ. We evaluate the curvature of the thermodynamical potential at the global minimum and for T, μ_B, μ_S at the freeze-out point. For parameter sets CI, CII, CIII we obtain $\xi_\sigma = 0.37$ fm, 0.41 fm, 0.40 fm, respectively. For the correlation length of the strange condensate we obtain $\xi_\zeta = 0.20$ fm in all three cases. The correlation lengths are not very much smaller than, say, the radius of a proton. Thus, even if the freeze-out point for high-energy *pp* collisions happens to be close to that for Au+Au collisions at RHIC energies, the transition from the symmetry restored to the broken phase might be different. Finally, we also note that the correlation lengths obtained from our effective potential are not larger than the thermal correlation length

$1/T$ at freeze-out, and so corrections beyond the mean-field approximation employed here should be analyzed in the future.

5 Conclusion

Hadron production and the characteristics of the expanding matter at RHIC are investigated in a chiral $SU(3)\,\sigma-\omega$ model. Particle ratios are compared with RHIC data for Au+Au at $\sqrt{130}$ AGeV and with noninteracting gas calculations. Since different versions of the chiral model show qualitatively different phase transition scenarios, we investigate how the particle production, the behaviour of the effective masses and the characteristics of the system during an adiabatic expansion depend on the models. Since we have shown that the current data on particle ratios are described by all three different phase transition scenarios and the ideal gas model, we can by this not favour or rule out any one scenario. In all interacting models the effective masses at freeze-out are shifted up to 15% from their vacuum values and are nearly identical for a given hadron. However, during the adiabatic expansion the effective masses for a given hadron differ considerably between the investigated models. The fitted chemical freeze-out temperatures and chemical potentials depend on the order of the phase transition. The cross-over case yields 15 MeV shifted T values as compared to the noninteracting gas model while the models with a first order phase transition yield more than 30 MeV lower temperatures. Furthermore, the fitted freeze-out points are located practically right on the phase transition boundary, in the first order phase transition scenarios, but T is always $\leq T_c$. *This suggests that at RHIC the system emerges after the chiral chiral phase transition. This of course is only true if a first order phase transition does actually occur in QCD at small chemical potentials and high T.* "Freezing" of the relative abundances of various hadrons in the symmetric phase (at $T > T_c$) is excluded. Finally, we are aware that the assumption of global freeze-out is not verified by a satisfactory description of the particle ratios, since e.g. nonequilibrium models showing continuous evaporation of hadrons from the expanding matter successfully describe particle ratios measured at SPS. Therefore, though very successful in describing particle ratios, the assumption of global freeze-out can only be regarded as one possible scenario for the hadronisation process in relativistic heavy ion collisions.

Acknowledgments

The authors are grateful to P. Braun-Munzinger, A. Dumitru, L. Gerland, M. Gorenstein, I. Mishustin, and K. Paech for fruitful discussions. This work is supported by Deutsche Forschungsgemeinschaft (DFG), Gesellschaft für Schwerionenforschung (GSI), Bundesministerium für Bildung und Forschung (BMBF), the Graduierenkolleg Theoretische und Experimentelle Schwerionenphysik and by the U.S. Department of Energy, Nuclear Physics Division (Contract No. W-31-109-Eng-38).

References

1. E. Fermi, Prog. Theor. Phys. **5**, 570 (1950).
2. L. D. Landau, Izv. Akad. Nauk. SSSR Ser. Fiz. **17**, 51 (1953).
3. G. D. Westfall *et al.*, Phys. Rev. Lett. **37**, 1202 (1976).
4. H. Stöcker, W. Greiner, and W. Scheid, Z. Phys. **A286**, 121 (1978).
5. L. P. Csernai and J. I. Kapusta, Phys. Rep. **131**, 223 (1986).
6. D. Hahn and H. Stöcker, Nucl. Phys. **A452**, 723 (1986).
7. P. Braun-Munzinger, J. Stachel, J. P. Wessels, and N. Xu, Phys. Lett. **B365**, 1 (1996).
8. J. Letessier and J. Rafelski, Int. J. Mod. Phys. **E9**, 107 (2000).
9. J. Rafelski, J. Letessier, and G. Torrieri, Phys. Rev. **C64**, 054907 (2001).
10. F. Becattini, J. Cleymans, A. Keranen, E. Suhonen, and K. Redlich, Phys. Rev. **C64**, 024901 (2001).
11. P. Braun-Munzinger, D. Magestro, K. Redlich, and J. Stachel, Phys. Lett. **B518**, 41 (2001).
12. P. Braun-Munzinger and J. Stachel, Nucl. Phys. **A638**, 3 (1998).
13. J. Cleymans and K. Redlich, Phys. Rev. **C60**, 054908 (1999).
14. P. Papazoglou, S. Schramm, J. Schaffner-Bielich, H. Stöcker, and W. Greiner, Phys. Rev. **C57**, 2576 (1998).
15. P. Papazoglou, D. Zschiesche, S. Schramm, J. Schaffner-Bielich, H. Stöcker, and W. Greiner, Phys. Rev. **C59**, 411 (1999).
16. D. Zschiesche, P. Papazoglou, S. Schramm, C. Beckmann, J. Schaffner-Bielich, H. Stöcker, and W. Greiner, Springer Tracts in Modern Physics **163**, 129 (2000).
17. J. Theis, G. Graebner, G. Buchwald, J. A. Maruhn, W. Greiner, H. Stöcker, and J. Polonyi, Phys. Rev. **D28**, 2286 (1983).
18. J. Schaffner, I. N. Mishustin, L. M. Satarov, H. Stöcker, and W. Greiner, Z. Phys. **A341**, 47 (1991).
19. G. E. Brown and M. Rho, Phys. Rep. **269**, 333 (1996).

20. M. Michalec, W. Florkowski, and W. Broniowski, Phys. Lett. **B520**, 213 (2001).
21. D. Zschiesche, S. Schramm, H. Stöcker, and W. Greiner, Phys. Rev. **C65**, 064902 (2002).
22. F. Karsch, E. Laermann, and A. Peikert, Nucl. Phys. **B605**, 579 (2001).
23. F. Karsch, hep-lat/9903031 (1998).
24. B. D. Serot and J. D. Walecka, Int. J. Mod. Phys. **E6**, 515 (1997).
25. J. Schechter, Phys. Rev. **D21**, 3393 (1980).
26. D. Zschiesche *et al.*, Nucl. Phys. **A681**, 34 (2001).
27. S. A. Bass *et al.*, Prog. Part. Nucl. Phys. **41**, 225 (1998).
28. UrQMD Collaboration, http://www.th.physik.uni-frankfurt.de/ urqmd/.
29. M. Michalec, Ph.D. thesis, Institute of Nuclear Physics, Crakow, 2001.
30. W. Florkowski, W. Broniowski, and M. Michalec, Acta Phys. Polon. **B33**, 761 (2002).
31. N. Xu and M. Kaneta, Nucl. Phys. **A698**, 306 (2002).
32. D. Zschiesche *et al.* (unpublished).
33. J. Cleymans and K. Redlich, Phys. Rev. Lett. **81**, 5284 (1998).
34. C. Spieles, H. Stöcker, and C. Greiner, Eur. Phys. J. **C2**, 351 (1998).
35. P. R. Subramanian, H. Stöcker, and W. Greiner, Phys. Lett. **B173**, 468 (1986).
36. S. A. Bass, P. Danielewicz, S. Pratt, and A. Dumitru, J. Phys. **G27**, 635 (2001).
37. S. A. Bass and A. Dumitru, Phys. Rev. **C61**, 064909 (2000).
38. R. Rapp and E. V. Shuryak, Phys. Rev. Lett. **86**, 2980 (2001).
39. C. Greiner and S. Leupold, J. Phys. **G27**, L95 (2001).
40. C. Greiner, J. Phys. **G28**, 1631 (2002).
41. C. Greiner, Nucl. Phys. **A698**, 591 (2002).
42. O. Scavenius, A. Dumitru, and J. T. Lenaghan, (2002), hep-ph/0201079.
43. F. Becattini and U. W. Heinz, Z. Phys. **C76**, 269 (1997).
44. F. Becattini, M. Gazdzicki, and J. Sollfrank, Nucl. Phys. **A638**, 403 (1998).
45. R. Stock, Nucl. Phys. **A661**, 282 (1999).
46. R. Stock, Phys. Lett. **B456**, 277 (1999).

ANALYTICAL APPROACHES TO QCD

FRIEDER LENZ
Institut für Theoretische Physik III Universität Erlangen-Nürnberg
Staudtstraße 7 D-91058 Erlangen, Germany
flenz@theorie3.physik.uni-erlangen.de

In this talk, theoretical analyzes of low energy, non-perturbative Quantum Chromodynamics are presented. The focus of the discussion is confinement, a phenomenon unique to strong interaction physics. The theoretical framework for these analyzes is set by the Faddeev-Popov quantization of gauge theories. Particular emphasis is put on obstructions occurring when implementing gauge conditions and on the concomitant emergence of compact variables and singular fields. A detailed analysis of the non-perturbative dynamics associated with such exceptional field configurations is given. After discussion of the role of instantons as the simplest examples of singular configurations, I will turn to a description of the confinement related properties of the dynamics of compact fields and monopoles. Implications for the phenomenology of confined and plasma phases of QCD are discussed.

1 Introduction

Quantum Chromodynamics is the theory of strong interactions. Low energy phenomena in strong interaction physics are determined by two fundamental mechanisms of QCD, confinement and chiral symmetry breaking. Chiral symmetry breaking is a property which is not specific to QCD. There is theoretical evidence that in QED magnetic fields of sufficient strength induce a spontaneous breakdown of chiral symmetry [1]. In a variety of theoretical models such as the NJL model [2] or the instanton model [3] and within different approximations such as the mean-field approximation or the quenched and rainbow approximations in QED [4] chiral symmetry breaking occurs. On the other hand, in none of these models or approximations does confinement occur. Field theoretic models which exhibit confinement are 1+1 and 2+1 dimensional gauge theories. Here confinement is essentially enforced by the low dimensions although in 2+1 dimensions, a change from a logarithmically to a linearly rising interaction energy of static charges takes place. Confinement has been shown to arise by condensation of monopoles in supersymmetric gauge theories [5]. The existence of these monopoles as solutions to classical field equations is intimately connected to supersymmetry and the relevance of this mechanism of monopole condensation for QCD remains to be understood.
The connection between these two non-perturbative phenomena of QCD is an open issue. Results of lattice calculation point to such a connection. Similar temperatures for the confinement-deconfinement phase transition in quenched and for the chiral symmetry restoration in unquenched QCD have been obtained. This is not to be

expected since comparison of the characteristic energies associated with these two phenomena points to a certain hierarchy of the dynamics. The energy scale of the fluctuations associated with chiral symmetry breaking can be characterized by the pion mass of 140 MeV or by the constituent quark mass of mean-field models of a few hundreds of MeV. On the other hand, the lowest glueball mass in pure Yang-Mills theory is of the order of 1.5 GeV. Similarly comparison of the contribution of the quark condensate via the quark mass term to the ground state energy of QCD (obtained from the Gell-Mann,Oakes,Renner relation [6])

$$\Delta\mathcal{H}_\chi = m_q\langle 0|\bar{\psi}\psi|0\rangle \approx \ 25\,\mathrm{MeV\ fm}^{-3}$$

with the contribution of the gluon condensate [7]

$$\Delta\mathcal{H}_c = \ \frac{1}{2}\langle 0|\mathbf{E}^2|0\rangle \ \approx \ 750\,\mathrm{MeV\ fm}^{-3}$$

suggests gluon dynamics to be associated with significantly larger energy scales. If nevertheless the two phase transitions take place at roughly the same temperature, one is led to conclude that chiral symmetry breaking must be induced by the confinement-deconfinement transition.
The main topic in this discussion of analytical approaches to QCD will be confinement and its relation to basic properties of non-abelian gauge theories. In both recent numerical and analytical studies one has attempted to characterize confinement by identifying particular classes of field configurations which are far from the small-amplitude fluctuating free fields and may give rise to confinement. I mention here

- Compact fields. These fields extend e.g. up to Gribov horizons and may dominate infrared properties of QCD [8]. Confinement mechanism due to compact fields is akin to that of particles in an infinite square well.
- Singular fields
  - Point singularities - Instantons, Merons. These are fields which become singular at a point in space-time - whence the name instanton. The action density is finite for instantons and infinite for merons [9].
  - Line singularities - Monopoles [10,11].
  - Sheet singularities - Vortex sheets -in particular center vortices [12].
  - Volume singularities. The surface of an MIT-bag can be thought of representing singular fields. For a recent study of the dynamics of domains see[13].

Condensation of such singular or compact fields in the non-perturbative ground state is assumed to provide the basic mechanism for confinement. The wealth of possible

mechanism for confinement is a consequence of the gauge structure of QCD with its intricate consequences for the physics. Gauge theories are theories with redundant variables - four components of gauge fields are present in the Lagrangian for description of two polarization states. Furthermore, in non-abelian gauge theories, with their non-linear relation between gauge potential and field strength, the action density of a large class of singular gauge fields is finite and their contribution to the partition function may therefore not be suppressed by large or divergent values of the action. This forces us in the first place to consider such configurations. In general, finite action singular fields arise in the elimination of redundant variables (gauge fixing). Due to geometrical obstructions in implementing global gauge conditions [14] such singular fields appear in gauge theories in much the same way as singular points appear when defining a global coordinate system (e.g. azimuthal and polar angles) on the surface of a sphere. It is clear that in such a case no physical observable should depend on the singular nature of a particular choice of a representation. This independence of the physics reflects however also the weakness in all these attempts to associate the phenomenon of confinement with singular fields which owe their existence to a particular gauge choice.
This introduction into analytical studies of the physics of confinement starts with a summary of classical chromodynamics followed by a discussion of the quantization within the path integral formalism. Here I will emphasize the non-perturbative elements which make their first appearance in the Faddeev-Popov procedure. After presenting some of the arguments underlying the instanton model of strong interaction I will describe within the context of a specific gauge, the axial gauge, the emergence of compact and singular fields and give an overview of their possible relevance for the gluon and quark dynamics.

2 Quantum Chromodynamics

2.1 The Classical Theory

In this introductory section I review concepts, definitions and basic properties of gauge theories with emphasis on QCD.
Gauge Fields
In non-abelian gauge theories, gauge fields are matrix-valued functions of space-time. In $SU(N)$ gauge theories they can be represented by the generators of the corresponding Lie algebra, i.e. gauge fields and their color components are related by

$$A_\mu(x) = A_\mu^a(x) \frac{\lambda^a}{2},$$

where the color sum runs over the $N^2 - 1$ generators of $SU(N)$. The generators are hermitian, traceless matrices whose commutation-relations are specified by the structure constants f^{abc}

$$\left[\frac{\lambda^a}{2}, \frac{\lambda^b}{2}\right] = i f^{abc} \frac{\lambda^c}{2}.$$

The normalization is chosen as

$$\text{tr}\left(\frac{\lambda^a}{2} \cdot \frac{\lambda^b}{2}\right) = \frac{1}{2}\delta^{ab}.$$

Most of our applications will be concerned with $SU(2)$ gauge theories; in this case the generators are the Pauli matrices

$$\lambda^a = \tau^a ,$$

with structure constants

$$f^{abc} = \epsilon^{abc}.$$

Covariant derivative, field strength tensor, and its color components are defined by

$$D_\mu = \partial_\mu + igA_\mu, \tag{1}$$

$$F_{\mu\nu} = \frac{1}{ig}[D_\mu, D_\nu], \quad F^a_{\mu\nu} = \partial_\mu A^a_\nu - \partial_\nu A^a_\mu - g f^{abc} A^b_\mu A^c_\nu. \tag{2}$$

The definition of electric and magnetic fields in terms of the field strength tensor is the same as in electrodynamics

$$E^{ia}(x) = -F^{0ia}(x) \quad , \quad B^{ia}(x) = -\frac{1}{2}\epsilon^{ijk} F^{jka}(x) \ . \tag{3}$$

The dimensions of gauge field and field strength in 4-dimensional space-time are

$$[A] = \ell^{-1}, \quad [F] = \ell^{-2}$$

and therefore in absence of a scale

$$A^a_\mu \sim M^a_{\mu\nu} \frac{x^\nu}{x^2} .$$

In general, the action of such fields exhibits infrared and ultraviolet logarithmic divergencies. In the following we will discuss

- Yang-Mills Theories
  Only gauge fields are present. The YM Lagrangian is

$$\mathcal{L}_{YM} = -\frac{1}{4} F^{\mu\nu a} F^a_{\mu\nu} = -\frac{1}{2}\text{tr}\,\{F^{\mu\nu} F_{\mu\nu}\} = \frac{1}{2}(\mathbf{E}^2 - \mathbf{B}^2). \tag{4}$$

- Quantum Chromodynamics
  QCD contains besides the gauge fields- the gluons -, fermion fields - the quarks. Quarks are in the fundamental representation, i.e. in $SU(2)$ they are represented by 2-component color spinors. The QCD Lagrangian is (flavor dependence suppressed)

$$\mathcal{L}_{QCD} = \mathcal{L}_{YM} + \mathcal{L}_m, \quad \mathcal{L}_m = \bar{\psi}\left(i\gamma^\mu D_\mu - m\right)\psi, \tag{5}$$

  with the action of the covariant derivative on the quarks given by

$$(D_\mu \psi)^i = (\partial_\mu \delta^{ij} + igA_\mu^{ij})\,\psi^j \quad i,j = 1,2\,.$$

- Georgi-Glashow Model
  In the Georgi-Glashow model [15], the gluons are coupled to scalar (Higgs) fields in the adjoint representation, i.e. in $SU(2)$, the scalar fields are represented as 3-component color-vectors. Lagrangian and action of the covariant derivative are

$$\mathcal{L}_{GG} = \mathcal{L}_{YM} + \mathcal{L}_m, \quad \mathcal{L}_m = \frac{1}{2} D_\mu \phi D^\mu \phi - V(\phi). \tag{6}$$

$$(D_\mu \phi)^a = [D_\mu, \phi]\;^a = (\partial_\mu \delta^{ac} - gf^{abc} A_\mu^b)\phi^c\,.$$

The principle of least action

$$\delta S = 0, \quad S = \int d^4x \mathcal{L}$$

yields when varying the gauge fields

$$\begin{aligned}\delta S_{YM} &= -\int d^4x \,\mathrm{tr}\left\{F_{\mu\nu}\delta F^{\mu\nu}\right\} = -\int d^4x \,\mathrm{tr}\left\{F_{\mu\nu}\frac{2}{ig}\left[D^\mu, \delta A^\nu\right]\right\} \\ &= 2\int d^4x \,\mathrm{tr}\left\{\delta A^\nu \left[D^\mu, F_{\mu\nu}\right]\right\}\end{aligned}$$

the inhomogeneous field equations

$$[D_\mu, F^{\mu\nu}] = j^\nu \tag{7}$$

with j^ν the color current associated with the matter fields

$$j^{a\nu} = \frac{\delta \mathcal{L}_m}{\delta A_\nu^a}\,. \tag{8}$$

For QCD and the Georgi-Glashow model these currents are given by

$$j^{a\nu} = g\bar{\psi}\gamma^\nu \frac{\tau^a}{2}\psi, \quad j^{a\nu} = gf^{abc}\phi^b (D^\nu \phi)^c \tag{9}$$

respectively. As in electrodynamics, the homogeneous field equations for the Yang-Mills field strength

$$\left[D_\mu, \tilde{F}^{\mu\nu}\right] = 0\,,$$

with the dual field strength tensor

$$\tilde{F}^{\mu\nu} = \frac{1}{2}\,\epsilon^{\mu\nu\sigma\rho} F_{\sigma\rho}\,, \tag{10}$$

are obtained as the Jacobi-identities of the covariant derivative

$$[D_\mu, [D_\nu, D_\rho]] + [D_\nu, [D_\rho, D_\mu]] + [D_\rho, [D_\mu, D_\nu]] = 0\,,$$

i.e. they follow from the mere fact that the field strength is represented in terms of gauge potentials.

Gauge-transformations

Gauge transformations change the color orientation of the matter fields locally, i.e. in a space-time dependent manner, and are defined as

$$U(x) = \exp\{ig\alpha(x)\} = \exp\left\{ig\alpha^a(x)\,\frac{\tau^a}{2}\right\}\,,$$

with the arbitrary gauge function $\alpha^a(x)$. Matter fields transform covariantly with U

$$\psi \to U\psi\,, \quad \phi \to U\phi U^\dagger. \tag{11}$$

The transformation property of A is chosen such that $D_\mu\psi$ transforms as ψ. As in electrodynamics, this requirement makes the gauge fields transform inhomogeneously

$$A_\mu(x) \to U(x)\left(A_\mu(x) + \frac{1}{ig}\partial_\mu\right)U^\dagger(x) = A_\mu^{[U]}(x) \tag{12}$$

resulting in a covariant transformation law for the field strength

$$F_{\mu\nu} \to UF_{\mu\nu}U^\dagger. \tag{13}$$

Under infinitesimal gauge transformations ($|g\alpha^a(x)| \ll 1$)

$$A_\mu^a(x) \to A_\mu^a(x) - \partial_\mu\alpha^a(x) - gf^{abc}\alpha^b(x)\,A_\mu^c(x)\,, \tag{14}$$

physical observables must be independent of the choice of gauge (coordinate system in color space). Local quantities such as the Yang-Mills action density $\mathrm{tr}\,F^{\mu\nu}(x)F_{\mu\nu}(x)$ or matter field bilinears like $\bar{\psi}(x)\psi(x), \phi^a(x)\phi^a(x)$ are gauge-invariant, i.e. their value does not change under local gauge transformations. One also introduces non-local quantities which transform homogeneously. In this construction a basic building block is the path ordered integral

$$\Omega\left(x,y,C\right)=P\exp\left\{-ig\int_{s_0}^{s}d\sigma\frac{dx^{\mu}}{d\sigma}A_{\mu}\left(x\left(\sigma\right)\right)\right\}=P\exp\left\{-ig\int_{C}dx^{\mu}A_{\mu}\right\}. \tag{15}$$

It describes a gauge string between the space-time points x, y. Ω satisfies the differential equation

$$\frac{d\Omega}{ds}=-ig\frac{dx^{\mu}}{ds}A_{\mu}\Omega.$$

Gauge transforming this differential equation yields the transformation property of Ω:

$$\Omega\left(x,y,C\right)\rightarrow U\left(x\right)\Omega\left(x,y,C\right)U^{\dagger}\left(y\right).$$

With the help of Ω, non-local, gauge invariant quantities like

$$\mathrm{tr}\{F^{\mu\nu}(x)\Omega\left(x,y,C\right)F_{\mu\nu}(y)\}\,,\quad \bar{\psi}(x)\Omega\left(x,y,C\right)\psi(y),$$

or closed gauge strings - Wilson loops

$$\mathrm{tr}\left\{\Omega\left(x,x,C\right)\right\}$$

can be constructed. While $\bar{\psi}(x)\Omega\left(x,y,C\right)\psi(y)$ is an operator which connects the vacuum with meson-states for $SU(2)$ and $SU(3)$, fermionic baryons appear only in $SU(3)$. In $SU(3)$ a point-like gauge invariant baryonic state is obtained by creating three quarks in a color antisymmetric state at the same space-time point

$$\psi\sim\epsilon^{abc}\psi^{a}(x)\psi^{b}(x)\psi^{c}(x).$$

Under gauge transformations

$$\begin{aligned}\psi &\rightarrow \epsilon^{abc}U_{a\alpha}(x)\psi^{\alpha}(x)U_{b\beta}(x)\psi^{\beta}(x)U_{c\gamma}(x)\psi^{\gamma}(x)\\ &= \det(U(x))\epsilon^{abc}\psi^{a}(x)\psi^{b}(x)\psi^{c}(x)\,.\end{aligned}$$

Operators that create finite size baryonic states must contain appropriate gauge strings as given by the following expression and illustrated by this workshop's logo

$$\psi \sim \epsilon^{abc}(\Omega(u,x,C_1)\psi(x))^a(\Omega(u,y,C_2)\psi(y))^b(\Omega(u,z,C_3)\psi(z))^c$$

Superposition of quarks and gluons

with exchange of color

Under gauge transformations

$$\Omega(u,x,C)\psi(x) \to U(u)\Omega(u,x,C)U^\dagger(x)U(x)\psi(x).$$

With ψ, a simple, non-local and gauge-invariant operator has been constructed which, with appropriate flavor and spin-projections included, will create baryonic states with quantum numbers e.g. of the proton. The correct proton creation operator will exhibit many additional complications, such as string crossing, knots,... as illustrated by the logo, and also virtual quark pairs. Such complications are not required by local gauge invariance - they are determined by the dynamics of quarks and gluons.

Canonical Formalism

The canonical formalism is developed in the same way as in electrodynamics. The most appropriate gauge is the Weyl gauge [16,17]

$$A_0^a = 0, \quad a = 1, ..., N^2 - 1. \tag{16}$$

Due to the antisymmetry of $F_{\mu\nu}$ the Lagrangian (4) does not contain the time derivative of A_0 which in the canonical formalism has to be treated as a constrained variable. In the Weyl gauge these constraint variables are eliminated and the standard procedure of canonical quantization can be employed. In a first step, the canonical momenta of gauge and matter fields (quarks and Higgs fields) are identified

$$\frac{\delta \mathcal{L}_{YM}}{\partial_0 A_i^a} = -E^{a\,i}, \quad \frac{\delta \mathcal{L}_{mq}}{\partial_0 \psi^\alpha} = i\psi^{\alpha\,\dagger}, \quad \frac{\delta \mathcal{L}_{mH}}{\partial_0 \phi^a} = \pi^a .$$

By Legendre transformation, the Hamiltonian densities of the gauge fields

$$\mathcal{H}_{YM} = \frac{1}{2}(\boldsymbol{E}^2 + \boldsymbol{B}^2), \tag{17}$$

and of the matter fields

$$\text{QCD}: \quad \mathcal{H}_m = \psi^\dagger \left(\frac{1}{i}\gamma^0\gamma^i D_i + \gamma^0 m\right)\psi,$$

$$\text{Georgi-Glashow model :} \quad \mathcal{H}_m = \frac{1}{2}\pi^2 + \frac{1}{2}(\boldsymbol{D}\phi)^2 + V(\phi)$$

are obtained. The gauge condition (16) does not fix uniquely the gauge, it still allows for time-independent gauge transformations. As a consequence the Hamiltonian exhibits a local symmetry. This residual gauge symmetry is taken into account by requiring physical states $|\Phi\rangle$ to satisfy the Gauss law, i.e. the 0-component of the equation of motion (cf. Eq.(7))

$$([D_i, E^i] + j^0)|\Phi\rangle = 0.$$

Implementation of the Gauss law cannot be carried out in general in closed form which severely limits the application of the canonical formalism. In axial gauge a complete formulation has been given in [18]. The connection of canonical to path-integral quantization is discussed in detail in [19].

2.2 *Quantization - The Faddeev-Popov Procedure*

Gauge copies

Gauge theories are formulated in terms of redundant variables. Only in this way, a covariant, local representation of the dynamics of gauge degrees of freedom is possible. For quantization of the theory both canonically or in the path integral, redundant variables have to be eliminated - this procedure is called gauge fixing. The procedure is not unique and the implications of a particular choice are generally not well understood. In the path integral a sum over all field configurations is performed. Gauge theories contain with a given field an infinity of gauge copies i.e. fields which are obtained from each other by gauge transformations. The set of fields

$$O = \left\{ A^{[U]} | U \epsilon G \right\} \tag{18}$$

with A fixed and U running over the set of gauge transformations is called the gauge orbit. The fields in a given gauge orbit differ from each other by gauge transformations only and describe therefore the same physical situation. Instead of summing in the path integral over formally different but physically equivalent fields the integration is performed over the equivalence classes of such fields i.e. over the corresponding gauge orbits. The value of the action is gauge invariant, i.e. the same for all members of a given gauge orbit,

$$S\left[A^{[U]}\right] = S\left[A\right].$$

Therefore the action is seen to be a functional defined on classes (gauge orbits). Also the integration measure

$$d\left[A^{[U]}\right] = d\left[A\right] \quad , \quad d\left[A\right] = \prod_{x,\mu,a} dA^a_\mu\left(x\right) ,$$

is gauge invariant since shifts and rotations of an integration variable do not change the value of an integral. Therefore in the naïve path integral

$$\tilde{Z} = \int d[A]\, e^{iS[A]} \propto \int \prod_x dU(x) \;,$$

a "volume" associated with the gauge transformations $\prod_x dU(x)$ can be factorized and thereby the integration be performed over the gauge orbits. To turn this property into a working algorithm redundant variables are eliminated by imposing a gauge condition

$$f[A] = 0, \tag{19}$$

which is supposed to eliminate all gauge copies of a certain field configuration A. In other words, the functional f has to be chosen such that for arbitrary field configurations the equation

$$f[A^{[U]}] = 0$$

determines uniquely the gauge transformation U. If successful, the set of all gauge equivalent fields, the gauge orbit, is represented by exactly one representative. In order to write down an integral over gauge orbits we insert into the integral the gauge-fixing δ-functional

$$\delta\left[f(A)\right] = \prod_x \prod_{a=1}^{N^2-1} \delta\left[f^a\left(A(x)\right)\right].$$

This modification of the integral however changes the value depending on the representative chosen, as the following elementary identity shows

$$\delta\left(g(x)\right) = \frac{\delta(x-a)}{|g'(a)|} \quad , \quad g(a) = 0. \tag{20}$$

This difficulty is circumvented with the help of the Faddeev-Popov determinant $\Delta_f[A]$ defined implicitly by

$$\Delta_f[A] \int d[U]\, \delta\left[f\left(A^{[U]}\right)\right] = 1.$$

Multiplication of the path integral $\tilde{Z}$ with the above and taking into account the gauge invariance of the various factors yields

$$\begin{aligned}\tilde{Z} &= \int d[U] \int d[A]\, e^{iS[A]} \Delta_f[A]\, \delta\left[f\left(A^{[U]}\right)\right] \\ &= \int d[U] \int d[A]\, e^{iS[A^{[U]}]} \Delta_f\left[A^{[U]}\right] \delta\left[f\left(A^{[U]}\right)\right] = \int d[U]\, Z.\end{aligned}$$

The gauge volume has been factorized and, being independent of the dynamics, can be dropped. In summary, the final definition of the generating functional for gauge theories

$$Z[J] = \int d[A]\, \Delta_f[A]\, \delta(f[A])\, e^{iS[A]+i\int d^4x J^\mu A_\mu} \tag{21}$$

is given in terms of a sum over gauge orbits.

Faddeev-Popov Determinant

For the calculation of $\Delta_f[A]$ we first consider the change of the gauge condition $f^a[A]$ under infinitesimal gauge transformations . Taylor expansion

$$\begin{aligned} f_x^a\left[A^{[U]}\right] &\approx f_x^a[A] + \int d^4y \sum_{b,\mu} \frac{\delta f_x^a[A]}{\delta A_\mu^b(y)} \delta A_\mu^b(y) \\ &= f_x^a[A] + \int d^4y \sum_b M(x,y;a,b)\, \alpha^b(y) \end{aligned}$$

with δA_μ^a given by infinitesimal gauge transformations (14), yields

$$M(x,y;a,b) = \left(\partial_\mu \delta^{b,c} + g f^{bcd} A_\mu^d(y)\right) \frac{\delta f_x^a[A]}{\delta A_\mu^c(y)}\,. \tag{22}$$

In the second step we compute the integral

$$\Delta_f^{-1}[A] = \int d[U]\, \delta\left[f\left(A^{[U]}\right)\right]$$

by expressing the integration $d[U]$ by integration over the gauge functions α. We finally change to the variables $\beta = M\alpha$

$$\Delta_f^{-1}[A] = |\det M|^{-1} \int d[\beta]\, \delta[f(A) - \beta]$$

and arrive at the final expression for the Faddeev-Popov determinant

$$\Delta_f[A] = |\det M|\,. \tag{23}$$

Examples

- Lorentz Gauge

$$f_x^a(A) = \partial^\mu A_\mu^a(x) - \chi^a(x)$$

$$M(x,y;a,b) = -\left(\delta^{ab}\Box - g f^{abc} A_\mu^c(y)\, \partial_y^\mu\right) \delta^{(4)}(x-y)\,.$$

- Coulomb Gauge

$$f_x^a(A) = \text{div}\vec{A}^a(x) - \chi^a(x)$$

$$M(x,y;a,b) = \left(\delta^{ab}\Delta + gf^{abc}\vec{A}^c(y)\vec{\nabla}_y\right)\delta^{(4)}(x-y)\,.$$

- Axial Gauge

$$f_x^a(A) = n^\mu A_\mu^a(x) - \chi^a(x)$$

$$M(x,y;a,b) = -\delta^{ab}n_\mu\partial_y^\mu\delta^{(4)}(x-y)\,.$$

We note that in axial gauge, the Faddeev-Popov determinant does not depend on the gauge fields and therefore changes the generating functional by an irrelevant factor.

Gribov-Horizons

As the elementary example (20) shows, a vanishing Faddeev-Popov determinant ($g'(a) = 0$) indicates the gauge condition to exhibit a quadratic or higher order zero. This implies that at this point in function space, the gauge condition is satisfied by at least two gauge equivalent configurations i.e. vanishing of $\Delta_f[A]$ implies the existence of zero modes associated with M (Eq.(22))

$$M\chi_0 = 0$$

and therefore the gauge choice is ambiguous. The (connected) spaces of gauge fields which make the gauge choice ambiguous

$$\mathcal{M}_H = \{A|\ \det M = 0\}$$

are called Gribov horizons [8]. Around Gribov horizons, pairs of infinitesimally close gauge equivalent fields exist which satisfy the gauge condition. If on the other hand two gauge fields satisfy the gauge condition and are separated by an infinitesimal gauge transformation these two fields are separated by a Gribov horizon. The region beyond the horizon thus contains gauge copies of fields inside the horizon. In general one therefore needs additional conditions to select exactly one representative of the gauge orbits. The structure of Gribov horizons and of the space of fields which contain no Gribov copies depends on the choice of the gauge. Without specifying further the procedure we associate an infinite potential energy $\mathcal{V}[A]$ with every gauge copy of a configuration which already has been taken into account, i.e. after gauge fixing, the action is supposed to contain implicitly this potential energy

$$S[A] \to S[A] - \int d^4x\,\mathcal{V}[A]. \tag{24}$$

With the above expression, and given a reasonable gauge choice, the generating functional is written as an integral over gauge orbits and can serve as starting point for further formal developments such as the canonical formalism [19] or applications e.g. perturbation theory.

The occurrence of Gribov horizons points to a more general problem in the gauge fixing procedure. Unlike in electrodynamics, global gauge conditions may not exist in non-abelian gauge theories [14]. In other words, it may not be possible to formulate a condition which in the whole space of gauge fields selects exactly one representative. This difficulty of imposing a global gauge condition is similar to the problem of global coordinate choice on e.g. S^2. In this case one either has to resort to some patching procedure and use more than one set of coordinates (like for the Wu-Yang Monopole) or deal with singular fields arising from these gauge ambiguities (Dirac Monopole). Gauge singularities are analogous to the coordinate singularities on non-trivial manifolds (azimuthal angle on north pole).

3 Instantons

Instantons are solutions of the classical field equations with distinguished topological properties [20]. Our discussion of instantons starts with a discussion of configurations of vanishing energy expected to be of particular importance for the quantum mechanical ground state [21,17]. As follows from the Yang-Mills Hamiltonian (17) in the Weyl gauge (16) static, zero-energy solutions of the equations of motion (7) satisfy

$$\mathbf{E} = 0 \quad \mathbf{B} = 0,$$

and are therefore of the following structure

$$\mathbf{A} = \frac{1}{ig} U(\mathbf{x}) \boldsymbol{\nabla} U^\dagger(x). \tag{25}$$

Vanishing of the electric and magnetic fields (cf. Eqs.(2), (3)) is easily verified. A gauge potential with vanishing field strength is called pure gauge. Pure gauges in electrodynamics are gradients of scalar time-independent functions. In $SU(2)$ Yang-Mills theory, the manifold of zero energy solutions is according to (25) given by the set of all $U(\mathbf{x}) \in SU(2)$. Topologically

$$SU(2) \sim S^3 , \tag{26}$$

since any $U \in SU(2)$ can be written as

$$U = a_0 + i\mathbf{a} \cdot \boldsymbol{\tau}$$

with the unitarity constraint

$$1 = U\,U^+ = (a_0 + i\mathbf{a} \cdot \boldsymbol{\tau})(a_0 - i\mathbf{a} \cdot \boldsymbol{\tau}) = a_0^2 + \mathbf{a}^2 .$$

We impose the requirement that at infinity, $U(\mathbf{x})$ approaches a unique value independent of the direction of $\mathbf{x}$

$$U(\mathbf{x}) \to \text{const.} \quad \text{for } |\mathbf{x}| \to \infty.$$

Thereby the configuration space becomes compact $R^3 \longrightarrow S^3$. In this way pure gauges define a map $U(\mathbf{x}) :\ S^3 \longrightarrow S^3$ to which a winding number can be assigned. This winding number counts how many times the 3-sphere of gauge transformations $U(\mathbf{x})$ is covered if $\mathbf{x}$ covers once the 3-sphere of the compactified configuration space and is given in terms of the gauge fields by

$$n_w = \frac{g^2}{16\pi^2} \int d^3x\, \epsilon^{ijk} \left(A_i^a\, \partial_j\, A_k^a - \frac{g}{3}\, \epsilon_{abc}\, A_i^a\, A_j^b\, A_k^c \right). \tag{27}$$

The expression on the right hand side yields an integer only if A is a pure gauge. Examples of gauge transformations giving rise to non-trivial winding (hedgehog solution for n=1) are

$$U_n(\mathbf{x}) = \exp\{i\pi n \frac{\mathbf{x}\boldsymbol{\tau}}{\sqrt{x^2+\rho^2}}\} \tag{28}$$

with winding number $n_w = n$. These topological considerations show that Yang-Mills theory considered as a classical system possesses an infinity of different lowest energy ($E = 0$) solutions which can be labeled by an integer n. They are connected to each other by gauge fields which cannot be pure gauges and which therefore produce a finite value of the magnetic field, i.e. of the potential energy. The schematic plot of the potential energy in Fig.1 shows that the ground state of QCD can be expected to exhibit similar properties as that of a particle moving in a periodic potential. In the quantum mechanical case too, an infinite degeneracy is present with the winding number in gauge theories replaced by the integer characterizing the equilibrium positions of the particle. "Classical vacua" are states with values of the coordinate of a mechanical system $x = n$ given by the equilibrium positions. Correspondingly, in gauge theories the classical vacua, the "n - vacua" are given by the pure gauges (Eqs.(25),(28)). To proceed from here to a description of the quantum-mechanical ground state, tunneling processes have to be included which, in such a semiclassical approximation, connect classical vacua with each other. Thereby the quantum mechanical ground state becomes a linear superposition of classical vacua. Such tunneling solutions are most easily obtained by changing to imaginary time with a concomitant change in the time component of the gauge potential

$$\to -it\,; \quad A_0 \to -iA_0. \tag{29}$$

The metric becomes Euclidean and there is no distinction between covariant and contravariant indices. Thereby the tunneling solutions become solutions of the classical field equations derived from the Euclidean action S_E i.e. the Yang Mills action

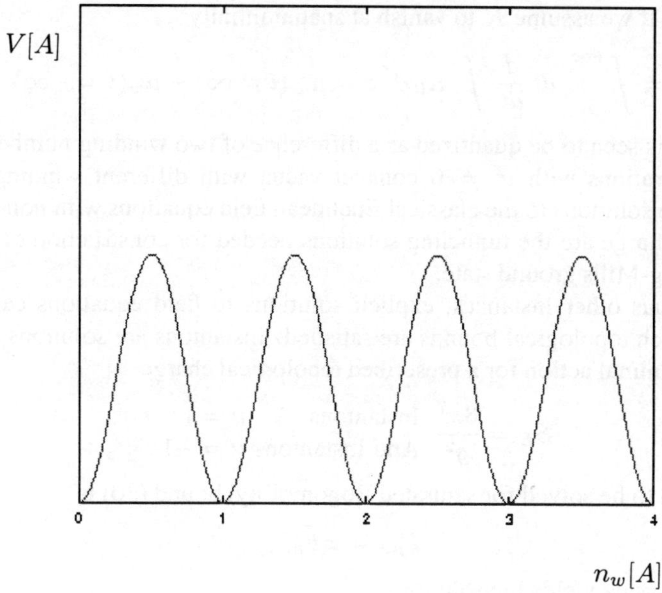

Fig. 1: Potential energy $V[A] = \int d^3x B[A]^2$ as a function of the winding number.

(cf. Eq.(4)) modified by the substitution (29). This action can be rewritten in the following form (cf. Eq.(10))

$$S_E = \frac{1}{4}\int d^4x\; F_{a\mu\nu}\, F_{a\mu\nu} = \frac{1}{4}\int d^4x \left(\pm F_{a\mu\nu}\tilde{F}_{a\mu\nu} + \frac{1}{2}\left(F_{a\mu\nu} \mp \tilde{F}_{a\mu\nu}\right)^2\right) \tag{30}$$

$$\geq \pm\frac{1}{4}\int d^4x\; F_{a\mu\nu}\, \tilde{F}_{a\mu\nu}\,. \tag{31}$$

This bound for S_E (Bogomoln'yi-bound) is determined by the 'topological charge', i.e. it can be rewritten as a surface term

$$\nu = \frac{g^2}{32\pi^2}\int d^4x\; F\,\tilde{F} = \int d\sigma_\mu\, K_\mu \tag{32}$$

of the topological current K^μ

$$K_\mu = \frac{g^2}{16\pi^2}\,\epsilon_{\mu\alpha\beta\gamma}\left(A_{a\alpha}\,\partial_\beta\, A_{a\gamma} - \frac{g}{3}\,\epsilon_{abc}\, A_{a\alpha}\, A_{b\beta}\, A_{c\gamma}\right). \tag{33}$$

Furthermore, if we assume K to vanish at spatial infinity

$$\nu = \int_{-\infty}^{+\infty} dt \, \frac{d}{dt} \int K_0 \, d^3 x = n_w (t = \infty) - n_w (t = -\infty) \qquad (34)$$

the charge ν is seen to be quantized as a difference of two winding numbers. Hence field configurations with $\nu \neq 0$ connect vacua with different winding number. Therefore the solutions to the classical Euclidean field equations with non-vanishing topological charge are the tunneling solutions needed for construction of the semi-classical Yang-Mills ground state.
Like in various other instances, explicit solutions to field equations can be constructed if such topological bounds are satisfied. Instantons are solutions in the Euclidean of minimal action for a prescribed topological charge

$$S_E = \frac{8\pi^2}{g^2} \begin{array}{ll} \text{Instantons} & \nu = 1 \\ \text{Anti Instantons} & \nu = -1 \, . \end{array}$$

The equation to be solved for saturated Bogomol'nyi bound (30) is

$$F_{\mu\nu} = \pm \tilde{F}_{\mu\nu} . \qquad (35)$$

A spherical ansatz yields the solutions

$$A_{a\mu} = -\frac{2}{g} \frac{\eta_{a\mu\nu} x_\nu}{x^2 + \rho^2} \qquad F^2_{\mu\nu} = \frac{1}{g^2} \frac{192 \rho^4}{(x^2 + \rho^2)^4} , \qquad (36)$$

with the 't Hooft symbol [22]

$$\eta_{a\mu\nu} = \begin{cases} \epsilon_{a\mu\nu} & \mu, \nu = 1, 2, 3 \\ \delta_{a\mu} & \nu = 0 \\ -\delta_{a\nu} & \mu = 0 \end{cases} .$$

The size of the instanton ρ can be freely chosen. Asymptotically, gauge potential and field strength behave as

$$A \underset{|x| \to \infty}{\longrightarrow} \frac{1}{x} \qquad F \underset{|x| \to \infty}{\longrightarrow} \frac{1}{x^4} .$$

The unexpected strong decrease in the field strength is the result of a partial cancellation of abelian and non-abelian contributions to $F_{\mu\nu}$ (Eq.(2)). For instantons, the asymptotics of the gauge potential is actually gauge dependent. By a gauge transformation, the asymptotics can be changed to x^{-3}. In this "singular" gauge, the gauge potential is given by

$$A_{a\mu} = -\frac{2\rho^2}{g x^2} \frac{\bar{\eta}_{a\mu\nu} x_\nu}{x^2 + \rho^2} , \quad \bar{\eta}_{a\mu\nu} (1 - 2\delta_{\mu,4})(1 - 2\delta_{\nu,4}) .$$

Instanton Gas - A model for low energy phenomena of QCD.

In the semiclassical approximation as sketched above, the non-perturbative QCD ground state is assumed to be given by topologically distinguished pure gauges and the instantons connecting the different classical vacuum configurations. In the instanton model for the description of low energy phenomena one replaces the QCD partition function (Eq.(21)), i.e. the weighted sum over all gauge fields by a sum over (singular gauge) instanton fields

$$A_\mu = \sum_{i=1}^{N} U(i)\, A_\mu(i)\, U^+(i)\,, \tag{37}$$

with

$$A_\mu(i) = -\bar{\eta}_{a\mu\nu} \frac{1}{g(x-z(i))^2} \frac{x_\nu - z_\nu(i)}{(x-z(i))^2 + \rho^2}\, \tau_a\,.$$

The gauge field is composed of N instantons with their centers located at the positions $z(i)$ and color orientations specified by the $SU(2)$ matrices $U(i)$. The instanton ensemble for calculation of n-point functions is obtained by summing over these positions and color orientations

$$Z[J] = \int \prod_{i=1}^{N} [dU(i)dz(i)]\, e^{-S_E[A]+i\int J\cdot A\, d^4x}.$$

Starting point of hadronic phenomenology in terms of instantons are the fermionic zero modes concentrated around each individual instanton

$$\not{D}\psi_0 = 0,$$

$$\psi_0 = \frac{\rho}{\pi\sqrt{x^2}} \frac{\gamma x}{(x^2+\rho^2)^{\frac{3}{2}}} \frac{1+\gamma_5}{2}\, \varphi_0.$$

Existence of zero-modes is guaranteed by the non-vanishing topological charge of instantons. In the instanton model, the functional integration over quarks too is truncated and replaced by a sum over the zero-modes in a given configuration of non-overlapping instantons. A successful description of hadronic properties has been achieved [3], although a dilute gas of instantons does not confine quarks and gluons. To describe confinement, merons have been proposed [9] as the relevant field configurations. Merons are solutions of classical equations of motion. They are literally half-instantons, i.e. up to a factor of $1/2$ they are identical with instantons (Eq.(36)) (in regular gauge)

$$A^M_{a\mu}(x) = \frac{1}{2} A^I_{a\mu}(x) = -\frac{1}{g} \frac{\eta_{a\mu\nu} x_\nu}{x^2}$$

and carry half a unit of topological charge. By this change of normalization the cancellation between abelian and non-abelian contributions to the field strength is upset and therefore, asymptotically

$$A \sim \frac{1}{x}, \quad F \sim \frac{1}{x^2}.$$

The action

$$S \sim \int d^4x \frac{1}{x^4}$$

exhibits a logarithmic infrared in addition to the ultraviolet divergence. Unlike instantons in singular gauge ($A \sim x^{-3}$) , merons always overlap - a dilute gas limit of an ensemble of merons does not exist, i.e. merons are strongly interacting.

4 Compact Variables and Singular Fields

Unlike instantons, most of the special field configurations considered to be important in generating confinement emerge only in the process of elimination of the redundant variables. Their possible relevance for confinement is therefore intimately linked to a particular gauge choice. In the following discussion I will present a detailed study of the emergence of compact and singular fields within the axial gauge [23] and discuss their possible role in the dynamics of the confined and plasma phase.

Kinematics For proper formulation of the axial gauge it is advantageous to assume that space-time has a compact direction. We choose this direction to be spacelike and of extension L and we require

$$0 \leq x_3 \leq L \quad x = (x_\perp, x_3),$$

with

$$x_\perp = (x^0, x^1, x^2).$$

We impose periodic boundary conditions for gauge and bosonic matter fields

$$A^a_\mu(x_\perp, x_3 + L) = A^a_\mu(x_\perp, x_3). \tag{38}$$

Fermion fields are subject to anti-periodic boundary conditions

$$\psi(x_\perp, x_3 + L) = -\psi(x_\perp, x_3). \tag{39}$$

We have chosen the compact direction to be spacelike in order to be able to apply the standard canonical formalism in the analysis of QCD. The results to be obtained can in a simple way be related to QCD at finite temperature. Covariance connects relativistic field theories at finite extension with the same systems at finite temperature provided the boundary conditions (38) and (39) are chosen. To illustrate this surprising connection [24,23] we compare energy density and pressure of the black-body radiation with the same quantities observed in the Casimir effect

Stefan-Boltzmann	Casimir
$\epsilon = \frac{\pi^2}{15} T^4$	$p = -\frac{\pi^2}{15} L^{-4}$
$p = \frac{\pi^2}{45} T^4$	$\epsilon = -\frac{\pi^2}{45} L^{-4}$. (40)

This result indicates that in relativistic theories, thermal and quantum fluctuations are closely related to each other. The connection is easily established by considering the partition function given in terms of the Euclidean form (cf. Eq.(29)) of the Lagrangian

$$Z = \int_{\text{period.}} D[...]e^{-\int_0^\beta dx_0 \int dx_1 dx_2 dx_3 \mathcal{L}_E[...]}$$

which describes a system of infinite extension at temperature $T = \beta^{-1}$.

The partition function

$$Z = \int_{\text{period.}} D[...]e^{-\int_0^L dx_3 \int dx_0 dx_1 dx_2 \mathcal{L}_E[...]}$$

describes the same dynamical system in its ground state at finite extension L in 3-direction.

As consequence, by interchanging the coordinate labels in the Euclidean form one easily derives

$$\mathcal{Z}(\beta, L) - \mathcal{Z}(L, \beta)$$

$$\epsilon(\beta, L) = -p(L, \beta). \qquad (41)$$

These relations hold irrespective of the dynamics of the system. In particular it implies that any phase transition taking place when heating up a system has as its counterpart a phase transition occurring when compressing the system. Critical temperature and critical length are related by

$$T_c = \frac{1}{L_c}.$$

For QCD with its supposed phase transition at about 150 MeV this relation implies the existence of a phase transition when compressing the system beyond 1.3 fm.

4.1 Gauge Fixing

We now carry out the elimination of redundant variables and attempt to eliminate the 3-component of the gauge field $A_3(x)$. Formally this can be achieved by applying the gauge transformation

$$\Omega(x) = P \exp ig \int_0^{x_3} dz\, A_3\left(x_\perp, z\right).$$

It is straightforward to verify that the gauge transformed 3-component of the gauge field indeed vanishes (cf. Eq.(12))

$$A_3\left(x\right) \to \Omega\left(x\right)\left(A_3\left(x\right) + \frac{1}{ig}\partial_3\right)\Omega^\dagger\left(x\right) = 0.$$

However this gauge transformation to axial gauge is not quite legitimate. The gauge transformation is not periodic

$$\Omega(x_\perp, x_3 + L) \neq \Omega(x_\perp, x_3).$$

In general, gauge fields do not remain periodic either under transformation with Ω. To correct for this shortcoming, we introduce the Polyakov loop

$$P(x_\perp) = P \exp\left\{ig \int_0^L dx_3\, A_3(x)\right\}, \tag{42}$$

a path ordered integral with identical end-points - or equivalently the Wilson loop along the compact direction (cf. Eq.(15)). As we have seen, the trace of these objects is gauge invariant and therefore with A_3 also gauge invariant objects are eliminated by Ω. Periodicity is preserved and the loop variables $\mathrm{tr}\, P(x_\perp)$ are restored with the following modified gauge transformation

$$\Omega_{ag}(x) = \Omega_D\left(x_\perp\right)\left(P^\dagger\left(x_\perp\right)\right)^{x_3/L}\Omega(x)\,. \tag{43}$$

The gauge fixing to axial gauge thus proceeds in three steps

- Elimination of the 3-component of the gauge field $A_3(x)$
- Restoration of the Polyakov loops $P(x_\perp)$
- Elimination of the gauge variant components of the Polyakov loops $P(x_\perp)$ by diagonalization

$$\Omega_D\left(x_\perp\right) P(x_\perp)\Omega_D^\dagger\left(x_\perp\right) = e^{igLa_3(x_\perp)\,\tau_3/2}\,. \tag{44}$$

Generating Functional

With the above procedure we have by explicit construction of the appropriate gauge transformations established that the 3-component of the gauge field indeed can be eliminated in favor of a diagonal, x^3-independent field $a_3(x_\perp)$. In the language of the Faddeev-Popov procedure the axial gauge condition (cf. Eq.(19)) therefore reads

$$f[A] = A_3 - \left(a_3 + \frac{\pi}{gL}\right)\frac{\tau_3}{2}.$$

The field $a_3(x_\perp)$ is compact.

$$a_3 = a_3(x_\perp), \qquad -\frac{\pi}{gL} \le a_3(x_\perp) \le \frac{\pi}{gL}.$$

It is interesting to compare QED and QCD in axial gauge in order to identify at this level already properties which are related to the non-abelian character of QCD. In QED the same procedure can be followed with the omission of the third step. Once more, a lower dimensional field has to be kept for periodicity and gauge invariance. However in QED, the integer part of $a_3(x_\perp)$ cannot be gauged away; as winding number of the mapping $S^1 \to S^1$ it is protected topologically. In QCD, the appearance of the compact variable is ultimately due to the elimination of the gauge field A_3, an element of the Lie algebra, in favor of $P(x_\perp)$, an element of the compact Lie group.

With the help of the auxiliary field $a_3(x_\perp)$, the generating functional for QCD in axial gauge is written as

$$Z[J] = \int d[a_3]d[A]\,\Delta_f[A]\,\delta\left(A_3 - \left(a_3 + \frac{\pi}{gL}\right)\frac{\tau_3}{2}\right)\, e^{iS[A]+i\int d^4x J^\mu A_\mu}. \quad (45)$$

This generating functional contains as dynamical variables the fields $a_3(x_\perp), A_\perp(x)$ with

$$A_\perp(x) = (A_0(x), A_1(x), A_2(x)).$$

It is one of the unique features of axial gauge QCD that the Faddeev-Popov determinant (cf. Eqs.(23),22))

$$\Delta_f[A] = |\det D_3|$$

can be evaluated in closed form

$$\frac{\det D_3}{(\det \partial_3)^3} = \prod_{x_\perp} \frac{1}{L^2}\cos^2 gLa_3(x_\perp)/2$$

and absorbed into the measure

$$Z[J] = \int D[a_3]d[A_\perp]\, e^{iS\left[A_\perp, a_3 - \frac{\pi}{gL}\right] + i\int d^4x JA}.$$

The measure

$$D[a_3] = \prod_{x_\perp} \cos^2\left(gLa_3(x_\perp)/2\right)\, \Theta\left[(a_3(x_\perp)^2 - (\pi/gL)^2\right] da_3(x_\perp)\,. \quad (46)$$

is nothing else than the Haar measure of the gauge group. It reflects the presence of variables (a_3) which are built from elements of the Lie group and not of the Lie algebra. The associated volume element of the first polar angle on S^3 (cf. Eq.(26)) naturally appears with, as will be seen, far reaching consequences.

4.2 Center Symmetry and Polyakov Loops

For preserving periodicity of the gauge fields, gauge transformations $U(x)$ do not have to be strictly periodic. In order to specify more precisely the conditions on $U(x)$ we consider the transformation of the Polyakov-loop variables $P(x_\perp)$

$$P(x_\perp) \to U(x_\perp, L)\, P(x_\perp) U^\dagger(x_\perp, 0)$$

with $x = (x_\perp, 0)$ and $x = (x_\perp, L)$ describing identical points. The periodicity properties remain unchanged if

$$U(x_\perp, L) = c_U \cdot U(x_\perp, 0)\ ,$$

with c_U in the center of the group. One therefore can classify gauge transformations according to the value of c_U (± 1 in SU(2)).

$$\mathrm{tr}(P(x_\perp)) \to \mathrm{tr}(c_U P(x_\perp)) \overset{\mathrm{SU(2)}}{=} \pm\mathrm{tr}(P(x_\perp)).$$

Gauge transformations with $c = 1$ are strictly periodic; an important example of an SU(2) gauge transformation u_- with $c = -1$ is

$$u_- = e^{i\pi\hat{\psi}\tau x_3/L} = \cos \pi x_3/L + i\hat{\psi}\tau \sin \pi x_3/L.$$

Here $\hat{\psi}$ is a constant unit vector in color space. Under u_-, gauge fields transform as

$$A_\mu^{[u_-]} = e^{i\pi\hat{\psi}\tau x_3/L} A_\mu e^{-i\pi\hat{\psi}\tau x_3/L} - \frac{\pi}{gL}\hat{\psi}\tau\delta_{\mu 3}$$

and indeed remain periodic. Gauge transformations $U(x)$ can be decomposed into different classes labeled by the values of c_U. In $SU(2)$, the gauge orbits $\mathcal{O}$ (cf. Eq.(18)) are decomposed according to $c = \pm 1$ into sub-orbits $\mathcal{O}_\pm$. Thus these suborbits are characterized by the sign of the Polyakov loop at some fixed $x_\perp^0$

$$A(x) \in \mathcal{O}_\pm \quad , \quad \text{if} \quad \pm\, \mathrm{tr}(P(x_\perp^0)) \geq 0. \quad (47)$$

Strictly speaking it is not the trace of the Polyakov loop rather only its modulus $|\mathrm{tr}(P(x_\perp)|$ which is invariant under all gauge transformations. Complete gauge fixing, i.e. a representation of gauge orbits $\mathcal{O}$ by exactly one representative is only possible if the gauge fixing transformations are not strictly periodic. In turn, if gauge fixing is carried out with strictly periodic gauge fixing transformations ($U, c_U = 1$) the resulting ensemble of gauge fields contains one representative for each of the suborbits (47). As a consequence, the ensemble must be symmetric under residual discrete gauge transformations which change the sign of the Polyakov loop

$$\text{tr}(P(x_\perp)) \to -\text{tr}(P(x_\perp)).$$

It is straightforward to construct the residual symmetry transformations in axial gauge.

Center Reflections

Center reflections Z are residual gauge transformations which in axial gauge are given by

$$Z = ie^{i\pi\tau_1/2}e^{i\pi\tau_3 x^3/L}.$$

Center reflections change the sign of the Polyakov loop

$$Z^2 = 1, \quad c_Z = -1,$$

transform the gauge fields, and leave the action invariant

$$Z: \quad a_3 \to -a_3, \quad A^3_\mu \to -A^3_\mu, \quad \Phi_\mu \to \Phi^\dagger_\mu, \quad S[A_\perp a_3] \to S[A_\perp a_3]\,. \tag{48}$$

The off-diagonal gluon fields have been represented in a spherical basis by the antiperiodic fields

$$\Phi_\mu(x) = \frac{1}{\sqrt{2}}(A^1_\mu(x) + iA^2_\mu(x))e^{-i\pi x^3/L}\,. \tag{49}$$

In such a basis Z acts as (abelian) charge conjugation with the charged fields $\Phi_\mu(x)$ and the "photons" described by the neutral fields $A^3_\mu(x), a_3(x_\perp)$.
Under center reflections, the trace of the Polyakov loop

$$\frac{1}{2}\,\text{tr}\,P(x_\perp) = -\sin\frac{1}{2}gLa_3(x_\perp) \tag{50}$$

changes sign

$$Z: \quad \text{tr}\,P(x_\perp) \to -\text{tr}\,P(x_\perp). \tag{51}$$

The center symmetry is a standard symmetry of the canonical formalism. Center reflection commutes with the Hamiltonian

$$[H, Z] = 0\,.$$

Stationary states in SU(2) Yang-Mills theory can therefore be classified according to their Z-Parity. Equivalently, in a center-symmetric ensemble the distribution of the

Polyakov-loop variables a_3 must be symmetric around $a_3 = 0$ (cf. Fig.2). The dynamics of the Polyakov loop is intimately connected to confinement. The Polyakov loop is associated with the free energy of a single heavy charge. In electrodynamics, the coupling of a heavy pointlike charge to an electromagnetic field is given by

$$\begin{aligned}\delta\mathcal{L} &= \int d^4x j^\mu A_\mu = \int d^4x \delta(\mathbf{x}-\mathbf{y})\, A_0(x) \\ &= \int_0^L dx_0 A_0(x_0,\mathbf{y}),\end{aligned}$$

which, in the Euclidean and after interchange of coordinate labels 0 and 3, reduces to the Polyakov loop. The property of the system to confine can be formulated as a symmetry property. The expected infinite free energy of a static color charge results in a vanishing ground state expectation value of the Polyakov loop

$$\langle 0|\mathrm{tr}\, P(x_\perp)|0\rangle = 0 \tag{52}$$

in the confined phase. This property is guaranteed if the vacuum is center symmetric. The vacuum expectation value

$$\langle 0|trP(x_\perp)|0\rangle \sim \int_{-\frac{\pi}{gl}}^{\frac{\pi}{gl}} da_3(x_\perp)\cos^2\left(gLa_3(x_\perp)/2\right)\,\psi^2[a_3,A_\perp]\,\sin\frac{1}{2}gLa_3(x_\perp)$$

vanishes, if the wavefunctional $\psi[a_3, A_\perp]$ is, up to a phase, reproduced under the center reflections (Eq.(48)). The interaction energy of two static charges is, up to an additive constant, given by the Polyakov-loop correlator

$$\langle 0|\mathrm{tr}P(x_\perp)\mathrm{tr}P(0)|0\rangle = e^{-LV(x_\perp)}. \tag{53}$$

Thus vanishing of the Polyakov loops in the center symmetric phase indicates an infinite free energy of static color charges, i.e. confinement. For non-zero Polyakov-loop expectation values the free energy of a static color charge is finite and the system is deconfined. A non-vanishing expectation value is possible only if center symmetry is broken. Thus in the transition from the confined to the plasma phase the center symmetry, i.e. a discrete part of the underlying gauge symmetry, must be spontaneously broken. In the broken phase, the distribution of the Polyakov-loop variables will not be symmetric anymore around $a_3 = 0$. We expect a distribution centered around the "perturbative" value $a_3 = -\frac{\pi}{gL}$ corresponding to $A_3 = 0$ (cf. Fig.2).

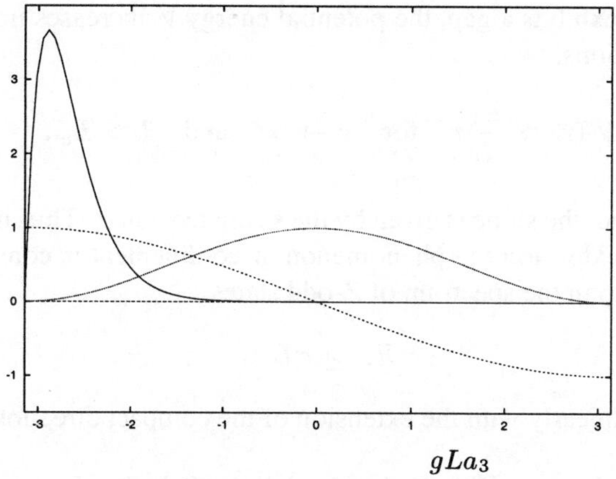

Fig.2: Polyakov loop (Eq.(50)), Haar measure (Eq.(46)) and schematic center symmetry breaking distribution.

4.3 The Spectrum of the SU(2) Yang-Mills Theory

In this section we will, independent of any approximation, characterize on the basis of confinement and covariance, the spectrum of the SU(2) Yang Mills theory in the confined, center symmetric phase.

- In the center-symmetric phase,

$$Z|0\rangle = |0\rangle$$

the vacuum expectation value of the Polyakov loop vanishes (Eq.(52)).

- The correlation function of Polyakov loops yields the interaction energy V of static color charges (in the fundamental representation)

$$\exp\left\{-LV\left(r\right)\right\} = \langle 0|T\left[\text{tr}\,P\left(x_{\perp}^{E}\right)\text{tr}\,P\left(0\right)\right]|0\rangle\,, \qquad r^2 = \left(x_{\perp}^{E}\right)^2. \tag{54}$$

- Due to the rotational invariance in Euclidean transverse space, $x_{\perp}^{E}$ can be chosen to point in the time direction. After insertion of a complete set of excited states

$$\exp\left\{-LV\left(r\right)\right\} = \sum_{n_-} \left|\langle n_-|\text{tr}\,P\left(0\right)|0\rangle\right|^2 e^{-E_{n_-}r}\,. \tag{55}$$

In the confined phase, the ground state does not contribute to this sum (Eq.(52)). The excited states contributing must be odd under center reflections

$$Z|n_-\rangle = -|n_-\rangle.$$

If the spectrum exhibits a gap, the potential energy V increases linearly with r for large separations,

$$V(r) \approx \frac{E_1}{L} r \quad \text{for} \quad r \to \infty \quad \text{and} \quad L > L_c \,. \tag{56}$$

- On the other hand, the slope is given by the string tension σ. Thus in Yang Mills theory at finite extension the phenomenon of confinement is connected to the presence of a gap in the spectrum of Z-odd states

$$E_- \geq \sigma L \,, \tag{57}$$

which increase linearly with the extension of the compact direction.

- As the Polyakov loops tr $P(x_\perp)$, also the *Adjoint Polyakov Loops*

$$P_{\text{ad}}(x_\perp) = \frac{1}{3}\left(1 - 2\cos gLa_3(x_\perp)\right)$$

are built from the fields $a_3(x_\perp)$. The correlation function of the adjoint Polyakov loops describes the static interaction energy of charges in the adjoint representation. These charges can be shielded by gluons. One therefore does not expect a linearly raising interaction energy.

- The adjoint Polyakov loops are even under center reflections and therefore the center symmetry does not force vanishing of the vacuum expectation value. We therefore must assume that in general

$$\langle 0 | \left(1 - 2\cos gLa_3(x_\perp)\right) | 0 \rangle \neq 0.$$

Applying similar arguments as above for the spectrum of Z-even states, yields

$$E_+ = m_{gb} \quad m_{gb} L \gg 1 \,,$$

i.e. an extension independent gap in the spectrum determined by a glueball mass.

- Summary : SU(2) Yang-Mills theory contains two sectors of excitations which, in the confined phase, are not connected by any physical process.

  - The hadronic sector, the sector of Z-even states with a mass gap $E_+ = m_{gb} \approx 1.5\text{GeV}$
  - The gluonic sector, the sector of Z-odd states with mass gap $E_- = \sigma L$.

When compressing the system, the gap in the Z=-1 sector decreases to about 650 MeV at $L_c \approx .75$fm ($T_c \approx 270$ MeV).
According to SU(3) lattice gauge calculations, when approaching the critical temperature $T_c \approx 220$ MeV, the lowest glueball mass decreases to $m_{gb}(T_c) = 770$ MeV [25,26] and a value of $\sigma(170\,\mathrm{MeV})/T_c \approx 670$ MeV is obtained for the gap in the Z-odd sector; a more recent calculation [27] yields the significantly higher value of 1250 MeV for the glueball mass at T_c.

At the phase transition, an increase in free energy occurs through coupling the two sectors by a spontaneous breakdown of the center symmetry. This suggests that the phase transition in the Yang-Mills theory transcends the process of melting of hadronic glueballs by unfreezing gluonic degrees of freedom not visible in the confined phase.

4.4 Perturbation Theory in the Center-Symmetric Phase

The center symmetry protects the highly excited Z-odd states from mixing with the low-lying hadronic Z-even states. Any approximation compatible with confinement has therefore to respect the center symmetry. I will describe some first steps towards the development of a perturbative, but center-symmetry preserving scheme. First we disregard the couplings of the Polyakov-loop variables $a_3(x_\perp)$ to the charged gluons Φ_μ (Eq.(49)). The system of decoupled Polyakov-loop variables is determined by the Hamiltonian

$$h = \int d^2x_\perp \left[-\frac{1}{2L}\frac{\delta^2}{\delta a_3(\mathbf{x}_\perp)^2} + \frac{L}{2}\left(\nabla a_3(\mathbf{x}_\perp)\right)^2 \right] . \tag{58}$$

and by the boundary conditions at $a_3 = \pm\frac{\pi}{gL}$ for the "radial" wave function

$$\hat{\psi}[a_3]|_{\text{boundary}} = 0 . \tag{59}$$

This system has a simple mechanical analogy. The Hamiltonian describes a 2 dimensional array of degrees of freedom interacting harmonically with their nearest neighbors (magnetic field energy of the Polyakov-loop variables). If we disregard for a moment the boundary condition, the elementary excitations are "sound waves" which run through the lattice. This is actually the model we would obtain in electrodynamics, with the sound waves representing the massless photons. Mechanically we can interpret the boundary condition as a result of an infinite square well in which each mechanical degree of freedom is trapped, as is illustrated in the following Figure.

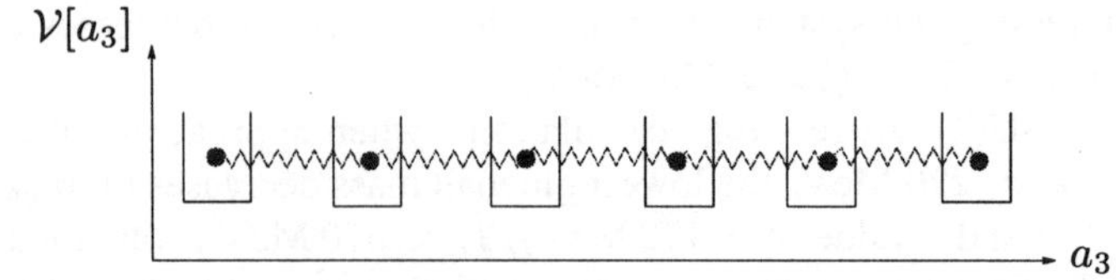

Fig.3: System of harmonically coupled Polyakov-loop variables (58) trapped by the boundary condition (59) in infinite square wells.

This infinite potential is of the same origin as the one introduced in Eq.(24) to suppress contributions of fields beyond the Gribov horizon. Considered classically, waves with sufficiently small amplitude and thus with sufficiently small energy can propagate through the system without being affected by the presence of the walls of the potential. Quantum mechanically this may not be the case. Already the zero point oscillations may be changed substantially by the infinite square well. With space discretized (lattice spacing ℓ) and rescaled dynamical variables

$$\tilde{a}_3(x_\perp) = gLa_3(x_\perp)/2\,,$$

it is seen that for $\ell \ll L$ the electric field (kinetic) energy dominates. Dropping the nearest neighbor interaction, the ground state wavefunctional is given by

$$\hat{\Psi}_0\left[\,\tilde{a}_3\right] = \prod_{\mathbf{x}_\perp}\left[\left(\frac{2}{\pi}\right)^{1/2}\cos\left(\tilde{a}_3(\mathbf{x}_\perp)\right)\right].$$

In the absence of the nearest neighbor interaction the system does not support waves. Excitations remain localized. States of lowest excitation energy are obtained by exciting a single degree of freedom at one site $\tilde{\mathbf{x}}_\perp$ into its first excited state

$$\cos\left(\tilde{a}_3(\tilde{\mathbf{x}_\perp})\right) \rightarrow \sin\left(2\tilde{a}_3(\tilde{\mathbf{x}_\perp})\right)$$

with excitation energy

$$\Delta E = \frac{3}{8}\frac{g^2 L}{\ell^2}\,. \tag{60}$$

Thus this perturbative calculation is in agreement with our general considerations and yields excitation energies rising with the extension L. From comparison with Eq.(57) the string constant

$$\sigma = \frac{3}{8}\frac{g^2}{\ell^2}$$

is obtained. This value coincides with the strong coupling limit of lattice gauge theory. However, unlike lattice gauge theory in the strong coupling limit, here no confinement like behavior is obtained in QED. Only in QCD the Polyakov-loop variables a_3 are compact and thereby give rise to localized excitations rather than waves.

Perturbative Coupling of Gluonic Variables

In the next step one may include coupling of the Polyakov-loop variables to each other via the nearest neighbor interactions. As a result of this coupling, the spectrum contains bands of excited states centered around the excited states in absence of the magnetic coupling [28]. The width of these bands is suppressed by a factor ℓ^2/L^2 as compared to the excitation energies (60) and can therefore be neglected in the continuum limit. Significant changes occur by the coupling of the Polyakov-loop variables to the charged gluons Φ_μ. We continue to neglect the magnetic coupling $(\partial_\mu a_3)^2$. The Polyakov-loop variables a_3 appearing at most quadratically in the action, can be integrated out in this limit and the following effective action is obtained

$$S_{\text{eff}}\left[A_\perp\right] = S\left[A_\perp\right] + \frac{1}{2}M^2 \sum_{a=1,2} \int d^4x A^a_\mu\left(x\right) A^{a,\mu}\left(x\right) . \qquad (61)$$

The antiperiodic boundary conditions of the charged gluons (cf. Eq.(49)) reflect the mean value of A_3 in the center-symmetric phase

$$A_3 = a_3 + \frac{\pi}{gL},$$

while the geometrical, (g-independent) mass

$$M^2 = \left(\frac{\pi^2}{3} - 2\right)\frac{1}{L^2} \qquad (62)$$

arises from their fluctuations. Antiperiodic boundary conditions describe the appearance of Aharonov-Bohm fluxes in the elimination of the Polyakov-loop variables. The original periodic charged gluon fields may be continued to be used if the partial derivative ∂_3 is replaced by the covariant one

$$\partial_3 \to \partial_3 + \frac{i\pi}{2L}[\tau_3, \quad .$$

Such a change of boundary conditions is a phenomenon well known in quantum mechanics. It occurs as a result of a magnetic flux affecting the motion of a particle on a circle. Similarly, the charged massive gluons move in a constant color neutral gauge field of strength $\frac{\pi}{gL}$, pointing in the spatial 3 direction. With x_3 compact, a color magnetic flux is associated with this gauge field,

$$\Phi_{\text{mag}} = \frac{\pi}{g} , \qquad (63)$$

corresponding to a magnetic field of strength

$$B = \frac{1}{gL^2} .$$

Also quark boundary conditions are changed under the influence of the color-magnetic fluxes

$$\psi\left(x\right)\rightarrow\exp\left[-ix_{3}\frac{\pi}{2L}\tau_{3}\right]\psi\left(x\right). \tag{64}$$

Depending on their color they acquire a phase $\pm i$ when transported around the compact direction. Within the effective theory, the Polyakov-loop correlator can be calculated perturbatively. As is indicated in the diagram below, Polyakov loops propagate only through their coupling to the charged gluons.

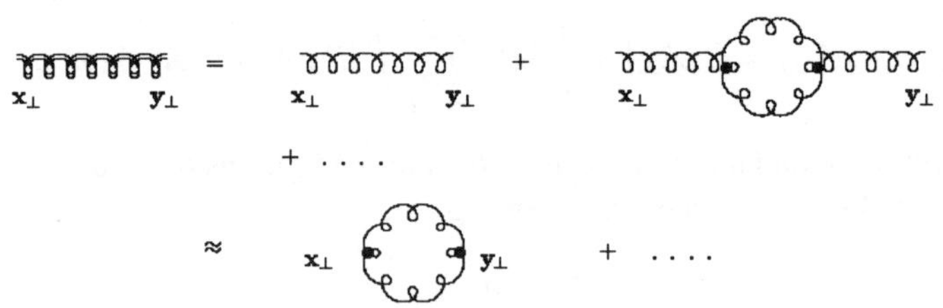

Fig.4: One loop contribution from charged gluons (curly lines) to propagator of Polyakov loops (wavy lines).

Confinement like properties are preserved when coupling the Polyakov loops to charged gluons. A linear rise of the interaction energy of fundamental charges and constant asymptotic value for adjoint charges are obtained. The value of the string constant is determined by the threshold for charged gluon pair production

$$\sigma_{\rm pt}=\frac{2}{L^{2}}\sqrt{\frac{4\pi^{2}}{3}-2}\,. \tag{65}$$

As a consequence of the coupling of the Polyakov loops to the charged gluons, the string constant has changed from the strong coupling value to a value which vanishes in the thermodynamic limit. This is a result of the perturbative treatment of the charged gluons. A realistic string constant can only be expected to arise if the threshold of a Z-odd pair of charged gluons increases linearly with the extension L (Eq.(57)); perturbatively (Eq.(65)), this threshold decreases like $1/L$.
Within this approximation also the effect of dynamical quarks on the Polyakov-loop variables can be calculated by including quark loops besides the charged gluon loop in the calculation of the Polyakov-loop propagator (cf. Fig.4). As a result of this coupling the interaction energy of static charges ceases to rise linearly; it saturates for asymptotic distances at a value

$$V(r)\approx 2m\,.$$

Thus string breaking by dynamical quarks is obtained. This is a remarkable and rather unexpected result. Even though perturbation theory has been employed, the asymptotic value of the interaction energy is independent of the coupling constant g in contradistinction to the e^4 dependence of the Uehling potential in QED. Furthermore, the quark loop contribution vanishes if calculated with anti-periodic or periodic boundary conditions. A finite result only arises with the boundary conditions (64) modified by the Aharonov-Bohm fluxes. The $1/g$ dependence of the strength of these fluxes (Eq.(63)) is responsible for the coupling constant independence of the asymptotic value of $V(r)$.

4.5 *Polyakov Loops in the Plasma Phase*

In the high temperature or small extension limit of the center symmetric phase charged gluons decouple because of the geometrical mass (62) and the antiperiodic boundary conditions (49) induced by the Bohm-Aharonov flux

$$\Delta E \approx \frac{\pi}{L} \to \infty .$$

Only neutral gluons contain zero modes and contribute to thermodynamic quantities at high temperature. This is in conflict with results of lattice gauge calculations [29] and we therefore will assume that the center symmetry is spontaneously broken for $L \leq L_c = 1/T_c$. In the high temperature phase, Aharonov-Bohm fluxes must be screened and the geometrical mass reduced. On the basis of this assumption, I now describe the development of a phenomenological treatment of the plasma phase[30]. For technical simplicity, I will neglect the space time dependence of a_3 and describe the results for vanishing geometrical mass M. For the description of the high-temperature phase it is more appropriate to use the variables

$$\chi = gLa_3 + \pi$$

with vanishing average Aharonov-Bohm flux. Charged gluons satisfy quasi-periodic boundary conditions

$$A_\mu^{1,2}(x_\perp, x_3 + L) = e^{i\chi} A_\mu^{1,2}(x_\perp, x_3). \tag{66}$$

Furthermore we will calculate the thermodynamic properties by evaluation of the energy density in the Casimir effect (cf. Eqs.(40), (41)). In the Casimir effect, the central quantity to be calculated is the ground state energy of gluons between plates on which the fields have to satisfy appropriate boundary conditions. In accordance with our choice of boundary conditions (Eq.(38)) we assume the enclosing plates to extend in the x^1, x^2 directions and to be separated in x^3 direction. The essential observation for the following phenomenological description is the dependence of the Casimir energy on the boundary conditions and therefore on the presence of

Aharonov-Bohm fluxes. The Casimir energy of charged gluons is obtained by summing, after regularization, the zero point energies

$$\varepsilon(L,\chi) = \frac{1}{2}\sum_{n=-\infty}^{\infty}\int \frac{d^2k_\perp}{(2\pi)^2}\left[\frac{\mathbf{k}_\perp^2 + (2\pi n+\chi)^2}{L^2}\right]^{1/2} = \frac{4\pi^2}{3L^4}B_4\left(\frac{\chi}{2\pi}\right) \quad (67)$$

with

$$B_4(x) = -\frac{1}{30} + x^2(1-x)^2.$$

Thermodynamic stability requires positive pressure at finite temperature and thus, according to Eq. (41), a negative value for the Casimir energy-density. This requirement is satisfied if

$$\chi \leq 1.51.$$

For complete screening ($\chi = 0$) of the Aharonov-Bohm fluxes the expression for the pressure in black-body radiation is obtained (the factor of two difference between Eqs. (40) and (67) accounts for the two charged gluonic states). Unlike QED, QCD is not stable for vanishing Aharonov-Bohm fluxes. In QCD the perturbative ground state energy can be lowered by spontaneous formation of magnetic fields. Magnetic stability can be reached if the strength of the Aharonov-Bohm fluxes does not decrease beyond a certain minimal value. By calculating the Casimir effect in the presence an external, homogeneous color-magnetic field

$$B_i^a = \delta^{a3}\delta_{i3}B$$

this minimal value can be determined. The energy of a single quantum state is given in terms of the oscillator quantum number m for the Landau orbits, in terms of the momentum quantum number n for the motion in the (compact) direction of the magnetic field and by a magnetic moment contribution ($s = \pm 1$)

$$E_{mns} = \left[2gH(m+1/2) + \frac{(2\pi n+\chi)^2}{L^2} + 2sgH\right]^{1/2}.$$

This expression shows that the destabilizing magnetic moment contribution $2sgH$ in the state with

$$s = -1, m = 0, n = 0$$

can be compensated by a non-vanishing Aharonov-Bohm flux χ of sufficient strength. For determination of the actual value of χ the sum over these energies has to be performed. After regularizing the expression, the Casimir energy density can be computed numerically. The requirement of magnetic stability yields a lower limit on χ which is shown in Figure 5.

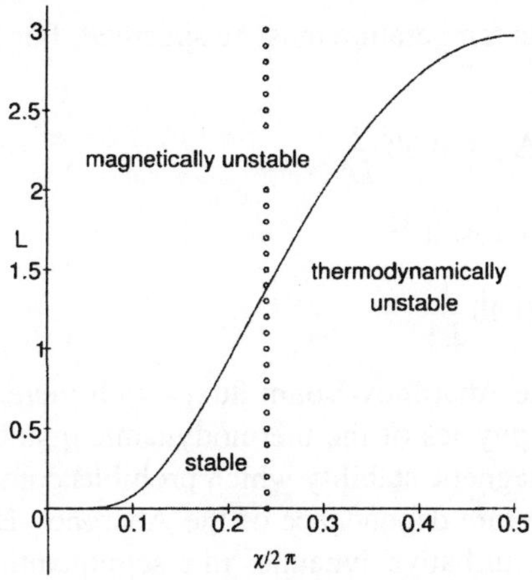

Fig.5: Regions of stability and instability in the (L, χ) plane. To the right of the circles, thermodynamic instability; above the solid line, magnetic instability.

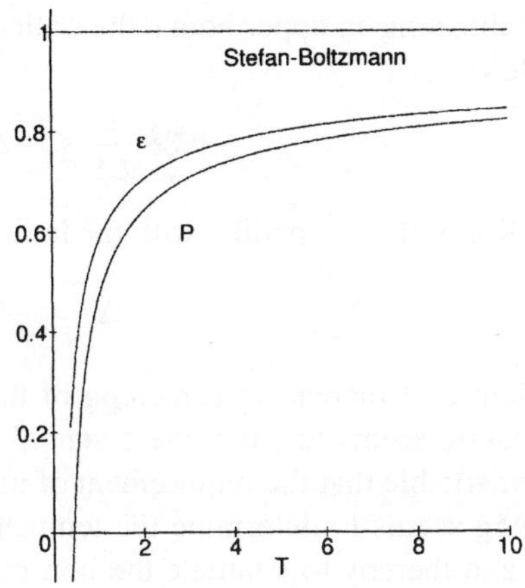

Fig. 6: Energy density and pressure normalized to Stefan-Boltzmann values vs. temperature in units of $\Lambda_{\rm MS}$.

The Figure shows that the Stefan-Boltzmann limit $\chi = 0$ is not compatible with magnetic stability for any value of the temperature. Identification of the Aharonov-Bohm flux with the minimal allowed values sets upper limits to energy density and pressure. These upper limits are shown in Figure 6. These results are reminiscent of lattice data [31] in the slow, logarithmic approach of energy density and pressure

$$\chi(T) \geq \frac{11}{12} g^2(T), \; T \rightarrow \infty$$

to the Stefan-Boltzmann limit. It appears that the finite value of the Aharonov-Bohm flux accounts for interactions present in the deconfined phase fairly well; qualitative agreement with lattice calculations is also obtained for the "interaction measure" $\epsilon - 3P$. Furthermore these limits on χ also yield a realistic estimate for the change in energy density $-\Delta\epsilon$ across the phase transition. The phase transition is accompanied by a change in strength of the Aharonov-Bohm flux from the center symmetric value π to a value in the stability region. The lower bound is determined by thermodynamic stability

$$-\Delta\epsilon \geq \epsilon(L_c, \chi = \pi) - \epsilon(L_c, \chi = 1.51) = \frac{7\pi^2}{180} \frac{1}{L_c^4}.$$

For establishing an upper bound the critical temperature must be specified. For $T_c \approx$ 270 MeV

$$0.38 \, \frac{1}{L_c^4} \leq -\Delta\epsilon \leq 0.53 \, \frac{1}{L_c^4}.$$

These limits are compatible with the lattice result [32]

$$\Delta\epsilon = -0.45 \frac{1}{L_c^4} \, .$$

The picture of increasing screening of the Aharonov-Bohm fluxes with increasing temperature seems to catch the essential physics of the thermodynamic quantities. It is remarkable that the requirement of magnetic stability which prohibits complete screening seems to determine the temperature dependence of the Aharonov-Bohm fluxes and thereby to simulate the non-perturbative dynamics in a semiquantitative way.

4.6 Monopoles

The discussion of the dynamics of the Polyakov loops has demonstrated the significant changes which arise if compact variables are present. The results discussed strongly suggest that confinement arises naturally in a setting where the dynamics is dominated by such compact variables. The analysis of the dynamics of the Polyakov loops as well as the development of of a framework for phenomenological treatment in particular of the plasma phase could be carried out by choosing judiciously the gauge. In axial gauge, the Polyakov-loop variables are elementary fields which derive their compact nature from the compactness of the gauge group.
The Polyakov-loop variables $a_3(x_\perp)$ constitute only a small set of degrees of freedom in gauge theories. In axial gauge, the remaining degrees of freedom $A_\perp(x)$ are standard fields which, with interactions neglected, describe freely propagating particles. As a consequence, the coupling of the compact variables to the other degrees of freedom, almost destroys the confinement present in the system of uncoupled Polyakov-loop variables. This can be prevented to happen only if the dynamics of all the gluon fields acquire infrared properties similar to those of the Polyakov loop variables. In particular for confinement with a realistic string tension to be obtained, a gap in the spectrum of all the Z-odd states has to be present which rises linearly with the extension of the system (cf. Eq.(57)). Mechanisms for producing such a gap have not been identified for variables other than the Polyakov-loop variables. In the axial gauge representation it is tempting to connect such mechanisms to the presence of monopoles whose existence is intimately linked to the compactness of the Polyakov-loop variables. In this last section I will sketch the emergence of such singular fields in axial gauge and discuss some elements of their dynamics.

Singular fields arise in the last step of the gauge fixing procedure (43) where, by diagonalization of the Polyakov loops, the variables characterizing the orientation of the Polyakov loops in color space are eliminated as redundant variables. The diagonalization of group elements is achieved by the unitary matrix

$$\Omega_D = e^{i\omega\tau} = \cos\omega + i\tau\hat{\omega}\sin\omega\,,$$

with $\omega(x_\perp)$ depending on the Polyakov loop $P(x_\perp)$ to be diagonalized. This diagonalization is ill defined if

$$P(x_\perp) = \pm 1\,, \tag{68}$$

i.e. if the Polyakov loop is an element of the center of the group. Diagonalization of an element in the neighborhood of the center of the group is akin to the definition of the azimuthal angle on the sphere close to the north or south pole. With Ω_D ill defined, the transformed fields

$$A'_\mu(x) = \Omega_{\rm D}(x_\perp)\left[A_\mu(x) + \frac{1}{ig}\partial_\mu\right]\Omega^\dagger_{\rm D}(x_\perp)$$

develop singularities. By analyzing this expression it can be seen that the singularities are monopoles with their strength quantized for topological reasons. The condition (68) specifies the color orientation of the Polyakov loop and therefore determines one point on S^3 (cf. Eq.(26)). As a consequence, the diagonalization is, in general, ill defined along a world-line. In axial gauge the monopole world-lines are parallel to the x_3- axis.

Properties of Singular Fields

- Dirac monopoles extended to include color constitute the simplest examples of singular fields (Euclidean $x_\perp = \mathbf{x}$)

$$\mathbf{A} \sim \frac{m}{2gr}\left\{\frac{1+\cos\theta}{\sin\theta}\,\hat{\varphi}\tau_3 + [(\hat{\varphi} + i\hat{\theta})e^{\ i\varphi}(\tau_1 - i\tau_2) + \mathrm{h.c}]\right\}.$$

  Apart from the pole at $r = 0$ the fields contain a string-in 3 space (here chosen along $\theta = 0$) and therefore a sheet-like singularity in 4 space which emanates from the monopole (word-line).

- Monopoles are characterized by 2 charges, the "north-south" charge for the two center elements of SU(2) (Eq.68)

$$z = \pm 1 \tag{69}$$

  and the quantized strength of the singularity

$$m = \pm 1, \pm 2, ... \tag{70}$$

- The topological charge (32) is determined by the two charges of the monopoles present in a given configuration [34,35,33]

$$\nu = \frac{1}{2}\sum_i m_i z_i .$$

Thus after elimination of the redundant variables topological charge resides exclusively in singular field configurations.

- The action of singular fields is in general finite and can be arbitrarily small for $\nu = 0$. The singularities in the abelian and non-abelian contributions to field strength cancel since by gauge transformations singularities in gauge covariant quantities cannot be generated.

To illustrate the importance of the connection between between topological charges and monopole charges we consider the singularity content of instantons in axial gauge [36]. To this end we calculate the Polyakov loop of instantons. Assuming the instantons size ρ to be small as compared to the extension L of the compact direction, the Polyakov loop of an instanton (36) located at the origin is given by

$$\frac{1}{2}\,\mathrm{tr}P(x_\perp = t, z, 0) = \frac{1}{2}\,\mathrm{tr}P\exp\Big\{ -i\int_0^L dx_3 \frac{\eta_{a3\mu}\tau^a x^\mu}{\sqrt{t^2+z^2+x_3^2+\rho^2}}\Big\} . \qquad (71)$$

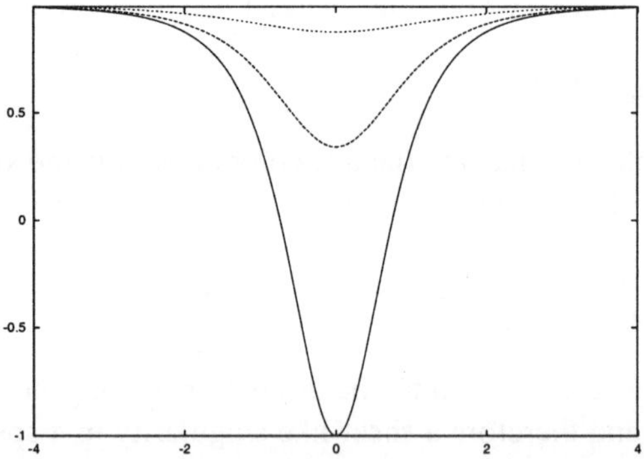

Fig.7: Polyakov loop (71) of an instanton (36) of size $\rho = 1$ for minimal distance to the center $z = 0, 1, 2$.

As this Figure illustrates, instantons contain a $z = -1$ monopole at the center and a $z = 1$ monopole at infinity; these monopoles carry the topological charge of the instanton. Furthermore tunneling processes represented by instantons connect field configurations of different winding number (cf. Eq.(34)) but with the same value for the Polyakov loop. In the course of the tunneling the Polyakov loop of the instanton may pass through or get close to the center element $z = -1$, it however always returns to its original value $z = +1$. Thus we conclude that an instanton ensemble in the dilute gas limit is not center symmetric and therefore cannot give rise to confinement. One cannot rule out that, with increasing instanton density, the $z = -1$ values of the Polyakov loop are encountered more and more frequently and a center-symmetric ensemble may finally be reached in the high density limit.

Elements of Monopole Dynamics

In axial gauge, instantons are composed of two monopoles. An instanton gas (Eq.(37)) of finite density n_{I} therefore contains field configurations with infinitely many monopoles. The instanton density in 4-space can be approximately converted to a monopole density in 3-space [33]

$$n_{\mathrm{M}} \sim (Ln_{\mathrm{I}}\rho)^{3/2} \quad , \quad \rho \ll L,$$

$$n_{\mathrm{M}} \sim Ln_{\mathrm{I}} \quad , \quad \rho \geq L\,.$$

At large extensions or equivalently low temperatures the monopole density diverges for constant instanton density. Nevertheless, the action density of an instanton gas remains finite. This is in accordance with our expectation that production of monopoles is not necessarily suppressed by large values of the action. Furthermore a finite or possibly even divergent density of monopoles as in the case of the dilute instanton gas does not imply confinement.

Beyond generation of monopoles via instantons, the system has the additional option of producing one type ($z = +1$ or $z = -1$) of poles and corresponding antipoles only. No topological charge is associated with such singular fields and their occurrence is not limited by the instanton bound (Eqs. (30), (35)) on the action as is the case for a pair of monopoles of opposite z-charge. Thus entropy favors production of such configurations. The entropy argument also applies in the plasma phase. For purely kinematical reasons a decrease in the monopole density must be expected as the above estimates within the instanton model show. This decrease is counteracted by the enhanced probability to produce monopoles when, with decreasing L, the Polyakov loop approaches more and more the center of the group, as has been discussed in the previous section (cf. Fig. 5). A finite density of singular fields is likely to be present also in the deconfined phase. In order for this to be compatible with the partially perturbative nature of the plasma phase and with dimensional reduction to QCD_{2+1}, poles and antipoles may have to be strongly correlated with each other and to form effectively a gas of dipoles.

Since entropy favors proliferate production of monopoles and monopoles may be produced with only a small increase in the total action, the coupling of the monopoles to the quantum fluctuations must ultimately prevent unlimited increase in the number of monopoles. A systematic study of the relevant dynamics has not been carried out. Monopoles are not solutions to classical field equations and therefore even on the level of bilinear terms in the action singular fields are mixed with quantum fluctuations. Nevertheless two mechanisms can be identified which might limit the production of monopoles.

- The 4-gluon vertex couples pairs of monopoles to charged and neutral gluons and can generate masses for all the color components of the gauge fields. A simple estimate yields

$$\delta m^2 = -\frac{\pi}{V} \sum_{\substack{i,j=1 \\ i<j}}^{N} m_i m_j |\mathbf{x}_{\perp i} - \mathbf{x}_{\perp j}|.$$

  with the monopole charges m_i and positions $\mathbf{x}_{\perp\, i}$. If operative also in the deconfined phase, this mechanism would give rise to a magnetic gluon mass.

- In general, fluctuations around singular fields generate infinite action. Finite action results only, if the fluctuations $\delta\phi, \delta A^3$ satisfy the boundary conditions

$$\delta\phi(x)\, e^{2i\varphi(x_\perp)} \quad \text{continuous along the strings}$$

$$\delta\phi(x)\Big|_{\text{at pole}} = \delta A^3\Big|_{\text{at pole}} = 0\,.$$

  For finite monopole density, long wave-length fluctuations cannot simultaneously satisfy boundary conditions related to monopoles or strings which are close to each other and one therefore might suspect quantum fluctuations with wavelengths

$$\lambda \geq \lambda_{\text{max}} = n_{\text{M}}^{-1/3}$$

  to be suppressed.

We note that both mechanisms would also suppress the propagators of the quantum fluctuations in the infrared. Thereby the decrease in the string constant by coupling Polyakov loops to charged gluons could be alleviated if not cured.

5 Conclusions

In this talk I have described theoretical efforts in the analysis of non-perturbative QCD. Common to many such efforts presently undertaken is the attempt to identify the relevant field configuration which incorporate the low-energy phenomena of chiral symmetry breaking and of confinement. With regard to the breakdown of chiral symmetry and its related phenomena, efforts have met success with the development of the instanton model. With its failure to account for confinement and with the expectation that the two phenomena are dynamically connected, this success is incomplete. The issue of identification of the field configurations which are relevant for confinement - or equivalently the issue of dynamical mechanism such as the Higgs mechanism for mass generation - remains to be settled. The focus in this presentation has been on confinement and I have emphasized the role of compact variables and of singular field configurations which both arise as a result of fixing the gauge. The appearance of such peculiar configurations reflects obstructions in imposing these gauge conditions.

The physics of compact fields and its consequences for confinement can be considered to be qualitatively understood. As I have illustrated in detail, the compactness of fields may give rise to a localization of excitations, i.e. a suppression of waves. No particles are associated to such fields and as we have seen, linearly rising interaction energies of static charges may result. Localization has clearly its origin in the non-abelian character of the gauge group, it does not occur in electrodynamics where, within the same sequence of approximations propagating excitations, the photons, are correctly described. Despite these qualitative insights into the physics of compact variables, a complete theoretical framework has not been developed. In particular, as we also have seen, the property of compactness may easily get lost by coupling compact to propagating degrees of freedom. This effective decompactification occurs irrespective of the coupling strength and therefore the coupling between localized and propagating excitations must be forbidden by symmetry. For the compact Polyakov loop variables this symmetry has been identified as the center symmetry. This identification is strongly supported by the string breaking mechanism which I have shown to occur as a symmetry breaking mechanism induced by dynamical fermions. Apart from decompactification also localization on small length scales has to be prevented to occur in order to comply with the requirement of asymptotic freedom.

Conceptual and technical difficulties continue to plague the development of a viable framework to deal with the dynamics of singular fields. Like the compact variables, singular fields have their origin in the non-abelian character of the gauge group. The possibility of cancellations between abelian and non-abelian contributions to the field strength is the origin for the appearance of singular gauge fields of finite action. Unlike instantons or the 't Hooft-Polyakov monopole, singular field configurations in general do not satisfy classical field equations. The action associated with such fields may be arbitrarily small. Semiclassical treatments are not directly applicable. Singular fields and fluctuations can mix. In general fluctuations around a singular field destroy the delicate balance of abelian and non-abelian contributions to the field strength resulting in an infinite action. There are no standard tools available to handle the difficult constraints imposed on the fluctuations by finiteness of the action. Essentially nothing is known about the treatment of singular fields in the canonical formalism. It is not obvious how to include them in an Ansatz for the QCD ground state wave-functional. Simple QED like structures for the wave-functional would exclude such fields by erroneously assigning to them an infinite energy. It seems unlikely that the issue of singular fields can be avoided. Smooth, but topologically non-trivial configurations must be in general expected to give rise to singular fields after gauge fixing. However, the occurrence of singular fields is by itself not sufficient for generating non-perturbative phenomena. In the context of the instanton model we have seen explicitly that a finite or possibly even divergent density of monopoles does not give rise to confinement. Furthermore, the geometry of gauge conditions like the axial of Lorentz-gauge conditions and the ensuing gauge fixing procedures are independent of the presence of matter in the adjoint representation, while depending on the dynamics of the matter field a confining or a Higgs phase may be realized. Understanding the mechanisms for confinement must also include understanding of their ineffectiveness in the Higgs phase [37]. Similar remarks apply to the appearance of singular fields in systems with varying extension.

Above all, it is the gauge dependence of the singular fields which so far has prevented a proper formulation of their dynamics. Existence of obstructions in implementing global gauge conditions may be the only property which is common to different gauge fixed formulations of QCD. The various gauge dependent manifestations of such obstructions seem to have very little in common. Instantons, for instance, possess in (regular) Lorentz-gauge a point singularity at infinity while in axial gauge, they exhibit in addition word-line and world-sheet singularities. Insights beyond perturbation theory into the interrelation between formulations of QCD in different gauges seem to be required for further progress in our understanding of the mechanisms of confinement in terms of gauge fixed singular fields.

Acknowledgments

I thank V.E. Herscovitz and C. Vasconcellos for the kind hospitality extended to me during the workshop and my stay at the Instituto de Fisica de UFRGS.

References

1. A.V. Kutznetsov and N.V. Mikheev, *Rev. Lett.* **89**, 011601 (2002).
2. Y. Nambu and G. Jona-Lasinio, *Phys. Rev.* **122**, 345 (1961); **124**, 246 (1961).
3. T. Schäfer and E.V. Shuryak, *Rev. Mod. Phys.* **70**, 323 (1998).
4. A. Bashir and M.R. Pennington, *Phys. Rev.* D **50**, 7679 (1994).

5. N. Seiberg, E. Witten, *Nucl. Phys.* B **426**, 19 (1994), B **431**, 484 (1995)
6. M. Gell-Mann, R.J. Oakes, and B.Renner, *Phys. Rev.* **175**, 2195 (1968).
7. O. Nachtmann, in: Lectures on QCD, Applications, edt. by F. Lenz et al., Springer, (1997).
8. V. N. Gribov, *Nucl. Phys.* B **139**, 1 (1978).
9. C.G. Callan, R.F. Dashen and D.J. Gross, *Phys. Rev.* D **17**, 2717 (1978).
10. S. Mandelstam, *Phys. Rep.* **67**, 109 (1980).
11. G. 't Hooft, *Nucl. Phys.* B **190**, 455 (1981).
12. H. Reinhardt, *Nucl. Phys.* B **628**, 133 (2002).
13. A.C. Kalloniatis and S.N. Nedelko, *Phys. Rev.* D **64**, 114025 (2001).
14. I. M. Singer, *Comm. Math. Phys.* **60**, 7 (1978).
15. H. Georgi and S. Glashow, *Phys. Rev. Lett.* **28**, 1494 (1972).
16. H. Weyl, *"Gruppentheorie und Quantenmechanik"*, Hirzel, Leipzig 1928.
17. R. Jackiw, *Rev. Mod. Phys.* **52**, 661 (1980).
18. F. Lenz, H.W.L. Naus and M. Thies, *Ann. Phys.* **233**, 317 (1994).
19. F. Lenz and S. Wörlen, in: "At the Frontier of Particle Physics, Handbook of QCD" edited by M.Shifman, Vol.2, p.762
20. A.A. Belavin, A.M. Polyakov, A.S. Schwartz and Yu.S. Tyupkin, *Phys. Lett.* B **59**, 85 (1975).
21. J.D. Bjorken, in: "Lectures on Lepton Nucleon Scattering and Quantum Chromodynamics", W. Atwood *et al.*, Birkhäuser (1982).
22. G. 't Hooft, *Phys. Rev.* D **14**, 3432 (1976).
23. F. Lenz and M. Thies, *Ann. Phys.* **268**, 308 (1998).
24. D.J. Toms, *Phys. Rev.* D **21**, 928 (1980).
25. J. Fingberg, U. Heller, and F. Karsch, *Nucl. Phys.* B **392**, 493 (1993).
26. B. Grossman, S. Gupta, U.M. Heller, and F. Karsch, *Nucl. Phys.* B **417**, 289 (1994).
27. M. Ishii, H. Suganuma and H. Matsufuru, *Phys. Rev.* D **66**, 014507 (2002).
28. F. Lenz, E.J. Moniz and M. Thies, *Ann. Phys.* **242**, 429 (1995).
29. T. Reisz, *Z. Phys.* C **53**, 169 (1992).
30. V.L. Eletsky, A.C. Kalloniatis, F. Lenz, and M. Thies, *Phys. Rev.* D **57**, 5010 (1998)
31. F. Karsch, E. Laermann, and A. Peikert, *Phys. Lett.* B **478**, 447 (2000).
32. J. Engels, F. Karsch and K. Redlich, *Nucl. Phys.* B**435**, 295 (1995).
33. O. Jahn and F. Lenz, *Phys. Rev.* D **58**, 85006 (1998).
34. M. Quandt, H. Reinhardt and A. Schafke, *Phys. Lett.* B **446**, 290 (1999).
35. C. Ford, T. Tok and A. Wipf, *Phys. Lett.* B **456**, 155 (1999).
36. F. Lenz, J.W. Negele, and M. Thies, to be published.
37. F. Lenz, J.W. Negele, L. O'Raifeartaigh and M. Thies, *Ann. Phys.* **285**, 25 (2000).

NUCLEAR AND HIGH-ENERGY ASTROPHYSICS

FRIDOLIN WEBER

Department of Physics, San Diego State University
5500 Campanile Drive San Diego, California 92182-1233, USA
fweber@sciences.sdsu.edu

There has never been a more exciting time in the overlapping areas of nuclear physics, particle physics and relativistic astrophysics than today. Orbiting observatories such as the Hubble Space Telescope, Rossi X-ray Timing Explorer (RXTE), Chandra X-ray satellite, and the X-ray Multi Mirror Mission (XMM) have extended our vision tremendously, allowing us to see vistas with an unprecedented clarity and angular resolution that previously were only imagined, enabling astrophysicists for the first time ever to perform detailed studies of large samples of galactic and extragalactic objects. On the Earth, radio telescopes (e.g., Arecibo, Green Bank, Parkes, VLA) and instruments using adaptive optics and other revolutionary techniques have exceeded previous expectations of what can be accomplished from the ground. The gravitational wave detectors LIGO, LISA VIRGO, and Geo-600 are opening up a window for the detection of gravitational waves emitted from compact stellar objects such as neutron stars and black holes. Together with new experimental forefront facilities like ISAC, ORLAND and RIA, these detectors provide direct, quantitative physical insight into nucleosynthesis, supernova dynamics, accreting compact objects, cosmic-ray acceleration, and pair-production in high energy sources which reinforce the urgent need for a strong and continuous feedback from nuclear and particle theory and theoretical astrophysics. In my lectures, I shall concentrate on three selected topics, which range from the behavior of superdense stellar matter, to general relativistic stellar models, to strange quark stars and possible signals of quark matter in neutron stars.

1 Introduction

A forefront area of research, both experimental and theoretical, concerns the exploration of the subatomic structure of superdense matter and the determination of the equation of state – that is, the relation between pressure P, temperature T and density ϵ – associated with such matter.[1] Knowing its properties is of key importance for our understanding of the physics of the early universe, its evolution in time to the present day, compact stars, various astrophysical phenomena, and laboratory physics. The high-temperature domain of the phase diagram of superdense matter is probed by relativistic heavy-ion colliders. Complementary to this, neutron stars contain cold superdense matter permanently in their centers (cf. Figure 1), which make them superb astrophysical laboratories for probing the low-density/high-density domain of the phase diagram of superdense matter[2,3].

Neutron stars are dense, neutron-packed remnants of massive stars that

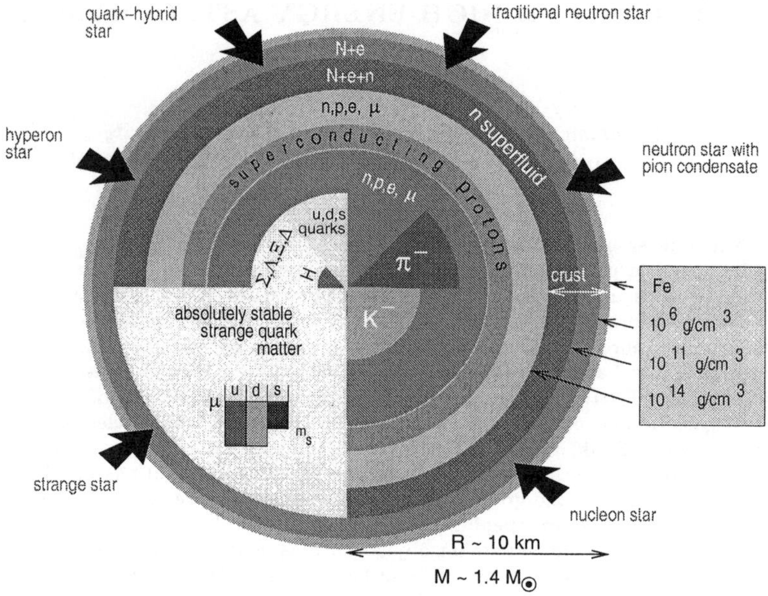

Figure 1. Competing structures and novel phases of subatomic matter predicted by theory to make their appearance in the cores ($R \lesssim 8$ km) of neutron stars.[3]

blew apart in supernova explosions. They are typically about 10 kilometers across and spin rapidly, often making several hundred rotations per second. The discovery rate of new rotating neutron stars, spotted as pulsars by radio telescopes [4,5], is rather high. To date about 1400 pulsars are known. Depending on star mass and rotational frequency, gravity compresses the matter in the core regions of pulsars up to more than ten times (~ 1.5 GeV/fm^3) the density of ordinary atomic nuclei, thus providing a high-pressure environment in which numerous subatomic particle processes plausibly compete with each other and novel phases of matter may exist. The most spectacular ones stretch from the generation of new baryonic particles (e.g., $\Sigma, \Lambda, \Xi, \Delta$) to quark (u, d, s) deconfinement to the formation of Boson condensates (π^-, K^-, H-matter), as illustrated in Figure 1. There are theoretical suggestions of even more exotic processes inside pulsars, such as the formation of absolutely stable quark matter, a configuration of matter even more stable than the most stable atomic nucleus, ^{56}Fe! In the latter event, pulsars would be largely composed of pure quark matter, eventually enveloped in thin nuclear crusts (bottom-left portion in Figure 1). No matter which physical processes are actually realized inside neutron stars, each one leads to fingerprints, some more pronounced than others though, in the observable stellar quantities. Paired with the unprecedented wealth of new observational pulsar data, it seems to be within reach for the first time ever to seriously explore the subatomic structure of matter in the high-density low-temperature portion of its phase diagram from observed pulsar data. To this aim Einstein's field equation of relativistic gravity,

$$G^{\mu\nu} \equiv R^{\mu\nu} - \frac{1}{2} g^{\mu\nu} R = 8\,\pi\, T^{\mu\nu}(\epsilon, P(\epsilon))\,, \qquad (1)$$

is to be solved in combination with the latest theories of the subatomic structure of matter[3]. The latter follows according to the scheme

$$\frac{\partial \mathcal{L}(\{\phi\})}{\partial \phi} - \partial_\mu \frac{\partial \mathcal{L}(\{\phi\})}{\partial(\partial_\mu \phi)} = 0 \quad \Rightarrow \quad P(\epsilon)\,, \qquad (2)$$

where $\mathcal{L}(\{\phi\})$ denotes a given stellar matter lagrangian.[3] In general, $\mathcal{L}$ is a complicated functional of the numerous hadron and quark fields, collectively written as $\{\phi\}$, that acquire finite amplitudes up to the highest densities reached in the cores of neutron stars. According to what has been said just above, plausible candidates for ϕ are the charged states of the SU(3) baryon octet, $p, n, \Sigma, \Lambda, \Xi$[6], the charged states of the Δ[7,8], π^- [9] and K^- [10–14] mesons, as well as u, d, s quarks[15,16,17]. The conditions of chemical equilibrium and electric charge neutrality of stellar matter require the presence of leptons too, in which case $\phi = e^-, \mu^-$. Theories of superdense matter enter Einstein's field equation (1) via the energy-momentum tensor $T^{\mu\nu}$, which contains the equation of state of the stellar matter, $P(\epsilon)$. Because of the rather uncertain behavior of the matter at supernuclear densities, the models derived for the equation of state differ considerably from each other. This has its origin in various sources such as: (1) the many-body technique used for the determination of the equation of state, (2) the model adopted for the nucleon–nucleon force, (3) assumptions about the fundamental building blocks of neutron star matter, (4) the inclusion of boson condensates, and (5) the considerations of a possible phase transition of confined hadronic matter into deconfined quark matter[3,18].

In general Eqs. (1) and (2) were to be solved simultaneously since the particles move in curved spacetime whose geometry, determined by Einstein's field equations, is coupled to the total mass energy ϵ of the matter. In the case of neutron stars, however, the long-range gravitational force can be cleanly separated from the short-range nuclear force so that Eqs. (1) and (2) decouple from each other. This simplifies the study of compact stellar objects considerably.

2 Models for the Equation of State of Superdense Neutron Star Matter

2.1 Non-relativistic Models

For non-relativistic models, the starting point are phenomenological nucleon-nucleon interactions, denoted V_{ij}, which fit the nucleon-nucleon scattering data and the deuteron properties. In order to achieve the correct binding energy of nuclear matter at the empirical saturation density ρ_0 ($\rho_0 = 0.15$ nucleons/fm^3 which corresponds to a mass density of $\epsilon = 2.5 \times 10^{14}$ g/cm^3), two-nucleon potentials are supplemented with three-nucleon interactions V_{ijk}. The hamiltonian is then of the form

$$\mathcal{H} = \sum_i \left(\frac{-\hbar^2}{2m} \right) \nabla_i^2 + \sum_{i<j} V_{ij} + \sum_{i<j<k} V_{ijk} \,. \tag{3}$$

A proven many-body method frequently adopted to solve the associated many-body Schrödinger equation $\mathcal{H}|\Psi >= E|\Psi >$ is the variational approach[19,20] where a variational trial function $|\Psi_v >$ is constructed from a symmetrized product of two-body correlation operators (F_{ij}) acting on an unperturbed ground-state, i.e.,

$$|\Psi_v > = \left[\hat{S} \prod_{i<j} F_{ij} \right] |\Phi > \,, \tag{4}$$

where $|\Phi >$ denotes the antisymmetrized Fermi-gas wave function,

$$|\Phi > = \hat{A} \prod_j \exp(i\, \boldsymbol{p}_j \cdot \boldsymbol{x}_j) \,. \tag{5}$$

The correlation operator contains variational parameters which are varied to minimize the energy per baryon for a given density ρ:[19,20]

$$E_v(\rho) = \min \left\{ \frac{< \Psi_v \,|\, \mathcal{H} \,|\, \Psi_v >}{< \Psi_v \,|\, \Psi_v >} \right\} \geq E_0 \,. \tag{6}$$

As indicated, E_v constitutes an upper bound to the ground-state energy E_0. The energy density $\epsilon(\rho)$ and pressure $P(\rho)$ are obtained from Eq. (6) through the following manipulations:

$$\epsilon(\rho) = \rho\,(E_v(\rho) + m)\,, \qquad P(\rho) = \rho^2 \frac{\partial}{\partial\rho} E_v(\rho)\,, \tag{7}$$

which can be combined to an equation of state of the form $P(\epsilon)$ applicable to the stellar structure calculations.

A representative collection of non-relativistic nuclear equations of state is listed in Tables 1 and 2, where the following abbreviations are used: N = pure neutron matter; NP = n, p, and leptons in chemical equilibrium; and K denotes the incompressibility (in MeV) of symmetric nuclear matter at saturation density. Several of these equations of state are shown in Figures 2 and 3 where the pressure is plotted as a function of total energy density (in units of the density of normal nuclear matter, $\epsilon_0 = 140\ \mathrm{MeV/fm^3}$). The Thomas-Fermi equation of state TF96 of Table 1 is based on the new Thomas-Fermi approach of Myers and Swiatecki[21].

Table 1. Non-relativistic models for the equation of state of neutron star matter.

EOS	Properties (see text)	References
WFF(UV_{14}+TNI)	NP, K=261	22
WFF(UV_{14}+UVII)	NP, K=202	22
WFF(AV_{14}+UVII)	NP, K=209	22
FP(V_{14}+TNI)	N, K=240	23
APR(A_{18}+UIX)	NP	24
TF96	N, K=234	25,26,27,28

The effective interaction v_{12} of this new approach,

$$v_{12} = -\frac{2\,T_0}{\rho_0}\,Y(r_{12})\Big\{\frac{1}{2}(1 \mp \xi)\,\alpha - \frac{1}{2}(1 \mp \zeta)\Big(\beta\Big(\frac{p_{12}}{k_{F_0}}\Big)^2 - \gamma\,\frac{k_{F_0}}{p_{12}} + \sigma\Big(\frac{2\bar{\rho}}{\rho_0}\Big)^{\frac{2}{3}}\Big)\Big\}\,, \tag{8}$$

consists of the Seyler-Blanchard potential[29], generalized by the addition of one momentum dependent and one density dependent term[21]. The upper (lower) sign in (8) corresponds to nucleons with equal (unequal) isospin. The quantities k_{F_0}, T_0 $(= k_{F_0}^2/2m)$, and ρ_0 are the Fermi momentum, the Fermi energy and the particle density of symmetric nuclear matter. The potential's radial dependence is described by the normalized Yukawa interaction

$$Y(r_{12}) = \frac{1}{4\,\pi\,a^3}\,\frac{e^{-\,r_{12}/a}}{r_{12}/a}\,. \tag{9}$$

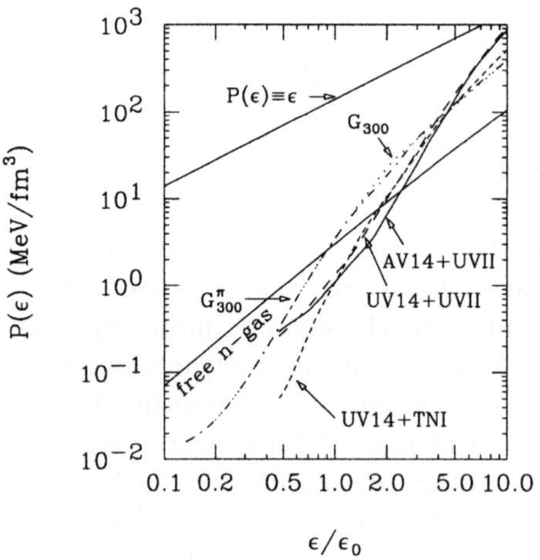

Figure 2. Graphical illustration of AV_{14}+ UVII, UV_{14}+UVII, UV_{14}+TNI, G_{300}, and G^{π}_{300} [3].

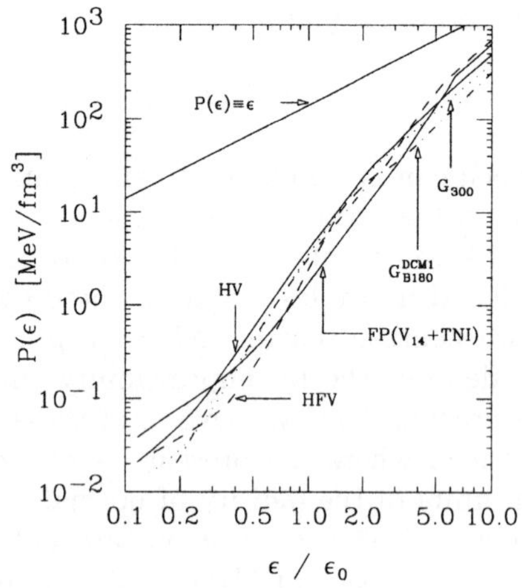

Figure 3. Graphical illustration of equations of state HV, HFV, FP(V_{14} + TNI), G_{300}, and G^{DCM1}_{B180}.

Table 2. Nuclear matter properties of the equations of state compiled in Table 1.

EOS	E/A (MeV)	ρ_0 (fm^{-3})	K (MeV)	M^* (MeV)	a_{sy} (MeV)
WFF(UV_{14}+TNI)	−16.6	0.157	261	0.65	30.8
WFF(UV_{14}+UVII)	−11.5	0.175	202	0.79	29.3
WFF(AV_{14}+UVII)	−12.4	0.194	209	0.66	27.6
FP(V_{14}+TNI)	−16.00	0.159	240	0.64	–
APR(A_{18}+UIX)	−16.00	0.16	-	-	–
TF96	−16.04	0.161	234	–	32.0

Its strength depends both on the magnitude of the particles' relative momentum, p_{12}, and on an average of the densities at the locations of the particles. The parameters ξ and ζ were introduced in order to achieve better agreement with asymmetric nuclear systems, and the behavior of the optical potential is improved by the term $\sigma(2\bar{\rho}/\rho_0)^{2/3}$. Here the average density is defined by $\bar{\rho}^{2/3} = (\rho_1^{2/3} + \rho_2^{2/3})/2$, where ρ_1 and ρ_2 are the relevant densities of the interacting particle (neutron or protons) at points 1 and 2. The potential's seven free parameters α, β, γ, σ, ξ, ζ, a are adjusted to the properties of finite nuclei, the parameters of the mass formula, and the behavior of the optical potential[27]. The nuclear matter properties at saturation obtained for TF96 are summarized in Table 2, where the listed quantities are: binding energy of normal nuclear matter at saturation density, E/A; compression modulus, K; effective nucleon mass, M^* ($\equiv m^*/m$ where m denotes the free nucleon mass); and the asymmetry energy, $a_{\rm sy}$. The new Thomas-Fermi force has the advantage over the standard Seyler-Blanchard interaction to not only reproduce the ground-state properties of finite nuclei and infinite symmetric nuclear matter, but also the optical potential and, as revealed by stellar structure calculations [27], the properties of neutron stars as well. These features render the new Thomas-Fermi model very attractive for investigations of the properties of dense nuclear matter.

2.2 Relativistic Field Theoretical Models

Relativistic field theories of dense nuclear matter and finite nuclei have enjoyed a renaissance in recent years, and they have the virtue of describing nuclear matter at saturation, many features of finite nuclei, both spherical and deformed, and they extrapolate causally to high density[30]. The starting point of relativistic theories of superdense neutron star matter is a lagrangian of the following type,[3]

$$\begin{aligned}\mathcal{L}(x) &= \sum_{B=p,n,\Sigma^{\pm,0},\Lambda,\Xi^{0,-},\Delta^{++,+,0,-}} \mathcal{L}^0_B(x) + \sum_{L=e^-,\mu^-} \mathcal{L}_L(x) \\ &+ \sum_{M=\sigma,\omega,\pi,\varrho,\eta,\delta,\phi} \left\{ \mathcal{L}^0_M(x) + \sum_{B=p,n,\ldots,\Delta^{++,+,0,-}} \mathcal{L}^{\rm Int}_{B,M}(x) \right\} . \qquad (10)\end{aligned}$$

The subscript B runs over all baryon species that may become populated in dense neutron star matter (Table 3). The nuclear forces are mediated by a collection of scalar, vector, and isovector mesons [31] which is compiled in Table 4. The equations of motion that follow from Eq. (10) for the baryon fields are given by[3]

$$\begin{aligned}(i\gamma^\mu\partial_\mu - m_B)\,\psi_B(x) &= g_{\sigma B}\,\sigma(x)\,\psi_B(x) \\ &+ \left\{ g_{\omega B}\gamma^\mu\omega_\mu(x) + \frac{f_{\omega B}}{4m_B}\sigma^{\mu\nu}F_{\mu\nu}(x)\right\}\psi_B(x) \\ &+ \left\{ g_{\rho B}\,\gamma^\mu\,\boldsymbol{\tau}\cdot\boldsymbol{\rho}_\mu(x) + \frac{f_{\rho B}}{4\,m_B}\,\sigma^{\mu\nu}\,\boldsymbol{\tau}\cdot\mathbf{G}_{\mu\nu}(x)\right\}\psi_B(x) \\ &+ \frac{f_{\pi B}}{m_\pi}\,\gamma^\mu\gamma^5\,\left(\partial_\mu\,\boldsymbol{\tau}\cdot\boldsymbol{\pi}(x)\right)\psi_B(x)\,, \qquad (11)\end{aligned}$$

Table 3. Masses, quantum numbers (spin J_B, isospin I_B, strangeness S_B, hypercharge Y_B, third component of isospin I_{3B}), and electric charges (q_B) of baryons.

Baryon (B)	m_B (MeV)	J_B	I_B	S_B	Y_B	I_{3B}	q_B
n	939.6	1/2	1/2	0	1	−1/2	0
p	938.3	1/2	1/2	0	1	1/2	1
Σ^+	1189	1/2	1	−1	0	1	1
Σ^0	1193	1/2	1	−1	0	0	0
Σ^-	1197	1/2	1	−1	0	−1	−1
Λ	1116	1/2	0	−1	0	0	0
Ξ^0	1315	1/2	1/2	−2	−1	1/2	0
Ξ^-	1321	1/2	1/2	−2	−1	−1/2	−1
Δ^{++}	1232	3/2	3/2	0	1	3/2	2
Δ^+	1232	3/2	3/2	0	1	1/2	1
Δ^0	1232	3/2	3/2	0	1	−1/2	0
Δ^-	1232	3/2	3/2	0	1	−3/2	−1

where the standard field theoretical notation of Ref.[32] is used. The meson fields in (11) obey the following field equations,[3]

$$(\partial^\mu \partial_\mu + m_\sigma^2)\,\sigma(x) = -\sum_B g_{\sigma B}\,\bar{\psi}_B(x)\psi_B(x) - m_N\, b_N\, g_{\sigma N}\,(g_{\sigma N}\sigma(x))^2 - c_N\, g_{\sigma N}\,(g_{\sigma N}\sigma(x))^3\,, \tag{12}$$

$$\partial^\mu F_{\mu\nu}(x) + m_\omega^2\,\omega_\nu(x) = \sum_B \Big\{ g_{\omega B}\,\bar{\psi}_B(x)\gamma_\nu\psi_B(x) - \frac{f_{\omega B}}{2m_B}\,\partial^\mu\left(\bar{\psi}_B(x)\sigma_{\mu\nu}\psi_B(x)\right)\Big\}\,, \tag{13}$$

$$\left(\partial^\mu \partial_\mu + m_\pi^2\right)\,\boldsymbol{\pi}(x) = \sum_B \frac{f_{\pi B}}{m_\pi}\,\partial^\mu\left(\bar{\psi}_B(x)\,\gamma_5\,\gamma_\mu\,\boldsymbol{\tau}\,\psi_B(x)\right)\,, \tag{14}$$

$$\partial^\mu \mathbf{G}_{\mu\nu}(x) + m_\rho^2\,\boldsymbol{\rho}_\nu(x) = \sum_B \Big\{ g_{\rho B}\,\bar{\psi}_B(x)\boldsymbol{\tau}\gamma_\nu\psi_B(x) - \frac{f_{\rho B}}{2m_B}\,\partial^\mu\left(\bar{\psi}_B(x)\boldsymbol{\tau}\sigma_{\mu\nu}\psi_B(x)\right)\Big\}\,, \tag{15}$$

with the field tensors $F^{\mu\nu}$ and $\mathbf{G}^{\mu\nu}$ defined as $F^{\mu\nu} = \partial^\mu\omega^\nu - \partial^\nu\omega^\mu$ and $\mathbf{G}^{\mu\nu} = \partial^\mu\boldsymbol{\rho}^\nu - \partial^\nu\boldsymbol{\rho}^\mu$. Equations (11) through (15) are to be solved simultaneously in combination with the neutron star matter constraints of electric charge

Table 4. Mesons and their quantum numbers.[33] The entries are: spin J_M, parity π, isospin I_M, and mass m_M of meson M.

Meson (M)	J_M^π	I_M	Coupling	Mass (MeV)
σ	0^+	0	scalar	550
ω	1^-	0	vector	783
$\pi^\pm$	0^-	1	pseudovector	140
π^0	0^-	1	pseudovector	135
ρ	1^-	1	vector	769
η	0^-	0	pseudovector	549
δ	0^+	1	scalar	983
ϕ	1^-	0	vector	1020
K^+	0^-	1/2	pseudovector	494
K^-	0^-	1/2	pseudovector	494

neutrality and chemical equilibrium. The former can be written as

$$\sum_B q_B\,(2J_B+1)\,\frac{p_{F,B}^3}{6\pi^2} - \sum_{\lambda=e,\mu}\frac{p_{F,\lambda}^3}{3\pi^2} - \varrho_M\,\Theta(\mu^M - m_M) = 0\,, \quad (16)$$

while chemical (β) equilibrium is expressed as

$$\mu^B = \mu^n - q_B\,\mu^e\,, \quad (17)$$

where μ^n and μ^e denote the chemical potentials of neutrons and electrons whose knowledge is sufficient to determine the chemical potential of any other baryon present in the system. The only mesons that may plausibly condense (density ρ_M) in neutron star matter are the π^- [9,34] or the K^- [10,35,36,37]. Equations (16) and (17) lead to constraints on the Fermi momenta of baryons and leptons, k_{F_B} and k_{F_L} respectively. The leptons in neutron star matter can be treated as free particles, which obey the following Dirac equation,

$$(i\gamma^\mu\partial_\mu - m_L)\,\psi_L(x) = 0\,, \quad (18)$$

with $L = e^-, \mu^-$. An elegant mathematical framework that enables one to solve the equations of motion (11) through (18) is the relativistic Green's function method, which is outlined next. The starting point is the Martin-Schwinger hierarchy of coupled Green's functions[38]. In the lowest order, the Martin-Schwinger hierarchy can be truncated by factorizing the four-point Green's function $g_2(1,2;1'2')$, where primed (unprimed) arguments refer to ingoing (outgoing) fermions, into a product of two-point Green's functions $g_1(1;1')$, as schematically shown in Figure 4. This leads to the well-known

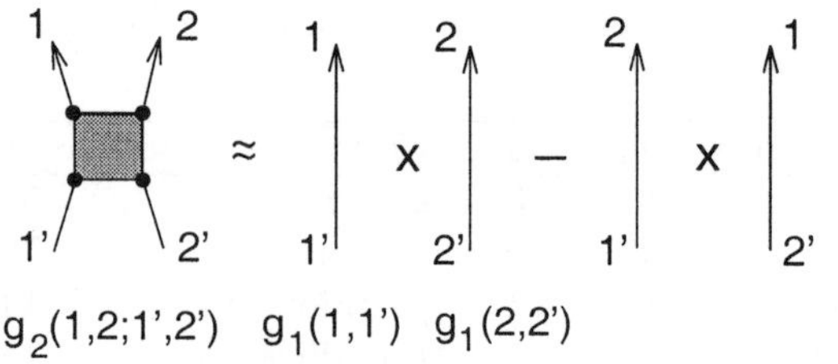

Figure 4. Factorization of four-point baryon Green function g_2 into antisymmetrized products of two-point baryon Green functions, $g_1 \times g_1$. Direct (Hartree) and exchange (Fock) contribution are shown. This factorization scheme truncates the many-body equations at the Hartree-Fock level.

relativistic Hartree (i.e., mean-field) and Hartree-Fock approximations. The T-matrix approximation, also known as ladder (Λ) approximation, goes beyond Hartree and Hartree-Fock by truncating the Martin-Schwinger hierarchy by factorizing the six-point Green's function $g_3(123; 1'2'3')$ into products of four-point and two-point functions by means of which dynamical two-particle correlations in matter, which are connected to a two-body boson-exchange potential v, are taken into account. The T matrix (effective two-particle interaction in matter) is computed from an integral equation of the following type,[3]

$$T = v - v^{\text{ex}} + \int v \, \Lambda \, T \,, \tag{19}$$

with Λ the baryon-baryon propagator in intermediate scattering states. In the framework of the so-called Λ^{00} approximation, the baryons in intermediate states propagate as free particles, leading to a $\Lambda \approx \Lambda^{00} \equiv i g_1^0 g_1^0$. This is in sharp contrast to the relativistic Brueckner-Hartree-Fock (RBHF) approximation where the intermediate baryons are coupled to the nuclear background and, therefore, are constrained to scattering states outside their respective Fermi seas[39,40,41]. The basic input quantity in Eq. (19) is the nucleon-nucleon interaction in free space. Representative models are, for instance, the Bonn meson-exchange model[31] and the boson-exchange potentials of Brockmann and Machleidt[42]. A characteristic feature of these potentials is that their parameters are adjusted to the relativistic two-nucleon scattering data and the properties of the deuteron. In this respect a parameter-free treatment of the many-body problem is achieved. The Born approximation to the T-matrix,

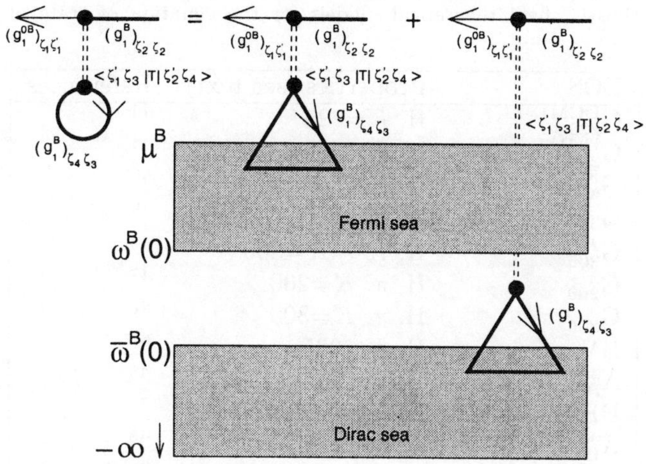

Figure 5. Graphical representation of $g_1^{0B}\Sigma^B g_1^B$ of Dyson's equation (22). The Fermi sea of each baryon is filled up to the highest single-particle energy state, μ^B. The infinite Dirac sea is filled with antibaryons, which too modify the motion of the baryons. The Fermi and Dirac-sea graphs are referred to as medium and vacuum polarization contributions, respectively.[30,43]

which merely sums numerous boson-exchange potentials,

$$< 12 \mid v \mid 1'2' > = \sum_{M=\sigma,\omega,\pi,\rho,\eta,\delta,\phi} \delta^4_{11'} \, \Gamma^M_{11'} \, \Delta^M_{12} \, \Gamma^M_{22'} \, \delta^4_{22'} \, , \qquad (20)$$

neglects dynamical correlations among baryons. It is this approximation which reduces the ladder approximation to the relativistic Hartree and Hartree-Fock approximations[9]. The symbols Γ^M in Eq. (20) stands for the various meson-nucleon vertices, and Δ^M denotes the free meson propagator of a meson of type M whose explicit forms are given in Ref.[3]. The baryon self-energy, or mass operator, is obtained from the T matrix according to[3]

$$\Sigma^B = i \sum_{B'=p,n,\Sigma^{\pm,0},\Lambda,\Xi^{0,-},\Delta^{++,+,0,-}} \int \left[\mathrm{tr}\left(T^{BB'} \, g_1^{B'}\right) - T^{BB'} \, g_1^{B'} \right] , \qquad (21)$$

where B' sums all the charged baryon states whose thresholds are reached in neutron star matter treated within this framework. The baryon Green's functions g_1^B in Eq. (21) are given as solutions of Dyson's equation,

$$g_1^B = g_1^{0\,B} + g_1^{0\,B} \, \Sigma^B(\{g_1^{B'}\}) \, g^B \, , \qquad (22)$$

Table 5. Relativistic, field theoretical models for the equation of state of neutron star matter.

EOS	Properties (see text)	References
G_{225}^{DCM1}	H, K=225	44
G_{265}^{DCM2}	H, K=265	44
G_{M78}^{K240}	H, K=240	45
G_{300}	H, K=300	46
$G_{300}^{K^-}$	H, K^-, K=300	46
G_{200}^{π}	H, π, K=200	47
G_{300}^{π}	H, π, K=300	46
HV	H, K=285	6,7
$\Lambda_{\rm Bonn}^{00}$ + HV	H, K=186	48
HFV	H, Δ, K=376	7
$\Lambda_{\rm HEA}^{00}$ + HFV	H, Δ, K=115	48
$\Lambda_{\rm BroB}^{\rm BHF}$ + HFV	H, Δ, K=249	39,40,41

which terminates the set of equations to be solved self-consistently. The diagrammatic representation of the second term of Dyson's equation is shown in Figure 5. It is this term which corrects the free Green's function for medium effects originating from the presence of Fermi seas of filled baryons. The equation of state follows from the stress-energy density tensor, $T_{\mu\nu}$, of the system as[3]

$$E(\rho) \;=\; (< T_{00} > / \rho\,) - m\,, \tag{23}$$

where

$$T_{\mu\nu}(x) \;=\; \sum_{\chi=B,L} \partial_\nu \psi_\chi(x)\, \frac{\partial \mathcal{L}(x)}{\partial\,(\partial^\mu\, \psi_\chi(x))} \;-\; g_{\mu\nu}\, \mathcal{L}(x)\,. \tag{24}$$

The pressure is obtained from $E(\rho)$ as in Eq. (7).

A broad collection of relativistic models for the equation of state of superdense neutron star matter is compiled in Table 5. The specific properties of these equations of state are described in Table 6, where the following abbreviations are used: N = pure neutron; NP = n, p, leptons; π = pion condensation; H = composed of n, p, hyperons ($\Sigma^{\pm,0}$, Λ, $\Xi^{0,-}$), and leptons; $\Delta = \Delta_{1232}$-resonance; K^- = condensate of negatively charged kaons; and K = incompressibility (in MeV). All equations of state of this collection account for neutron star matter in full baryonic equilibrium. A few selected equations of state of this collection are shown in Figure 3. An inherent feature

Table 6. Nuclear matter properties of the equations of state compiled in Table 5.

EOS	E/A (MeV)	ρ_0 (fm^{-3})	K (MeV)	M^* (MeV)	a_{sy} (MeV)
$\mathrm{G}_{225}^{\mathrm{DCM1}}$	−16.0	0.16	225	0.796	32.5
$\mathrm{G}_{265}^{\mathrm{DCM2}}$	−16.0	0.16	265	0.796	32.5
$\mathrm{G}_{\mathrm{M78}}^{\mathrm{K240}}$	−16.3	0.153	240	0.78	32.5
G_{300}	−16.3	0.153	300	0.78	32.5
$\mathrm{G}_{300}^{K^-}$	−16.3	0.153	300	0.78	32.5
G_{200}^{π}	−15.95	0.145	200	0.8	36.8
G_{300}^{π}	−16.3	0.153	300	0.78	32.5
HV	−15.98	0.145	285	0.77	36.8
$\Lambda_{\mathrm{Bonn}}^{00}$ + HV	−11.9	0.134	186	0.79	34
HFV	−15.54	0.159	376	0.62	30
$\Lambda_{\mathrm{HEA}}^{00}$ + HFV	−8.7	0.132	115	0.82	29
$\Lambda_{\mathrm{BroB}}^{\mathrm{BHF}}$ + HFV	−15.73	0.172	249	0.73	34.3

of relativistic equations of state is that they do not violate causality, i.e., the velocity of sound given by $v_s = \sqrt{dP/d\epsilon}$ is smaller than the velocity of light at all densities. This is not the case for most non-relativistic models for the equation of state. Among the latter, WFF(UV_{14} + TNI), APR(A_{18} + UIX), and the TF96 models for the equation of state do not violate causality up to densities relevant for the construction of models of neutron stars from them.

2.3 Baryon-Lepton Composition of Neutron Star Matter

At densities lower than the density of normal nuclear matter, $\epsilon_0(=140\ \mathrm{MeV/fm}^3)$, neutron star matter consists of only p, n, e^-, and μ^- whose densities are determined by the matter equations (16) through (22). At densities larger than ϵ_0 the more massive baryon states $\Sigma^{\pm,0}, \Lambda, \Xi^{0,-}, \Delta^{++,\pm,0}$ become populated since their chemical potentials, μ^B, become larger than the lowest-lying energy eigenstates $\omega^B(\boldsymbol{p}=0)$ of these particles,

$$\mu^B \equiv \mu^n - q_B\, \mu^e \geq \omega^B(\boldsymbol{p}=0)\,. \tag{25}$$

In relativistic field theory, the energy eigenstates are given by[3]

$$\omega^B(\boldsymbol{p}) = \Sigma_0^B(\omega^B(\boldsymbol{p}),\boldsymbol{p}) + I_{3B}\, \Sigma_{03}^B(\omega^B(\boldsymbol{p}),\boldsymbol{p}) + [m_B + \Sigma_S^B(\omega^B(\boldsymbol{p}),\boldsymbol{p})]\,, \tag{26}$$

while in the non-relativistic case one has

$$\omega^B(\boldsymbol{p}) = \frac{1}{2\,m_B}\,\boldsymbol{p}^2 + \Sigma^B(\boldsymbol{p})\,, \tag{27}$$

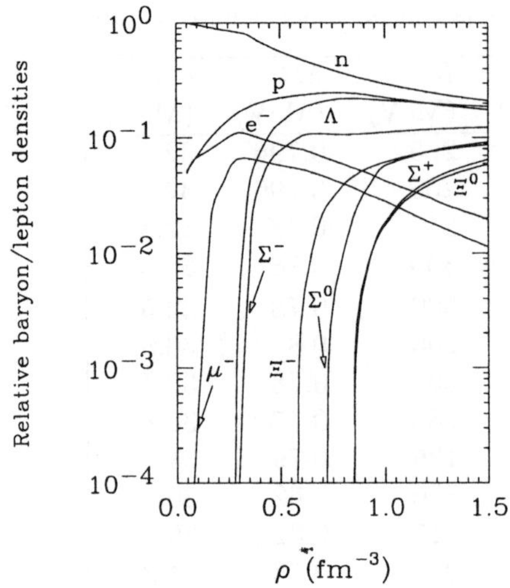

Figure 6. Baryon-lepton composition, normalized to the total baryon density ρ, of neutron star matter computed for HV.

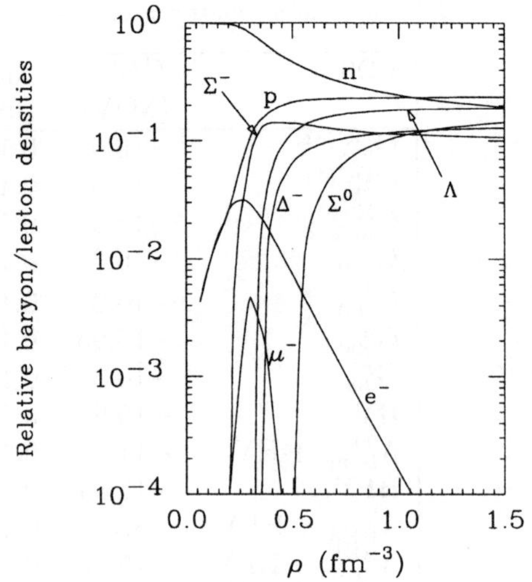

Figure 7. Same as Figure 6, but computed for HFV.

where Σ^B denotes the non-relativistic one-particle potential felt by a baryon in matter. Since the chemical potentials μ^n and μ^e are positive and the self-energies Σ^B_{03} are negative (Σ^B_{03} is proportional to the difference of proton and neutron densities, which is negative for neutron star matter because of its neutron excess), it follows from Eq. (25) that (i) negatively charged baryons are charge-favored, and (ii) baryons having the opposite isospin orientation as the neutron (i.e., $I_{3B} > 0$) are isospin-favored.[6] The physical reason behind item (i) is that negatively charged baryons replace electrons with high Fermi momenta, while (ii) is making the matter more isopin symmetric. Both items enable neutron star matter to settle down in an energetically more favorable state. Figures 6 through 9 illustrate possible particle compositions of neutron star matter computed for different many-body techniques as well as different nuclear forces. One sees that in relativistic Hartree (HV) the Λ hyperon, which remains unaffected by the effects (i) and (ii) since its electric charge $q_\Lambda = 0$ and the third component of isospin $I_{3\Lambda} = 0$, has the lowest threshold of all hyperons. In the case of relativistic Hartree-Fock (HFV), the charge favored but isospin unfavored Σ^- possesses the lowest threshold, and the Δ makes its appearance too. The latter is also the case for relativistic Brueckner-Hartree-Fock type calculations.[8] The important net effect of all these new degrees of

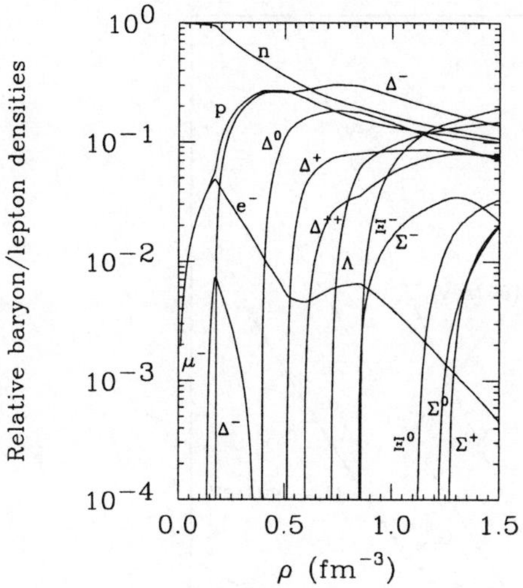

Figure 8. Same as Figure 6, but for RBHF 1 parameter set (see Ref. [3]). The hyperons are non-universally coupled, the Δ's universally.

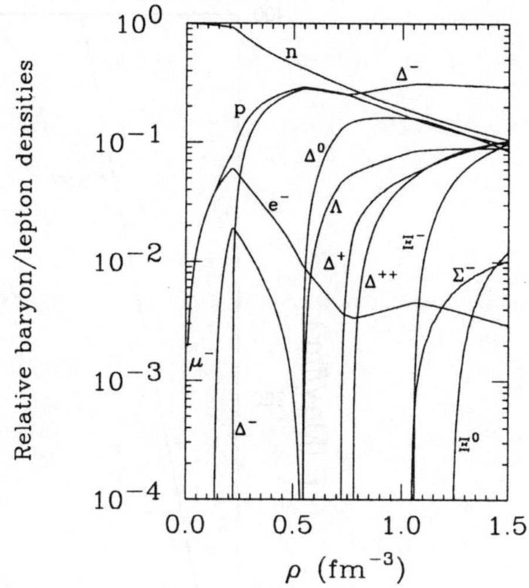

Figure 9. Same as Figure 8, but for non-universally coupled hyperons and Δ's.[8]

freedom is a softening of the equation of state at supernuclear densities, which reduces the maximum mass of a neutron star.

2.4 H-dibaryons

A novel particle that may make its appearance in the center of a neutron star is the H-dibaryon, a doubly strange six-quark composite with spin and isospin zero, and baryon number two.[49] Since its first prediction in 1977, the H-dibaryon has been the subject of many theoretical and experimental studies as a possible candidate for a strongly bound exotic state. In neutron stars, which may contain a significant fraction of Λ hyperons (Figures 6 through 9), the Λ's could combine to form H-dibaryons, which could give way to the formation of H-matter at densities somewhere between $3\,\epsilon_0$ and $6\,\epsilon_0$ depending on the in-medium properties of the H-dibaryon.[50,51,52] This could cause a significant softening of the equation of state as shown in Figure 10 (curves labeled U_H). In Ref.[50] it was shown that H-dibaryons with a vacuum mass of about 2.2 GeV and a moderately attractive potential in the medium of about $U_H = -30$ MeV could go into a Bose condensate in the cores of neu-

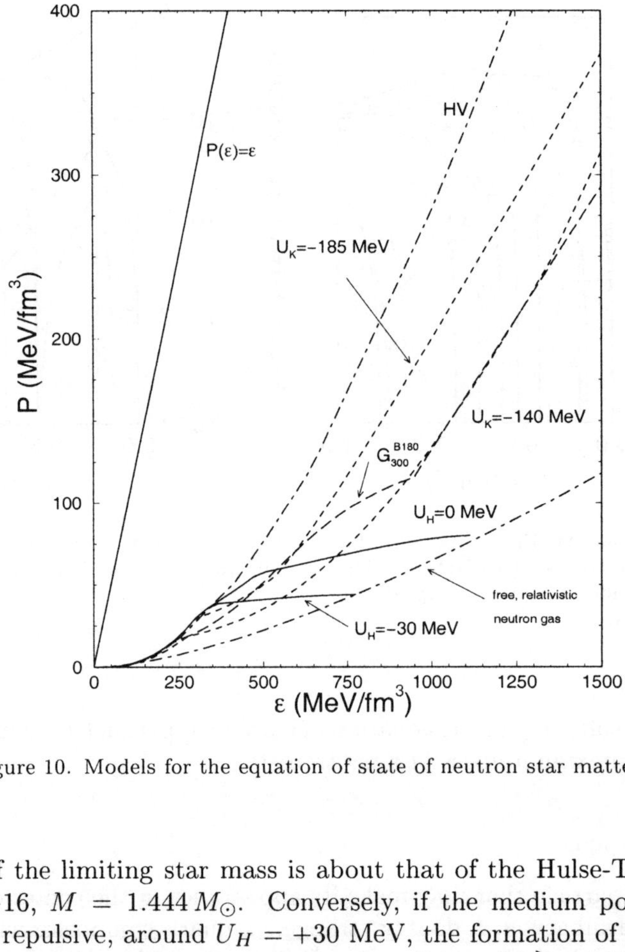

Figure 10. Models for the equation of state of neutron star matter.[3]

tron stars if the limiting star mass is about that of the Hulse-Taylor pulsar PSR 1913+16, $M = 1.444\, M_\odot$. Conversely, if the medium potential were moderately repulsive, around $U_H = +30$ MeV, the formation of H-dibaryons may only take place in heavier neutron stars of mass $M \gtrsim 1.6\, M_\odot$. If indeed formed, however, H-matter may not remain dormant in neutron stars but, because of its instability against compression could trigger the conversion of neutron stars into hypothetical strange stars[52,53,54].

2.5 Condensation of K^- Mesons

Once the reaction

$$e^- \rightarrow K^- + \nu \tag{28}$$

becomes possible in a neutron star, it becomes energetically advantageous

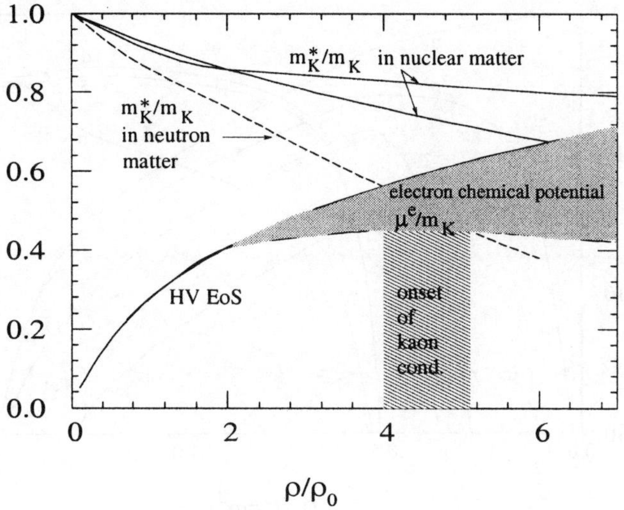

Figure 11. Effective kaon mass in nuclear and neutron star matter[55-59].

for the star to replace the fermionic electrons with the bosonic K^- mesons. Whether or not this actually happens depends on how quickly the K^- drops with density in dense matter (Figure 11). An analysis of the Kaon EOS data[60] shows that the attraction from nuclear matter may bring the K^- mass down to $m^*_{K^-} \simeq 200$ MeV at $\rho \sim 3\,\rho_0$. For neutron-rich matter, the K^- mass may drop as[11-14]

$$m^*_{K^-}(\rho) \simeq m_{K^-}\left(1 - 0.2\,\frac{\rho}{\rho_0}\right), \tag{29}$$

with $m_K = 495$ MeV the K^- vacuum mass. Combining this relation with the outcome for μ^e shown in Figure 11 shows that the threshold condition for the onset of K^- condensation, $\mu^e = m^*_K$ would be fulfilled at densities $\rho \gtrsim 3\rho_0$, which are easily reached in neutron stars.

2.6 Quark Deconfinement

The phase transition between confined hadronic matter and quark matter is characterized by the conservation of baryon charge and electric charge. The Gibbs condition for phase equilibrium then is that the two associated chemical potentials μ^n and μ^e, corresponding to baryon and electric charge

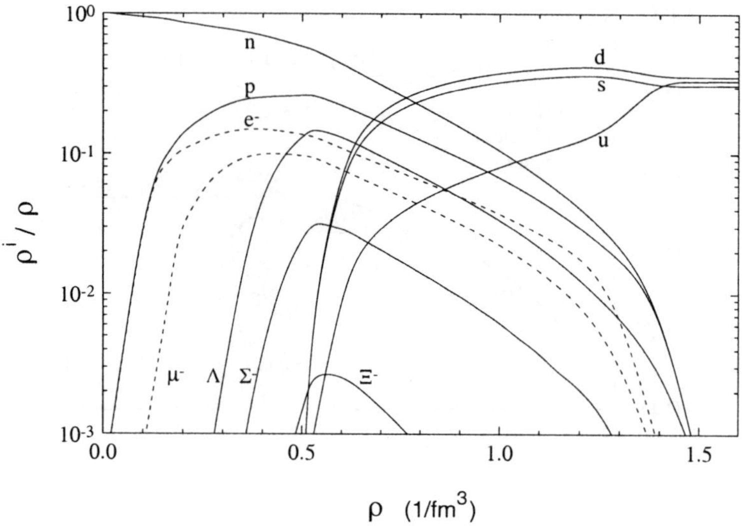

Figure 12. Sample composition of chemically equilibrated neutron (quark-hybrid) star matter[3,62].

conservation, and the pressure in the two phases be equal,

$$P_{\rm H}(\mu^n, \mu^e, \{\phi\}, T) = P_{\rm Q}(\mu^n, \mu^e, T)\,. \tag{30}$$

Here $P_{\rm H}$ denotes the pressure of hadronic matter computed for a hadronic lagrangian $\mathcal{L}(\{\phi\})$ as given in Eq. (10). The pressure of quark matter, $P_{\rm Q}$, is obtainable from the bag model[17]. The quark chemical potentials μ^u, μ^d, μ^s are related to the baryon and charge chemical potentials as

$$\mu^u = \frac{1}{3}\,\mu^n - \frac{2}{3}\,\mu^e\,, \qquad \mu^d = \mu^s = \frac{1}{3}\,\mu^n + \frac{1}{3}\,\mu^e\,. \tag{31}$$

Equation (30) is to be supplemented with the two global relations for conservation of baryon charge and electric charge within an unknown volume V containing A baryons[61]. The first one is given by

$$\rho \equiv \frac{A}{V} = (1-\chi)\,\rho_{\rm H}(\mu^n, \mu^e, T) + \chi\,\rho_{\rm Q}(\mu^n, \mu^e, T)\,, \tag{32}$$

where $\chi \equiv V_{\rm Q}/V$ denotes the volume proportion of quark matter, $V_{\rm Q}$, in the unknown volume V, and ρ_i $(i = {\rm H}, {\rm Q})$ is the particle number density. Global neutrality of electric charge within the volume V can be written as

$$0 = \frac{Q}{V} = (1 - \chi)\, q_{\rm H}(\mu^n, \mu^e, T) + \chi\, q_{\rm Q}(\mu^n, \mu^e, T) + q_{\rm L}\,, \tag{33}$$

with q_i the electric charge densities of hadrons, quarks, and leptons. For a given temperature T, Eqs. (30) through (33) serve to determine the two independent chemical potentials and the volume V for a specified volume fraction χ of the quark phase in equilibrium with the hadronic phase. After completion $V_{\rm Q}$ is obtained as $V_{\rm Q} = \chi V$. Through Eqs. (30) to (33), the chemical potentials obviously depend on the proportion χ of the phases in equilibrium, and hence so also all properties that depend on them, the energy densities, baryon and charge densities of each phase, and the common pressure. For the mixed phase, the volume proportion of quark matter varies from $0 \le \chi \le 1$, and the energy density is the linear combination of the two phases,[61]

$$\epsilon = (1 - \chi)\, \epsilon_{\rm H}(\mu^n, \mu^e, \{\phi\}, T) + \chi\, \epsilon_{\rm Q}(\mu^n, \mu^e, T)\,. \tag{34}$$

By solving the models of confined and deconfined phases in both pure phases and in the mixed phase, we can compute the baryon, lepton and quark populations in neutron star matter from Eqs. (30) through (33). One such sample outcome is shown in Figure 12. Three features emerge immediately from this population. Firstly, one sees that the transition from pure hadronic matter to the mixed phase occurs at rather low density of about $3\,\rho_0$. Depending on the bag constant and the underlying nuclear many-body approximation, threshold values even as small as about $2\,\rho_0$ can be obtained.[61,62] Secondly, the lepton density saturates as soon as deconfined quark matter is generated, since charge neutrality can be achieved more economically among the baryon-charge carrying particles themselves. Thirdly, the presence of quark matter enables the hadronic regions of the mixed phase to arrange themselves to be more isospin symmetric than in the pure phase by transferring charge to the quark phase in equilibrium with it. Symmetry energy will be lowered

Table 7. Models for the equation of state of neutron star matter accounting for quark deconfinement.

EOS	Properties (see text)	Reference
$G_{\rm B180}^{\rm DCM1}$	H,Q, K=225, $B^{1/4} = 180$	61,63
$G_{\rm B180}^{\rm DCM2}$	H,Q, K=265, $B^{1/4} = 180$	61,63
$G_{\rm B180}^{\rm K240}$	H,Q, K=240, $B^{1/4} = 180$	45
$G_{\rm B180}^{\rm K300}$	H,Q, K=300, $B^{1/4} = 180$	45

Table 8. Nuclear matter properties of the equations of state compiled in Table 7.

EOS	E/A (MeV)	ρ_0 (fm^{-3})	K (MeV)	M^* (MeV)	a_{sy} (MeV)
G^{DCM1}_{B180}	−16.0	0.16	225	0.796	32.5
G^{DCM2}_{B180}	−16.0	0.16	265	0.796	32.5
G^{K240}_{B180}	−16.3	0.153	240	0.78	32.5
G^{K300}_{B180}	−16.3	0.153	300	0.70	32.5

thereby at only a small cost in rearranging the quark Fermi surfaces. Thus the mixed phase region consists of positively charged nuclear matter and negatively charged quark matter. Models for the equation of state of neutron star matter which account for quark deconfinement are listed in Tables 7 and 8. A graphical illustration of these equations of state is shown in Figure 13.

3 Observed Neutron Star Properties

Orbiting observatories such as the Hubble Space Telescope, the Chandra X-ray satellite, and the X-ray Multi Mirror Mission have extended our vision tremendously, allowing us to look at neutron stars with an unprecedented clarity and angular resolution that previously were only imagined, The properties of such objects such as masses, rotational frequencies, radii, moments of inertia, redshifts, and temperatures are known to be sensitive to the adopted microscopic model for the nucleon-nucleon interaction and thus to the nuclear equation of state[3,64]. Hence by means of comparing theoretically determined values for these quantities with observed ones one may be able to constrain the behavior of superdense matter from the (unprecedented wealth of new) observational pulsar data. In the following we briefly summarize several important star properties.

3.1 Masses

The gravitational mass is of special importance since it can be inferred directly from observations of X-ray binaries and binary radio pulsar systems. The masses of the latter appear to be centered about 1.35 $M_\odot$[65]. The mass of the Hulse-Taylor radio pulsar PSR 1913+16 [66], given by $1.444 \pm 0.003)\, M_\odot$, is somewhat higher than this value. Examples of neutron stars in X-ray binaries are Vela X-1 and the burster Cygnus X-2. Their respective masses are $M = 1.87^{+0.23}_{-0.17}\, M_\odot$ [67,68] and $M = (1.78 \pm 0.23)\, M_\odot$ [69]. Indications for

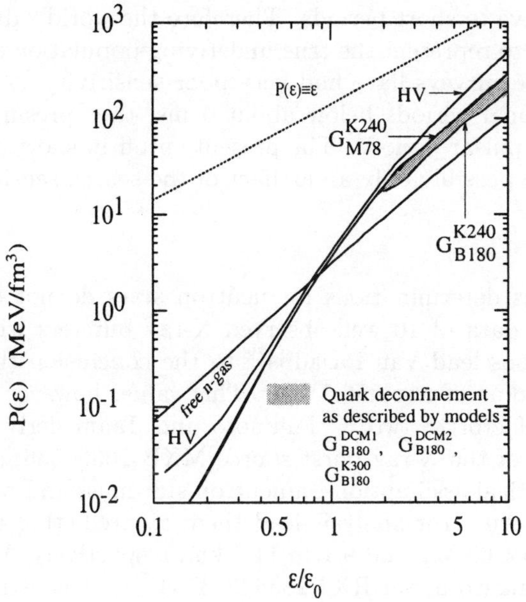

Figure 13. Models for the equations of state of neutron star matter. The shaded area marks quark deconfinement taken into account in models G^{K300}_{B180}, G^{DCM2}_{B180}, G^{K240}_{B180}, G^{DCM162}_{B180}.

the possible existence of very heavy neutron stars, with masses around $2\,M_\odot$ (e.g., neutron star 4U 1636–536), may also come from the observation of quasi-periodic oscillations in luminosity in low-mass X-ray binaries[70]. These mass values, which constitute upper boundaries on the star mass, are larger than the typical $1.35\,M_\odot$ masses found in neutron star binaries, presumably due to accreted matter.

3.2 Rotational Frequencies of Fast Pulsars

The rotational periods of fast pulsars provide constraints on the equation of state when combined with the mass constraint. The fastest pulsars observed so far have rotational periods of 1.6 ms, which corresponds to a rotational frequency of 620 Hz. The successful model for the nuclear equation of state must therefore account for rotational neutron star periods of at least $P =$ 1.6 ms as well as for the masses discussed in Section 3.1.

Observationally, the searches for rapidly rotating radio pulsars are biased, being less sensitive to short periods. Therefore the world's data of radio pulsars are unlikely to represent the true underlying population of fast pulsars[71]. Most of the large surveys have had very poor sensitivity to millisecond pulsars with rotational periods below about 4 ms, thus presumably distorting the statistics on pulsar periods. The present cutoff in short periods at about 1 ms is therefore possibly only an artifact of the search sensitivity.

3.3 Radii

Direct radius determinations for neutron stars do not exist. However, combinations of data of 10 well-observed X-ray bursters with special theoretical assumptions lead Van Paradijs[72] to the conclusion that the emitting surface has a radius of about 8.5 km. This value, however, may be underestimated by a factor of two[73]. Fujimoto and Taam derived from the observational data of the X-ray burst source MXB 1636–536, under somewhat uncertain theoretical assumptions, a neutron star mass and radius of 1.45 $M_\odot$ and 10.3 km[74]. An error analysis lead them to predicting mass and radius ranges of 1.28 to 1.65 $M_\odot$ and 9.1 to 11.3 km, respectively. Very recently the nearby, isolated neutron star RXJ 185635–3754 [75,76] has attracted particular attention because of its interpretation as a possible strange quark star[77]. (The physics of strange quark stars will be discussed in Section 5.) The observed X-ray flux and temperature from this object, if originating from a uniform blackbody, appears to correspond to an effective stellar radius significantly smaller than $\sim$ 10 km, the canonical neutron star radius. This follows from the relation

$$R_\infty = R\left(1 - \frac{2M}{R}\right)^{-1/2} \approx 7\,\frac{D}{120\ \mathrm{pc}}\ \mathrm{km}\,, \tag{35}$$

where R is the neutron star radius, M its mass, and D the distance to the star. One major uncertainty in (35) is buried in the numerical coefficient, which could increase by 50% or more if the star's surface temperature is not uniform. Moreover, the existence of an atmosphere on this neutron star could alter the coefficient probably even more strongly[78]. The other major uncertainty in obtaining a radius is the star's distance which, however, has now been determined to be $D = 117 \pm 12$ pc [79], nearly double the originally published distance. Taking these considerations into account, the interpretation of RXJ 185635–3754 as a quark star appears questionable[79].

3.4 Moment of Inertia

Another global neutron star property is the moment of inertia, I. Early estimates of the energy-loss rate from pulsars spanned a wide range of I, i.e., $7 \times 10^{43} < I < 7 \times 10^{44}$ g/cm^3[80]. From the luminosity of the Crab nebula ($\sim 2 - 4 \times 10^{38}$ erg/sec), several authors have found a lower bound on the moment of inertia of the pulsar given by $I \gtrsim 4 - 8 \times 10^{44}$ g cm^2.[81,82,83]

3.5 Redshift

Next we mention the neutron star redshift, z, which is given by

$$z = \left(1 - \frac{2\,M}{R}\right)^{-1/2} - 1\,, \tag{36}$$

with M and R the star's mass and radius. Liang has considered the neutron star redshift data base provided by measurements of γ-ray burst redshifted annihilation lines in the range 300 – 511 keV[84]. These bursts have widely been interpreted as gravitationally redshifted 511 keV $e^{\pm}$ pair annihilation lines from the surfaces of neutron stars. From this he showed that there is tentative evidence to support a neutron star redshift range of $0.2 \le z \le 0.5$, with the highest concentration in the narrower range $0.25 \le z \le 0.35$. A particular role plays the source of the 1979 March 5 γ-ray burst source, which has been identified with SNR N49 by its position. From the interpretation of its emission, which has a peak at ~ 430 keV, as the 511 keV $e^{\pm}$ annihilation line the resulting gravitational redshift has a value of $z = 0.23 \pm 0.05$.

3.6 Magnetic Fields

Neutron stars are highly magnetized objects[85]. There is a general plausibility argument to the effect that, if internal magnetic flux in a star is conserved owing to high conductivity, then the surface magnetic field increases as R^{-2} where R is the stellar radius. Thus, if the sun were to collapse to neutron star dimensions (roughly a factor of 7×10^4 in radius), one would magnify the general 1 G solar field to about 5×10^9 G. This value, while being in good agreement with the magnetic field strength of $B \sim 4 - 6 \times 10^{11}$ G estimated for the X-ray pulsar 1 E 2259+586[86], the central source in the supernova remnant G 109.1–1.0, is a bit shy of 10^{12} G inferred from line features in the pulsed hard X-ray spectrum of Her X–1 and 4U 0115–63[87,88,89]. Which physical mechanisms might be important to create the extraordinarily strong magnetic fields of neutron stars is an open issue. One possible mechanism may be differential rotation of the stellar core, as proposed for planetary magnetic fields. Besides that, ferromagnetism has been examined as a possible magnetic field source, while differential rotation between say superfluid protons and normal electrons has been another suggestion[85]. Effects of strong magnetic fields on

Table 9. Overview of neutrino emitting processes relevant for neutron star cooling[3].

Name	Particle processes	Efficiency
Modified Urca	$n + n \to n + p + e^- + \bar{\nu}_e$	slow
	$n + p + e^- \to n + n + \nu_e$	
Direct Urca	$n \to p + e^- + \bar{\nu}_e$	fast
	$p + e^- \to n + \nu_e$	
Quark modified Urca	$d + u + e^- \to d + d + \nu_e$	slow
	$u + u + e^- \to u + d + \nu_e$	
	$d + u + e^- \to d + s + \nu_e$	
	$u + u + e^- \to u + s + \nu_e$	
Quark direct Urca	$d \to u + e^- + \bar{\nu}_e$	fast
	$u + e^- \to d + \nu_e$	
	$s \to u + e^- + \bar{\nu}_e$	
	$u + e^- \to s + \nu_e$	
π^- condensate	$n + < \pi^- > \to n + e^- + \bar{\nu}_e$	fast
K^- condensate	$n + < K^- > \to n + e^- + \bar{\nu}_e$	fast
Quark bremsstrahlung	$Q_1 + Q_2 \to Q_1 + Q_2 + \nu + \bar{\nu}$	slow
Core bremsstrahlung	$n + n \to n + n + \nu_e + \bar{\nu}_e$	slow
	$n + p \to n + p + \nu_e + \bar{\nu}_e$	
	$e^- + p \to e^- + p + \nu_e + \bar{\nu}_e$	
Crust bremsstrahlung	$e^- + (A, Z) \to e^- + (A, Z) + \nu_e + \bar{\nu}_e$	slow

the structure of compact stars were studied, for instance, in Refs.[90,91]. Particularly striking are so-called magnetars, which are interpreted as neutron stars with superstrong surface magnetic fields on the order of $\sim 10^{15}$ G.[92] This translates to interior magnetic fields of up to $\sim 10^{18}$ G, which is about the highest possible value a star can sustain.

3.7 Cooling Data

The detection of thermal photons from a stellar surface via X-ray observatories like ASCA and ROSAT serves as the principal window on the properties of a star. The surface temperatures of stars are derivable from the measured flux and spectrum. The cooling rate of a hot, young neutron star is primarily dependent at early times (first several thousand years) on the neutrino emissivity of the core's composition. The possible existence of Bose-Einstein condensates or quark matter enhance the neutrino emissivity from the core,

Table 10. Ages τ (in years) and luminosities L (in erg s^{-1}) / temperatures T (in K) of pulsars.[3]

Pulsar	SNR	$\log_{10}\tau$	L or T	Refs.
1706–44		4.25	$\log_{10} L = 32.8 \pm 0.7$	93
1823–13		4.50	$\log_{10} L = 33.2 \pm 0.6$	94
2334+61		4.61	$\log_{10} L = 33.1 \pm 0.4$	95
0531+21	Crab	3.09	$\log_{10} L = 35.5 \pm 0.3$	96
1509–58	MSH 15-52	3.19	$\log_{10} L = 33.6 \pm 0.4$	97,98
0540–69		3.22	$\log_{10} L = 36.2 \pm 0.2$	99
1951+32	CTB 80	5.02	$\log_{10} L = 33.8 \pm 0.5$	100
1929+10		6.49	$\log_{10} L = 28.9 \pm 0.5$	101,102
0950+08		7.24	$\log_{10} L = 29.6 \pm 1.0$	103
J0437–47		8.88	$\log_{10} L = 30.6 \pm 0.4$	104
0833–45	Vela	4.05	$\log_{10} L = 32.9 \pm 0.2$	105
0656+14		5.04	$\log_{10} L = 32.6 \pm 0.3$	106
0630+18	Geminga	5.51	$\log_{10} L = 31.8 \pm 0.4$	107
1055–52		5.73	$\log_{10} L = 33.0 \pm 0.6$	108
J0205+6449	3C58	2.91	$\log_{10} T = 6.0334$	109

leading to a more rapid early cooling (Table 9). Superfluidity, on the other hand, has the opposite effect on cooling. Quantitative constraints have been hampered by the relatively small number of young pulsars known, the complication that several of them also display non-thermal, beamed X-ray emission from their magnetospheres, and uncertainties in distance and interstellar absorption. Table 10 gives an overview of observed neutron star temperatures. Because of its unusually low temperature which appears to fall well below predictions from standard cooling models[109], the neutron star in supernova remnant (SNR) 3C58 received much attention recently (see Section 6).

4 Properties of Neutron Star Models

4.1 Non-rotating Star Models

For many studies of neutron star properties it is sufficient to treat neutron star matter as a perfect fluid. The energy-momentum tensor of such a fluid is given by

$$T^{\mu\nu} = u^\mu \, u^\nu \left(\epsilon + P \right) + g^{\mu\nu} \, P \,, \tag{37}$$

where u^μ and u^ν are four-velocities defined as

$$u^\mu \equiv \frac{dx^\mu}{d\tau}\,, \qquad u^\nu \equiv \frac{dx^\nu}{d\tau}\,. \tag{38}$$

They are the components of the macroscopic velocity of the stellar matter with respect to the actual coordinate system that is being used to derive the stellar equilibrium equations. The production of curvature by the star's mass is specified by Einstein's field equations,

$$G_{\mu\nu} = 8\,\pi\,T_{\mu\nu}\,, \quad \text{where} \quad G_{\mu\nu} \equiv R_{\mu\nu} - \frac{1}{2}\,g_{\mu\nu}\,R \tag{39}$$

is the Einstein tensor. The Ricci tensor $R_{\mu\nu}$ is obtained from the Riemann tensor $R^\tau{}_{\mu\sigma\nu}$ by contraction, that is, $R_{\mu\nu} = R^\tau{}_{\mu\sigma\nu}\,g^\sigma{}_\tau$ which leads to

$$R_{\mu\nu} = \frac{\partial}{\partial x^\nu}\,\Gamma^\sigma_{\mu\sigma} - \frac{\partial}{\partial x^\sigma}\,\Gamma^\sigma_{\mu\nu} + \Gamma^\kappa_{\mu\sigma}\,\Gamma^\sigma_{\kappa\nu} - \Gamma^\kappa_{\mu\nu}\,\Gamma^\sigma_{\kappa\sigma}\,, \tag{40}$$

with $\Gamma^\sigma_{\mu\nu}$ the Christoffel symbol defined as

$$\Gamma^\sigma_{\mu\nu} \equiv \frac{1}{2}\,g^{\sigma\lambda}\left(\frac{\partial}{\partial x^\nu}\,g_{\mu\lambda} + \frac{\partial}{\partial x^\mu}\,g_{\nu\lambda} - \frac{\partial}{\partial x^\lambda}\,g_{\mu\nu}\right)\,. \tag{41}$$

The scalar curvature of space-time R in Eq. (39), also known as Ricci scalar, follows from Eq. (40) as

$$R = R_{\mu\nu}\,g^{\mu\nu}\,. \tag{42}$$

Finally, we need to specify the metric of a non-rotating body in general relativity theory. Assuming spherical symmetry, the metric has the form

$$ds^2 = -e^{2\,\Phi(r)}\,dt^2 + e^{2\,\Lambda(r)}\,dr^2 + r^2\,d\theta^2 + r^2\,\sin^2\theta\,d\phi^2\,, \tag{43}$$

where $\Phi(r)$ and $\Lambda(r)$ are radially varying metric functions. Introducing the covariant components of the metric tensor,

$$g_{tt} = -\,e^{2\,\Phi}\,, \; g_{rr} = e^{2\,\Lambda}\,, \; g_{\theta\theta} = r^2\,, \; g_{\phi\phi} = r^2\,\sin^2\theta\,, \tag{44}$$

the non-vanishing Christoffel symbols of a spherically symmetric body are

$$\Gamma^r_{tt} = e^{2\,\Phi - 2\,\Lambda}\,\Phi'\,, \; \Gamma^t_{tr} = \Phi'\,, \; \Gamma^r_{rr} = \Lambda'\,, \Gamma^\theta_{r\theta} = r^{-1}\,, \; \Gamma^\phi_{r\phi} = r^{-1}\,,$$

$$\Gamma^r_{\theta\theta} = -\,r\,e^{-2\,\Lambda}\,, \Gamma^\phi_{\theta\phi} = \frac{\cos\theta}{\sin\theta}\,, \; \Gamma^r_{\phi\phi} = -\,r\,\sin^2\theta\,e^{-2\,\Lambda}\,, \; \Gamma^\theta_{\phi\phi} = -\,\sin\theta\,\cos\theta\,, \tag{45}$$

where accents denote differentiation with respect to the radial coordinate. From Eqs. (37), (39) and (45) one derives the structure equations of spherically symmetric neutron stars known as Tolman-Oppenheimer-Volkoff equations,[110,111]

$$\frac{dP}{dr} = -\frac{\epsilon(r)\, m(r)}{r^2}\, \frac{[1 + P(r)/\epsilon(r)]\, [1 + 4\pi r^3 P(r)/m(r)]}{1 - 2m(r)/r} \,. \tag{46}$$

Note that we use units for which the gravitational constant and velocity of light are $G = c = 1$ so that $M_\odot = 1.5$ km. The boundary condition to (46) is $P(r = 0) \equiv P_c = P(\epsilon_c)$, where ϵ_c denotes the energy density at the star's center, which constitutes an input parameter. The pressure is to be computed out to that radial distance where $P(r = R) = 0$ which determines the star's radius R. The mass contained in a sphere of radius r ($\leq R$), denoted by $m(r)$, follows as

$$m(r) = 4\pi \int_0^r dr'\, r'^2\, \epsilon(r') \,. \tag{47}$$

The star's total gravitational mass is thus given by $M \equiv m(R)$.

Figure 14 exhibits the gravitational masses of non-rotating neutron stars as a function of central energy density for a sample of equations of state of Tables 1 and 5. Each star sequence is shown up to densities that are slightly larger than those of the maximum-mass star (indicated by tick marks) of each sequence. Stars beyond the mass peak are unstable against radial oscillations and thus cannot exist stably (collapse to black holes) in nature. One sees that all equations of state are able to support non-rotating neutron star models of gravitational masses $M \geq M(\text{PSR } 1913 + 16)$. On the other hand, rather massive stars of say $M \gtrsim 2\, M_\odot$ can only be obtained for those models of the equation of state that exhibit a rather stiff behavior at supernuclear densities. The largest maximum mass value, $M = 2.2\, M_\odot$, is obtained for HFV, which is caused by the stiffening exchange term. Knowledge of the maximum mass value is of great importance for two reasons. Firstly, because the largest known neutron star mass imposes a lower bound on the maximum mass of a theoretical model. The current lower bound is about $1.60\, M_\odot$ [neutron star 4U 0900–40 ($\equiv$ Vela X–1)] which does not set too stringent a constraint on the nuclear equation of state. The situation could easily change if an accurate future determination of the mass of this neutron star should result in a value that is close to its present upper bound of $2.1\, M_\odot$. In this case most of the equations of state of our collection would be ruled out. The second reason is that the maximum mass is essential in order to identify black hole candidates.[112,113] For example, if the mass of a compact companion of an optical star is determined to exceed the maximum mass of a neutron star

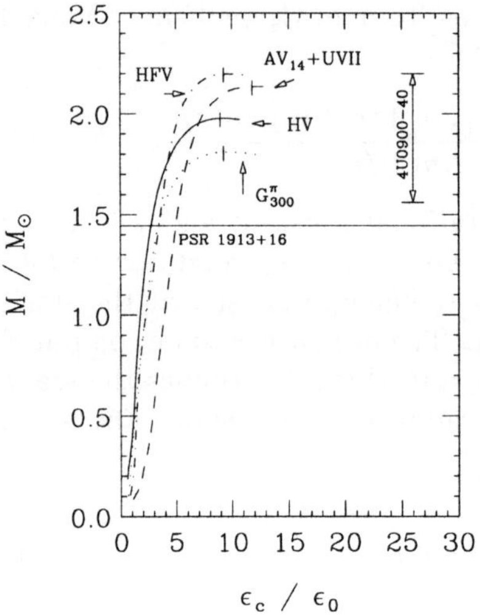

Figure 14. Non-rotating neutron star mass as a function of central density.[3]

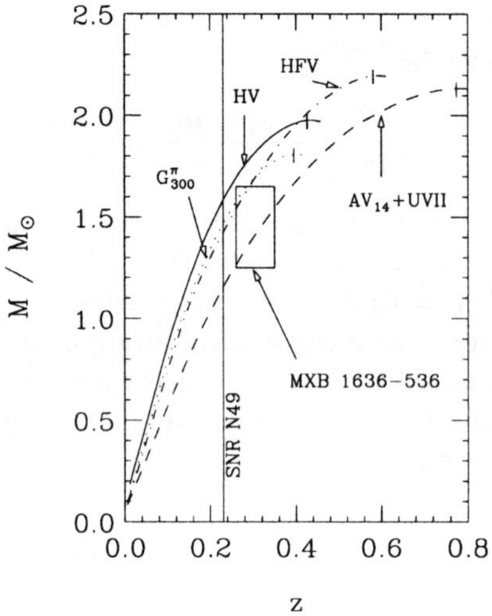

Figure 15. Non-rotating neutron star mass as a function of redshift.[3]

it must be a black hole. Since the maximum mass of stable neutron stars studied here is $\sim 2.2\,M_\odot$, compact companions being more massive than that value are predicted to be black holes.

The neutron star mass as a function of gravitational redshift, defined in Eq. (36), is shown in Figure 15. Maximum-mass stars have redshifts in the range $0.4 \lesssim z \lesssim 0.8$, depending on the stiffness of the equation of state. Neutron stars of typically $M \approx 1.5\,M_\odot$ (e.g., PSR 1913+16) are predicted to have redshifts in the considerably narrower range $0.2 \le z \le 0.32$. The rectangle covers masses and redshifts in the ranges of $1.30 \le M/M_\odot \le 1.65$ and $0.25 \le z \le 0.35$, respectively. The former range has been determined from observational data on X-ray burst source MXB 1636–536[74], while the latter is based on the neutron star redshift data base provided by measurements of gamma-ray burst pair annihilation lines.[84] From the redshift value of SNR N49 one would expect a neutron mass star of mass $1.1 \lesssim M/M_\odot \lesssim 1.6$. Figure 16 displays the radius as a function of gravitational redshift.

Figure 17 shows the moment of inertia of neutron stars, given by[114]

$$I(\Omega) = 4\,\pi \int_0^{\pi/2} d\theta \int_0^{R(\theta)} dr\, e^{\lambda+\mu+\nu+\psi}\, \frac{\epsilon + P(\epsilon)}{e^{2\nu-2\psi} - (\Omega-\omega)^2}\, \frac{\Omega-\omega}{\Omega}\,, \qquad (48)$$

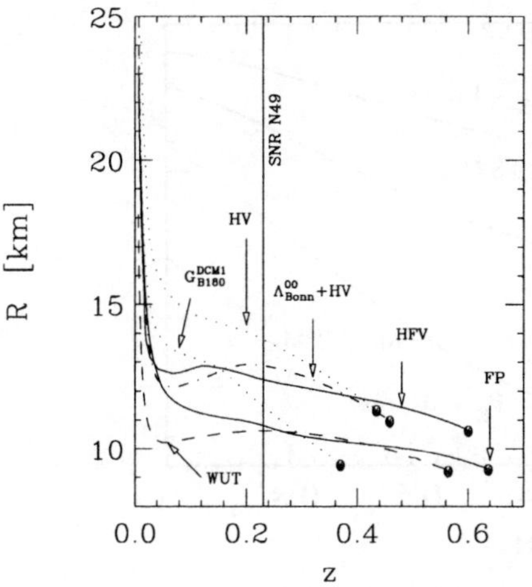

Figure 16. Radius as a function of redshift for a collection of equations of state.[3]

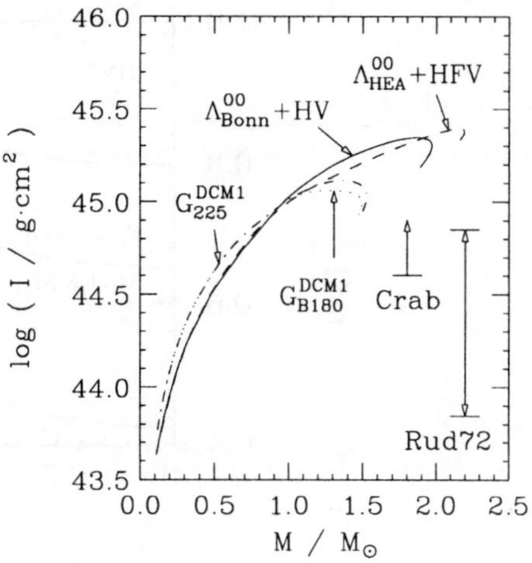

Figure 17. Moment of inertia as a function of mass for a collection of equations of state.[3]

as a function of gravitational mass. In newtonian mechanics one has for a sphere of uniform density $I \propto R^2 M$. The general relativistic expression for the moment of inertia, given in Eq. (48), is considerably more complicated as it accounts for the dragging effect of the local inertial frames (see Figure 18) and the curvature of space-time.[3] Nevertheless the qualitative dependence of I on mass and radius as expressed in the classical expression remains valid. Estimates for the upper and lower bounds on the moment of inertia of the Crab pulsar derived from the pulsar's energy loss rate (labeled Rud72), and the lower bound on the moment of inertia derived from the luminosity of the Crab nebula (labeled Crab) are shown in Figure 17. The arrows refer only to the value of the moment of inertia of the Crab pulsar and not to its mass, which is not known.

4.2 Rotating Star Models

The stellar equations describing rotating compact stars are considerably more complicated than those of non-rotating compact stars.[3] These complications have their cause in the deformation of rotating stars plus the general relativistic effect of the dragging of local inertial frames. This reflects itself in

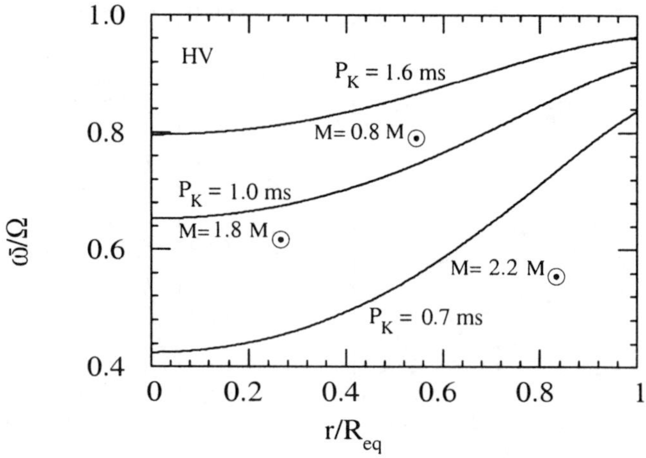

Figure 18. Dragging of the local inertial frames inside rotating neutron stars, from the center to the equator, calculated for the HV of Table 5. $P_{\rm K}$ and M denote Kepler period, defined in (50), and gravitational mass.[62]

a metric of the form[3,115]

$$ds^2 = -\, e^{2\,\nu}\, dt^2 + e^{2\,\psi}\, \big(d\phi - \omega\, dt\big)^2 + e^{2\,\mu}\, d\theta^2 + e^{2\,\lambda}\, dr^2\,, \tag{49}$$

where each metric function ν, ψ, μ and λ depends on the radial coordinate r, polar angle θ, and implicitly on the star's angular velocity Ω. The quantity ω denotes the angular velocity of the local inertial frames, which are dragged along in the direction of the star's rotation. This frequency too depends on r, θ and Ω. Of particular interest is the relative frame dragging frequency $\bar{\omega}$ defined as $\bar{\omega}(r, \theta, \Omega) \equiv \Omega - \omega(r, \theta, \Omega)$, which typically increases from about 15% at the surface to about 60% at the center of a neutron star that rotates at its Kepler frequency[3].

4.3 Mass Shedding versus Gravity Wave Emission

The Kepler frequency, $\Omega_{\rm K}$, is the maximum frequency a star can have before mass loss (mass shedding) at the equator sets in. It sets an absolute upper limit on stable rapid rotation. In classical mechanics the expression for $\Omega_{\rm K}$, determined by the equality between centrifuge and gravity, is readily

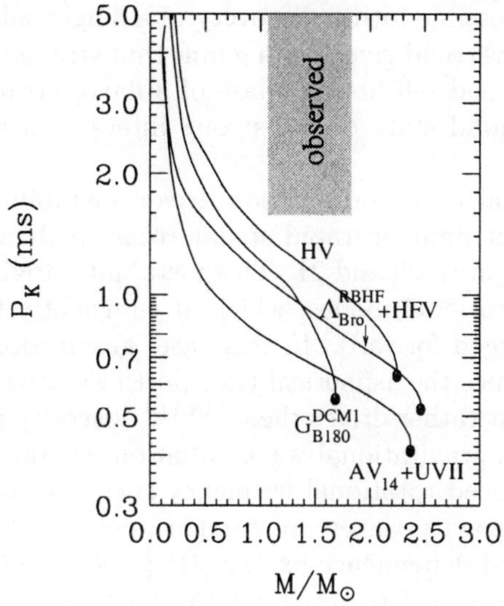

Figure 19. Onset of mass shedding from rapidly spinning neutron stars, computed for a collection of equations of state[3]. The Kepler period is defined in Eq. (50).

obtained as $\Omega_K = \sqrt{M/R^3}$. Its general relativistic counterpart is given by[3,115]

$$\Omega_K = \omega + \frac{\omega'}{2\psi'} + e^{\nu-\psi}\sqrt{\frac{\nu'}{\psi'} + \left(\frac{\omega'}{2\psi'}e^{\psi-\nu}\right)^2}\,, \qquad P_K \equiv \frac{2\pi}{\Omega_K}\,. \tag{50}$$

The primes denote derivatives with respect to the Schwarzschild radial coordinate. In order to construct stellar models that rotate at their respective Kepler frequencies, equation (50) is to be solved self-consistently together with Einstein's field equations for a given model for the equation of state.[3,115]

Figure 19 shows Ω_K as a function of rotating star mass. The rectangle indicates both the approximate range of observed neutron star masses as well as the observed rotational periods which, currently, are $P \geq 1.6$ ms. One sees that all pulsars so far observed rotate below the mass shedding frequency and so can be interpreted as rotating neutron stars. Half-millisecond periods or even smaller ones are excluded for neutron stars of mass $1.4\,M_\odot$.[3,115,116] The situation appears to be very different for stars made up of self-bound strange quark matter, the so-called strange stars which will be introduced in Section 5. Such stars can withstand stable rotation against mass shedding down to rotational periods in the half-millisecond regime or even below.[117]

Consequently, the possible future discovery of a single sub-millisecond pulsar spinning at say 0.5 ms could give a strong hint that strange stars actually exist, and that the deconfined self-bound phase of 3-flavor strange quark matter is in fact the true ground state of the strong interaction rather than nuclear matter.

The gravitational radiation-reaction driven instability in compact stars sets a more stringent limit on rapid stellar rotation than mass shedding, as can be seen from Figures 20 and 21. This instability originates from counter-rotating surface vibrational modes, which at sufficiently high rotational star frequencies are dragged forward. In this case, gravitational radiation which inevitably accompanies the aspherical transport of matter does not damp the instability modes but rather drives them.[118,119] Viscosity plays the important role of damping such gravitational-wave radiation reaction-driven instabilities at a sufficiently reduced rotational frequency such that the viscous damping rate and power in gravity waves are comparable[120]. The instability modes are taken to have the dependence $\exp[i\omega_m(\Omega)t + im\phi - t/\tau_m(\Omega)]$, where ω_m is the frequency of the surface mode which depends on the angular velocity Ω of the star, ϕ denotes the azimuthal angle, and τ_m is the time scale for the mode which determines its growth or damping. The rotation frequency Ω at which it changes sign is the critical frequency for the particular mode, m (=2,3,4,...). It is conveniently expressed as the frequency, denoted by Ω^{ν}_{m}, that solves[121]

$$\Omega^{\nu}_{m} = \frac{\omega_m(0)}{m} \left[\tilde{\alpha}_m(\Omega^{\nu}_{m}) + \tilde{\gamma}_m(\Omega^{\nu}_{m}) \left(\frac{\tau_{g,m}}{\tau_{\nu,m}} \right)^{\frac{1}{2m+1}} \right] , \tag{51}$$

where

$$\omega_m(0) \equiv \sqrt{\frac{2m(m-1)}{2m+1} \frac{M}{R^3}} \tag{52}$$

is the frequency of the vibrational mode in a non-rotating star. The time scales for the gravitational radiation reaction to grow, $\tau_{g,m}$, and for viscous damping to set in, $\tau_{\nu,m}$, are given by[3]

$$\tau_{g,m} = \frac{2}{3} \frac{(m-1)\,[(2m+1)!!]^2}{(m+1)\,(m+2)} \left(\frac{2m+1}{2m(m-1)} \right)^m \left(\frac{R}{M} \right)^{m+1} R, \tag{53}$$

$$\tau_{\nu,m} = \frac{R^2}{(2m+1)\,(m-1)} \frac{1}{\nu} . \tag{54}$$

The shear viscosity, ν, depends on a star's temperature according to $\nu(T) \propto T^{-2}$. It is small in very hot ($T \approx 10^{10}$ K) and therefore young stars but

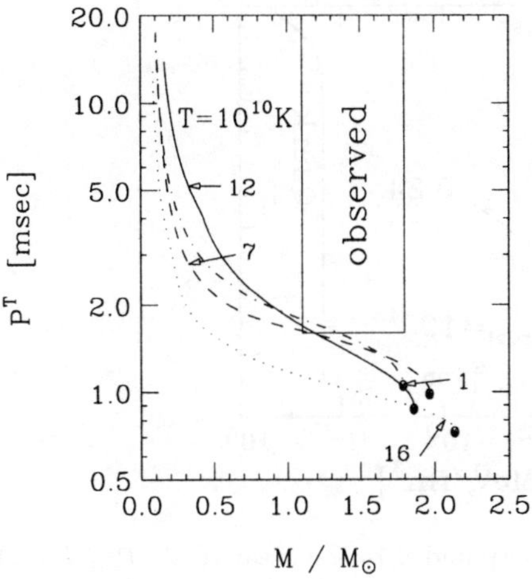

Figure 20. Gravitational radiation-reaction instability period P^T versus mass for newly born stars of temperature $T = 10^{10}$ K.[3]

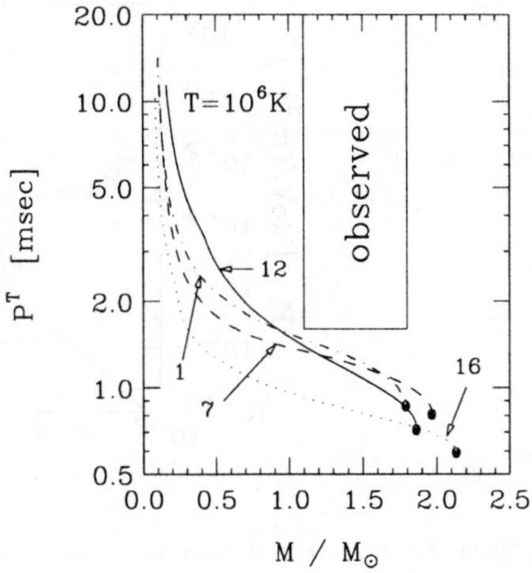

Figure 21. Gravitational radiation-reaction instability period P^T versus mass for old stars of temperature $T = 10^6$ K.[3]

considerably larger in cold ones ($T \approx 10^6$ K). Information about the pulsations of the rotating star is contained in the functions $\tilde{\alpha}_m$ and $\tilde{\gamma}_m$.[121,122] A characteristic feature of Eqs. (51) through (54) is that Ω_m^ν merely depends on radius and mass (R and M) of the spherical star model, which renders solving these equations rather straightforward.

Figure 20 shows the critical rotational periods at which the emission of gravity waves sets in in hot ($T \approx 10^{10}$ K) pulsars newly born in supernova explosions. Figure 21 is the analog to Figure 20, but for old and therefore cold pulsars of temperature $T = 10^6$ K, like neutron stars in binary systems that are being spun up by mass accretion from a companion. One sees that the limiting rotational periods P^T ($\equiv 2\pi/\Omega_m^\nu$) are the smaller the more massive (the smaller the radius) the star (cf. Figure 16). A comparison between Figures 20 and 21 shows that the instability periods are shifted toward smaller values the colder the star due to the larger viscosity in colder objects. Consequently, the instability modes of neutron stars in binary systems are excited at smaller rotational periods than for hot, newly born pulsars in supernovae. This conclusion hinges on the role of bulk viscosity of neutron star matter which may become very large at high temperatures. Sawyer has pointed out that is proportional to the sixth power of the temperature as compared with a

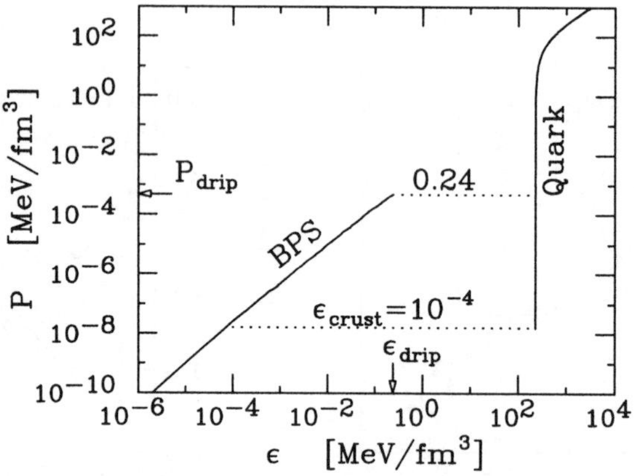

Figure 22. Equation of state of a strange star surrounded by a nuclear crust. $P_{\rm drip}(\epsilon_{\rm drip})$ denotes the pressure at the maximum possible inner crust density determined by neutron drip, $\epsilon_{\rm crust} = 0.24$ MeV/fm^3 (4.3×10^{11} g/cm^3). Any inner crust value smaller than that is possible. As an example, we show the equation of state for $\epsilon_{\rm crust} = 10^{-4}$ MeV/fm^3 (10^8 g/cm^3).[114]

T^{-2} dependence for the shear viscosity[123]. This means that at temperatures $T \gtrsim 10^9$ K the bulk viscosity may dominate over the shear viscosity and regulate the gravitational radiation-reaction driven instability in rapidly rotating neutron stars, pushing their critical rotational periods toward smaller values, possibly even as small as the Kepler period.

The dependence of P^T on the equation of state is shown too in Figures 20 and 21. One sees that the lower limits on P^T are set by non-relativistic equations of state due to the relatively small radii obtained for stars constructed from them. The rectangles in Figures 20 and 21 labeled "observed" cover both the range of observed neutron star masses, $1.1 \lesssim M/M_\odot \lesssim 1.8$ as well as observed pulsar periods, $P \geq 1.6$ ms. One sees that even the most rapidly rotating pulsars so far observed have rotational periods larger that those at which the gravitational radiation reaction-driven instability sets in. Thus all observed pulsars can be understood as rotating neutron stars. The observation of pulsars possessing masses in the observed range but rotational periods that are smaller than say ~ 1 ms, depending on temperature and thus on the pulsar's history, would be in contradiction to our collection of equations of state.

Finally, we mention the r-mode instability which has attracted a great deal of attention over the last several years. Like the f-modes discussed above,

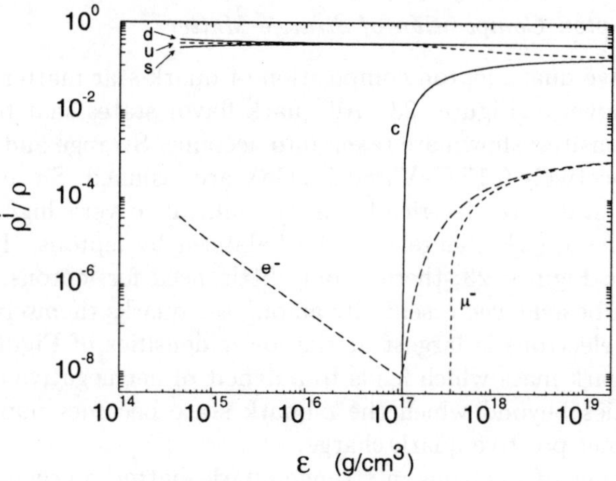

Figure 23. Relative quark and lepton densities in cold quark star matter as a function of energy density for a bag constant of $B^{1/4} = 145$ MeV.[17]

this instability too is produced by the Chandrasekhar, Friedman and Schutz (CFS) mechanism and render every star made of a perfect fluid unstable at all rates of rotation.[124,125] The question whether or not it provides a much more severe constraint on the stable rotation of viscous stars, however, appears to be still an open issue.

5 Strange Quark Matter Stars

5.1 The Strange Matter Hypothesis

The theoretical possibility that strange quark matter may be absolutely stable with respect to nuclear matter, that is, the energy per baryon of such matter is less than 930 MeV, has been pointed out independently by Bodmer, Terazawa, and Witten.[126,127,128] This so-called strange matter hypothesis constitutes one of the most startling possibilities regarding the behavior of superdense matter, which, if true, would have implications of fundamental importance for cosmology, the early universe, its evolution to the present day, astrophysical compact objects. Even to the present day there is no sound scientific basis on which one can either confirm or reject the hypothesis so that it remains a serious possibility of fundamental significance for various phenomena[3,129].

5.2 Quark-lepton Composition of Strange Matter

The relative quark-lepton composition of quark star matter at zero temperature is shown in Figure 23. All quark flavor states that become populated at the densities shown are taken into account. Strange and charm quark masses of respectively 0.15 GeV and 1.2 GeV are assumed. Since stars in their lowest energy state are electrically charge neutral to very high precision[130], any net positive quark charge must be balanced by leptons. In general, as can be seen in Figure 23, there is only little need for leptons, since charge neutrality can be achieved essentially among the quarks themselves. The concentration of electrons is largest at the lower densities of Figure 23 due to the finite s-quark mass which leads to a deficit of net negative quark charge, and at densities beyond which the c-quark state becomes populated which increases the net positive quark charge.

The presence of electrons in strange quark matter is crucial for the possible existence of a nuclear crust on such objects, as will be discussed below. Recently it has been argued that strange quark matter is a color superconductor (see Section 7) which, at extremely high densities, is in the Color-Flavor-Locked (CFL) phase. This phase is rigorously electrically neutral with no electrons required[131]. If the CFL phase would extend all the way to the surface of a strange star, then strange stars would not be able to carry nuclear crusts because of the missing electric dipole layer. However, for sufficiently large strange quark masses, the "low" density regime of strange matter is rather expected to form a 2-flavor color superconductor (2SC) in which electrons are present[132,133].

As shown in Refs.[17,134,135], the electrons, because they are bound to strange matter by the Coulomb force rather than the strong force, extend several hundred fermi beyond the surface of the strange star. Associated with this electron displacement is an electric dipole layer which can support, out of contact with the surface of the strange star, a crust of nuclear material, which it polarizes[134,135]. The maximal possible density at the base of the crust (inner crust density) is determined by neutron drip, which occurs at about 4.3×10^{11} g/cm^3.

5.3 Equation of State of Strange Stars with Crust

The somewhat complicated situation of the structure of a strange star with crust described above can be represented by a proper choice of equation of state which consists of two parts[114]. At densities below neutron drip it is represented by the low-density equation of state of charge-neutral nuclear matter, for which we use the Baym-Pethick-Sutherland equation of state[136,137].

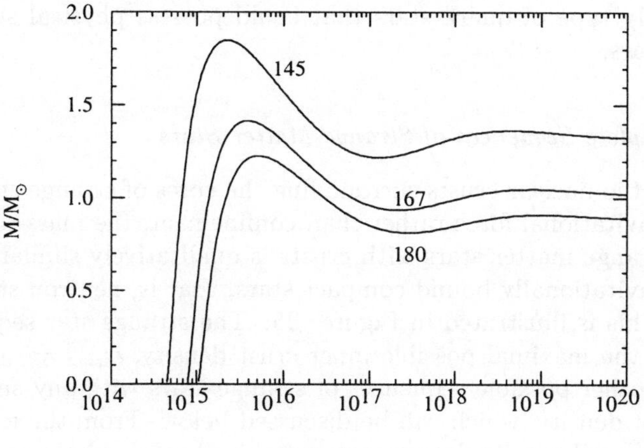

Figure 24. Gravitational mass of strange and charm stars for different $B^{1/4}$ values.

The star's strange matter core is described by the bag model. The graphical illustration of such an equation of state is shown in Figure 22. Notice that there is a discontinuity in energy density between strange quark matter and hadronic matter across the electron surface (dipole gap) inside the star where the pressure of the hadronic crust at its base equals the pressure of the strange core at its surface[114].

5.4 Strange and Charm Stars

Figure 24 shows the mass of quark matter stars as a function on central star density for different, representative bag constants. Stars with central densities smaller than a few times 10^{15} g/cm^3 are composed of up, down and strange quarks only and are referred to as strange stars. Charm quarks are present at central densities higher than $\sim 10^{17}$ g/cm^3. Stars with densities larger than that are thus denoted charm stars. Only the sequences computed for $B^{1/4} = 145$ MeV consist of matter that is absolutely stable at zero external pressure. The other two sequences correspond to strange matter whose energy per baryon number is about 960 MeV ($B^{1/4} = 167$ MeV) and 1030 MeV ($B^{1/4} = 180$ MeV). A stability analysis shows that charm stars are unstable against radial oscillations and thus cannot exist stably in the universe.[138]

So the only type of quark stars that could possess physical significance are strange stars.

5.5 *Complete Sequences of Strange Matter Stars*

Since the nuclear crusts surrounding the cores of strange stars are bound by the gravitational force rather than confinement, the mass-radius relationship of strange matter stars with crusts is qualitatively similar to the one of purely gravitationally bound compact stars, that is, neutron stars and white dwarfs. This is illustrated in Figure 25. The strange star sequence is computed for the maximal possible inner crust density, $\epsilon_{\rm crust} = \epsilon_{\rm drip}$. Of course there are other possible sequences of strange stars with any smaller value of inner crust density, which will be discussed below. From the maximum mass star, the central density decreases monotonically through the sequence in each case. The neutron star sequence is computed for HFV of Table 5, a representative model for the equation of state of neutron star matter, which has been combined at subnuclear densities with the Baym-Pethick-Sutherland equation of state. Hence the white dwarfs shown in Figure 25 are computed for the latter equation of state. Those gravitationally bound stars with radii $\lesssim 200$ km and $\gtrsim 3000$ km represent stable neutron stars and white dwarfs, respectively. The fact that strange stars with crust possess smaller radii than neutron stars leads to smaller mass shedding periods, as already indicated by the classical expression $P_{\rm K} = 2\pi\sqrt{R^3/M}$. Of course the general relativistic expression (50), which is to be applied to neutron and strange stars, is considerably more complicated. However the qualitative dependence of $P_{\rm K}$ on mass and radius remains valid.[139] Due to the smaller radii of strange stars, the complete sequence of such objects (and not just those close to the mass peak, as is the case for neutron stars) can sustain extremely rapid rotation.[140] In particular, a strange star with a typical pulsar mass of $\sim 1.45\, M_\odot$ can rotate at Kepler periods as small as $0.55 \lesssim P_{\rm K}/{\rm msec} \lesssim 0.8$, depending on crust thickness and bag constant[114,140]. This range is to be compared with $P_{\rm K} \sim 1$ ms obtained for neutron stars of the same mass[141].

The minimum-mass configuration of the strange star sequence, labeled "a" in Figure 25,has a mass of about $\sim 0.017\, M_\odot$ (about 17 Jupiter masses). Depending on the chosen value of inner crust density, strange matter stars can be even by orders of magnitude lighter than this value.[140] If abundant enough in our Galaxy, such low-mass strange stars could be seen by the gravitational microlensing searches. Strange stars located to the right of "a" consist of small strange cores ($\lesssim 3$ km) surrounded by a thick nuclear crust made up of white dwarf material. Such objects are thus called strange dwarfs. Their

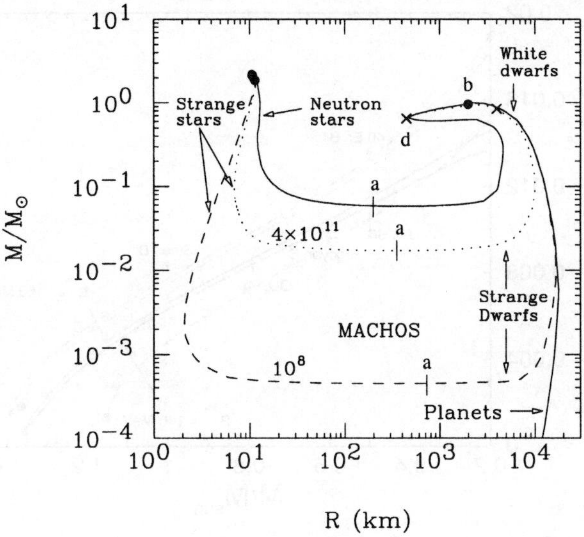

Figure 25. Mass versus radius of strange stars with nuclear crusts (dashed and dotted curves) and neutron stars and white dwarfs (solid curve). The strange stars carry nuclear crusts with chosen inner densities of $\epsilon_{\rm crust} = 4 \times 10^{11}$ g/cm^3 and 10^8 g/cm^3. The two crosses denote the termination points of the strange star sequences where the quark matter cores have shrunk to zero. Dots refer to maximum mass stars; minimum mass stars are located at the vertical bars labeled "a"[3].

cores have shrunk to zero at the points labeled "X". What is left is an ordinary white dwarf with a central density equal to the inner crust density of the former strange dwarf.[140,142] A detailed stability analysis of strange stars against radial oscillations shows that the strange dwarfs between "b" and "d" are unstable against the fundamental eigenmode[140]. Hence such objects cannot exist stably in nature. However all other stars of this sequence are stable against oscillations. So, in contrast to neutron stars and white dwarfs, the branches of strange stars and strange dwarfs are stably connected with each other.[140,143]

Until recently, only rather vague tests of the theoretical mass-radius relation of white dwarfs were possible. This has changed dramatically because of the availability of new data emerging from the Hipparcos project.[144] These data allow the first accurate measurements of white dwarf distances and, as a result, establishing the mass-radius relation of such objects empirically. Figure 26 shows a comparison of several data from the Hipparcos project with the

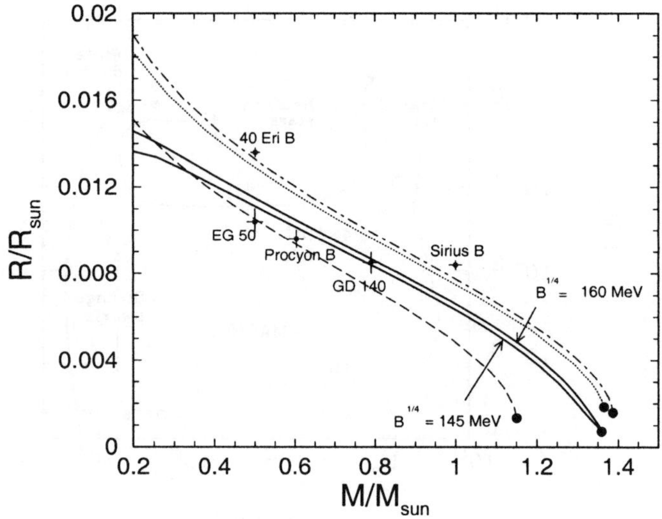

Figure 26. Comparison of theoretical mass-radius relationships of strange dwarfs ($\epsilon_{\rm crust} = \epsilon_{\rm drip}$) and several ordinary white dwarf stars (^{12}C dwarfs: dot-dashed; ^{24}Mg: dotted; ^{56}Fe: dashed double-dotted) with data from the Hipparcos project[145].

mass-radius relationships of strange dwarfs (solid lines) and ordinary white dwarfs computed for different compositions. The outcome could suggest that there is evidence for a bimodality in the white-dwarf populations one of which may contain strange-matter cores.

5.6 *Hadronic Crust on Strange Stars and Pulsar Glitches*

Of considerable relevance for the viability of the strange matter hypothesis is the question of whether strange stars can exhibit glitches in rotation frequency. From the study performed in Refs. [114,146] it is known that the ratio of the crustal moment of inertia to the total moment of inertia, $I_{\rm crust}/I_{\rm total}$, varies between 10^{-3} and $\sim 10^{-5}$. If the angular momentum of the pulsar is conserved in a stellar quake then the relative frequency change and moment of inertia change are equal, and one arrives for the change of the star's frequency at[114]

$$\frac{\Delta\Omega}{\Omega} = \frac{|\Delta I|}{I_0} > \frac{|\Delta I|}{I} \equiv f\frac{I_{\rm crust}}{I} \sim (10^{-5} - 10^{-3})\, f, \text{ with } \; 0 < f < 1\,. \qquad (55)$$

Here I_0 denotes the moment of inertia of that part of the star whose frequency is changed in the quake. It might be that of the crust only, or some fraction, or all of the star. The factor f in Eq. (55) represents the fraction of the crustal moment of inertia that is altered in the quake, i.e., $f \equiv |\Delta I|/I_{\rm crust}$. Since the observed glitches have relative frequency changes $\Delta\Omega/\Omega = (10^{-9} - 10^{-6})$, a change in the crustal moment of inertia of $f \lesssim 0.1$ would cause a giant glitch even in the least favorable case.[114] Moreover it turns out that the observed range of the fractional change in the spin-down rate, $\dot{\Omega}$, is consistent with the crust having the small moment of inertia calculated and the quake involving only a small fraction f of that, just as in Eq. (55). For this purpose we write[114]

$$\frac{\Delta\dot{\Omega}}{\dot{\Omega}} = \frac{\Delta\dot{\Omega}/\dot{\Omega}}{\Delta\Omega/\Omega}\frac{|\Delta I|}{I_0} = \frac{\Delta\dot{\Omega}/\dot{\Omega}}{\Delta\Omega/\Omega}\, f\, \frac{I_{\rm crust}}{I_0} > (10^{-1} \text{ to } 10)\, f\,, \qquad (56)$$

where use of Eq. (55) has been made. Equation (56) yields a small f value, i.e., $f < (10^{-4} \text{ to } 10^{-1})$, in agreement with $f \lesssim 10^{-1}$ established just above. Here measured values of the ratio $(\Delta\Omega/\Omega)/(\Delta\dot{\Omega}/\dot{\Omega}) \sim 10^{-6}$ to 10^{-4} for the Crab and Vela pulsars, respectively, have been used.

6 Thermal Evolution of Neutron and Strange Stars

The predominant cooling mechanism of hot (temperatures of several $\sim 10^{10}$ K) newly formed neutron stars immediately after formation is neutrino emission, with an initial cooling time scale of seconds. Already a few minutes after birth, the internal neutron star temperature drops to $\sim 10^9$ K. Photon emission overtakes neutrinos only when the internal temperature has fallen to $\sim 10^8$ K, with a corresponding surface temperature roughly two orders of magnitude smaller. Neutrino cooling dominates for at least the first 10^3 years, and typically for much longer in standard cooling (modified Urca) calculations. Being sensitive to the adopted nuclear equation of state, the neutron star mass, the assumed magnetic field strength, the possible existence of superfluidity, meson condensates and quark matter, theoretical cooling calculations, as summarized in Figure 27, provide most valuable information about the interior matter and neutron star structure. The thermal evolution of a neutron star also yields information about such temperature sensitive properties as transport coefficients, transition to superfluid states, crust solidification, and internal pulsar heating mechanisms such as frictional dissipation at the crust-superfluid interfaces[147]. The stellar cooling tracks in Figure 27 are computed for the broad collection of equations of state introduced in Section 2. Recent Chandra observations have identified pulsar J0205+6449 at the center of the young Crab-like supernova remnant 3C58. Historical evidence suggests an association of the remnant with supernova SN 1181, which makes

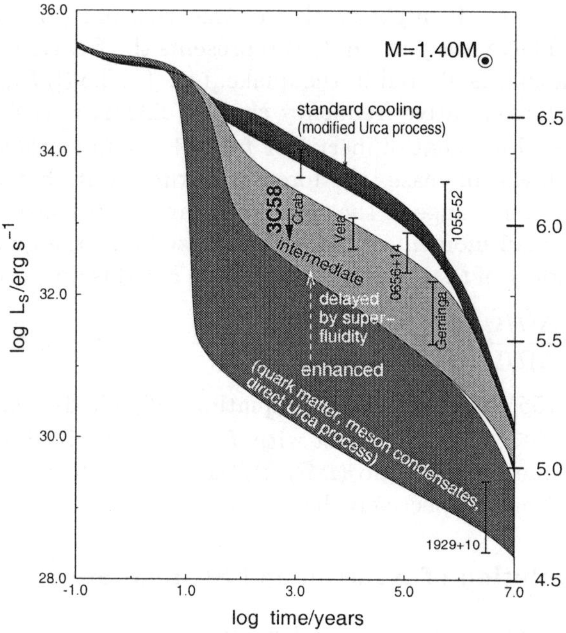

Figure 27. Cooling behavior of a 1.4 $M_\odot$ neutron star based on competing assumptions about the behavior of superdense matter. Three distinct cooling scenarios, referred to as "standard", "intermediate", and "enhanced" can be distinguished (cf. Section 3.7). The band-like structures reflect the uncertainties inherent in the equation of state of superdense matter[3].

3C58 younger (see Table 10) than Crab. The temperature of J0205+6449 was recently determined to be $\log_{10} T = 6.0334$ [109] which, as can be seen from Figure 27, falls well below predictions of standard cooling calculations[109]. Processes which speed up cooling considerably range from the presence of meson condensates [13,148], to quark matter [149,150], to the direct Urca process [151] (see Table 9 for an overview of neutron star cooling mechanisms).

7 Color Superconductivity of Quark Matter

There has been much recent progress in our understanding of quark matter, culminating in the discovery that if quark matter exists it will be in a color superconducting state[132,133,152,153]. The phase diagram of such matter is very complex[132,133]. At asymptotic densities the ground state of QCD with a vanishing strange quark mass is the color-flavor locked (CFL) phase. This phase is electrically neutral in bulk for a significant range of chemical potentials and strange quark masses[131]. If the strange quark mass is heavy enough to be ignored, then up and down quarks may pair in the two-flavor superconducting (2SC) phase. Other possible condensation patters are the CFL–K^0 phase [154] and the color-spin locked (2SC+s) phase[155]. The magnitude of the gap energy lies between $\sim$ 50 and 100 MeV. Color superconductivity thus modifies the equation of state at the order $(\Delta/\mu)^2$ level, which is only a few percent. Such small effects can be safely neglected in present determinations of models for the equation of state of neutron stars and strange quark matter stars, as is the case here. There has been much recent work on how color superconductivity in neutron stars could affect their properties. (See Refs.[132,133,159–159] and references therein.) These studies reveal that possible signatures include the cooling by neutrino emission, the pattern of the arrival times of supernova neutrinos, the evolution of neutron star magnetic fields, rotational (r-mode) instabilities, and glitches in rotation frequencies.

Aside from neutron star properties, an additional test of color superconductivity may be provided by upcoming cosmic ray space experiments such as AMS [160] and ECCO[161]. As shown in Ref.[162], finite lumps of color-flavor locked strange quark matter (strangelets), which should be present in cosmic rays if strange matter is the ground state of the strong interaction, turn out to be significantly more stable than strangelets without color-flavor locking for wide ranges of parameters. In addition, strangelets made of CFL strange matter obey a charge-mass relation of $Z/A \propto A^{-1/3}$, which differs significantly from the charge-mass relation of strangelets made of "ordinary" strange quark matter. In the latter case, Z/A would be constant for small baryon numbers A and $Z/A \propto A^{-2/3}$ for large A[129,162,163]. This difference may allow an experimental test of CFL locking in strange quark matter[162].

8 Particle Thresholds and Quark Deconfinement in Rotating Neutron Stars

Figures 28 and 29 reveal that the weakening of centrifuge accompanied by the slowing-down of a rotating neutron star causes a significant increase of its central density, which is accompanied by a rearrangement of the particle population and the creation of new phases of matter. From Figure 29, for instance, one reads off that the central density of a neutron star of mass $M = 1.42\, M_{\odot}$ increases from about 450 MeV/fm^3 for rotation at $\Omega_{\rm K}$ to more than 1500 MeV/fm^3 for $\Omega = 0$, which is a $\sim$ 66% effect. Of course, this effect is smaller for the lighter stars, since for them the changes of the gravitational

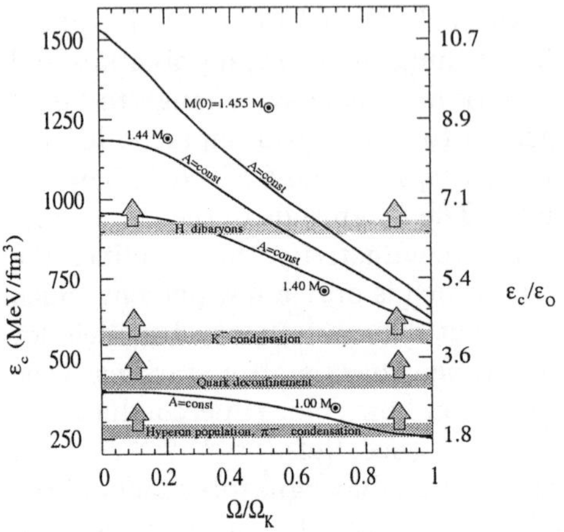

Figure 28. Central density versus rotational frequency of neutron stars of constant baryon number, A, computed for G_{B180}^{K300}. Threshold densities of various possible phases of superdense matter are indicated. $M(0)$ is the non-rotating star mass, Ω_K stands for the Kepler frequency[62].

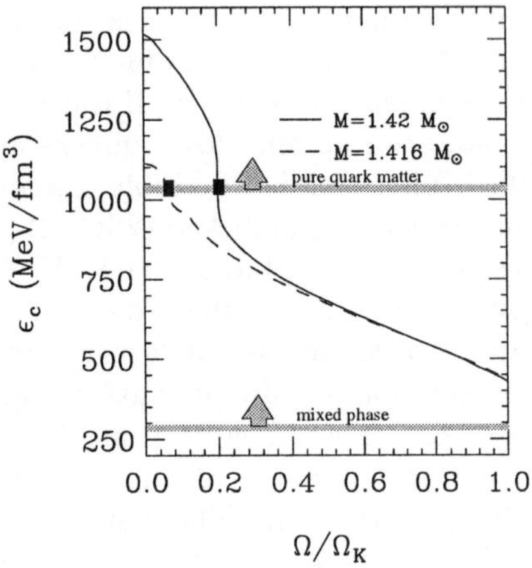

Figure 29. Central star density versus rotational frequency for two sample stars computed for equation of state G_{B180}^{K300}.

pull are weaker. Such dramatic changes in the interior density of slowing-down neutron stars imply profound changes of their interior composition, because of the changing particle composition with density. Figures 30 through 33 illustrate how these changes carry over to the internal structure of conventional neutron stars. In each case the star's rotational frequency covers the maximal possible range $0 \leq \Omega \leq \Omega_K$. It is evident that as rotating stars spin down they become significantly less deformed, which leads to closer equality between the polar and equatorial radii. At the same time the central density rises from below to above the threshold densities of heavier baryons (i.e., Σ, Λ, Ξ). For some pulsars the mass and initial rotational frequency may be such that the central density rises from below the critical density for dissolution of baryons into their quark constituents, examples of which are shown in Figures 34 through 37. The evolution of the central mass densities of these stellar models can be inferred in reference with Figure 29. For rotational frequencies below $\sim 1250\ s^{-1}$ the quark-hybrid star models consist of an inner region of pure quark matter (q) surrounded by a few kilometer thick shell of mixed phase of hadronic and quark matter. This phase consists of structures like hadronic drops (hd), hadronic rods (hr), hadronic slabs (hs), quark slabs (qs), quark

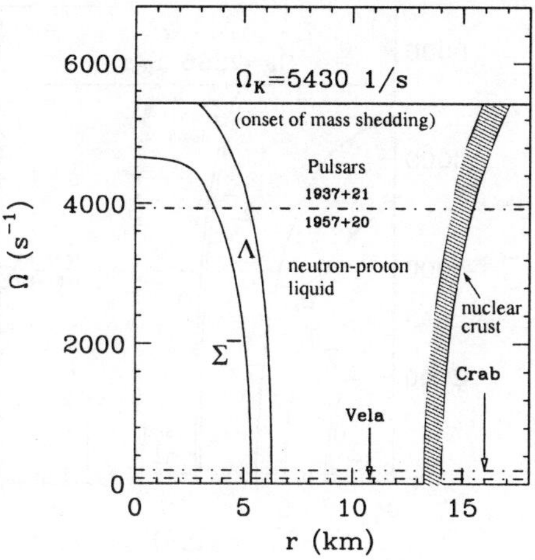

Figure 30. Frequency dependence of hyperon thresholds in equatorial neutron star direction computed for HV. The star's non-rotating mass is 1.40 $M_\odot$.[3]

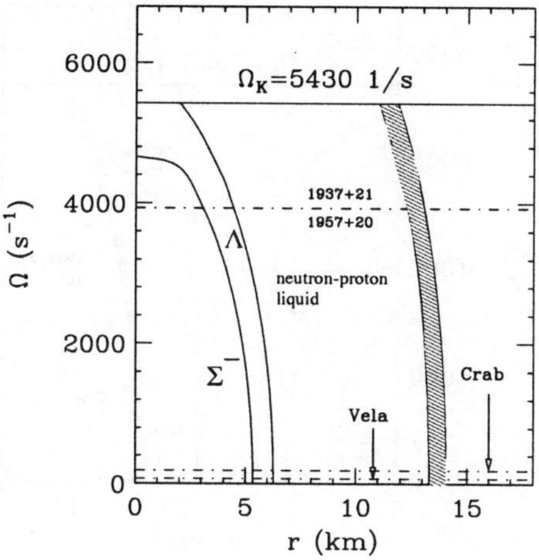

Figure 31. Same as Figure 30, but in polar direction.[3]

rods (qr), and quark drops (qd), in each case arranged in a lattice structure[61]. This shell is surrounded by a sphere of hadronic liquid (hl), itself enveloped in a thin nuclear crust of heavy ions. These quark-hadron structures are a consequence of the competition of the Coulomb and surface energies of the hadronic and quark matter phase[61]. They may have dramatic effects on pulsar observables including transport properties and the theory of glitches.

If the mass and initial rotational frequency of a pulsar is such that during its slowing-down phase the interior density rises from below to above the critical density for the quark-hadron phase transition, first at the center where the density is highest (Figures 28 and 29) and then in a region expanding in the radial outward direction away from the star's center, matter will be converted from the relatively incompressible nuclear matter phase to the highly compressible quark matter phase. The tremendous weight of the overlaying layers of nuclear matter tends to compress the quark matter core, which causes the entire star to shrink on a macroscopic length scale, as shown in Figures 34 through 37. The mass concentration in the core will be further enhanced by the increasing gravitational attraction of the quark core on the overlaying nuclear matter. The moment of inertia thus decreases anomalously with decreasing rotational frequency as the new phase slowly engulfs a grow-

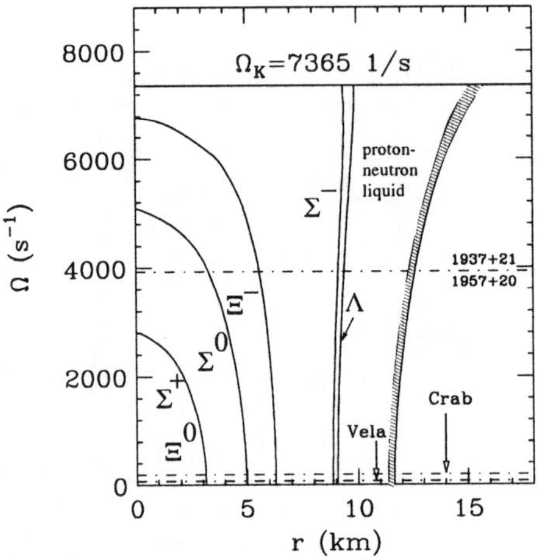

Figure 32. Frequency dependence of hyperon thresholds in equatorial neutron star direction computed for HV. The star's non-rotating mass is 1.978 $M_\odot$.[3]

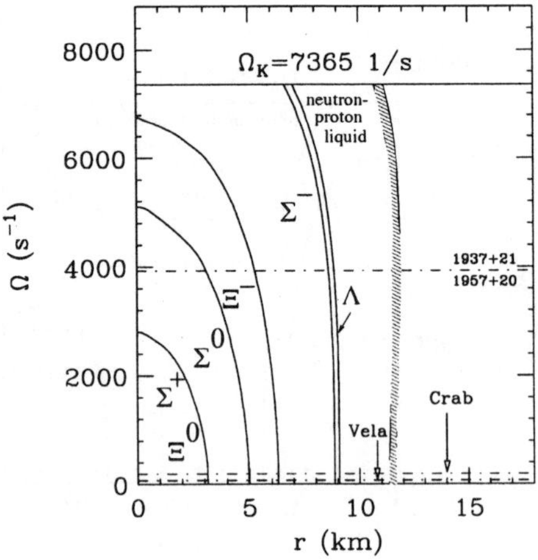

Figure 33. Same as Figure 32, but in polar direction[3].

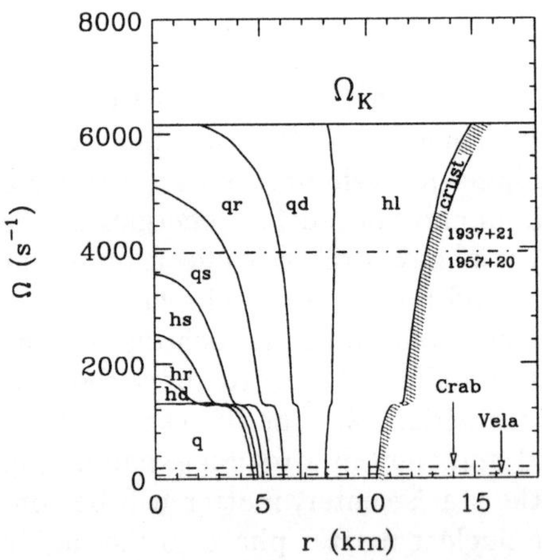

Figure 34. Frequency dependence of quark structure in equatorial neutron star direction computed for G_{B180}^{K300} and a non-rotating star mass of 1.42 $M_\odot$[3].

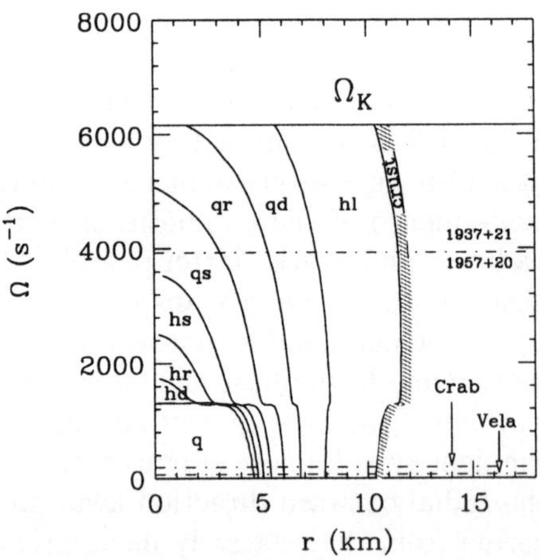

Figure 35. Same as Figure 34, but in polar direction.[3]

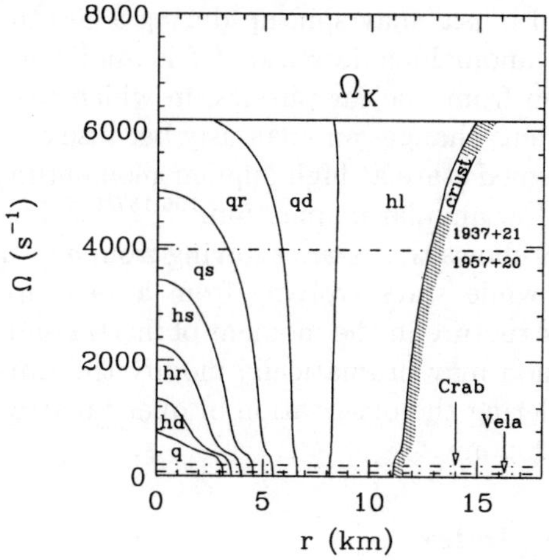

Figure 36. Frequency dependence of quark structure in equatorial direction computed for G^{K300}_{B180} and a non-rotating star mass of $1.416\,M_{\odot}$[3].

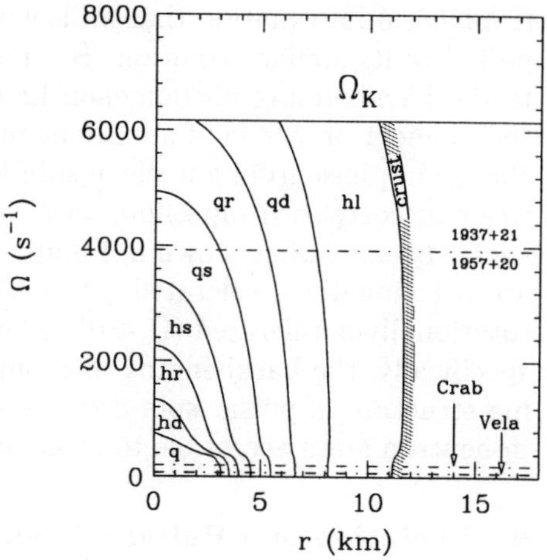

Figure 37. Same as Figure 36, but in polar direction[3].

ing fraction of the star[164], as can be seen from Figures 38 and 39. Figure 38 shows the moment of inertia, I, computed self-consistently from Eq. (48) for several sample stars having the same baryon number but different internal constitutions[165]. The curve labeled $M = 1.420\,M_{\odot}$, computed for G^{K300}_{B180}, shows the moment of inertia of the quark-hybrid star of Figures 34 and 35. The other curves correspond to a standard hyperon star constructed for G^{K240}_{M78} and a standard neutron star where hyperons are ignored. In accordance with what has been said just above, the shrinkage of quark-hybrid stars driven by the development of quark matter cores is the less pronounced the smaller the quark matter cores which are being built up in their centers during spin-down. Correspondingly the reduction of I weakens with decreasing star mass, as shown in Figure 38 for several sample star masses.

The decrease of the moment of inertia caused by the quark-hadron phase transition shown in Figure 38 is superimposed on the response of the stellar shape to a decreasing centrifugal force as the star spins down due to the loss of rotational energy. In order to conserve angular momentum not carried off by particle radiation from the star, the deceleration rate $\dot{\Omega}$ (< 0) must respond correspondingly by decreasing in absolute magnitude. More than that, $\dot{\Omega}$ may even change sign, as shown in Figures 39[164], which carries the important astrophysical information that an isolated pulsar may spin up during a certain period of its stellar evolution. Such an anomalous decrease of I is analogous to the backbending phenomenon known from nuclear physics, in which case the moment of inertia of an atomic nucleus changes anomalously because of a change in phase from a nucleon spin aligned state at high angular momentum to a pair correlated superfluid phase at low angular momentum.[166,167,168] The stellar backbending shown in Figure 39 shows that stars evolving from $b \to a$ are rotationally accelerated ($\dot{\Omega} > 0$), while stars evolving from $a \to b$ are rotationally decelerated ($\dot{\Omega} < 0$). The structure in the moment of inertia and, specifically, the backbending phenomenon may dramatically modify the timing structure of pulsar spin-down, rendering the observation of quark matter in neutron stars accessible to radio astronomy.

9 Evolution of a Pulsar's Braking Index

Pulsars are identified by their periodic signal believed to be due to a strong magnetic field fixed in the star and oriented at an angle from the rotation axis. The star's angular velocity of rotation decreases slowly but measurably over time, and usually the first and occasionally second time derivative can also be measured. Various energy loss mechanisms could be at play such as the magnetic dipole radiation, part of which is detected on each revolution, as well as other losses such as ejection of charged particles.[169] Without loss of generality, one can assume that pulsar slow-down is governed by a single mechanism, or several mechanisms having the same power law. The energy balance equation can then be written in the form

$$\frac{dE}{dt} = \frac{d}{dt}\left(\frac{1}{2}\, I(\Omega)\, \Omega^2\right) = -\, C\; \Omega^{n+1} \,. \tag{57}$$

In the case of magnetic dipole radiation, the constant C is equal to $C = \frac{2}{3}\mu^2 \sin^2 \alpha$ (μ denotes the star's magnetic moment), and $n = 3$ if I is kept constant during spin-up (down). If, as is customary, the star's angular velocity Ω is regarded as the only time dependent quantity, one obtains the usual formula for the rate of change of the pulsar frequency, given by

$$\dot{\Omega} = -\, K\; \Omega^n \,, \tag{58}$$

with K a constant and n the braking index. As known from Figures 38 and 39, the moment of inertia is not constant in time but responds to changes in rotational frequency. This response changes the value of the braking index in a frequency dependent manner, that is $n = n(\Omega)$. Thus during any epoch of observation, the braking index will be measured to be different from

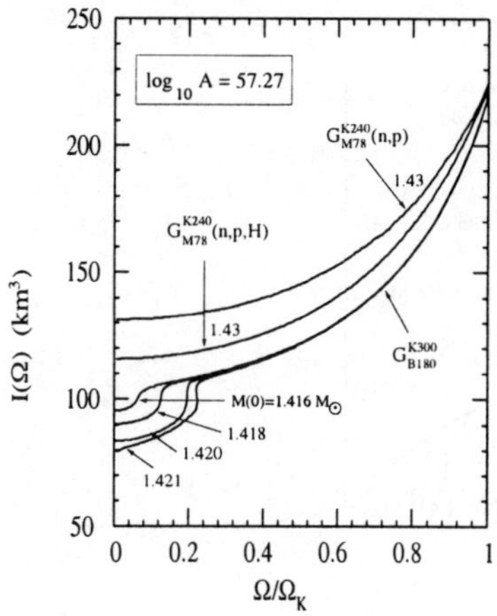

Figure 38. Moment of inertia versus frequency of neutron stars with the same baryon number, A, but different constitutions. The dips at low Ω's are caused by quark deconfinement[62].

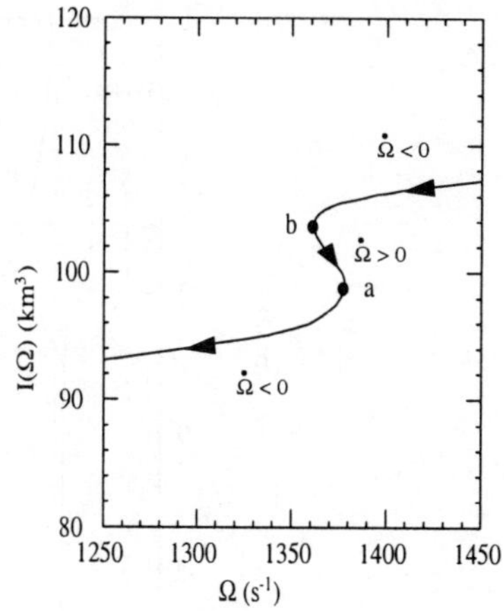

Figure 39. Enlargement of the lower-left portion of Figure 38 for the star of mass $M = 1.421\,M_\odot$, which is characterized by a backbending of I for frequencies between a and b.[3,164]

it canonical value $n = 3$ by a certain amount. How different depends, for any given pulsar, on its rotational frequency and for different pulsars of the same frequency, on their mass and on their internal constitution. When the frequency response of the moment of inertia is taken into account, Eq. (58) is to be replaced with[3,164]

$$\dot\Omega = -2\,I\,K\,\frac{\Omega^n}{2\,I+\frac{dI}{d\Omega}\,\Omega} = -K\,\Omega^n\left(1-\frac{\frac{dI}{d\Omega}}{2\,I}\,\Omega+\left(\frac{\frac{dI}{d\Omega}}{2\,I}\,\Omega\right)^2+\cdots\right), \qquad (59)$$

where $K = C/I$. This explicitly shows that the frequency dependence of $\dot\Omega$ corresponding to any mechanism that absorbs (or deposits) rotational energy cannot be a simple power law as given in equation (58) (with K a constant). It must depend on the mass and internal constitution of the star through the response of the moment of inertia to rotation as expressed in (59). It is interesting to note that Eq. (59) can be represented in the form of (58) but now with a frequency dependent prefactor, by evaluating

$$n(\Omega) \equiv \frac{\Omega\,\ddot\Omega}{\dot\Omega^2} = 3 - \frac{3\,\frac{dI}{d\Omega}\,\Omega+\frac{d^2I}{d\Omega^2}\,\Omega^2}{2\,I+\frac{dI}{d\Omega}\,\Omega}\,. \qquad (60)$$

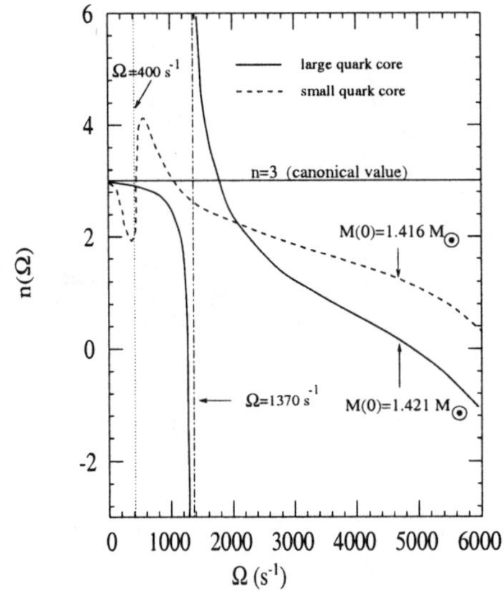

Figure 40. Braking index, n, of quark-hybrid stars of Figures 38 and 39. The anomalies in n at $\Omega \sim 400\ \mathrm{s}^{-1}$ and $\Omega \sim 1370\ \mathrm{s}^{-1}$ are caused by quark deconfinement. The overall reduction of n below 3 is caused by the frequency dependence of the moment of inertia.[62]

One sees that this effective braking index depends explicitly and implicitly on Ω and reduces to the canonical expression $n = 3$ only if I is independent of frequency.

As an example, we show in Figure 40 the change of the braking index with frequency for two selected quark-hybrid stars. For illustration purposes we assume dipole radiation. Because of the structure in the moment of inertia, driven by the phase transition of baryonic matter into deconfined quark matter, the braking index deviates dramatically from $n = 3$ at those rotational frequencies where quark deconfinement leads to the built-up of pure quark matter cores in the centers of these stars. Such anomalies in $n(\Omega)$ are not obtained for conventional neutron stars or hyperon stars because their moments of inertia increase smoothly with Ω, as known from Figure 38. The observation of such an anomaly in the timing structure of pulsar spin-down could thus be interpreted as a possible signal of quark deconfinement in the center of a pulsar. Of course, because of the extremely small temporal change of a pulsar's rotational period, one cannot measure the shape of the curve which is in fact not necessary. Just a single anomalous value of n that differed significantly from the canonical value of $n = 3$ would suffice.[164,170]

As a final but very important point on this subject, we estimate the typical duration over which the braking index is anomalous if deconfinement is well pronounced. The time span can be estimated from

$$\Delta T \simeq -\frac{\Delta\Omega}{\dot{\Omega}} = \frac{\Delta P}{\dot{P}}, \tag{61}$$

where $\Delta\Omega$ is the frequency interval of the anomaly. The range over which $n(\Omega)$ is smaller than zero and larger than six (Figure 40) is $\Delta\Omega \approx -100\ \mathrm{s}^{-1}$, or $\Delta P \approx -2\pi\Delta\Omega/\Omega^2 \approx 3 \times 10^{-4}$ s at $\Omega = 1370\ \mathrm{s}^{-1}$. So, for a millisecond pulsar whose period derivative is typically $\dot{P} \simeq 10^{-19}$ one has $\Delta T \simeq 10^8$ years, as illustrated in Figure 41. The dipole age of such pulsars is about 10^9 years. So, as a rough estimate one may expect about 10% of the 25 presently known solitary millisecond pulsars to be in the transition epoch during which pure quark matter cores are gradually being built up in their centers. These pulsars could be signaling the ongoing process of quark deconfinement in their cores. Last but not least we note that the spin-up time (region *b–a* in Figure 39) is about 1/5 of the time span ΔT, or about 1/50 of the dipole age $\tau = -\Omega/(\dot{\Omega}(n-1))$.

It is probably needless to say that the observation of a pulsar with an anomalously large braking index, signaling the existence of quark matter in the centers of pulsars, would be a momentous discovery with most far-reaching consequences for nuclear and particle physics. It would help to clarify how quark matter behaves, and give a boost to theories about the early Universe as well as laboratory searches for the production of quark matter in heavy-ion collisions, which has become a forefront area of modern physics. On an even more fundamental level, such a discovery would prove that the essentially free quark state predicted for matter at very high energy densities actually exists, and give us a picture of an early phase of the Universe that is based on radio pulsar observation, which may be coined quark astronomy.

10 Quark Matter in Neutron Stars in Low-Mass X-ray Binaries

The signal of quark deconfinement described in Section 9 applies only to isolated neutron stars, where deconfinement is driven by the gradual stellar contraction as the star spins down. The situation is reversed in neutron stars in binary systems, which experience a spin-up torque because of the transfer of angular momentum carried by the matter picked up by the star's magnetic field from the surrounding accretion disk. The spin-up torque causes a change

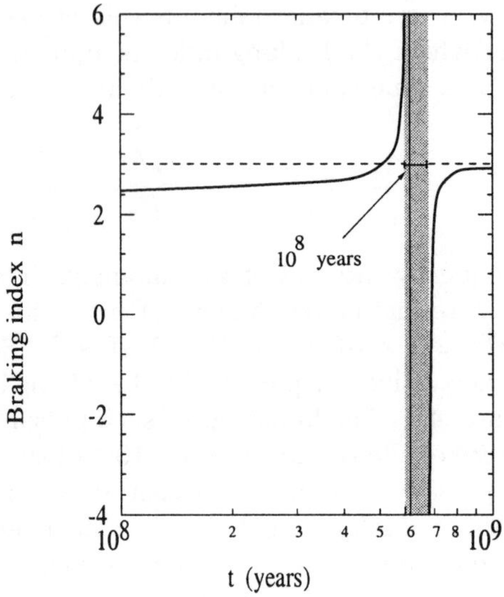

Figure 41. Braking index versus time for the quark-hybrid star of mass $M = 1.421\,M_\odot$ of Figure 40. The epoch over which n is anomalous because of quark deconfinement, $\sim 10^8$ years, is indicated by the shaded area.[3]

in the stars' angular momentum according to the relation[171]

$$\frac{dJ}{dt} = \dot{M}\tilde{l}(r_{\rm m}) - N(r_{\rm c}) \,, \tag{62}$$

where $\dot{M}$ denotes the accretion rate and

$$\tilde{l}(r_{\rm m}) = \sqrt{M r_{\rm m}} \tag{63}$$

is the angular momentum added to the star per unit mass of accreted matter. The quantity N stands for the magnetic plus viscous torque term,

$$N(r_{\rm c}) = \kappa\, \mu^2\, r_{\rm c}^{-3} \,, \tag{64}$$

with $\mu \equiv R^3 B$ the star's magnetic moment. The quantities $r_{\rm m}$ and $r_{\rm c}$ denote the radius of the inner edge of the accretion disk and the co-rotating radius, respectively, and are given by ($\xi \sim 1$)

$$r_{\rm m} = \xi\, r_{\rm A} \,, \tag{65}$$

and

$$r_{\rm c} = \left(M \Omega^{-2}\right)^{1/3} . \tag{66}$$

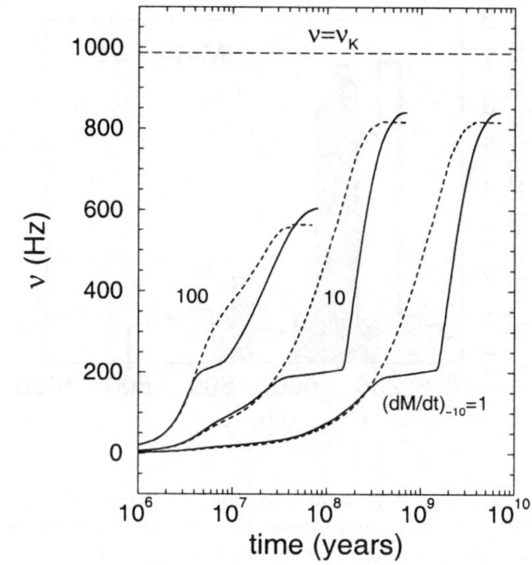

Figure 42. Evolution of spin frequencies of accreting X-ray neutron stars with (solid curves) and without (dashed curves) quark deconfinement.[171] The spin plateau around 200 Hz signals the ongoing process of quark re-confinement in the stellar centers.

Accretion will be inhibited by a centrifugal barrier if the neutron star's magnetosphere rotates faster than the Kepler frequency at the magnetosphere. Hence $r_{\rm m} < r_{\rm c}$, otherwise accretion onto the star will cease. The Alfén radius $r_{\rm A}$ in Eq. (65) is defined by

$$r_{\rm A} = \left(\frac{\mu^4}{2M\dot{M}^2} \right)^{1/7} . \tag{67}$$

The rate of change of a star's angular frequency Ω ($\equiv J/I$) then follows from Eq. (62) as

$$I(t)\frac{d\Omega(t)}{dt} = \dot{M}\tilde{l}(t) - \Omega(t)\frac{dI(t)}{dt} - \kappa\mu(t)^2 r_{\rm c}(t)^{-3} , \tag{68}$$

with the explicit time dependences as indicated. There are two terms on the right-hand-side of Eq. (68) that grow linearly respectively quadratically with Ω. Ignoring the linear term shows that mass transfer can spin up a neutron

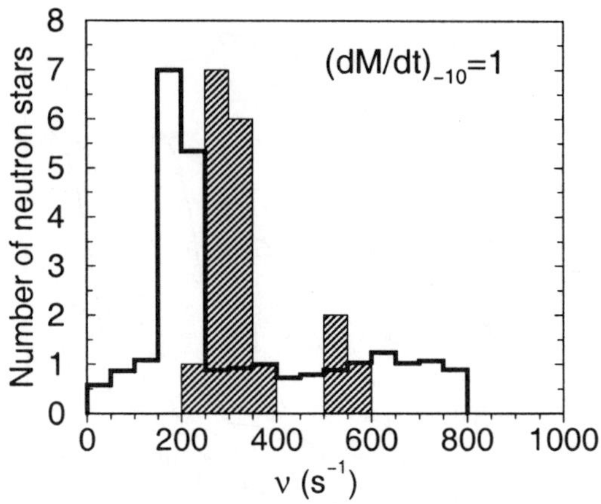

Figure 43. Calculated spin distribution of X-ray neutron stars. The spike in the calculated distribution (unshaded diagram) corresponds to the spinout of the quark matter phase. Otherwise the spike would be absent. The shaded histogram displays the observed data.[171]

star to an equilibrium period ($P = 2\pi/\Omega$) of[172]

$$P_{\rm eq} = 2.4 \text{ ms} \left(\frac{\dot{M}}{\dot{M}_{\rm Edd}}\right)^{-3/7} \left(\frac{M}{M_\odot}\right)^{-5/7} R_6^{15/7}\, B_9^{6/7}\,, \tag{69}$$

where R_6 and B_9 are the star's radius and its magnetic field in units of 10^6 cm and 10^9 G, respectively. $\dot{M}_{\rm Edd}$ in Eq. (69) denotes the maximum possible accretion rate, known as the Eddington limit, at which the accretion luminosity equals the luminosity at which the radiation pressure force on ionized hydrogen plasma near the star balances the gravitational acceleration force exerted by the star on the plasma. This condition leads to an Eddington accretion rate of $\dot{M}_{\rm Edd} = 1.5 \times 10^{-8} R_6 M_\odot/{\rm yr}$. For a typical accretion rate of $\dot{M}_{-10} \equiv \dot{M}/(10^{-10} M_\odot/{\rm yr})$, the Eddington rate can be expressed as $\dot{M}_{\rm Edd} = 150\, R_6\, \dot{M}_{-10}^{-1}\, \dot{M}$. The low-mass X-ray binaries (LMXBs) observed with the Rossi X-ray Timing Explorer are divided into Z sources and A(toll) sources, and appear to accrete at rates of $\dot{M}_{-10} \sim 200$ and $\dot{M}_{-10} \sim 2$, respectively.

The temporal change of the moment of inertia of accreting neutron stars which undergo phase transitions is crucial for their spin evolution.[171] This temporal change, on the other hand, renders the computation of the moment of inertia, defined in Eq. (48), very cumbersome since each quantity on the right-hand-side of (48) varies accordingly during stellar spin-up. The solution of Eq. (68) in combination with Eq. (48) is shown in Figure 42. The result is most striking. One sees that quark matter remains relatively dormant in the stellar core until the star has been spun up to frequencies at which the central density is about to drop below the threshold density at which quark matter exists. As known from Figure 39, this manifests itself in a significant increase of the star's moment of inertia. The angular momentum added to a neutron star during this phase of evolution is therefore consumed by the star's expansion, inhibiting a further spin-up until the quark matter has been converted into a mixed phase of matter made up of hadrons and quarks. Such accreters, therefore, tend to spend a greater length of time in the critical frequencies than otherwise. There will be an anomalous number of accreters that appear at or near the same frequency, as shown in Figure 43. This is what was actually found recently with the Rossi X-ray Timing Explorer (shown by the shaded area in Figure 43). Quark deconfinement constitutes a most striking explanation for this anomaly[171,173,174] though alternative explanations were suggested too.[175,176]

Acknowledgments

I wish to thank the organizers of Hadron 2002, Prof. Cesar A. Z. Vasconcellos and Prof. Victoria E. Herscovitz, for creating a very stimulating and enjoyable atmosphere at this meeting as well for their outstanding hospitality.

References

1. Proceedings of the International Conference on Nuclear Physics at the Turn of the Millennium: Structure of Vacuum & Elementary Matter, ed. by H. Stöcker, A. Gallmann, and J. H. Hamilton (World Scientific, Singapore, 1997).
2. R. D. Blandford, A. Hewish, A. G. Lyne and L. Mestel (eds.), *Pulsars as Physics Laboratories*, Phil. Trans. R. Soc. Lond. A **A341** (1992) 1–192.
3. F. Weber, *Pulsars as Astrophysical Laboratories for Nuclear and Particle Physics*, High Energy Physics, Cosmology and Gravitation Series (IOP Publishing, Bristol, Great Britain, 1999).
4. R. N. Manchester and J. H. Taylor, *Pulsars* (Freeman and Company, San Francisco, 1977).
5. A. G. Lyne and F. Graham-Smith, *Pulsar Astronomy* (Cambridge University Press, Cambridge, Great Britain, 1990).
6. N. K. Glendenning, Astrophys. J. **293** (1985) 470.
7. F. Weber and M. K. Weigel, Nucl. Phys. **A505** (1989) 779.

8. H. Huber, F. Weber, M. K. Weigel, and Ch. Schaab, Int. J. Mod. Phys. **E7**, No. 3 (1998) 301.
9. H. Umeda, K. Nomoto, S. Tsuruta, T. Muto, and T. Tatsumi, *Neutron Star Cooling and Pion Condensation*, in: The Structure and Evolution of Neutron Stars, ed. by D. Pines, R. Tamagaki, and S. Tsuruta (Addison-Wesley, New York, 1992) p. 406.
10. D. Kaplan and A. Nelson, Phys. Lett. **B175** (1986) 57.
11. G. Q. Li, C.-H. Lee, and G. E. Brown, Nucl. Phys. **A625** (1997) 372.
12. G. Q. Li, C.-H. Lee, and G. E. Brown, Phys. Rev. Lett. **79** (1997) 5214.
13. G. E. Brown, *Supernova Explosions, Black Holes and Nucleon Stars*, in: Proceedings of the Nuclear Physics Conference – INPC '95, ed. by S. Zuxun and X. Jincheng (World Scientific, Singapore, 1996) p. 623.
14. G. E. Brown, Phys. Bl. **53** (1997) 671.
15. H. Fritzsch, M. Gell–Mann, and H. Leutwyler, Phys. Lett. **B47** (1973) 365.
16. G. Baym, *Neutron Stars and the Physics of Matter at High Density*, in: Nuclear Physics with Heavy Ions and Mesons, Vol. 2, Les Houches, Session XXX, ed. by R. Balian, M. Rho and G. Ripka (North-Holland, Amsterdam, 1978) p. 745.
17. Ch. Kettner, F. Weber, M. K. Weigel, and N. K. Glendenning, Phys. Rev. **D51** (1995) 1440.
18. H. Heiselberg and V. Pandharipande, Ann. Rev. Nucl. Part. Sci. **50** (2000) 481.
19. V. R. Pandharipande, Proc. 16th Solvay Conf. on Astrophysics and Gravitation, Université de Bruxelles, p. 177 (1974).
20. V. R. Pandharipande and R. B. Wiringa, Rev. Mod. Phys. **51** (1979) 821.
21. W. D. Myers and W. J. Swiatecki, Nucl. Phys. **A601** (1996) 141.
22. R. B. Wiringa, V. Fiks, and A. Fabrocini, Phys. Rev. **C38** (1988) 1010.
23. B. Friedman and V. R. Pandharipande, Nucl. Phys. **A361** (1981) 502.
24. A. Akmal, V. R. Pandharipande, and D. G. Ravenhall, Phys. Rev. **C58** (1998) 1804.
25. W. D. Myers and W. J. Swiatecki, Ann. Phys. (N. Y.) **204** (1990) 401.
26. W. D. Myers and W. J. Swiatecki, Ann. Phys. (N. Y.) **211** (1991) 292.
27. K. Strobel, F. Weber, M. K. Weigel, and Ch. Schaab, Int. J. Mod. Phys. **E6**, No. 4 (1997) 669.
28. K. Strobel, F. Weber, and M. K. Weigel, Z. Naturforsch. **A54** (1999) 83.
29. R. G. Seyler and C. H. Blanchard, Phys. Rev. **124** (1961) 227; **131** (1963) 355.
30. B. D. Serot and J. D. Walecka, Adv. Nucl. Phys. **16** (1986) 1.
31. R. Machleidt, K. Holinde, and Ch. Elster, Phys. Rep. **149** (1987) 1.

32. J. D. Bjorken and S. D. Drell, *Relativistic Quantum Fields* (Mc Graw-Hill, New York, 1965).
33. Particle Data Group, Rev. Mod. Phys. **56** (1984) S1.
34. T. Tatsumi and T. Muto, in: Nuclei in the Cosmos, ed. by H. Oberhummer and C. Rolfs (Springer-Verlag, Berlin, 1991).
35. G. E. Brown, K. Kubodera, and M. Rho, Phys. Lett. **B192** (1987) 273.
36. C.-H. Lee and M. Rho, *Kaon Condensation in Dense Stellar Matter*, Proc. of the International Symposium on Strangness and Quark Matter", ed. by G. Vassiliadis, A. Panagiotou, B. S. Kumar, and J. Madsen (World Scientific, Singapore, 1995) p. 283.
37. J. A. Pons, J. A. Miralles, M. Prakash, and J. M. Lattimer, Astrophys. J. **553** (2001) 382.
38. L. Wilets, *Green's Functions Method for the Relativistic Field Theory Many-Body Problem*, in: Mesons in Nuclei, Vol. III, ed. by M. Rho and D. Wilkinson (North-Holland, Amsterdam, 1979) p. 791.
39. H. Huber, F. Weber, and M. K. Weigel, Phys. Lett. **B317** (1993) 485.
40. H. Huber, F. Weber, and M. K. Weigel, Phys. Rev. **C50** (1994) R1287.
41. H. Huber, F. Weber, and M. K. Weigel, Phys. Rev. **C51** (1995) 1790.
42. R. Brockmann and R. Machleidt, Phys. Rev. **C42**, (1990) 1965.
43. C. J. Horowitz and B. D. Serot, Nucl. Phys. **A464** (1987) 613.
44. N. K. Glendenning, F. Weber, and S. A. Moszkowski, Phys. Rev. **C45** (1992) 844.
45. N. K. Glendenning, *Compact Stars, Nuclear Physics, Particle Physics, and General Relativity* (Springer-Verlag, New York, 1997).
46. N. K. Glendenning, Nucl. Phys. **A493** (1989) 521.
47. N. K. Glendenning, Phys. Rev. Lett. **57** (1986) 1120.
48. F. Weber, N. K. Glendenning, and M. K. Weigel, Astrophys. J. **373** (1991) 579.
49. R. L. Jaffe, Phys. Lett. **38** (1977) 195.
50. N. K. Glendenning and J. Schaffner-Bielich, Phys. Rev. **C58** (1998) 1298.
51. R. Tamagaki, Prog. Theor. Phys. **85** (1991) 321.
52. T. Sakai, J. Mori, A. J. Buchmann, K. Shimizu, and K. Yazaki, Nucl. Phys. **A625** (1997) 192.
53. A. Faessler, A. J. Buchmann, M. I. Krivoruchenko, and B. V. Martemyanov, Phys. Lett. **B391** (1997) 255.
54. A. Faessler, A. J. Buchmann, and M. I. Krivoruchenko, Phys. Rev. **C56** (1997) 1576.
55. G. Mao, P. Papazoglou, S. Hofmann, S. Schramm, H. Stöcker, and W. Greiner, Phys. Rev. **C59** (1999) 3381.
56. T. Waas, M. Rho, and W. Weise, Nucl. Phys. **A617** (1997) 449.

57. A. Gal, Nucl. Phys. **A691** (2001) 268.
58. A. Cieply, E. Friedman, A. Gal, and J. Mares, Nucl. Phys. **A696** (2001) 173.
59. N. Kaiser and W. Weise, Phys. Lett. **B512** (2001) 283.
60. R. Barth et al., Phys. Rev. Lett. **78** (1997) 4027.
61. N. K. Glendenning, Phys. Rev. **D46** (1992) 1274.
62. F. Weber, J. Phys. G: Nucl. Part. Phys. **25** (1999) R195.
63. N. K. Glendenning, Nucl. Phys. B (Proc. Suppl.) **B24** (1991) 110.
64. *Physics of Neutron Star Interiors*, ed. by D. Blaschke, N. K. Glendenning, and A. Sedrakian, Lecture Notes in Physics **578** (Spring-Verlag, Berlin, 2001).
65. S. E. Thorsett and D. Chakrabarty, Astrophys. J. **512** (1999) 288.
66. J. H. Taylor and J. M. Weisberg, Astrophys. J. **345** (1989) 434.
67. M. H. Van Kerkwijk, *Neutron Star Mass Determinations*, in: Black Holes in Binaries and Galactic Nuclei, ed. by L. Kaper, E. P. J. van den Heuvel, and P. A. Woudt (Springer-Verlag, Berlin, 2001).
68. O. Barziv, L. Kaper, M. H. van Kerkwijk, J. H. Telting, and J. van Paradijs, Astron. & Astrophys. **377** (2001) 925.
69. J. A. Orosz and E. Kuulkers, Mon. Not. R. Astron. Soc. **304** (1999) 132.
70. M. van der Klis, Ann. Rev. Astron. Astrophys. **38** (2000) 717.
71. L. Burderi et al., Astrophys. J. **560** (2001) L71.
72. J. Van Paradijs, Astrophys. J. **234** (1978) 609.
73. S. L. Shapiro and S. A. Teukolsky, *Black Holes, White Dwarfs, and Neutron Stars* (Wiley & Sons, New York, 1983).
74. M. Y. Fujimoto and R. E. Taam, Astrophys. J. **305** (1986) 246.
75. F. M. Walter, S. J. Wolk, and R. Neuhäuser, Nature **379** (1996) 233.
76. F. M. Walter and L. D. Matthews, Nature **389** (1997) 358.
77. J. Drake et al., *Is RXJ1856.5-3754 a Quark Star?*, (astro-ph/0204159).
78. J. A. Pons, F. M. Walter, J. M. Lattimer, M. Prakash, R. Neuhäuser, and P. An, Astrophys. J. **564** (2002) 981.
79. F. W. Walter and J. Lattimer, *A Revised Parallax and its Implications for RXJ185635-3754*, (astro-ph/0204199).
80. M. Ruderman, Ann. Rev. Astron. Astrophys. **10** (1972) 427.
81. G. Baym and C. Pethick, Ann. Rev. Nucl. Sci. **25** (1975) 27.
82. V. Trimble and M. Rees, Astrophys. Lett. **5** (1970) 93.
83. G. Börner and J. M. Cohen, Astrophys. J. **185** (1973) 959.
84. E. P. Liang, Astrophys. J. **304** (1986) 682.
85. F. C. Michel, Rev. Mod. Phys. **54** (1982) 1.
86. K. Iwasawa, K. Koyama, and J. P. Halpern, *Cyclotron Lines and the Pulse Period Change of X-ray Pulsar 1E2259+586*, in: The Structure

and Evolution of Neutron Stars, ed. by D. Pines, R. Tamagaki, and S. Tsuruta (Addison-Wesley, New York, 1992) p. 203.
87. J. Trümper, W. Pietsch, C. Reppin, W. Voges, R. Staubert, and E. Kendziorra, Astrophys. J. Lett. **219** (1978) L105.
88. W. Voges, W. Pietsch, C. Reppin, J. Trümper, E. Kendziorra, and R. Staubert, Astrophys. J. **263** (1982) 803.
89. W. Wheaton, Nature **282** (1979) 240.
90. S. Chakrabarty, D. Bandyopadhyay, and S. Pal, Phys. Rev. Lett. **78** (1997) 2898.
91. C. Y. Cardall, M. Prakash, and J. M. Lattimer, Astrophys. J. **554** (2001) 322.
92. R. C. Duncan and C. Thompson, Astrophys. J. **392** (1992) L9.
93. W. Becker, P. Predehl, J. Trümper and H. Ögelman, *IAU Circular* **5554** (1992), p. 1.
94. J. Finley and H. Ögelman, *IAU Circular* **5787** (1993).
95. W. Becker, *IAU Circular* **5805** (1993).
96. W. Becker and B. Aschenbach, in: *The Lives of Neutron Stars*, ed. by M. A. Alpar, Ü. Kiziloglu, and J. Van Paradijs (Kluwer, Dordrecht, 1995) p. 47.
97. F. Seward, F. Harnden, P. Murdin, and D. Clark, Astrophys. J. **267** (1983) 698.
98. E. Trussoni, W. Brinkmann, H. Ögelman, G. Hasinger, and B. Aschenbach, Astron. & Astrophys. **234** (1990) 403.
99. J. P. Finley, H. Ögelman, G. Hasinger, and J. Trümper, Astrophys. J. **410** (1993) 323.
100. S. Safi-Harb and H. Ögelman, in: *The Lives of Neutron Stars*, ed. by M. A. Alpar, Ü. Kiziloglu, and J. Van Paradijs (Kluwer, Dordrecht, 1995) p. 53.
101. H. Ogelman, *X-Ray Observations of Cooling Neutron Stars*, in: *The Lives of Neutron Stars*, ed. by M. A. Alpar, Ü. Kiziloglu and J. Van Paradijs (Kluwer, Dordrecht, 1995) p. 101.
102. S. D. Yancopoulos, T. D. Hamilton, and D. J. Helfland, Bull. American Astron. Soc. **25** (1993) 912.
103. F. Seward and Z.-R. Wang, Astrophys. J. **332** (1988) 199.
104. W. Becker and J. Trümper, Nature **365** (1993) 528.
105. H. Ögelman, J. Finley and H. Zimmermann, Nature **361** (1993) 136.
106. J. P. Finley, H. Ögelman, and Ü. Kiziloglu, Astrophys. J. **394** (1992) L21.
107. J. Halpern and M. Ruderman, Astrophys. J. **415** (1993) 286.
108. H. Ögelman and J. Finley, Astrophys. J. **413** (1993) L31.

109. P. Slane, D. J. Helfand, and S. S. Murray, *New Constraints on Neutron Star Cooling from Chandra Observations of 3C58*, (astro-ph/0204151).
110. J. R. Oppenheimer and G. M. Volkoff, Phys. Rev. **55** (1939) 374.
111. R. C. Tolman, Phys. Rev. **55** (1939) 364.
112. G. E. Brown and H. A. Bethe, Astrophys. J. **423** (1994) 659.
113. H. A. Bethe and G. E. Brown, Astrophys. J. **445** (1995) L129.
114. N. K. Glendenning and F. Weber, Astrophys. J. **400** (1992) 647.
115. J. L. Friedman, J. R. Ipser, and L. Parker, Astrophys. J. **304** (1986) 115.
116. J. L. Friedman, J. R. Ipser, and L. Parker, Phys. Rev. Lett. **62** (1989) 3015.
117. N. K. Glendenning, Mod. Phys. Lett. **A5** (1990) 2197.
118. S. Chandrasekhar, Phys. Rev. Lett. **24** (1970) 611.
119. J. L. Friedman, Phys. Rev. Lett. **51** (1983) 11.
120. L. Lindblom and S. L. Detweiler, Astrophys. J. **211** (1977) 565.
121. L. Lindblom, Astrophys. J. **303** (1986) 146.
122. C. Cutler and L. Lindblom, Astrophys. J. **314** (1987) 234.
123. R. F. Sawyer, Phys. Rev. **D39** (1989) 3804.
124. N. Andersson, Astrophys. J. **502** (1998) 708.
125. J. L. Friedman and S. M. Morsink, Astrophys. J. **502** (1998) 714.
126. A. R. Bodmer, Phys. Rev. **D4** (1971) 1601.
127. H. Terazawa, INS-Report-338 (INS, Univ. of Tokyo, 1979); J. Phys. Soc. Japan, **58** (1989) 3555; **58** (1989) 4388; **59** (1990) 1199.
128. E. Witten, Phys. Rev. **D30** (1984) 272.
129. *Strange Quark Matter in Physics and Astrophysics*, Proc. of the International Workshop, ed. by J. Madsen and P. Haensel, Nucl. Phys. B (Proc. Suppl.) **24B** (1991).
130. N. K. Glendenning, Phys. Lett. **B114** (1982) 392.
131. K. Rajagopal and F. Wilczek, Phys. Rev. Lett. **86** (2001) 3492.
132. K. Rajagopal and F. Wilczek, *The Condensed Matter Physics of QCD*, At the Frontier of Particle Physics / Handbook of QCD, ed. M. Shifman, (World Scientific) (2001), (hep-ph/0011333).
133. M. Alford, Ann. Rev. Nucl. Part. Sci. **51** (2001) 131.
134. C. Alcock, E. Farhi, and A. V. Olinto, Astrophys. J. **310** (1986) 261.
135. C. Alcock and A. V. Olinto, Ann. Rev. Nucl. Part. Sci. **38** (1988) 161.
136. G. Baym, C. Pethick, and P. Sutherland, Astrophys. J. **170** (1971) 299.
137. G. Baym, H. A. Bethe, and C. J. Pethick, Astrophys. J. **175** (1971) 225.
138. Ch. Kettner, F. Weber, M. K. Weigel, and N. K. Glendenning, *Stability of Strange Quark Stars with Nuclear Crust against Radial Oscillations*, Proc. of the International Symposium on Strangeness and Quark Matter, ed. by G. Vassiliadis, A. D. Panagiotou, B. S. Kumar, and J. Madsen

(World Scientific, Singapore, 1995) p. 333.
139. N. K. Glendenning and F. Weber, Phys. Rev. **D50** (1994) 3836.
140. N. K. Glendenning, Ch. Kettner, and F. Weber, Astrophys. J. **450** (1995) 253.
141. F. Weber and N. K. Glendenning, *Hadronic Matter and Rotating Relativistic Neutron Stars*, Proc. of the Nankai Summer School, "Astrophysics and Neutrino Physics", ed. by D. H. Feng, G. Z. He, and X. Q. Li (World Scientific, Singapore, 1993) p. 64–183.
142. P. F. C. Romanelli, *Strange Star Crusts*, B.S. Thesis, MIT (1986).
143. N. K. Glendenning, Ch. Kettner, and F. Weber, Phys. Rev. Lett. **74** (1995) 3519.
144. J. L. Provencal, H. L. Shipman, E. Hog, and P. Thejll, Astrophys. J. **494** (1998) 759.
145. G. J. Mathews, B. C. O'Gorman, K. Otsuki, I. Suh, and F. Weber, *Evidence for White Dwarfs with Strange-Matter Cores*, Univ. of Notre Dame preprint (2002).
146. J. L. Zdunik, P. Haensel and E. Gourgoulhon, Astron. & Astrophys. **372** (2001) 535.
147. Ch. Schaab, A. Sedrakian, F. Weber, and M. K. Weigel, Astron. & Astrophys. **346** (1999) 465.
148. D. Page, in: *The Many Faces of Neutron Stars*, ed. by R. Buccheri, J. van Paradijs, and M. A. Alpar (Kluwer Academic Publishers, Dordrecht, 1998), p. 539.
149. D. Blaschke, H. Grigorian, and D. N. Voskresensky, Astron. & Astrophys. **368** (2001) 561.
150. D. Page and V. V. Usov, *Thermal Evolution and Light Curves of Young Bare Strange Stars*, (astro-ph/0204275).
151. J. M. Lattimer, C. J. Pethick, M. Prakash, and P. Haensel, Phys. Rev. Lett. **66** (1991) 2701.
152. M. Alford, K. Rajagopal, and F. Wilczek, Phys. Lett. **B422** (1998) 247.
153. R. Rapp, T. Schäfer, E. V. Shuryak, and M. Velkovsky, Phys. Rev. Lett. **81** (1998) 53.
154. P. F. Bedaque and T. Schäfer (hep-ph/0105150).
155. T. Schäfer, Phys. Rev. **D62** (2000) 094007.
156. K. Rajagopal, Acta Physica Polonica **B31** (2000) 3021.
157. M. Alford, J. Bowers, and K. Rajagopal, Phys. Rev. **D63** (2001) 074016.
158. M. Alford, J. A. Bowers, and K. Rajagopal, J. Phys. **G27** (2001) 541.
159. D. Blaschke, D. M. Sedrakian, and K. M. Shahabasyan, Astron. & Astrophys. **350** (1999) L47.
160. The AMS home page is `http://ams.cern.ch`.

161. Information about ECCO can be found at: `http://ultraman.berkeley.edu.`
162. J. Madsen, Phys. Rev. Lett. **87** (2001) 172003.
163. J. Madsen, Lecture Notes in Physics **516** (1999) 162.
164. N. K. Glendenning, S. Pei, and F. Weber, Phys. Rev. Lett. **79** (1997) 1603.
165. F. Weber, N. K. Glendenning, and S. Pei, *Signal for the Quark-Hadron Phase Transition in Rotating Hybrid Stars*, Proc. of the 3rd International Conference on Physics and Astrophysics of Quark-Gluon Plasma, ed. by B. C. Sinha, D. K. Srivastava, and Y. P. Viyogi (Narosa Publishing House, New Delhi, 1998) p. 237.
166. B. R. Mottelson and J. G. Valatin, Phys. Rev. Lett. **5** (1960) 511.
167. F. S. Stephens and R. S. Simon, Nucl. Phys. **A183** (1972) 257.
168. A. Johnson, H. Ryde, and S. A. Hjorth, Nucl. Phys. **A179** (1972) 753.
169. M. A. Ruderman, in: *High Energy Phenomena around Collapsed Stars*, ed. by F. Pacini (D. Reidel Publishing Company, Dodrecht, 1987).
170. N. K. Glendenning, Nucl. Phys. **A638** (1998) 239c.
171. N. K. Glendenning and F. Weber, Astrophys. J. **559** (2001) L119.
172. D. Bhattacharya and E. P. J. van den Heuvel, Phys. Rep. **203** (1991) 1.
173. N. K. Glendenning and F. Weber, *Signal of Quark Deconfinement in Millisecond Pulsars and Reconfinement in Accreting X-ray Neutron Stars*, Lecture Notes in Physics **578**, (Springer-Verlag, Berlin, 2001), p. 305.
174. G. Poghosyan, H. Grigorian, D. Blaschke, Astrophys. J. **551** (2001) L73.
175. L. Bildsten, Astrophys. J. **501** (1998) L89.
176. N. Andersson, D. I. Jones, K. D. Kokkotas, and N. Stergioulas, Astrophys. J. **534** (2000) L75.

SEMINARS

LATTICE SIMULATIONS FOR THE RUNNING COUPLING CONSTANT OF QCD

A. CUCCHIERI

IFSC—São Paulo University, Caixa Postal 369, 13560-970 São Carlos SP, Brazil
attilio@if.sc.usp.br

The strong coupling constant $\alpha_s(\mu_0)$, taken at a fixed reference scale μ_0, is the single free parameter of QCD and should be known to the highest available precision. The value of α_s should also be determined with good accuracy over as large a range of scales as possible, in order to reveal potential anomalous running in the strength of the strong interaction. Lattice QCD is now able to calculate α_s with accuracy comparable to or better than experiment. We review the status of such lattice calculations in quenched and full QCD.

1 QCD coupling and β-function

The scale dependence of the QCD coupling $\alpha_s = g_s^2/(4\pi)$ is controlled by the β-function:

$$\mu \frac{\partial \alpha_s}{\partial \mu} = \beta(\alpha_s) = -\left(\frac{\beta_0}{2\pi}\alpha_s^2 + \frac{\beta_1}{4\pi^2}\alpha_s^3 + \frac{\beta_2}{64\pi^3}\alpha_s^4 + \ldots \right) , \qquad (1)$$

where $\beta_0 = 11 - 2N_f/3$, $\beta_1 = 51 - 19N_f/3$ and N_f is the number of flavors of quarks with mass less than the energy scale μ. The coefficients β_n for $n \geq 2$ are renormalization-scheme-dependent.

By solving the differential equation for α_s, a constant of integration is introduced. This constant is the one fundamental constant of QCD. Usually one chooses for this constant the value of α_s at a fixed reference scale μ_0. Equivalently one can consider the dimensional parameter Λ, defined for example as

$$\Lambda(N_f) = \mu \exp\left[\frac{-2\pi}{\beta_0\,\alpha_s(\mu)}\right] \left[\frac{\beta_0\,\alpha_s(\mu)}{2\pi}\right]^{-\beta_1/\beta_0^2}$$
$$\times \exp\left\{ - \int_0^{\alpha_s(\mu)} du \left[\frac{1}{\beta(u)} + \frac{2\pi}{\beta_0 u^2} - \frac{\beta_1}{\beta_0^2 u} \right] \right\} . \qquad (2)$$

Note that, by using this relation at the leading order, one gets that $\Lambda(N_f)$ is known with an accuracy of about 5% if $\alpha_s(\mu)$ is given with an accuracy of about 1%.

When two different renormalization schemes are considered one can write

$$\alpha_s^{(1)}(\mu) = \alpha_s^{(2)}(\mu)\left\{1 + a\frac{\alpha_s^{(2)}(\mu)}{4\pi} + b\left[\frac{\alpha_s^{(2)}(\mu)}{4\pi}\right]^2 + \ldots\right\} \tag{3}$$

and check that

$$\beta_2^{(1)} = \beta_2^{(2)} - 4a\beta_1 + 2(b - a^2)\beta_0 \tag{4}$$

$$\Lambda^{(1)}(N_f) = \Lambda^{(2)}(N_f)\exp\left(\frac{a}{2\beta_0}\right). \tag{5}$$

Using the Λ parameter we can also obtain a parametrization of the μ dependence of the QCD coupling[a]

$$\alpha_s(\mu) = \frac{2\pi}{\beta_0\log(\mu/\Lambda)}\left[1 - \frac{\beta_1}{\beta_0^2}\frac{\log(\log(\mu/\Lambda))}{\log(\mu/\Lambda)} + \ldots\right]. \tag{6}$$

Clearly the coupling decreases as the scale μ increases, corresponding to the property of asymptotic freedom. However, due to the logarithmic dependence one needs to consider a large range of scales μ in order to see a variation of $\alpha_s(\mu)$ by a factor two or three (see Fig. 9.2 in Ref. [2]).

1.1 Experimental results for $\alpha_s(\mu)$

The value for a quantity A from an experiment can be used to extract a value of $\alpha_s(\mu)$ by considering the following perturbative expansion (in some renormalization scheme)

$$A = c_0 + c_1\alpha_s(\mu) + c_2\alpha_s^2(\mu) + \ldots\ . \tag{7}$$

If the scale involved is low one should also try to take into account non-perturbative contributions to A. Clearly the physical result A does not depend on the chosen renormalization scheme. However, the truncated perturbative series does exhibit renormalization-scheme dependence. In particular, the finite coefficients c_i ($i \geq 2$) depend on the renormalization scale μ and, implicitly, on the chosen renormalization scheme. Therefore, for a given scheme, one should find the best choice[b] for the scale μ in order to reduce the scale ambiguity in the expansion (7).

In order to compare the values of $\alpha_s(\mu)$ from various experiments[3,4], these values must be evolved — using the renormalization group — to a common

[a]Note that this parametrization is not unique but depends on the choice of the constant of integration (see for example problem 13.11 in Ref. [1]).
[b]See Ref. [3] for various methods proposed for choosing the scale μ.

scale (usually the mass of the Z^0 boson) and related to a common scheme (usually the $\overline{MS}$ scheme). The present world-wide average[3] in the $\overline{MS}$ scheme for $\alpha_s(M_Z)^{(N_f=5)}$ is 0.1172 ± 0.002.

2 Numerical evaluation of $\alpha_s(\mu = q)$

One can perform the experiment also on a computer, using Lattice QCD.[5] For example, one can evaluate numerically the plaquette $W_{1,1} = < \mathrm{Tr}\, U_{\mu\nu}/3 >$ — which corresponds to the elementary gluonic action on the lattice — and compare the result with a perturbative expansion. Also in this case one should be careful with the choice of the scheme used for the perturbative expansion and with the choice of the scale μ. For example, it is well known[6] that the expansion of the plaquette $W_{1,1}$ in terms of the lattice bare coupling $\alpha_0 = g_0^2/4\pi$ yields a very poor result. In fact, for $g_0^2 = 1$ one-loop perturbation theory with $N_f = 0$ predicts $-\log W_{1,1} \approx 0.4228$ while the corresponding Monte Carlo result is about 0.5214.

A better expansion[6] is obtained if one considers an effective coupling defined in terms of a physical quantity. For example, if $V(q)$ is the inter-quark potential one can define

$$V(q) \;=\; -\frac{16\pi\alpha_V(q^2)}{3q^2} \tag{8}$$

and work in the α_V scheme. In this scheme one gets, at the one-loop level and with $N_f = 0$,

$$-\log W_{1,1} \;=\; c_1\,\alpha_V(q^2)\,\left[1 \,+\, \tilde{c}_2\,\alpha_V^2(q^2) \,+\, \ldots\right] , \tag{9}$$

where $\tilde{c}_2 = \frac{11}{4\pi}\log\left(\frac{\pi^2}{a^2q^2}\right) + \tilde{d}$ and $\tilde{d} \approx 1.329$. In order to fix the value of the scale q the idea introduced in Ref. [6] is to impose the relation

$$\alpha_V(\bar{q}^2)\,c_1 \;=\; \alpha_V(\bar{q}^2)\int d^4q\, f(q) \;=\; \int d^4q\,\alpha_V(q^2)\, f(q)\, , \tag{10}$$

namely $\bar{q}$ is the average momentum entering in the one-loop calculation. By expanding both $\alpha_V(\bar{q}^2)$ and $\alpha_V(q^2)$ in terms of $\alpha_V(\mu^2)$ one gets

$$\log(\bar{q}^2)\int d^4q\, f(q) \;=\; \int d^4q\, \log(q^2)\, f(q)\, . \tag{11}$$

In the case of the plaquette this prescription gives $\bar{q} \approx 3.4018/a$ where a is the lattice spacing. Clearly, this procedure can be repeated for different Wilson loops $W_{r,t}$, i.e. from the numerical values for $W_{r,t}$ one can obtain $\alpha_V(\bar{q}^2_{r,t})$ using a perturbative expansion like eq. (9). As can be seen from Fig.

8 in Ref. [6], the coupling $\alpha_V(q^2)$ obtained from such calculations runs with q in agreement with a two-loop formula (6).

In order to compare these results with experimental data one needs to fix the lattice spacing a, i.e. the scale $\bar{q}$ should be given in physical units. This can be done by comparing lattice data with an experimental input, as has been done[7] using the Υ meson spectrum, known experimentally with very good accuracy and for which accurate simulations can be done. Using results for $N_f = 0$ and $N_f = 2$ dynamical fermions, which allows extrapolation to physical case $N_f = 3$, the authors of Ref. [7] obtained a final result — in the $\overline{MS}$ scheme and evolved perturbatively to the Z^0 mass scale — of 0.1174 ± 0.0024. Similar results have been obtained by other groups, using different lattice setups, and have been included by the Particle Data Group[3] in the evaluation of the world-wide average of $\alpha_s(M_Z)^{(N_f=5)}$.

2.1 Ingredients for a numerical evaluation of the QCD running coupling constant

From the previous analysis it is clear that in order to evaluate the QCD running coupling constant using lattice numerical simulations one needs:[c]

- a good definition for α_s allowing high accuracy, control over finite-size effects and discretization effects,
- an accurate determination of the energy scale (i.e. the lattice spacing a),
- a perturbative relation with $\alpha_{\overline{MS}}$,
- the possibility of extending the procedure to dynamical fermions (unquenched simulations).

3 Running coupling in a finite box

The Alpha collaboration[9] studies the QCD running coupling using finite boxes (with physical size $L = Na$), a definition for α_s that explicitly depends on the scale L, and a non-perturbative finite-size scaling technique.[10] To this end one could consider, for example, the force between static quarks separated by a distance $r = L/2$, given by

$$\alpha_{q\bar{q}}(L) = \frac{3}{4} r^2 F_{q\bar{q}}(r, L)\bigg|_{r=L/2}, \tag{12}$$

[c]See Ref. [8] for a previous review of the numerical determination of α_s.

or a correlation function[11] for a separation $r = L/2$. Most of the simulations by the Alpha collaboration have been carried out using as definition for α_s the response of the system to a change of the boundary conditions in the Schrödinger functional approach (for details see Ref. [12]). In all cases the physical size L of the lattice is the only physical scale on which these couplings depend, i.e. the box size may be regarded as a reference scale, similar to the renormalization scale μ. Thus, one has a coupling running with $\mu = 1/L$.

The key point in considering these couplings is the use of a non-perturbative renormalization group technique. More specifically, one defines the step-scaling function $\sigma(u)$ through the relation $\alpha_s(2L) = \sigma(\alpha_s(L))$. Clearly this function provides the value of the coupling in a box of size $2L$ given a value for the same coupling in a box of size L. Thus, $\sigma(u)$ can be regarded as an "integrated" β-function. The step-scaling function can be evaluated non-perturbatively using numerical simulations by considering several values of L, as shown in Fig. 4 in Ref. [13]. Notice that for each data point in the plot one should carefully check for possible systematic effects, especially discretization errors. In fact, each point is the result of an extrapolation[d] to zero lattice spacing of data obtained from several simulations, done with the same physical size L but different lattice spacing a and number of sites per direction N.

Once the step-scaling function is known, i.e. after fitting the numerical data, one can evaluate the running coupling in a very large box $L_0 = L_{max}$ and then find the value of the coupling for boxes of sizes $L_i = L_0/2^i$, using a recursive procedure and a numerical inversion of the step-scaling function. [In order to keep errors under control one should avoid using the fit for the step-scaling function outside the range of the numerical data, i.e. one should avoid extrapolating the step-scaling function $\sigma(u)$ to very small values of u.] In this way one obtains the values $\alpha_s(\mu_i)$ for the running coupling at the scales $\mu_i = 1/L_i = 2^i/L_{max}$.

Since all the scales μ_i are given in terms of L_{max}, the last step consists in setting the scale for the largest lattice, i.e. finding the value of L_{max} in physical units by comparing lattice data with an experimental input. Thus, the use of a non-perturbative finite-size scaling technique avoids the requirement of a non-perturbative evaluation of the scale L_i for very small values of L_i: this is in general complicated by finite-size effects and by the fact that small values of L_i correspond to large value of β and that simulations at large β and small physical volumes are essentially "perturbative".

The final value[13] obtained by the Alpha collaboration in pure QCD is

[d]See Ref. [14] for possible problems with this extrapolation.

$\Lambda_{\overline{MS}}^{(N_f=0)} = 251 \pm 21$ MeV. Preliminary data with dynamical fermions ($N_f = 2$) have been presented last year[15].

4 Numerical determination of Λ

Several groups study the running of the QCD coupling (in some scheme) over a range of energies and then try to fit the data using a two- or three-loop expression for $\alpha(q)$ as in eq. (6), considering Λ as a fitting parameter. Alternatively, one can evaluate Λ as a function of $\alpha(q)$ and look for a plateau at high scale [note that in eq. (2) the left-hand side is constant, for a given value of N_f, even though the right-hand side depends explicitly on the scale μ].

4.1 Inter-quark potential

One can evaluate the QCD coupling using the static potential $V(r)$, which can be evaluated for several values of r with good accuracy using Wilson loop data and the relation[5]

$$W(r,t) = C(r) e^{[-tV(r)]} + \dots \; . \tag{13}$$

This formula should in general be considered in the limit of large t, but using[e] smearing and noise-reduction techniques one can increase the overlap of the evaluated quantity with the ground state, i.e. increase the value of $C(r)$, and obtain a good signal already for small values of t. Then, from $V(r)$ one can obtain the force $F(r)$ [and the running coupling $\alpha_{q\bar{q}}(1/r)$, see eq. (12)] using a finite-difference approximation for the derivative $dV(r)/dr$. In the quenched case[17] one obtains $\Lambda_{\overline{MS}}^{(N_f=0)} = 246 \pm 7 \pm 3$ MeV, in excellent agreement with the result from the Alpha collaboration. Unquenched simulations (also interesting to study string breaking[18]) are under way[19].

4.2 Three-gluon vertex

In this case one evaluates the three-gluon vertex[20] in the $\widetilde{MOM}$ scheme (in Landau gauge). For this quantity perturbative results are known to three loops[21] (and the corresponding beta function to four loops). In order to obtain good data a careful analysis of breaking of rotational symmetry is required[22]. From quenched simulations[23] one obtains the value $\Lambda_{\overline{MS}}^{(N_f=0)} = 295 \pm 5 \pm 15$ MeV, somewhat larger than the previous results. Preliminary unquenched

[e] For a review of this method see for example Ref. [16].

data[24] have been presented last year and the authors quote $\Lambda_{\overline{MS}}^{(N_f=2)} = 264 \pm 27$ MeV. A similar study has started recently using the quark-gluon vertex[25] yielding $\Lambda_{\overline{MS}}^{(N_f=0)} = 300\,^{+150}_{-180} \pm 55 \pm 30$ MeV, in agreement with the result obtained using the three-gluon vertex.

4.3 Gluon propagator

In this case[26] one evaluates the gluon propagator $D(q)$ (in Landau gauge) and defines $Z_3(q) = q^2 D(q)$. Then, the gluon data can be fitted by considering the two coupled differential equations

$$\frac{d\log Z_3(q)}{d\log q^2} = \Gamma(\alpha) = -\left[\frac{\gamma_0}{4\pi}\alpha + \frac{\gamma_1}{(4\pi)^2}\alpha^2 + \dots\right] \tag{14}$$

$$\frac{d\alpha}{d\log q} = \beta(\alpha) = -\left[\frac{\beta_0}{2\pi}\alpha + \frac{\beta_1}{(2\pi)^2}\alpha^2 + \dots\right], \tag{15}$$

where $\Gamma(\alpha)$ is the anomalous dimension of the gluon renormalization constant in the MOM scheme. At the lowest order these equations imply the well-known result $D(q^2) \sim 1/q^2 \log^c(q^2/\Lambda^2)$ with $c = \frac{\gamma_0}{2\beta_0} = 13/22$.

A solution for these two equations depends on the values $Z_3(\mu)$ and $\alpha(\mu)$ given at some scale μ, and a direct fit of the data provides a value for these two constants. The value $\alpha_s(\mu)$ obtained from the fit can then be evolved, using the renormalization group, to the Z^0 scale and related to the $\overline{MS}$ scheme, using the perturbative relation between the two schemes. The authors[26] quote $\Lambda_{\overline{MS}}^{(N_f=0)} = 319 \pm 14\,^{+10}_{-20}$ MeV as final result in the quenched case, in agreement with the results quoted in Sec 4.2.

4.4 Search for power corrections

Recently there have been various attempts of finding power corrections to the perturbative running of α_s, following theoretical predictions[27]. In particular data from the inter-quark potential[28], the three-gluon vertex[29,30] and the gluon propagator[30] have been reanalyzed. In all cases the authors found that a nonzero contribution proportional to $1/p^2$ improves the fit of the data. In particular, the result obtained in this way for the Λ parameter using the three-gluon vertex[29], i.e. $\Lambda_{\overline{MS}}^{(N_f=0)} = 237 \pm 3\,^{0}_{-10}$ MeV, is in much better agreement with the results quoted in Sections 3 and 4.1 than the original value (see Section 4.2).

5 Running couplings at the IFSC-USP

We now describe the simulations being performed since July 2001 at the Institute of Physics of the University of São Paulo, São Carlos (IFSC–USP). In connection with a grant from FAPESP ("Projeto Jovem Pesquisador", together with Tereza Mendes), we have set up a dedicated PC cluster at the IFSC–USP. The system has 16 nodes with 866 MHz Pentium III CPU and 256 MB RAM memory (working at 133 MHz), a server with 866 MHz Pentium III CPU and 512 MB RAM memory (working at 133 MHz) and is operating with Debian Linux. The machines are connected with a 100 Mbps full-duplex network and in total there are 130 GB in `/tmp` directories and 24.8 GB for the `/home` directory.

We are carrying out simulations considering two definitions of the running coupling constant. In the Coulomb gauge[f] we consider[31]

$$g_C^2(\mathbf{p}) \equiv \frac{11}{12}\mathbf{p}^2 V(\mathbf{p}) \equiv \frac{11}{12} g_0^2\, \mathbf{p}^2\, D_{44}(\mathbf{p})\,, \tag{16}$$

where $V(\mathbf{p})$ is the color-Coulomb potential, $D_{44}(\mathbf{p})$ is the time-time component (at equal time) of the gluon propagator and the momenta are three-dimensional[g]. Clearly, if $V(\mathbf{p})$ is governed by a string tension at large distances, i.e. goes like $1/\mathbf{p}^4$ at small momenta, then we should find $g_C^2(\mathbf{p}) \sim 1/\mathbf{p}^2$ in the infrared limit. In Landau gauge[h] we consider[32]

$$g_L^2(p) \equiv \frac{g_0^2}{4\pi}\left[p^2\, D(p)\right]\left[p^2\, G(p)\right]^2\,, \tag{17}$$

where $D(p)$ is the transverse gluon propagator and $G(p)$ is the ghost propagator. This expression enters directly in the quark Schwinger-Dyson equation[33] and can be interpreted as an effective interaction strength between quarks.[i] In this case one obtains $g_L^2(p) \sim p^{-2}$ if, for example, $D(p) \sim$ const and $G(p) \sim p^{-4}$ in the infrared limit. On the contrary, if the gluon propagator goes to 0 in the infrared limit and the ghost propagator blows up not faster than p^{-4} then $g_L^2(p)$ has an infrared fixed point. The last behavior has also been obtained[32] by solving (approximatively) a coupled set of Schwinger-Dyson equations.

---

[f] Work in collaboration with D. Zwanziger, New York University (NYU). Some of our simulations are done on a PC cluster at NYU.

[g] Remarkably, this definition of the running coupling constant is renormalization-group-invariant[31].

[h] Work in collaboration with K. Langfeld and J. C. R. Bloch, Tübingen University.

[i] This is also a renormalization-group-invariant quantity since (in Landau gauge) $\widetilde{Z}_1 = 1$.

In our project we are also interested in a numerical verification of Gribov's *confinement* scenarios — written in terms of the infra-red behavior for the propagators considered above — for these two gauges. Theoretical studies[34,35] predict that in Landau gauge, there should be a strong suppression of the (unrenormalized) transverse gluon propagator $D(p)$ in the infrared limit and an enhancement of the ghost propagator $G(p)$ in the same limit. These results clearly indicate the absence of a massless gluon from the physical spectrum and provide an indication of a long-range effect in the theory that may result in color confinement. The confinement scenario is particularly simple in the minimal Coulomb gauge where the ghost propagator determines directly the Coulomb interaction[34,31]. In fact, in this case, confinement of color, i.e. the enhancement at long range of the color-Coulomb potential $V(R)$, is due to the enhancement of ghost propagator $G(\mathbf{p})$ at small momenta. At the same time, the disappearance of gluons from the physical spectrum is manifested by the suppression at $\mathbf{p} = \vec{0}$ of the propagator $D_{ij}(\mathbf{p}, p_4)$ of 3-dimensionally transverse would-be physical gluons.

Preliminary results in the pure $SU(2)$ case for the two running coupling constants and Gribov's confinement scenarios have been presented in Refs. [36].

Acknowledgments

The author thanks the organizers of Hadron Physics 2002 for the invitation, and Tereza Mendes for her help in writing this manuscript. This work has been supported by FAPESP, Brazil (Project No. 00/05047-5).

References

1. *Quantum and statistical field theory*, M. Le Bellac (Oxford University Press, Oxford, 1995).
2. `http://pdg.lbl.gov/2001/figures.html`.
3. Review on "Quantum chromodynamics" in `http://pdg.lbl.gov/`.
4. P. N. Burrows et al., hep-ex/9612012.
5. See for example *Quarks, gluons and lattices*, M. Creutz (Cambridge University Press, Cambridge, 1983).
6. G. P. Lepage and P. B. Mackenzie, *Phys. Rev.* **D48** (1993) 2250.
7. C. T. H. Davies et al., *Phys. Rev.* **D56** (1997) 2755.
8. P. Weisz, *Nucl. Phys.* (Proc. Suppl.) **B47** (1996) 71.
9. See `http://www.ifh.de/computing/projects/parallel/ape/alpha/alpha.html`
10. M. Lüscher, P. Weisz and U. Wolff, *Nucl. Phys.* **B358** (1991) 221.

Also, see for example R. Sommer, hep-ph/9711243, in Schladming 1997 "Computing particle properties", and M. Lüscher, hep-lat/9802029.
11. G. M. de Divitiis et al., *Nucl. Phys.* **B422** (1994) 382; *Nucl. Phys.* **B433** (1995) 390.
12. M. Lüscher et al., *Nucl. Phys.* **B389** (1993) 247.
13. S. Capitani et al., *Nucl. Phys.* (Proc. Suppl.) **B63** (1998) 153.
14. A. Patrascioiu and E. Seiler, hep-lat/0009005.
15. A. Bode et al., *Phys. Lett.* **B515** (2001) 49.
16. G. S. Bali, *Phys. Rept.* **343** (2001) 1.
17. G. S. Bali and K. Schilling, *Phys. Rev.* **D47** (1993) 661.
18. B. Bolder et al., *Phys. Rev.* **D** (2001) 63; C. W. Bernard et al., *Phys. Rev.* **D64** (2001) 074509.
19. G. S. Bali et al., *Phys. Rev.* **D62** (2000) 054503.
20. C. Parrinello, *Phys. Rev.* **D50** (1994) 4247.
21. K. G. Chetyrkin and A. Retey, hep-ph/0007088.
22. P. Boucaud et al., *JHEP* **9810** (1998) 017.
23. P. Boucaud et al., *JHEP* **9812** (1998) 004.
24. P. Boucaud et al., *JHEP* **0201** (2002) 046.
25. J. Skullerud, A. Kizilersu and A. G. Williams, *Nucl. Phys.* (Proc.Suppl.) **B106** & 107 (2002) 841; J. Skullerud and A. Kizilersu, hep-ph/0205318.
26. D. Becirevic et al., *Phys. Rev.* **D60** (1999) 094509; *Phys. Rev.* D61 (2000) 114508.
27. V. I. Zakharov, *Nucl. Phys.* (Proc. Suppl.) **74** (1999) 392.
28. G. S. Bali, *Phys. Lett.* **B460** (1999) 170.
29. P. Boucaud et al., *JHEP* **0004** (2000) 006.
30. D. Becirevic et al., *Nucl. Phys.* (Proc. Suppl.) **B106** & **B107** (2002) 867.
31. D. Zwanziger, *Nucl. Phys.* **B518** (1998) 237.
32. L. von Smekal, R. Alkofer and A. Hauck, *Phys. Rev. Lett.* **79** (1997) 3591.
33. J. C. R. Bloch, hep-ph:/0202073.
34. V. N. Gribov, *Nucl. Phys.* **B139** (1978) 1.
35. D. Zwanziger, *Nucl. Phys.* **B364** (1991) 127; *Nucl. Phys.* **B412** (1994) 657.
36. A. Cucchieri and D. Zwanziger, *Phys. Rev.* D65 (2002) 014001; A. Cucchieri, T. Mendes and D. Zwanziger, *Nucl. Phys.* (Proc. Suppl.) **B106** & **B107** (2002) 697; talks to be presented at Lattice 2002 by D. Zwanziger (*Gluon propagator in minimal Coulomb gauge*) and A. Cucchieri (*Running coupling constant and propagators in $SU(2)$ Landau gauge*).

THE PROTON

CARLOS GARCIA CANAL
Laboratorio de Física Teórica, Departamento de Física, Universidad Nacional de La Plata, C.C. 67 La Plata, (1900), Argentina
garcia@venus.fisica.unlp.edu.ar

RODOLFO SASSOT
Departamento de Física, Universidad de Buenos Aires, Ciudad Universitaria, Pab.1 (1428) Buenos Aires, Argentina
sassot@df.uba.ar

In this talk we present a collection of selected topics concerning the structure of the proton and the fundamental interactions as seen inside it. These topics have been thoroughly covered by high energy experiments with ever increasing precision in recent years and beautifully illustrate our present knowledge of the standard model.

1 Introduction

The proton is, no doubt, the prototype for hadronic structure and consequently, the more precisely studied hadronic object. In the process of unveiling its structure, the proton allows to cross-check our understanding of the same interactions used to examine it; the standard model of fundamental interactions.

In this seminar we intend to present the proton as a window opened not only towards that fundamental hadronic structure but also towards the knowledge of the basic interactions in Nature. In this way we will see through the proton the most recent developments of Quantum Chromodynamics (QCD) and also of the electroweak interactions (EW) as a paradigm of unification.

Even more, almost all the present and future particle physics is based upon our comprehension of the fundamental interactions that involve the proton in the high energy regime. The precise knowledge of QCD at the highest energy scales is not only interesting in itself but is necessary for the analysis of W, Z and t-quark properties. Moreover, Higgs boson searches, and needless to mention SUSY particles and other non-standard particles searches, both at Tevatron and LHC, reckon on the precise determination of the corresponding cross sections. Consequently, to achieve a good precision in our knowledge of the most profound behavior of matter and in particular in the search of new physics, the proton turns out to be a fundamental issue.

The so called naive quark model provides the elementary structure of the proton in terms of three spin 1/2 electrically charged constituents: the quarks. In the sixties, it was understood that the known hadrons could be classified into representations of the unitary group called $SU(3)_F$.[1] The empirical evidence has shown that only some specific representations were present. In fact, the fundamental representations 3 and $\bar{3}$ are not realized in Nature in terms of hadrons. This led to postulate that the known hadrons were composite particles made up of three quarks in the case of baryons or of a quark-antiquark pair in the case of mesons. These quarks were characterized by a new quantum number called flavor with notation u, d and s that comprises the electric charge and baryon number. The first two flavors carry the quantum numbers of isospin ($SU(2)$), while the third one carries strangeness. The electric charge of these three quarks, measured in terms of the proton charge, are $Q_u = 2/3$; $Q_d = Q_s = -1/3$, respectively. So, in order to cope with the proton charge, this hadron has to have two u quarks and one d quark ($Q_p = 2/3+2/3-1/3 = 1$) whose masses can be immediately estimated from the uncertainty principle, namely

$$\begin{aligned} m_q &\sim \sqrt{(\Delta p_x)^2 + (\Delta p_y)^2 + (\Delta p_z)^2} \sim \sqrt{3}\,\Delta p_x \\ &\sim \frac{\sqrt{3}}{\Delta x} \sim \frac{\sqrt{3}}{1\,fm}\,200\,MeV\,fm \sim 340\,MeV\,. \end{aligned}$$

This description of the proton in terms of quarks was initially an effective picture in which these objects were considered as mere building blocks instead of real constituents. At the end of the sixties, a new generation of Rutherford like experiments called deep inelastic scattering (DIS) clearly show the presence of point-like constituents inside the proton.[2,3] They were initially named partons, and the observed properties allowed their identification with the quarks of the naive model but with almost vanishing masses. However, quarks were not alone in the proton; quarks can exchange neutral spin 1 bosons, a second kind of partons named gluons, that collectively carry half of the proton momentum and are ultimately responsible for the strong interactions. At the end of the seventies, a quantum field theory: Quantum Chromodynamics (QCD), provided a formal framework to explain the interactions among quarks and gluons.[4] This theory presents the property of asymptotic freedom and opened a new road in our understanding of the structure of matter. Even though the standing of the theory, with more than 25 years with us, is today well established, further theoretical refinements and the corresponding experimental validation have been constantly renewing the original enthusiasm on the subject of the high energy physics community.

During the eighties and nineties, particle physicists confirmed, in a rich variety of experiments and theoretical developments, the outstanding precision of the other paradigm of a gauge field theory, namely the electroweak (EW) theory.[5]

This lecture intends to provide an overview of the more recent results of high energy collisions related to the proton structure. First, we briefly review the essentials of the Standard Model: QCD + EW, based on Gauge Field Theory, including the main features of partons: quarks and gluons.

2 EW + QCD

The interactions among leptons, quarks and the mediators: photons, weak bosons and gluons, are described by the Standard Model based on the gauge group $SU(3)_C \times SU(2)_L \times U(1)_Y$. This is a quantum gauge field theory with a Lagrangian of the general type[5]

$$\mathcal{L} = -\frac{1}{4} F^a_{\mu\nu} F^{\mu\nu}_a + \bar{\psi}_i \, i \, \gamma^\mu D_\mu \psi_i \tag{1}$$

where

$$F^a_{\mu\nu} = \partial_\mu A^a_\nu - \partial_\nu A^a_\mu + g \, f^{abc} A_{b\mu} A_{c\nu} \tag{2}$$

stands for the corresponding gauge field tensor, ψ_i are the matter (lepton or quark) fields and the covariant derivative is defined by

$$D_\mu = \partial_\mu - i \, g \, T_a \, A^a_\mu \,. \tag{3}$$

The gauge coupling is represented by g and indices are summed over $a = 1, ..., 8$ for $SU(3)$, $a = 1, 2, 3$ for $SU(2)$ and $a = 1$ for $U(1)$, while i includes all quark quantum numbers. Finally, T_a and f_{abc} are the corresponding group generators and structure constants, respectively, which are related through

$$[T_a, T_b] = i \, f_{abc} \, T_c \,.$$

In the particular case of EW interactions, the gauge fields A^a_μ represent the photon (γ) and the weak bosons ($W^\pm$, Z). In QCD, A^a_μ stands for the gluons (g). Remember that γ and g are strictly massless particles while $W's$ and Z get their large masses through the Higgs mechanism. The masses of $W's$ and Z not only make difficult their production but also are responsible for the weakness of the weak interactions. In any gauge theory, interactions with fermions come from the covariant derivative (3) that is formally identical for any interaction. This fact conducted the present understanding of fundamental interactions in terms of a common mechanism which is the exchange of a gauge boson.

Beyond the leading order in perturbation theory, most of the pertinent amplitudes develop divergences.[6] The procedure used to deal consistently with the divergences, shows that the different couplings depend effectively on the energy scale of the process. Nevertheless, the dynamics of each particular interaction has distinctive features coming from the characteristics of the particular gauge group involved. In the case of electromagnetic interactions, part of EW, the coupling α_{em} connected to $U(1)$, that has the value $\sim 1/137$ at low momentum transfer, grows to a value $\sim 1/128$ at $Q^2 \sim m_Z^2$. On the other hand in the case of QCD, related to $SU(3)$, the coupling vanishes logarithmically, a feature known as *asymptotic freedom.* Consequently, in this regime perturbation theory is valid for the analysis of quark-gluon interactions and tests against experimental data can be performed in terms of hadrons.

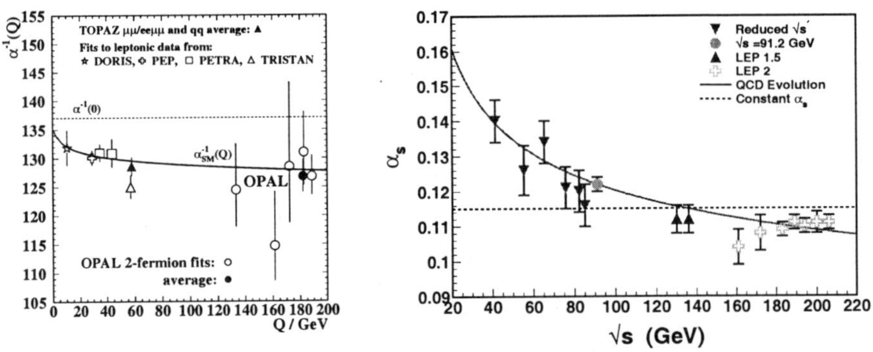

Figure 1: The running of α_{em} (a) and α_s (b).

In Figure 1, the Standard Model prediction for the running of both α_{em} and α_s is shown against recent data.[7,8] Notice that although the electromagnetic coupling slightly rises as the energy increases, the strong coupling falls logarithmically.

3 Deeply Inside the Proton

Deep Inelastic Scattering (DIS), the process in which a highly energetic lepton strikes a proton (or in general a nucleon, a hadron, or a nucleus) exchanging a highly virtual photon or weak boson with one of its constituents, has proved to be one of the most useful tools for unveiling the short distance structure of hadrons as well as many of the properties of the interactions relevant there. In the following we sketch the main features of DIS relevant for the comprehension of the most recent data. For a more detailed treatment see reference.[9]

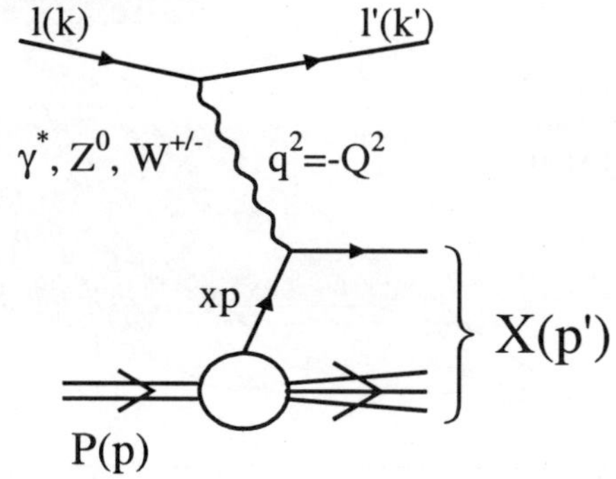

Figure 2: Deep inelastic scattering.

The scattering of a lepton from a proton at high momentum transfer, can be viewed as the (elastic) scattering of the lepton from a quark or antiquark inside the proton mediated by the exchange of the corresponding virtual vector boson $\gamma^{\star}$, Z^0 or $W^{+/-}$, as shown in Figure 2. The virtuality Q^2 of the vector boson, customarily defined as minus the boson four momentum squared, sets the resolving power of a DIS experiment. Clearly, the size d one can resolve inside the nucleon becomes smaller for a shorter photon, or gauge boson, wavelength, namely

$$d \sim \frac{\hbar c}{Q} \simeq 0.2 \frac{GeV\, fm}{Q}\,.$$

For example, $Q^2 = 4\,GeV^2$ implies $d \simeq 10^{-14}\,cm$ while $Q^2 = 40,000\,GeV^2$ means $d = 10^{-16}\,cm$. The virtuality Q^2 also constitutes the relevant scale at which the QCD coupling has to be evaluated and tells us how far we are from the QCD asymptotic limit. From a more phenomenological point of view, this variable can be directly related to the the scattering angle θ of the outgoing lepton, defined with respect to the incoming lepton direction in the laboratory frame

$$\begin{aligned} Q^2 &= -q^2 = -(k' - k)^2 \\ &\simeq 2\,E\,E'(1 - \cos\theta) > 0\,. \end{aligned} \tag{4}$$

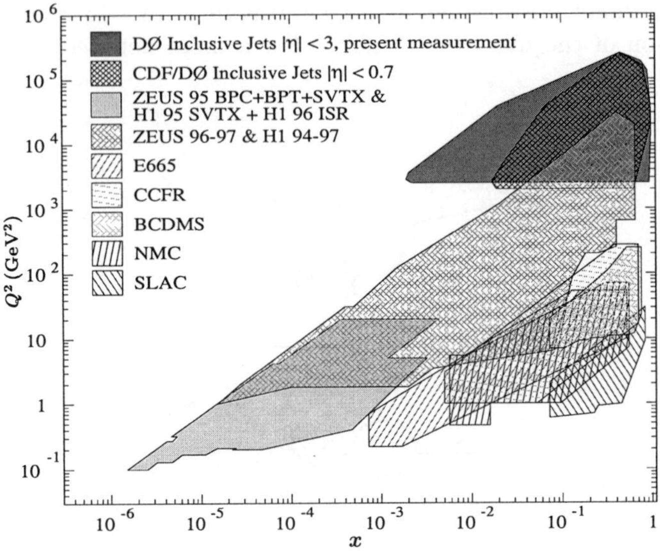

Figure 3: Experimental coverage of the (x,Q^2) plane.

If the scattering were fully elastic, i.e. with just a proton emerging as final state, then from energy-momentum conservation, the cross section would only depend on one variable, besides a trivial dependence around the incoming lepton axis. As the process we are considering is deeply inelastic, the cross section develops dependence in another invariant variable x, the well known Bjorken scale variable, defined by

$$x = \frac{Q^2}{2\,p \cdot q} \tag{5}$$

which, from energy-momentum conservation, happens to be the fraction of the proton momentum carried by the struck quark when measured in the infinite momentum frame. Both x and Q^2 are easily related to the invariant mass squared of the final hadronic state $W^2 = (p+q)^2$ as

$$W^2 = Q^2\,\frac{1-x}{x}\,, \tag{6}$$

showing that small values of x imply increasing W. In Figure 3 we show the updated coverage of the (x,Q^2) plane by different experiments[10]. Notice that present data span about six orders of magnitude both in x and Q^2.

Another variable of interest in terms of which the DIS cross section can be studied is the inelasticity of the scattered lepton

$$y = \frac{p \cdot q}{p \cdot k} = 1 - \cos^2\left(\frac{\theta^*}{2}\right) \tag{7}$$

related to the scattering angle θ^* in the lepton-quark center of mass system, and that is relevant for the analysis of helicity dependent interactions. Ignoring masses, the y variable is related to x and Q^2 through $Q^2 = x\,y\,s$.

In the naive parton model, interactions among partons are neglected because the proton is viewed as a collection of incoherent punctual constituents, or in dynamical words, due to asymptotic freedom. Consequently, the lepton-nucleon cross section is written as a convolution between the lepton-parton point-like cross section and the probability for finding a parton in the nucleon

$$\sigma_{lp} = \sigma_{lq} \otimes f_q(x) \tag{8}$$

being σ_{lq} the point-like cross section, fully predictable from EW theory, while the parton probability functions $f_q(x)$, of non perturbative origin, are taken from experiment.

The addition of QCD corrections to this picture implies taking into account the emission and absorption of gluons by the interacting quarks, thus modifying the point-like cross section in a scale (Q^2) dependent way. It is customary to shift the QCD driven Q^2-dependence from the point like cross section to redefined parton probability functions. These new scale dependent probability distributions also have to be measured once at some scale, but then the scale dependence is fully predicted by QCD.

Skipping some technicalities, the double differential DIS cross-sections both for neutral (NC) and charged current (CC) are finally written as[5]

$$\frac{d^2\sigma_{NC}}{dQ^2\,dx} = \frac{2\,\pi\,\alpha^2}{x\,Q^4}\,\mathcal{F}_{NC}(x,Q^2) \tag{9}$$

$$\frac{d^2\sigma_{CC}}{dQ^2\,dx} = \frac{G_F^2}{2\,\pi\,x}\left(\frac{M_W^2}{M_W^2+Q^2}\right)^2 \mathcal{F}_{CC}(x,Q^2)\,. \tag{10}$$

The first factor in the r.h.s. of both Eqs. (9) and (10), is the relic of the point-like boson quark cross section, and clearly includes the boson propagator. Here, α is the electromagnetic coupling, G_F the Fermi weak coupling and M_W the W weak boson mass. In the case of neutral currents, although the cross section is dominated by the photon exchange, the contribution coming from Z exchange and the photon-Z interference should also be included.

However in this first step we omit these contributions for simplicity. Inside the reduced cross section $\mathcal{F}$ one includes the incoherent contributions of each quark flavor, weighted by the corresponding (electric or weak) charge and probability density, and also some spin dependent factors. Up to the leading order in QCD perturbation theory, the $\mathcal{F}$ functions can be approximated by

$$\mathcal{F}^{\ell\pm}_{NC}(x,Q^2) \sim \left[1+(1-y)^2\right]\left[\frac{4}{9}\,x\,u(x,Q^2)+\frac{1}{9}\,x\,d(x,Q^2)+\cdots\right] \quad (11)$$

$$\mathcal{F}^{\ell+}_{CC}(x,Q^2) \sim x\,\bar{u}(x,Q^2)+(1-y)^2\,x\,d(x,Q^2)+\cdots \quad (12)$$

$$\mathcal{F}^{\ell-}_{CC}(x,Q^2) \sim x\,u(x,Q^2)+(1-y)^2\,x\,\bar{d}(x,Q^2)+\cdots \quad (13)$$

where the dots refer to contributions from other quark flavors. Notice that in addition to the *uud* quark densities, related to the naive picture providing the quantum numbers of the proton, the so called *valence quarks*, we have also included contributions from $u\bar{u}$, $d\bar{d}$, and $s\bar{s}$ pairs, the radiatively generated *sea* quark content of the proton.

The availability of different kinds of leptonic probes (electrons, positrons, positively and negatively charged muons, neutrinos, etc.) and their respective way of interacting with quarks, allows a first check of the physical content of the notion of quark or quark density, usually known as *universality*. In simple words, in order to make sense, the QCD parton model must be based in quark densities that do not depend on the leptonic probe or interaction mechanism. Furthermore, parton densities obtained in a given experiment must fit any other, and the Q^2 dependences must be the one predicted by QCD. This property has been thoroughly tested for the last 25 years in a large number of experiments with remarkable precision. These include neutral and charged DIS (as shown in Figures 4a, 4b, and 4c)[11,12,13], jet production in $p\bar{p}$ (Figure 5a)[10], and even polarized DIS (Figure 5b)[14].

The precision attained by these experiments allows very accurate determinations of parton distributions through which we have learned rather unexpected features of the proton structure such as flavor asymmetry at sea level, and the tiny contribution of quarks to the proton spin.

Although global QCD analysis of thousands of data are necessary for the accurate extraction of modern parton distributions, and the corresponding predictions for the next generation of experiments, somehow they seem to hide the simplicity once sought in the quark model. Nevertheless, high precision charged and neutral current experiments can easily take us back to that picture.

The distinctive features of the different electroweak exchanges (NC and CC) allow an interesting test related to the handedness of quarks, specifically

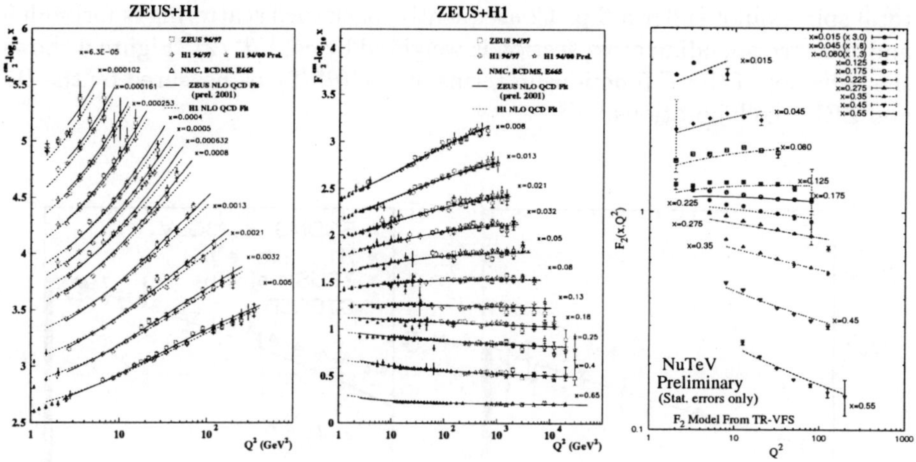

Figure 4: Structure function measurements.

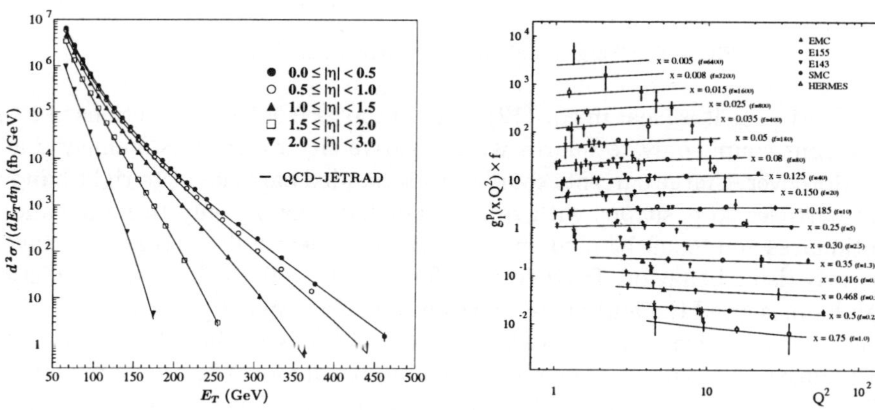

Figure 5: Inclusive jet production and polarized DIS.

confirming their spin 1/2 nature and -to the present accuracy of experiments- their point-like interaction with gauge bosons. Notice that in eqs. (10-12), the factor $(1-y) = \cos^2(\theta^*/2)$ distinguishes between quark and antiquark contributions. If the spin of both incoming particles add up to 0, i.e., both particles have the same handedness, any scattering angle θ^* is allowed. On the contrary, if quark and lepton have opposite handedness, i.e., add up to 1, this total spin cannot suffer a flip. Consequently, backward scattering is forbidden and the corresponding cross-section is weighted by $cos^4(\theta^*/2)$. Figure 6 shows the behavior of the $\mathcal{F}$ functions in terms of $cos^4(\theta^*/2)$ as measured by the H1 and ZEUS collaborations.[15,16]

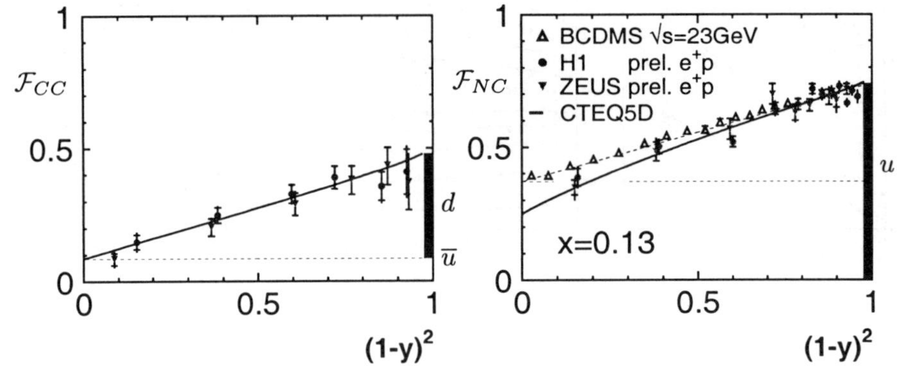

Figure 6: Charged and Neutral Current DIS measurements.

Notice, as it is clear in eq. (12), that in the case of CC interactions in $e^+ p$, positrons couple only to quarks with negative electric charges, namely d and $\bar{u}$. Moreover, charge currents couple to left handed particles or to right handed antiparticles so positrons with right handedness couple only to right-handed antiquarks and to left handed quarks. Figure 6a clearly shows that data have a linearly rising behavior. Furthermore, for $(1-y)^2 \to 0$, the backward direction can be measured the antiquark contribution. Consequently, the rise is mainly due to the d valence quark contribution at the given value of x (in the case of Figure 6a, $x = 0.13$). The corresponding results for NC, are also shown in Figure 6b. This data is mainly sensitive to the u valence quark distribution in the proton at the given value of x (again $x = 0.13$). In conclusion, the comparison of NC and CC measurements reveals that the quark model uud valence structure of the proton is by no means lost, even at the high scales of HERA.

4 EW Interactions Revisited

Now that we feel confident with our knowledge of the proton structure, we can use it as a benchmark for testing EW Interactions.

In Figure 7, the differential cross sections for $e^{\pm} p$ NC and CC DIS at

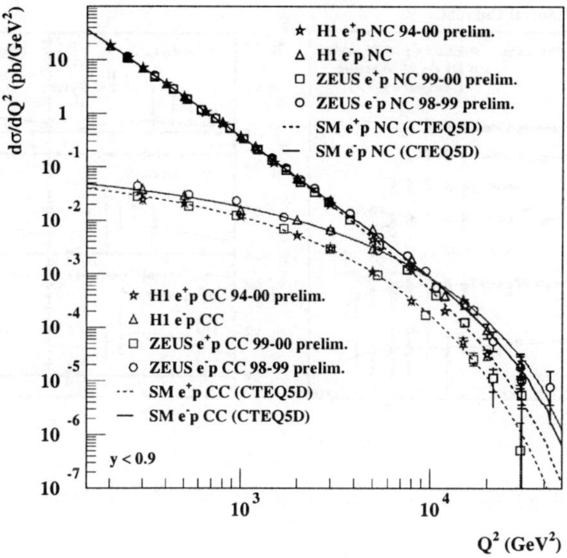

Figure 7: Differential cross sections for $e^{\pm}\,p$ NC and CC DIS.

high Q^2, as measured at HERA [15,16], are presented. The NC cross section falls by 6 orders of magnitude in the range $200\,GeV^2 < Q^2 < 10000\,GeV^2$. For $Q^2 < 1000\,GeV^2$, the NC cross section is approximately equal for both electron and positron, because in this region the NC processes are dominated by photon exchange (the Q^{-4} behavior coming from the photon propagator).

For $Q^2 \sim M_Z^2$, the positron cross section begins to stay below that of $e^-\,p$, because at these values of Q^2, besides the photon, one has the Z^0 exchange contributions that interfere positively in the case of the electron and suppress the $e^+\,p$ cross section. In fact, in this high Q^2 region,

$$\mathcal{F}_{NC}^{\ell^\pm}(x,Q^2) = \left[1+(1-y)^2\right]\,\tilde{F}_2(x,Q^2) - y^2\,\tilde{F}_L(x,Q^2) \mp \left[1-(1-y)^2\right]\,x\,\tilde{F}_3(x,Q^2) \tag{14}$$

with

$$\tilde{F}_2(x,Q^2) = F_2(x,Q^2) - v\,\frac{\kappa\,Q^2}{(Q^2-M_Z^2)}\,F_2^{\gamma\,Z} + (v^2+a^2)\left(\frac{\kappa\,Q^2}{(Q^2-M_Z^2)}\right)^2 F_2^Z\,. \tag{15}$$

Here $F_2^{\gamma\,Z}$ and F_2^Z are the contributions due to $\gamma - Z$ interference and to Z exchange respectively. The function $\tilde{F}_2$ is related to the sum of quark and

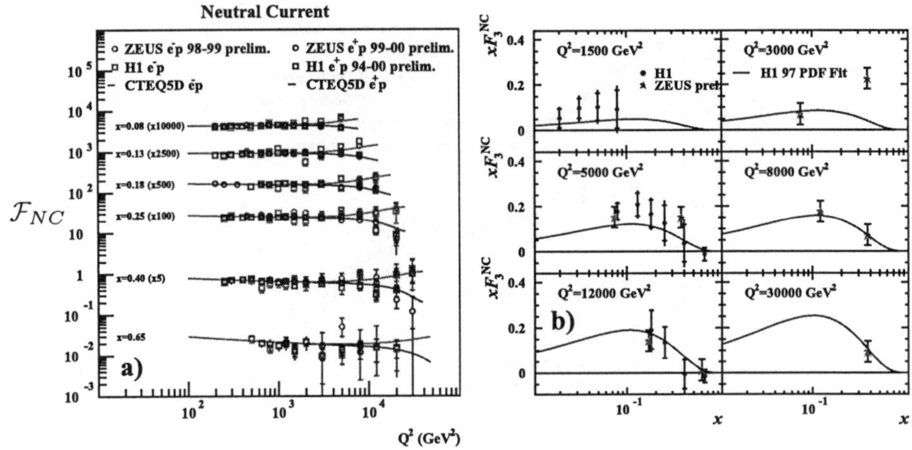

Figure 8: a) $\mathcal{F}_{NC}$ for $e^+\,p$ and $e^-\,p$ DIS; b) F_3^{NC} as extracted from data in a).

antiquark distributions, while $\tilde{F}_3$ to their difference. Clearly, v and a are the vector and the axial couplings of e to Z and $\kappa = 1/(4\,\sin^2\theta_W\,\cos^2\theta_W)$. The standard model gives a very good description of experimental data.

Referring to CC $e^\pm\,p$ DIS cross sections, they fall slowly for values of Q^2 up to $\sim 1000\,GeV^2$ and then more rapidly as Q^2 approaches the value M_W^2. The electron - proton CC cross section is always larger than the corresponding one to positron because CC interactions distinguish between flavors and the electron case picks out the u type quarks and the d type antiquarks, while the positron, the other way around. On the other hand, in the proton there are more u quarks than d quarks and CC suppresses the contribution of d quarks in $e^+\,p$ due to its chiral nature. Finally, for $Q^2 > 10000\,GeV^2$ the four different cross sections have a similar magnitude. This is another impressive verification of the EW unification provided by the standard model.

Figure 8a compares theoretical expectations for $\mathcal{F}_{NC}$ for $e^+\,p$ and $e^-\,p$ DIS against HERA data[15,16]. Notice that

$$\mathcal{F}_{NC}^{\mp} = F_2^{NC} \pm \frac{[1-(1-y)^2]}{[1+(1-y)^2]}\,F_3^{NC} \tag{16}$$

so that the F_3^{NC} contribution, coming from the $\gamma^* - Z^0$ interference adds for the electron case and substracts for that of the positron. Figure 8b shows F_3^{NC} data obtained from those of Figure 8a.

5 Transition Region in Q^2

Far from being closed, the description of the short distance structure of the proton remains a very active ground. One of the most interesting lines of current research concerns the kinematical domain where the QCD coupling begins to grow, eventually spoiling the perturbative treatment, opposite to that where we tested EW interactions, i.e. the low Q^2 domain.

The interesting point here is that although perturbative QCD desert us, Regge theory offers a hand at the other end of the passageway, leaving our inventive to fill the gap. The main point here is to analyze the behavior of the structure functions at low Q^2 and to determine how much one can push down the correctness of QCD perturbation predictions, and figure out what other mechanisms begin to play before we reach the safety on the other end.

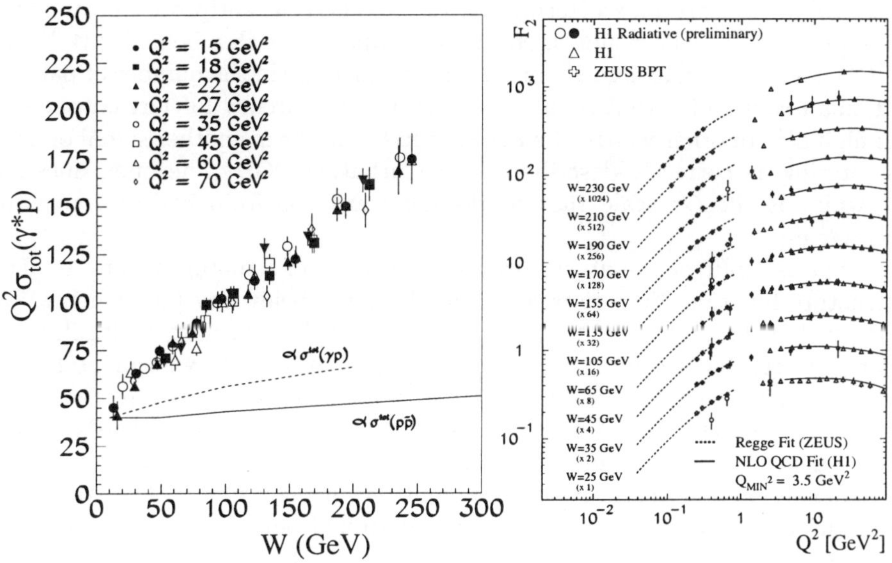

Figure 9: a) Virtual photon-proton cross section; b) F_2 through the transition region.

Neglecting the Z-contribution, the NC, DIS cross section can be written in terms of the flux of virtual photons times the virtual photon-proton cross section $\sigma^T_{\gamma^* p}$. This last cross section can be expressed as the sum of cross sections for the scattering of transverse and longitudinal (notice that they are virtual) photons, in an entirely similar way as in the case of real photons, as soon as the lifetime of the virtual photon is large compared to the interaction time. This implies $x \ll 0.1$. For this range of x, the total cross section can be written

$$\sigma^T_{\gamma^* p} \approx \frac{4\pi\alpha}{Q^2} F_2(W, Q^2)$$

where $W^2 \approx Q^2/x$ is the $\gamma^* - p$ c.m. energy. In Figure 9a, $Q^2\,\sigma^T_{\gamma^* p}$ data are shown as a function of W as measured by ZEUS.[17] These data rise almost linearly with W. By the way, if the $\gamma^* - p$ cross section is parametrized in terms of a pomeron-like trajectory, namely

$$\sigma^T_{\gamma^* p} \sim \left(W^2\right)^{\alpha_P(0)-1}$$

the intercept giving the best fit result $\alpha_P(0) \approx 1.45$. Figure 9a also includes the real photon-proton scattering cross section and the antiproton-proton one. Both cross sections are almost constant, compatible with a behavior $W^{0.2}$. In QCD, the rise of $\sigma^T_{\gamma^* p}$ comes from the strong increase of the number of partons at small values of x that reduce the proton transparency for virtual photons of high Q^2. In other words, the rise seen in the figure is another signal of hard scattering as correctly described by perturbative QCD. The open question is which is the Q^2 scale that marks the transition from hadronic to hard scattering.

Let us rephrase the above discussion in terms of structure functions. The structure function F_2^{NC} rises quite rapidly for x below say 0.01 when Q^2 is larger than around $2\,GeV^2$ and consequently QCD, at next to leading order perturbation theory, provides a good description of experimental facts in this regime. On the other hand, F_2^{NC}, for much lower Q^2 (in the range $0.045 < Q^2 < 0.65\,GeV^2$) does not rise with x falling. In other words, the x-dependence of the structure function undergoes a cross-over at $Q^2 \sim 1\,GeV^2$. This experimental observation is summarized in Figure 9b.[18] The transition to non-perturbative regime, from partonic to hadronic degrees of freedom at around $Q^2 \sim 1\,GeV^2$ seems to be smooth even if the details of the dynamics lying beneath are still a debt of the theory.

In summary it was verified that the Regge theory works at low Q^2, the photoproduction region . On the other hand, in the DIS regime, perturbative QCD describes inclusive NC cross section in terms of F_2. This is sketched in Figure 10.

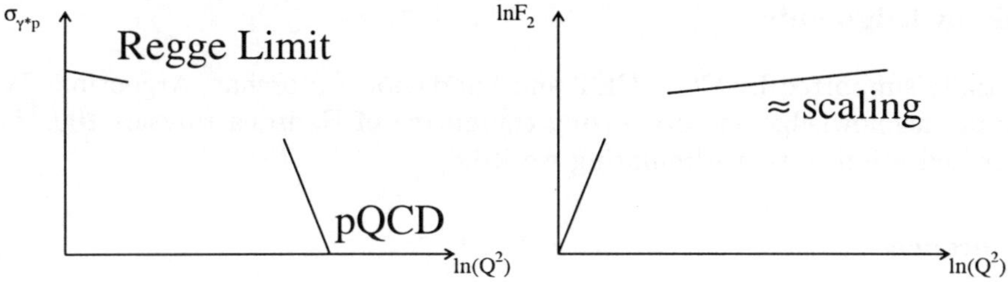

Figure 10: Range of validity of the different approaches.

6 Epilogue

Since its début in modern physics, the proton has played the role of the leading actor. We can start by mentioning for example, the proton spin as responsible for the hyperfine 21 *cm* spectroscopic line, of great interest in astrophysics, and go on through many of the most important breakthroughs in a variety of fields. The detection of the approximate symmetry between proton and neutron provided the first hint of a unitary internal symmetry. This was the starting point to the development of flavor symmetries from one side and later the birth of the new gauge Yang-Mills theories era from the other. Afterwards, DIS on protons opened the road to a new scale of elementariness and to the formalization of the theory of strong interactions of quarks and gluons. The most recent experiments at HERA allow the neutral and charged current weak interactions to be measured within a single experiment. As we have seen, these data provide a direct demonstration of the unification of electromagnetic and weak interactions.

The new advances of our knowledge of the non-perturbative region of QCD (low Q^2), seem to indicate that we are at the threshold of new developments that were hidden due to theoretical technical difficulties.

Last but not least, the everyday increasing precision in our knowledge of the proton structure is mandatory when trying to predict the potentialities of new experimental facilities in detecting the long waited Higgs.

In conclusion, it is not exaggerated to affirm that most of our present definitive knowledge of fundamental particles and interactions is closely related to the understanding of the proton. For this reason the proton is a "classic". Classic in the sense used for books by Italo Calvino: "a book that never ends saying what it has to say".

Acknowledgments

Partially supported by CONICET and Fundación Antorchas, Argentina. We warmly acknowledge the organizing committee of Hadrons Physics 2002 for their invitation to this stimulating workshop.

References

1. F. Close, *An Introduction to Quarks and Partons*, (Academic Press, London, 1979).
2. R. Feynman, *Photon-Hadron Interactions*, (Benjamin, New York, 1972).
3. F. Halzen, A. D. Martin, *Quarks and Leptons: An Introductory Course in Modern Particle Physics* (John Wiley & Sons, 1984).
4. F. J. Yndurain, *The Theory of Quark and Gluon Interactions* (Springer Verlag, Berlin, 1999).
5. E. Leader, E. Predazzi, *An Introduction to Gauge Theories and Modern Particle Physics*, (Cambridge University Press, 1996).
6. G. Sterman, *Quantum Field Theory*, (Cambridge University Press, 1993).
7. L3 Collaboration, L3 Note 2670 submitted to the International Europhysics Conference on high Energy Physics, Budapest, Hungary, 2001.
8. OPAL Collaboration, G. Abiendi *et al.*, *Eur. Phys. Jour.* C **13** (2000) 553.
9. C.A. Garcia Canal, R. Sassot, *Int. Jour. Mod. Phys.* A **15**, (2000) 3587.
10. D0 Collaboration, B. Abbot *et al.*, *Phys. Rev. Lett.* **86**, 1701 (2001).
11. H1 Collaboration, C. Adlof *et al.*, *Eur. Phys. Jour.* C **21** (2001) 33.
12. ZEUS Collaboration, S. Chekanov *et al.*, *Eur. Phys. Jour.* C **21** (2001) 433.
13. NUTEV Collaboration, D. Naples, International Europhysics Conference on high Energy Physics, Budapest, Hungary, 2001.
14. D. de Florian, O. Sampayo, R. Sassot, *Phys. Rev.* D **57**, 5803 (1998); D. de Florian, R. Sassot, *Phys. Rev.* D **62**, 094025 (2000).
15. H1 Collaboration, XX International Symposium on Lepton Photon Interactions at High Energies, Roma, Italy, 2001.
16. ZEUS Collaboration, International Europhysics Conference on high Energy Physics, Budapest, Hungary, 2001.
17. G. Wolf, VII Blois Workshop on Elastic and Diffractive Scattering, Seul, Korea 1996.
18. C. Issever, 9TH International Workshop on Deep Inelastic Scattering, Bologna, Italy, 2001.

LIGHT SCALAR MESONS $\sigma(500)$, $f_0(980)$ AND κ IN CHARM MESON DECAYS

IGNACIO BEDIAGA

Representing the Fermilab E791 Collaboration

Centro Brasileiro de Pesquisas Físicas, Rua Xavier Sigaud 150, 22290, Rio de Janeiro, Brazil

bediaga@cbpf.br

We present recent results on scalar light mesons based on Dalitz plot analysis of charm decays from Fermilab experiment E791. Scalar mesons are found to have large contributions to the decays studied, $D^+ \to K^-\pi^+\pi^+$ and $D^+, D_s^+ \to \pi^-\pi^+\pi^+$. From the first decay, we find good evidence for the existence of the light and broad κ meson and we measure its mass and width. We find strong evidence for the $\sigma(500)$ meson from $D^+ \to \pi^-\pi^+\pi^+$ decay and measure its mass and width. We also present the results obtained for the $f_0(980)$ parameters through the $D_s^+ \to \pi^-\pi^+\pi^+$ decay. These results demonstrate the importance of charm decays as a new environment for the study of light meson physics.

1 Introduction

The decays of charm mesons are currently a new source of information for the study of light meson spectroscopy, with the advantages of having well defined initial state (the D meson, a 0^- state with defined mass). This new information is complementary to that from scattering experiments and can be particularly relevant to the understanding of the scalar sector, due to the well-known difficulties of scattering experiments to observe light and broad scalar resonances.

Here we present an overview of the results we obtained analyzing the decays $D_s^+ \to \pi^-\pi^+\pi^+$ (see [1]), $D^+ \to \pi^-\pi^+\pi^+$ (see [2]) and $D^+ \to K^-\pi^+\pi^+$ (see [3]), using data collected in 1991/92 by Fermilab experiment E791 from $500 GeV/c$ π^--nucleon interactions. (For details see [4]).

For the $D^+ \to \pi^-\pi^+\pi^+$ decays, we find that a model with only known $\pi\pi$ resonances plus a non-resonant (NR) channel is not able to describe the data adequately. Thus we include a new amplitude in our fit function and find strong evidence for the presence of a light and broad scalar resonance, the $\sigma(500)$. The channel involving this scalar meson is responsible for half of the decay rate. We measure the mass and the width of this scalar meson to be $478^{+24}_{-23} \pm 17$ MeV$/c^2$ and $324^{+42}_{-40} \pm 21$ MeV$/c^2$, respectively. Other experiments have presented controversial evidence for this low-mass $\pi\pi$ resonance in partial wave analysis[5,6,7], with ambiguous interpretations of the characteristics of such particles[8,9].

We found a similar situation for the $D^+ \to K^-\pi^+\pi^+$ analysis. When we include all known $K\pi$ resonant channels plus a NR contribution, we find that this model is not able to describe the data. By including an extra scalar resonant state, with unconstrained mass and width, we obtain a fit which is substantially superior to that without this state. The values for its mass and width are found to be $797 \pm 19 \pm 42$ MeV/c^2 and $410 \pm 43 \pm 85$ MeV/c^2 respectively. We refer to this state as the κ. We also obtain new measurements for the mass and the width of the $K_0^*(1430)$ resonance.

From our analysis of $D_s^+ \to \pi^-\pi^+\pi^+$ decays we found that the isoscalar intermediate states are dominant in this decay as well. The largest contribution to this final state comes from the decay involving the scalar meson $f_0(980)$ whose nature is a long-standing puzzle. It has been described as a $q\bar{q}$ state , a $K\bar{K}$ molecule, a glueball, and a multiquark state [8,9]. We obtain new measurements for the $f_0(980)$ and $f_0(1370)$ masses and widths.

To obtain these results, we had to introduce a new approach in Dalitz plot analysis in order to extract the mass and width of the scalar resonances by using them as floating parameters in the fit. We begin this paper presenting the general method, applied in the $D^+ \to K^-\pi^+\pi^+$ Dalitz-plot analysis, then we discuss the $D_s^+ \to \pi^-\pi^+\pi^+$ and $D^+ \to \pi^-\pi^+\pi^+$ studies using the same procedure.

2 The $D^+ \to K^-\pi^+\pi^+$ Dalitz-Plot Analysis

From the original 2×10^{10} events collected by E791, and after reconstruction and selection criteria, we obtained the $D^+ \to K^-\pi^+\pi^+$ sample shown in Figure 1(a). The filled area represents the level of background; besides the combinatorial, the other main source of background comes from the reflection of the decay $D_s^+ \to K^-K^+\pi^+$ (through $\bar{K}^*K^+$ and $\phi\pi^+$). The cross-hatched region contains the events selected for the Dalitz-plot analysis. There are 15090 events in this sample, of which 6% are background.

Figure 1(b) shows the Dalitz-plot for these events. The two axes are the squared invariant-mass combinations for $K\pi$, and the plot is symmetrized with respect to the two identical pions. The plot presents a rich structure, where we can observe the clear bands from $\bar{K}^*(890)\pi^+$, and an accumulation of events at the upper edge of the diagonal, due to heavier resonances. To study the resonant substructure, we perform an unbinned maximum-likelihood fit to the data, with probability distribution functions (PDF's) for both signal and background sources. In particular, for each candidate event, the signal PDF is written as the square of the total physical amplitude $\mathcal{A}$ and it is weighted

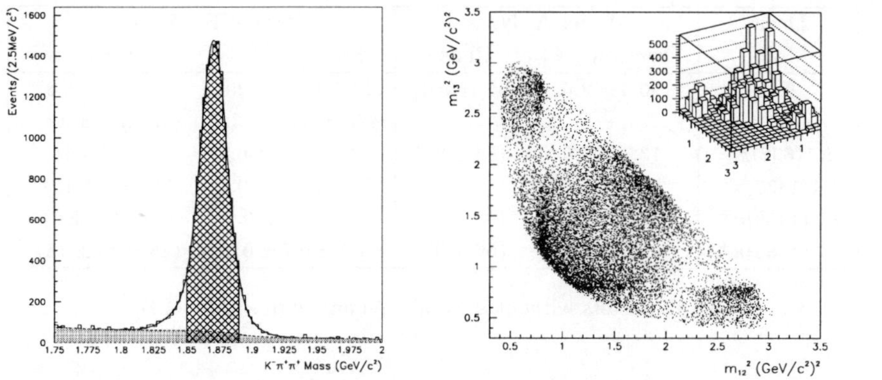

Figure 1. (a) The $K^-\pi^+\pi^+$ invariant mass spectrum. The filled area represents the background. (b) Dalitz plot corresponding to the events in the cross-hatched area of (a).

for the acceptance across the Dalitz plot (obtained by Monte Carlo (MC)) and by the level of signal to background for each event, as given by the line shape of Figure 1(a). The background PDF's (levels and shapes) are fixed for the Dalitz-plot fit, according to MC and data studies.

We begin describing our first approach to fit the data, which represents the conventional Dalitz-plot analysis including the known $K\pi$ resonant amplitudes ($\mathcal{A}_n$, $n \geq 1$), plus a constant non-resonant contribution. The signal amplitude, a Breit-Wigner parameterization times a Blatt-Weisskopf damping factor, is described in[3].

Using this model with well known resonances (Model A), we find contributions from the following channels: the non-resonant, responsible for more than 90% of the decay rate, followed by $\bar{K}_0^*(1430)\pi^+$, $\bar{K}^*(892)\pi^+$, $\bar{K}^*(1680)\pi^+$ and $\bar{K}_2^*(1430)\pi^+$. The decay fractions and relative phases are shown in Table 1. These values are in accordance with previous results from E691[10] and E687[11]. We thus confirm a high non-resonant contribution according to this model, which is totally unusual in D decays. Besides, there is an important destructive interference pattern, since all fractions add up to 140 %.

To evaluate the fit quality, we compute a χ^2 from binned, two-dimensional distributions of data and decay model events. The χ^2 comes from the differences in the binned numbers of events between the data and the model (from a fast MC simulation). We obtain $\chi^2/\nu = 2.7$ (ν being the number of degrees of

Decay Mode	Model A: No κ Fraction (%)	Phase	Model B: With κ Fraction (%)	Phase
NR	90.9 ± 2.6	$0°$ (fixed)	$13.0 \pm 5.8 \pm 2.6$	$(349 \pm 14 \pm 8)°$
$\kappa\pi^+$	–	–	$47.8 \pm 12.1 \pm 3.7$	$(187 \pm 8 \pm 17)°$
$\bar{K}^*(892)\pi^+$	13.8 ± 0.5	$(54 \pm 2)°$	$12.3 \pm 1.0 \pm 0.9$	$0°$ (fixed)
$\bar{K}_0^*(1430)\pi^+$	30.6 ± 1.6	$(54 \pm 2)°$	$12.5 \pm 1.4 \pm 0.4$	$(48 \pm 7 \pm 10)°$
$\bar{K}_2^*(1430)\pi^+$	0.4 ± 0.1	$(33 \pm 8)°$	$0.5 \pm 0.1 \pm 0.2$	$(306 \pm 8 \pm 6)°$
$\bar{K}^*(1680)\pi^+$	3.2 ± 0.3	$(66 \pm 3)°$	$2.5 \pm 0.7 \pm 0.2$	$(28 \pm 13 \pm 15)°$

Table 1. Results without κ (Model A) and with κ (Model B).

freedom), with a corresponding confidence level (CL) of 10^{-11}. In Figure 2(a) we show the $K\pi$ low and high squared-mass projections for data (error bars) and model (solid line). The discrepancies are evident in the very low-mass region for $m^2(K\pi)_{low}$ and near 2.5 $(\text{GeV}/c^2)^2$ for $m^2(K\pi)_{high}$. These regions of disagreement are the same observed previously by E687 [11]. We thus conclude that a model with the known $K\pi$ resonances, plus a non-resonant amplitude, is not able to describe the $D^+ \to K^-\pi^+\pi^+$ Dalitz plot satisfactorily.

A similar pattern – bad fit quality with large NR fraction – is found in the analysis of the decay $D^+ \to \pi^-\pi^+\pi^+$ when allowing only the established $\pi\pi$ resonances [2]. There we find that the inclusion of an extra scalar resonance improves the fit substantially, giving strong evidence for the $\sigma(500)$. (See the section on $D^+ \to \pi^-\pi^+\pi^+$ below.) Thus, we are led to try an extra scalar resonance in our fit model here.

This second fit model, Model B, is constructed by the inclusion of an extra scalar state, with unconstrained mass and width. For consistency, the mass and width of the other scalar state, the $K_0^*(1430)$, are also free parameters of the fit. We adopt a better description for these scalar states by introducing gaussian-type form-factors [12] to take into account the finite size of the decaying mesons. Two extra floating parameters are the meson radii r_D and r_R introduced for these meson sizes.

Using this model, we obtain the values of $797 \pm 19 \pm 42$ MeV/c^2 for the mass and $410 \pm 43 \pm 85$ MeV/c^2 for the width of the new scalar state (first error statistical, second error systematic), referred to here as the κ. The values of mass and width obtained for the $K_0^*(1430)$ are respectively $1459 \pm 7 \pm 6$ MeV/c^2 and $175 \pm 12 \pm 12$ MeV/c^2, appearing heavier and narrower than presented by the PDG [8]. The decay fractions and relative phases for Model B, with statistical errors, are given in Table 1. Compared to the results of Model A (without κ), the non-resonant mode drops from over 90% to 13 $\pm$

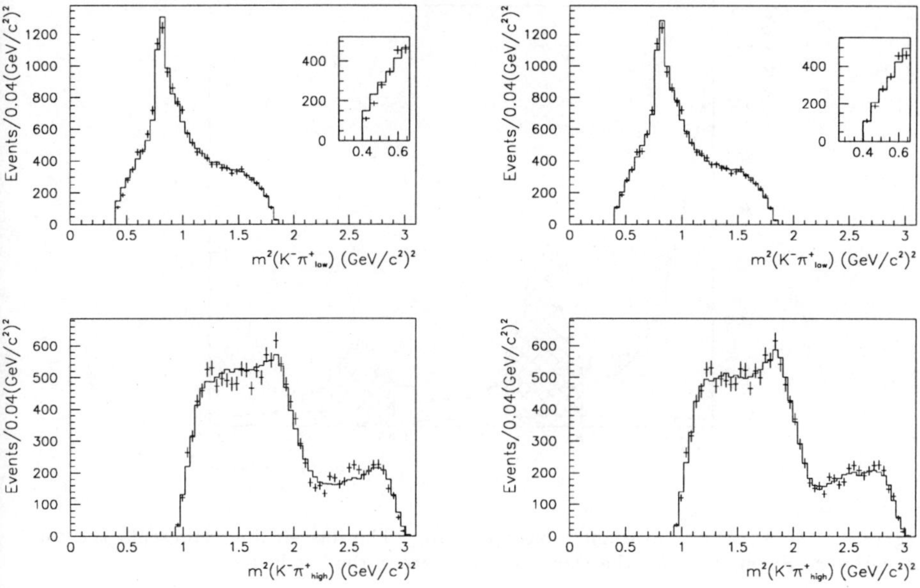

Figure 2. $m^2(K\pi_{\text{low}})$ and $m^2(K\pi_{\text{high}})$ projections for data (error bars) and fast MC (solid line): (a) fit to Model A, without κ, and (b) fit to Model B, with κ.

6%. The $\kappa\pi^+$ state is now the dominant channel with decay fraction about 50%. The meson radii r_D and r_R are found to be respectively 5.0 ± 0.5 GeV^{-1} and 1.6 ± 1.3 GeV^{-1}, in agreement with values used by other groups [13,14].

Moreover, the fit quality of Model B is substantially superior to that of Model A. The χ^2/ν is now 0.73 with a CL of 95%. The very good agreement between the model and the data can be seen in the projections of Figure 2(b). A number of studies were done to check these results. They are described in detail in reference [3].

3 The $D_s^+ \to \pi^-\pi^+\pi^+$ Results

In Figure3 we show the $\pi^-\pi^+\pi^+$ invariant mass distribution after reconstruction and selection criteria[1,2] for the sample collected by E791. Besides combinatorial background, reflections from the decays $D^+ \to K^-\pi^+\pi^+$ and $D^0 \to K^-\pi^+$ (plus one extra π^+ track) and $D_s^+ \to \eta'\pi^+$, $\eta' \to \rho^0(770)\gamma$ are all taken into account. The hatched regions in Figure 3 show the samples

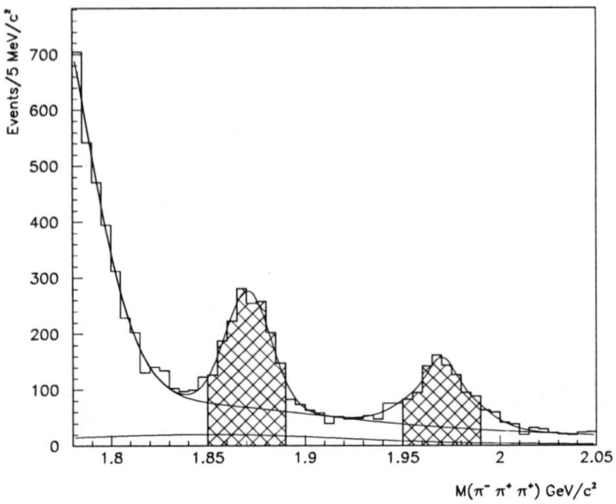

Figure 3. The $\pi^-\pi^+\pi^+$ invariant mass spectrum. The dashed line represents the total background. Events used for the Dalitz analyses are in the hatched areas.

used for the Dalitz-plot analysis. There are 1686 and 937 candidate events for D^+ and D_s^+ respectively, with a signal to background ratio of about 2:1. The Dalitz plots for these events are shown in Figure 4, the axes corresponding to the two $\pi^-\pi^+$ invariant-masses squared.

For the $D_s^+ \to \pi^-\pi^+\pi^+$ events in Figure 4(a), the signal amplitude includes all channels with well-known di-pion resonances [8]: $\rho^0(770)\pi^+$, $f_0(980)\pi^+$, $f_2(1270)\pi^+$, $f_0(1370)\pi^+$, $\rho^0(1450)\pi^+$ and the non-resonant, assumed constant across the Dalitz plot.

For the $f_0(980)\pi^+$ amplitude, instead of a simple Breit-Wigner form, we use a coupled-channel Breit-Wigner function [15],

$$BW_{f_0(980)} = \frac{1}{m_{\pi\pi}^2 - m_0^2 + im_0(\Gamma_\pi + \Gamma_K)}\,, \tag{1}$$

$$\Gamma_\pi = g_\pi\sqrt{m_{\pi\pi}^2/4 - m_\pi^2},\ \Gamma_K = \frac{g_K}{2}\left(\sqrt{m_{\pi\pi}^2/4 - m_{K^+}^2} + \sqrt{m_{\pi\pi}^2/4 - m_{K^0}^2}\right). \tag{2}$$

The $D_s^+ \to \pi^-\pi^+\pi^+$ Dalitz plot is fitted to obtain not only the decay fractions and phases of the possible sub-channels, but also the parameters of the $f_0(980)$

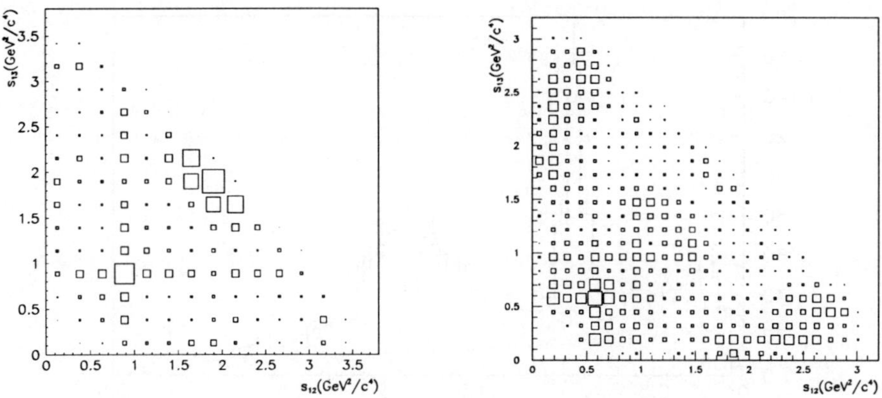

Figure 4. (a) The $D_s^+ \to \pi^-\pi^+\pi^+$ Dalitz plot and (b) the $D^+ \to \pi^-\pi^+\pi^+$ Dalitz plot. Since there are two identical pions, the plots are symmetrized.

Decay Mode	Fraction (%)	Phase
$f_0(980)\pi^+$	$56.5 \pm 4.3 \pm 4.7$	$0°$ (fixed)
NR	$0.5 \pm 1.4 \pm 1.7$	$(181 \pm 94 \pm 51)°$
$\rho^0(770)\pi^+$	$5.8 \pm 2.3 \pm 3.7$	$(109 \pm 24 \pm 5)°$
$f_2(1270)\pi^+$	$19.7 \pm 3.3 \pm 0.6$	$(133 \pm 13 \pm 28)°$
$f_0(1370)\pi^+$	$32.4 \pm 7.7 \pm 1.9$	$(198 \pm 19 \pm 27)°$
$\rho^0(1450)\pi^+$	$4.4 \pm 2.1 \pm 0.2$	$(162 \pm 26 \pm 17)°$

Table 2. Dalitz fit results for $D_s^+ \to \pi^-\pi^+\pi^+$.

state, g_π, g_K, and m_0, as well as the mass and width of the $f_0(1370)$. The other resonance masses and widths are taken from the PDG[8]. The resulting fractions and phases are given in Table 2. The measured $f_0(980)$ parameters are $m_0 = 977 \pm 3 \pm 2$ MeV/c^2, $g_\pi = 0.09 \pm 0.01 \pm 0.01$ and $g_K = 0.02 \pm 0.04 \pm 0.03$. By fitting the Dalitz plot using a simple Breit-Wigner function, for the $f_0(980)$ we find $m_0 = 975 \pm 3$ MeV/c^2 and $\Gamma_0 = 44 \pm 2 \pm 2$ MeV/c^2, and the results for fractions and phases are indistinguishable.

The confidence level of the fit for $D_s^+ \to \pi^-\pi^+\pi^+$ is 35%. In Figure 5 we show the $\pi^-\pi^+$ mass-squared projections for data (points) and model (solid lines, from fast-MC).

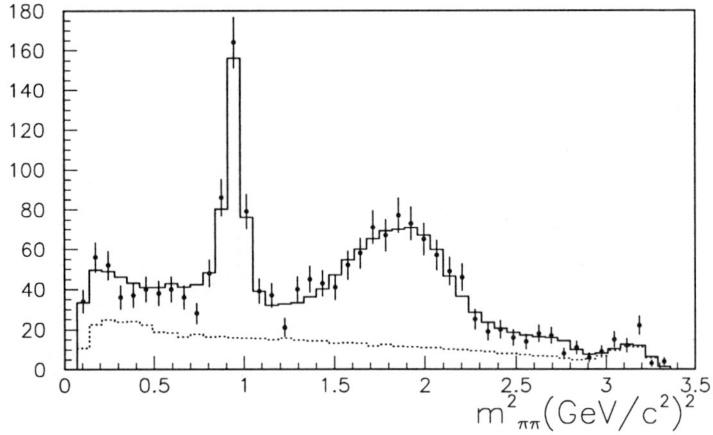

Figure 5. s_{12} and s_{13} ($m^2_{\pi\pi}$) projections for $D_s^+ \to \pi^-\pi^+\pi^+$ data (dots) and our best fit (solid). The area under the dashed curve corresponds to background.

4 The $D^+ \to \pi^-\pi^+\pi^+$ Results

In a first approach, we try to fit the $D^+ \to \pi^-\pi^+\pi^+$ Dalitz plot of Figure 4(b) with the same amplitudes used for the $D_s^+ \to \pi^-\pi^+\pi^+$ analysis. Using this model, the non-resonant, the $\rho^0(1450)\pi^+$, and the $\rho^0(770)\pi^+$ amplitudes are found to dominate, as shown in Table 3, and in agreement with previous reported analysis[16,17]. However, this model does not describe the data satisfactorily, especially at low $\pi^-\pi^+$ mass squared, as can be seen from Figure 6(a). The χ^2/ν obtained from the binned Dalitz plot for this model is 1.6, with a CL less than 10^{-5}.

To investigate the possibility that another $\pi^-\pi^+$ resonance contributes to the $D^+ \to \pi^-\pi^+\pi^+$ decay, we add an extra scalar resonance amplitude to the signal PDF, with mass and width as floating parameters in the fit.

We find that this model improves our fit substantially. The mass and the width of the extra scalar state are found to be $478^{+24}_{-23} \pm 17$ MeV/c^2 and $324^{+42}_{-40} \pm 21$ MeV/c^2, respectively. Referring to this state as the $\sigma(500)$, we obtain that $\sigma(500)\pi^+$ channel produces the largest decay fraction, as shown in Table 3. The non-resonant amplitude, which is dominant in the model

Decay Mode	Fit without $\sigma(500)\pi^+$ Fraction (%)	Phase	Fit with $\sigma(500)\pi^+$ Fraction (%)	Phase
$\sigma(500)\pi^+$	$--$	$--$	$46.3 \pm 9.0 \pm 2.1$	$(206 \pm 8 \pm 5)°$
$\rho^0(770)\pi^+$	20.8 ± 2.4	$0°$ (fixed)	$33.6 \pm 3.2 \pm 2.2$	$0°$ (fixed)
NR	38.6 ± 9.7	$(150 \pm 12)°$	$7.8 \pm 6.0 \pm 2.7$	$(57 \pm 20 \pm 6)°$
$f_0(980)\pi^+$	7.4 ± 1.4	$(152 \pm 16)°$	$6.2 \pm 1.3 \pm 0.4$	$(165 \pm 11 \pm 3)°$
$f_2(1270)\pi^+$	6.3 ± 1.9	$(103 \pm 16)°$	$19.4 \pm 2.5 \pm 0.4$	$(57 \pm 8 \pm 3)°$
$f_0(1370)\pi^+$	10.7 ± 3.1	$(143 \pm 10)°$	$2.3 \pm 1.5 \pm 0.8$	$(105 \pm 18 \pm 1)°$
$\rho^0(1450)\pi^+$	22.6 ± 3.7	$(46 \pm 15)°$	$0.7 \pm 0.7 \pm 0.3$	$(319 \pm 39 \pm 11)°$

Table 3. Dalitz fit results for $D^+ \to \pi^-\pi^+\pi^+$. First errors are statistical, second systematics (only for fit with $\sigma(500)\pi^+$ mode).

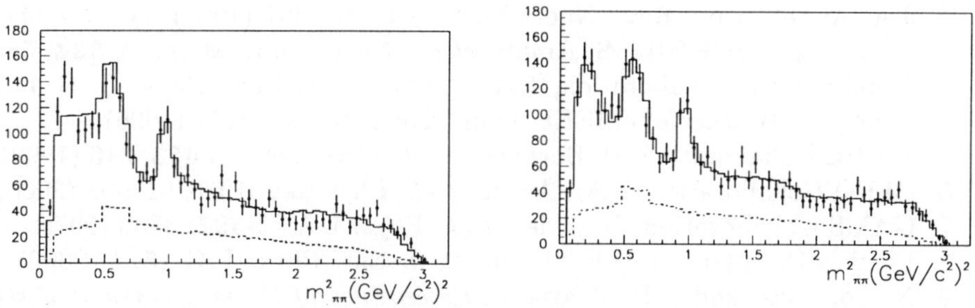

Figure 6. s_{12} and s_{13} ($m^2_{\pi\pi}$) projections for $D^+ \to \pi^-\pi^+\pi^+$ data (dots) and our best fit (solid) for models (a) without and (b) with $\sigma(500)\pi^+$ amplitude. The dashed distributions correspond to the expected background levels.

without $\sigma(500)\pi^+$, drops substantially. This model describes the data much better, as can be seen by the $\pi\pi$ mass squared projection in Figure 6(b). The χ^2/ν is now 0.9, with a corresponding confidence level of 91%.

5 Conclusion

From the data of the Fermilab E791 experiment, we studied the Dalitz plots of the decays $D^+ \to K^-\pi^+\pi^+$, $D_s^+ \to \pi^-\pi^+\pi^+$ and $D^+ \to \pi^-\pi^+\pi^+$. In these three final states, the scalar intermediate resonances were found to give the main contribution to the decay rates. We obtained strong evidence for the existence of $\sigma(500)$ and κ scalar mesons, measuring their masses and widths. We also obtained new measurements for masses and widths of the other scalars studied, $f_0(980)$, $f_0(1430)$ and $K_0^*(1430)$.

The results presented here show the potential of D meson decays for the study of light meson spectroscopy, in particular in the scalar sector.

References

1. E791 Collaboration, E.M. Aitala *et al.*, Phys. Rev. Lett. **86**, 765 (2001).
2. E791 Collaboration, E.M. Aitala *et al.*, Phys. Rev. Lett. **86**, 770 (2001).
3. C. Göbel, in *Proceedings of Heavy Quarks at Fixed Target* , Frascati Physics Series Vol. XX (Laboratory Nazionali de Frascati, Frascati (Roma), Italy, 2001), p. 373, edited by I.Bediaga, J. Miranda and A. Reis, hep-ex/0012009. And E791 Collaboration, E.M. Aitala *et al.* hep-ex/0204018, submitted to Phys. Rev. Lett.
4. J.A. Appel, Ann. Rev. Nucl. Part. Sci. **42**, 367 (1992); D. Summers *et al.*, hep-ex/0009015; S. Amato *et al.*, Nucl. Instr. Meth. A **324**, 535 (1993); E791 Collaboration, E.M. Aitala *et al.*, Eur. Phys. J. C **4**, 1 (1999). S. Bracker *et al* IEEE Trans. Nucl. Sci **43**, 2457 (1996).
5. WA102 Collaboration, D. Barberis, *et al.* Phys. Lett. B **453**, 316 (1999).
6. CLEO Collaboration, D.M. Asner *et al.*, Phys Rev D **61**, 012002 (2000)
7. GAMS Collaboration, D. Alde, *et al.*, Phys. Lett. B**397**, 350 (1997).
8. Particle Data Group, D.E. Groom *et al.* Eur. Phys. J. C **15**, 1 (2000).
9. N. Törnqvist and A.D. Polosa, in *Proceedings of Heavy Quarks at Fixed Target*, Frascati Physics Series Vol. XX (Frascati National Laboratory, Frascati (Roma), Italy, 2001), p. 385 edited by I.Bediaga, J. Miranda and A. Reis, hep-ph/0011107. And F.E. Close and N. Törnqvist, hep-ph/0204205.
10. E691 Collaboration, J.C. Anjos *et al.*, Phys. Rev. D **48**, 56 (1993).
11. E687 Collaboration, P.L. Frabetti *et al.*, Phys. Lett. B **331**, 217 (1994).
12. N.A. Törnqvist, Z. Phys. C **68**, 647 (1995).
13. ARGUS Collaboration, H. Albrecht *et al.*, Phys. Lett. B **308**, 435(1993).
14. CLEO Collaboration, S. Kopp *et al.*, Phys. Rev. D **63**, 092001 (2001).
15. WA76 Collaboration, T.A. Armstrong *et al.*, Z. Phys. C **51**, 351 (1991).
16. E691 Collaboration, J.C. Anjos *et al.*, Phys. Rev. Lett. **62**, 125 (1989).
17. E687 Collaboration, P.L. Frabetti *et al.*, Phys. Lett. B **407**, 79 (1997).

DIRECT URCA PROCESSES IN SUPERDENSE CORES OF NEUTRON STARS

L. B. LEINSON
Institute of Terrestrial Magnetism, Ionosphere and Radio Wave Propagation RAS, 142190 Troitsk, Moscow Region, Russia
leinson@izmiran.rssi.ru

The neutrino emission due to formation and breaking of Cooper pairs of protons in super-conducting cores of neutron stars is considered with allowing for the electromagnetic coupling of protons to ambient electrons. Collective response of electrons to the proton quantum transition contributes coherently to the complete interaction with a neutrino field and enhances the neutrino-pair production. Our calculation shows that the contribution of the vector weak current to the neutrino-pair emissivity of protons is much larger than that calculated without taking into account the plasma effects. Partial contribution of the pairing protons to the total neutrino radiation from the neutron star core is very sensitive to the critical temperatures for the proton and neutron pairing. We show domains of these parameters where the neutrino radiation, caused by a singlet-state pairing of protons is dominating.

The simplest neutrino-emitting processes by which neutron stars can loose their thermal energy are beta decay of neutrons and electron capture on protons $n \to p + e^- + \bar{\nu}_l, \quad p + e^- \to n + \nu_l$. These reactions, widely known as the direct Urca processes, are a central point of any modern scenarios of evolution of neutron stars because the neutrino energy losses caused by the direct Urca processes lead to a rapid cooling of the degenerate neutron star core. In the non-relativistic approximation, the energy losses caused by the direct Urca processes have been calculated more than ten years ago[1]

$$Q_{nr} = \frac{457\pi}{10080} G_F^2 C^2 \left(C_V^2 + 3C_A^2\right) T^6 M_n^* M_p^* \mu_e \Theta\left(p_e + p_{\rm p} - p_{\rm n}\right) . \qquad (1)$$

Here G_F is the Fermi weak coupling constant; $C = \cos\theta_C = 0.973$ is the Cabibbo factor; $C_V = 1$ and $C_A \simeq 1.26$ are the nucleon weak coupling constants; M_n^* and M_p^* are the effective nucleon masses, and μ_e is the electron chemical potential; $\Theta(x) = 1$ if $x \geq 0$ and zero otherwise.

The energy exchange in the matter goes naturally on the temperature scale $\sim T$, which is small compared to typical kinetic energies of degenerate particles. Therefore the momenta of in-medium fermions are fixed at their values at Fermi surfaces, which we denote as $p_{\rm n}$, $p_{\rm p}$ for the nucleons and p_e for electrons respectively. Then the "triangle" condition, $p_{\rm p} + p_e > p_{\rm n}$, required by the step-function, is necessary for conservation of the total momentum in the reaction and, together with the condition of charge neutrality of the medium, results in the threshold dependence on the proton concentration.

In spite of widely adopted importance of the direct Urca reactions, the corresponding neutrino energy losses are not well investigated yet. The simple formula (1) has been derived in the non-relativistic manner by neglecting all interactions between nucleons. Actually, however, the super-threshold proton fraction, necessary for the direct Urca processes to operate in the degenerate nuclear matter, appears at large densities, where the Fermi momenta of participating nucleons are comparable with their effective mass. Moreover, according to modern numerical simulations, the central density of the star can be up to eight times larger than the nuclear saturation density. This implies a substantially relativistic motion of nucleons in the super-dense neutron star core. The equation of state for such a matter is usually derived in the relativistic approach, and the relevant neutrino energy losses must be consistent with the used relativistic equation of state.

I propose the totally relativistic consideration of the direct Urca processes on nucleons. Let us consider first the relativistic kinematics of the neutron beta decay $n \to p + l + \bar{\nu}_l$ in the degenerate matter under beta equilibrium. In what follows we consider massless neutrinos of energy and momentum $k_1 = (\omega_1, \mathbf{k}_1)$ with $\omega_1 = |\mathbf{k}_1|$. The energy-momentum of the final lepton $l = e^-, \mu^-$ of mass m_l is denoted as $k_2 = (\omega_2, \mathbf{k}_2)$ with $\omega_2 = \sqrt{\mathbf{k}_2^2 + m_l^2}$. Thus, the energy and momentum conservation in the beta decay is given by the following equations

$$\begin{aligned} E_{\mathrm{n}}(\mathbf{p}) - E_{\mathrm{p}}(\mathbf{p}') - \omega_1 - \omega_2 = 0 \\ \mathbf{p} - \mathbf{p}' - \mathbf{k}_1 - \mathbf{k}_2 = 0\,, \end{aligned} \tag{2}$$

where $E_{\mathrm{n}}(\mathbf{p})$ and $E_{\mathrm{p}}(\mathbf{p}')$ are the in-medium energies of the neutron and the proton respectively.

Since the anti-neutrino energy is $\omega_1 \sim T$, and the anti-neutrino momentum $|\mathbf{k}_1| \sim T$ is much smaller than the momenta of other particles, we can neglect the neutrino contributions in Eqs. (2). Then the momentum conservation, $\mathbf{p}_{\mathrm{p}} + \mathbf{p}_l = \mathbf{p}_{\mathrm{n}}$, implies the well known "triangle" condition, $p_{\mathrm{p}} + p_l > p_{\mathrm{n}}$, necessary for the direct Urca processes to operate. However, the momentum conservation is necessary but insufficient condition for the direct beta decay to occur. The energy conservation requires a one more condition

$$E_{\mathrm{n}}(p_{\mathrm{n}}) - E_{\mathrm{p}}(p_{\mathrm{p}}) = \sqrt{(\mathbf{p}_{\mathrm{n}} - \mathbf{p}_{\mathrm{p}})^2 + m_l^2}\,. \tag{3}$$

Squaring of both sides of this equation gives

$$(E_{\mathrm{n}}(p_{\mathrm{n}}) - E_{\mathrm{p}}(p_{\mathrm{p}}))^2 - (\mathbf{p}_{\mathrm{n}} - \mathbf{p}_{\mathrm{p}})^2 = m_l^2. \tag{4}$$

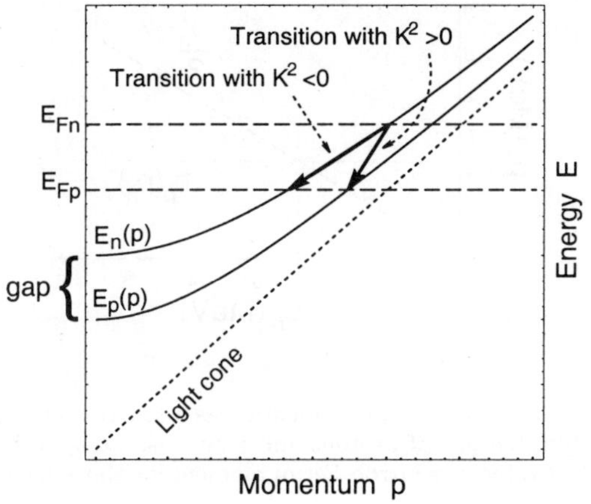

Figure 1. The nucleon transition caused by the beta decay. If the neutron and proton energy spectra are identical (the upper curve), the momentum transfer due to such a transition is space-like. The energy gap between protons and neutrons allows the time-like momentum transfer to the leptons.

If we denote as $K = (E_{\rm n} - E_{\rm p}, \mathbf{p}_{\rm n} - \mathbf{p}_{\rm p})$, the energy-momentum transferred from the nucleon, this equation can be readily recognized as the following condition $K_\mu^2 = m_l^2$. This condition naturally arises from the time-like momentum of the final lepton pair, $K = k_1 + k_2 \simeq k_2$, and can be satisfied only in the presence of the energy gap between spectra of protons and neutrons. Really, consider the neutron decay in the degenerate matter. In this case the neutron from the neutron Fermi surface E_{Fn} undergoes transition into the proton at the proton Fermi energy E_{Fp}, as it is shown in Fig.1. If the neutron and proton energy spectra are identical (the upper curve, for example), the momentum transfer due to such a transition is space-like contrary to the time-like momentum of the final lepton pair. In this case the neutron decay is kinematically forbidden. In contrast, the energy gap between protons and neutrons allows the time-like momentum transfer to the leptons, and thus opens the direct neutron decay. Therefore the possibility of the direct neutron decay depends substantially on the model of nuclear matter.

Consider, for example, a degenerate free gas consisting of neutrons, protons, and electrons under beta equilibrium. In this case the energy gap exists only due to the mass difference of the neutron and the proton. If we denote the masses as $M_{\rm n}$ and $M_{\rm p}$ respectively, then the corresponding Fermi energies

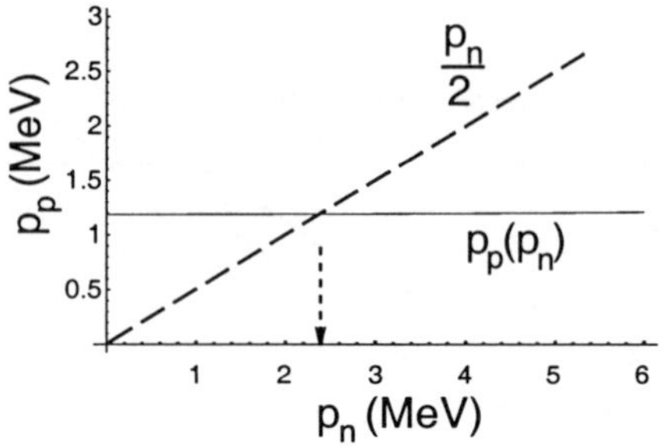

Figure 2. Solid line is the proton Fermi momentum as a function of neutron Fermi momentum in the beta-equilibrated gas of protons and neutrons. Long-dashed line is $p_{\rm n}/2$. The arrow shows the critical value of neutron Fermi momentum above which the direct neutron decay is forbidden.

are of the following relativistic form

$$E_{\rm n}\left(p_{\rm n}\right)=\sqrt{M_{\rm n}^2+p_{\rm n}^2}, \qquad E_{\rm p}\left(p_{\rm p}\right)=\sqrt{M_{\rm p}^2+p_{\rm p}^2}.$$

Due to charge neutrality the number density of electrons, $n_e \propto p_e^3$, equals to the number density of protons, $n_{\rm p} \propto p_{\rm p}^3$. This implies equality of the proton and electron Fermi momenta, $p_{\rm e}=p_{\rm p}$, so that the "triangle" condition reads $2p_{\rm p} \geq p_{\rm n}$. Neutrinos and anti-neutrinos freely escape from the neutron star. Therefore the chemical potential of neutrinos equals zero, $\mu_\nu = 0$, and the condition of chemical equilibrium can be written as $\mu_p + \mu_e = \mu_n$. Chemical potentials of degenerate particles can be approximated by their individual Fermi energies, yielding

$$\sqrt{m_{\rm e}^2+p_{\rm p}^2}+\sqrt{M_{\rm p}^2+p_{\rm p}^2}=\sqrt{M_{\rm n}^2+p_{\rm n}^2}. \tag{5}$$

Solution of this equation

$$p_{\rm p}\left(p_{\rm n}\right)=\frac{1}{2\sqrt{M_{\rm n}^2+p_{\rm n}^2}}\sqrt{\left(p_{\rm n}^2+M_{\rm n}^2-\left(M_{\rm p}+m_{\rm e}\right)^2\right)\left(p_{\rm n}^2+M_{\rm n}^2-\left(M_{\rm p}-m_{\rm e}\right)^2\right)}, \tag{6}$$

gives the proton Fermi momentum as a function of neutron Fermi momentum in the beta-equilibrated gas of protons and neutrons. This function is shown by solid line. As required by the "triangle" condition, $2p_{\rm p} \geq p_{\rm n}$, the direct neutron decay in such a medium is open when the proton Fermi momentum, $p_{\rm p}$, is larger than one half of the neutron Fermi momentum, $p_{\rm n}/2$. In Fig.2, this function is shown by the long-dashed line. We can see, the direct neutron decay is allowed to the left of the crossing point and is forbidden at the right-hand side. The crossing point can be found as solution of the following equation

$$p_{\mathrm{p}}\left(p_{\mathrm{n}}\right)=\frac{p_{\mathrm{n}}}{2}\,. \tag{7}$$

That is the critical value of the neutron Fermi momentum above which the direct neutron decay is forbidden

$$p_{\mathrm{n}}^{c}=\frac{\sqrt{\left(\left(M_p+M_n\right)^2-m_l^2\right)\left(\left(M_n-M_p\right)^2-m_l^2\right)}}{\sqrt{2M_p^2+2m_l^2-M_n^2}}=2.381\,1\,MeV\,. \tag{8}$$

Thus the direct neutron decay is forbidden if the number density of neutrons exceeds the following critical value

$$n_n^c=\frac{\left(p_{\mathrm{n}}^c\right)^3}{3\pi^2}=5.932\times 10^{31}\,cm^{-3}\,. \tag{9}$$

This number density is much smaller than that typical for neutron star cores, therefore, in the cooling neutron stars, the direct neutron decay can be kinematically allowed only due to strong interactions.

Notice, some of models of strong interaction also forbid the direct neutron decay. Consider, for example, a simple model for the baryon matter, which contains fields for baryons and neutral scalar (σ) and vector (ω_μ) mesons. The isoscalar mesons equally interact with protons and neutrons, which therefore have identical energy spectra. As discussed, in this case, the energy and the momentum in the direct neutron decay cannot be conserved simultaneously.

To allow the direct neutron decay, the model of nuclear matter must be generalized to include some additional degrees of freedom and couplings, which are able to create a large energy gap between possible energies of protons and neutrons. For this purpose, besides isoscalar mesons σ and ω, the model should include also isovector mesons. The isovector meson couples differently to protons and neutrons, thus creating the energy gap necessary for the direct neutron decay.

In the lowest order in the Fermi weak coupling constant G_F, the matrix element of the neutron beta decay is found to be

$$\mathcal{M}_{fi}=-i\frac{G_F C}{\sqrt{2}}\bar{u}_l\left(k_2\right)\gamma_\mu\left(1+\gamma_5\right)\nu\left(-k_1\right)\,\times \tag{10}$$

$$\times\bar{u}_{\mathrm{p}}\left(P'\right)\left[C_V\gamma^\mu+\frac{1}{2M}C_M\sigma^{\mu\nu}q_\nu+C_A\left(\gamma^\mu\gamma_5+F_q\,q^\mu\gamma_5\right)\right]u_{\mathrm{n}}\left(P\right)\,.$$

In contrast to the non-relativistic expression, this matrix element includes also the terms caused by weak magnetism and pseudoscalar interaction. In the mean field approximation, we assume $C_V = 1$, $C_M = \lambda_p - \lambda_n \simeq 3.7$, $C_A = 1.26$, and

$$F_q = -\frac{2M^*}{(m_\pi^2 - q^2)}\,. \tag{11}$$

In the above λ_p and λ_n are the anomalous magnetic moments of the proton and the neutron respectively. Note that the matrix element (10) is of the same form as that for the neutron decay in a free space, but with the total momentum transfer replaced with the kinetic momentum transfer $q = (\varepsilon - \varepsilon, \mathbf{p} - \mathbf{p})$.

We consider the total energy which is emitted into neutrino and antineutrino per unit volume and time. Within beta equilibrium, the inverse reaction $p + l \to n + \nu_l$ corresponding to a capture of the lepton l, gives the same emissivity as the beta decay, but in neutrinos. Thus, the total energy losses Q for the Urca processes is twice more than that caused by the beta decay. Taking this into account by Fermi's "golden" rule we find [4]:

$$\begin{aligned}
Q = & \frac{457\pi}{10\,080} G_F^2 C^2 T^6 \Theta\left(p_l + p_\mathrm{p} - p_\mathrm{n}\right) \left\{ \left(C_A^2 - C_V^2\right) M^{*2} \mu_l \right. \\
& + \frac{1}{2}\left(C_V^2 + C_A^2\right)\left[4\varepsilon_\mathrm{n}\varepsilon_\mathrm{p}\mu_l - \left(\varepsilon_\mathrm{n} - \varepsilon_\mathrm{p}\right)\left(\left(\varepsilon_\mathrm{n} + \varepsilon_\mathrm{p}\right)^2 - p_l^2\right)\right] \\
& + C_V C_M \frac{M^*}{M}\left[2\left(\varepsilon_\mathrm{n} - \varepsilon_\mathrm{p}\right) p_l^2 - \left(3\left(\varepsilon_\mathrm{n} - \varepsilon_\mathrm{p}\right)^2 - p_l^2\right)\mu_l\right] \\
& + C_A\left(C_V + 2\frac{M^*}{M} C_M\right)\left(\varepsilon_\mathrm{n} + \varepsilon_\mathrm{p}\right)\left(p_l^2 - \left(\varepsilon_\mathrm{n} - \varepsilon_\mathrm{p}\right)^2\right) \\
& + C_M^2 \frac{1}{4M^2}\left[8M^{*2}\left(\varepsilon_\mathrm{n} - \varepsilon_\mathrm{p}\right)\left(p_l^2 - \left(\varepsilon_\mathrm{n} - \varepsilon_\mathrm{p}\right)\mu_l\right)\right. \\
& + \left(p_l^2 - \left(\varepsilon_\mathrm{n} - \varepsilon_\mathrm{p}\right)^2\right)\left(2\varepsilon_\mathrm{n}^2 + 2\varepsilon_\mathrm{p}^2 - p_l^2\right)\mu_l \\
& \left. - \left(p_l^2 - \left(\varepsilon_\mathrm{n} - \varepsilon_\mathrm{p}\right)^2\right)\left(\varepsilon_\mathrm{n} + \varepsilon_\mathrm{p}\right)^2\left(2\varepsilon_\mathrm{n} - 2\varepsilon_\mathrm{p} - \mu_l\right)\right] \\
& \left. - C_A^2 M^{*2}\Phi\left(1 + m_\pi^2\Phi\right)\left[\mu_l\left(\left(\varepsilon_\mathrm{n} - \varepsilon_\mathrm{p}\right)^2 + p_l^2\right) - 2\left(\varepsilon_\mathrm{n} - \varepsilon_\mathrm{p}\right) p_l^2\right]\right\} .
\end{aligned} \tag{12}$$

In the above, ε_n and ε_p are the kinetic Fermi energies of neutrons and protons respectively, and the last term, with

$$\Phi = \frac{1}{m_\pi^2 + p_l^2 - \left(\varepsilon_\mathrm{n} - \varepsilon_\mathrm{p}\right)^2}, \tag{13}$$

represents the contribution of the pseudoscalar interaction.

In the following we consider a self-consistent relativistic model of nuclear matter in which baryons, $B = n, p, \Sigma^-, \Sigma^0, \Sigma^+, \Lambda$, interact via exchange of σ, ω, and ρ mesons [2]. The Lagrangian density,

$$\mathcal{L} = \sum_B \bar{B} \left[\gamma_\mu \left(i\partial^\mu - g_{\omega B}\omega^\mu - \frac{1}{2} g_{\rho B} \mathbf{b}^\mu \cdot \tau \right) - (M_B - g_{\sigma B}\sigma) \right] B$$
$$- \frac{1}{4} F_{\mu\nu} F^{\mu\nu} + \frac{1}{2} m_\omega^2 \omega_\mu \omega^\mu - \frac{1}{4} \mathbf{B}_{\mu\nu} \mathbf{B}^{\mu\nu} + \frac{1}{2} m_\rho^2 \mathbf{b}_\mu \mathbf{b}^\mu$$
$$+ \frac{1}{2} \left(\partial_\mu \sigma \partial^\mu \sigma - m_\sigma^2 \sigma^2 \right) - U(\sigma) + \bar{l} \left(i\gamma_\mu \partial^\mu - m_l \right) l, \qquad (14)$$

includes the interaction of baryon fields B with a scalar field σ, a vector field ω_μ and an isovector field $\mathbf{b}_\mu$ of ρ-meson. In the above, B are the Dirac spinor fields for baryons, $\mathbf{b}_\mu$ is the isovector field of ρ-meson. We denote as τ the isospin operator, which acts on the baryons of the bare mass M_B. The leptons are represented only by electrons and muons, $l = e^-, \mu^-$, which are included in the model as non-interacting particles. The field strength tensors for the ω and ρ mesons are $F_{\mu\nu} = \partial_\mu \omega_\nu - \partial_\nu \omega_\mu$ and $\mathbf{B}_{\mu\nu} = \partial_\mu \mathbf{b}_\nu - \partial_\nu \mathbf{b}_\mu$, respectively. The potential $U(\sigma)$ represents the self-interactions of the scalar field and is taken to be of the form

$$U(\sigma) = \frac{1}{3} bM \left(g_{\sigma N} \sigma \right)^3 + \frac{1}{4} c \left(g_{\sigma N} \sigma \right)^4 . \qquad (15)$$

The parameters of the model are chosen as suggested in Ref. [3] to reproduce the nuclear matter equilibrium density, the binding energy per nucleon, the symmetry energy, the compression modulus, and the nucleon effective mass at saturation density $n_0 = 0.16\ fm^{-3}$.

In the mean field approximation, when the contribution of mesons reduce to classical condensate fields $\langle \sigma \rangle = \sigma_0$, $\langle \omega^\mu \rangle = \omega_0 \delta^{\mu 0}$, $\langle \mathbf{b}^\mu \rangle = (0, 0, \rho_0)\, \delta^{\mu 0}$, only the baryon fields must be quantized. This procedure yields the following linear Dirac equation for the nucleon

$$\left(i\partial_\mu \gamma^\mu - g_\omega \gamma^0 \omega_0 - \frac{1}{2} g_\rho \gamma^0 \rho_0 \tau_3 - (M - g_\sigma \sigma_0) \right) \Psi(x) = 0 . \qquad (16)$$

The stationary and uniform condensate fields equally shift the effective masses

$$M^* = M - g_\sigma \sigma_0 \qquad (17)$$

but lead to different potential energies of protons and neutrons

$$U_\mathrm{n} = g_\omega \omega_0 - \frac{1}{2} g_\rho \rho_0, \quad U_\mathrm{p} = g_\omega \omega_0 + \frac{1}{2} g_\rho \rho_0, \qquad (18)$$

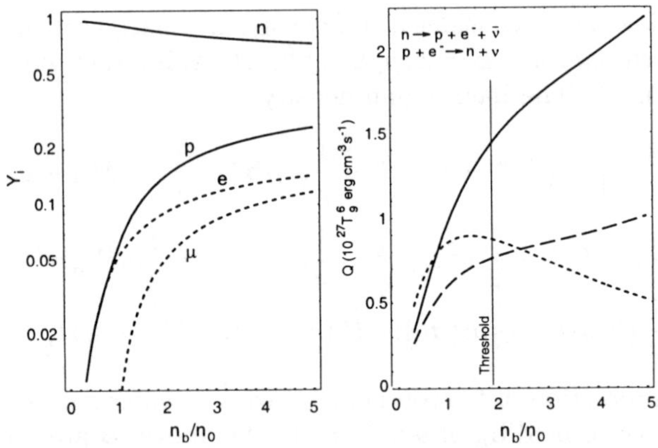

Figure 3. The left panel shows the composition of neutrino-free matter in beta equilibrium among nucleons, electrons and muons. The fractions are shown versus the baryon number density, in units of saturation density. Right panel represents the emissivity of the direct Urca processes, as given by our formula (solid curve) in comparison with the emissivity calculated without the effects of weak magnetism and pseudo scalar interaction (long-dashed curve). The short-dashed curve is the non-relativistic emissivity. All the emissivities are given in units $10^{27}T_9^6\ erg\,cm^{-3}s^{-1}$, where the temperature $T_9 = T/10^9\,K$.

thus creating the energy gap $U_{\rm n} - U_{\rm p} = -g_\rho \rho_0$ between possible energies of protons and neutrons.

Neutrino energy losses caused by the direct Urca on nucleons depend essentially on the composition of beta-stable nuclear matter. In order to examine numerically the relativistic effects in the direct Urca processes, we consider first a simplified model for degenerate nuclear matter of the standard composition consisting on neutrons, protons, electrons, and muons under beta-equilibrium. The left panel of Figure 3 shows the composition of neutrino-free matter in beta equilibrium among nucleons, electrons and muons. The fractions of the constituents are shown versus the baryon number density, in units of saturation density. The right panel represents the emissivity of the direct Urca process, as given by our formula in comparison with the emissivity calculated without the effects of weak magnetism and pseudo scalar interaction. The short-dashed curve is the non-relativistic emissivity. All the emissivities are given in units $10^{27}T_9^6\ erg\,cm^{-3}s^{-1}$, where the temperature $T_9 = T/10^9\,K$. Thus the curves demonstrate how the emissivity varies along with the matter density.

We can observe that the relativistic effects dramatically modify the emis-

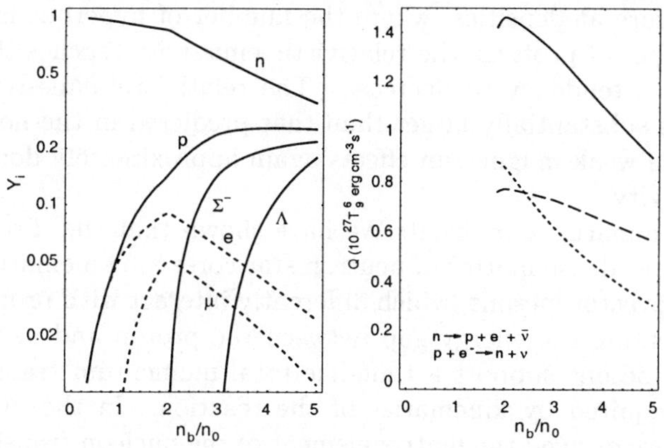

Figure 4. The same as in Fig. 3 but for the matter composition including hyperons.

sivity. The non-relativistic approximation approaches our result only at densities much smaller than the threshold density, indicated by the vertical line. Due to the decrease of the nucleon effective mass, the non-relativistic formula predicts a decreasing of the emissivity as density increases above the threshold. In contrast, our formula shows a substantial increase of the neutrino energy losses as we go to larger densities. The long-dashed curve demonstrates the energy losses obtained from the relativistic formula by formally setting C_M equal zero and Φ equal zero. This eliminates the weak magnetism and pseudo scalar contributions. A comparison of the long-dashed and solid curve demonstrates that the weak magnetism effects approximately double the neutrino energy losses.

To examine the dependence of energy losses on the matter composition we consider the model, which includes nucleon and hyperon degrees of freedom. The composition of neutrino-free matter in beta equilibrium among nucleons, hyperons, electrons and muons is shown in Figure 4 versus the baryon number density (left panel). The right panel represents the emissivity of the direct Urca processes, as given by our formula in comparison with the emissivity calculated without the effects of weak magnetism and pseudo scalar. The short dashed curve is again the non-relativistic emissivity. We see that the neutrino energy losses caused by the direct Urca processes on nucleons depend essentially on the composition of beta-stable nuclear matter. Appearance of hyperons in the system suppresses the nucleon fractions and lepton abundance. Therefore at densities, where the number of hyperons is comparable with the number of protons, the relativistic emissivity reaches the maximum and then has a tendency to decrease. The relativistic emissivity, however, is found to be substantially larger than that predicted in the non-relativistic approach. The weak magnetism effects again approximately double the relativistic emissivity.

Let us summarize our results.We have shown that the direct Urca processes in a super-dense matter of neutron star cores are kinematically allowed only due to isovector mesons, which differently interact with protons and neutrons. By creating the energy gap between the proton and neutron spectra the isovector mesons support a time-like total momentum transfer from the nucleon, as required by kinematics of the reaction. In the mean field approximation, we derived the matrix element of the nucleon transition current, which is found to be a function of the space-like kinetic momentum transfer. In the mean field approximation, we have calculated the neutrino energy losses caused by the direct Urca processes on nucleons. Our Equation (12) for neutrino energy losses exactly incorporates the effects of nucleon recoil, parity violation, weak magnetism, and pseudoscalar interaction. To quantify the relativistic effects we consider a self-consistent relativistic model, widely used in the theory of relativistic nuclear matter. The relativistic energy losses are up to four times larger than those obtained in the non-relativistic approach. In our analysis, we pay special attention to the effects of weak magnetism and pseudoscalar interaction in the neutrino energy losses. We found that weak magnetism approximately doubles the neutrino emissivity. The efficiency of the direct Urca processes involving different kinds of baryons depends essentially on the composition of the beta-stable nuclear matter. The energy losses in the standard nuclear matter increase along with the density. In contrast, the total energy losses in the multi-component medium have a tendency to decrease.

References

1. J. M. Lattimer, C.J. Pethick, M. Prakash, and P. Haensel, *Phys. Rev. Lett.* **66** (1991) 2701.
2. B. D. Serot and J. D. Walecka, *Adv. Nucl. Phys.* **19**, eds. J.W. Negele and E. Vogt, (Plenum, New York, 1986); B.D. Serot, *Rep. Prog. Phys.* **55** (1992) 1855.
3. N. K.Glendenning and S. A. Moszkowski, *Phys. Rev. Lett.* **67** (1991) 2414.
4. L. B. Leinson, *Phys. Lett.* **B 532** (2002) 267;

DIFFRACTIVE PHYSICS AT HERA

L. FAVART
Chercheur Qualifié FNRS
Université Libre de Bruxelles, 1050 Brussels, Belgium
lfavart@ulb.ac.be

A great interest for diffraction has been generated these last years by the HERA data. They give us the first opportunity to understand high energy diffractive physics in terms of a fundamental theory, i.e. QCD. A review of the main results from H1 and ZEUS in this field is presented and the relation with the proton structure function at small x-Bjorken values is discussed.

Introduction

Presently, one of the most important tasks in particle physics is the understanding of the strong force. For this purpose, the Quantum Chromodynamics theory (QCD) seems to be the best candidate. An important characteristic of this theory is that the coupling constant α_S is larger than 1 when the interaction between the different actors, quarks and gluons, does not take place at distances small enough. This means that to be calculable within a perturbative approach, the interaction between these constituents requires the presence in the process of a "hard" scale, i.e. a large transverse momentum or a large mass.

The lepton beam of the high energy $e-p$ collider HERA (reaching an energy in the center of mass system $\sqrt{s} = 300$ GeV) is a prolific source of photons in a large virtuality range such that the study of $\gamma^* p$ interactions provides a completely new and deep insight into the QCD dynamics.

A major discovery at HERA is the observation of the strong rise of the total cross section at high energy in the deep inelastic scattering (DIS) i.e. $\gamma^* p$ interactions with large Q^2 values (Q^2 being the negative of the squared four-momentum of the exchanged photon). This is inconsistent with the case of the photoproduction ($Q^2 \simeq 0$), which shows a soft dependence in the total hadronic energy, W, similar to the hadron-hadron interaction case and well described by the Regge phenomenological theory.

HERA is thus a unique device to test QCD in the perturbative regime and to study the transition between perturbative and non-perturbative domains. One of the remarkable successes of this theory is the correct prediction of the evolution of the proton structure function with Q^2, for $Q^2 > 1$ GeV^2, now measured with a very good acuracy[1]. This evolution allows the extraction of the gluon density in the proton, which is not directly measurable. The fast rise of the gluon density for decreasing x, shown in Fig. 1 reflects the fast rise of the cross section with W.

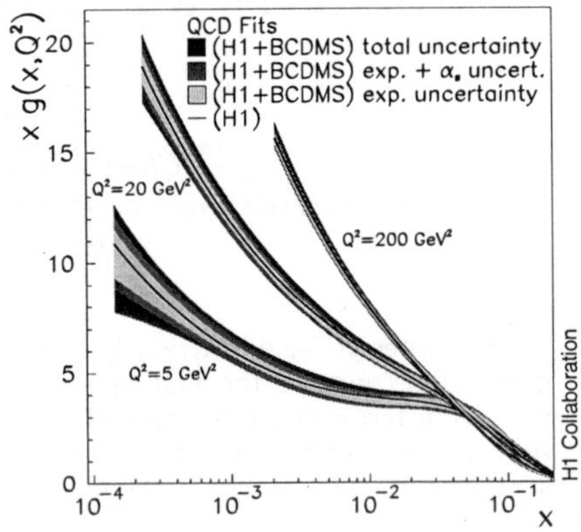

Figure 1. The gluon density in the proton measured by H1 as a function of x, for different values of Q^2.

The other main window opened at HERA for the understanding of the strong force is the study of diffractive interaction. Diffraction has been successfully described, already more than 30 years ago, via the introduction of an exchanged object carrying the vacuum quantum numbers, called the Pomeron ($I\!P$). Whilst Regge-based models give a unified description of all pre-HERA diffractive data, this approach is not linked to an underlying fundamental theory.

The second major result at HERA is the observation that 8 – 10% of the events in the deep inelastic regime ($Q^2 > 1$ GeV^2) present the charateristics of diffractively produced events[2], i.e. a large rapidity gap (LRG) without hadronic activity between the two hadronic low mass sub-systems, X and Y, as illustrated in Fig. 2.The gaps being significantly larger than implied by particle density fluctuation during the hadronisation process, these events are attributed to "hard" diffraction, i.e. to the exchange of a colourless object at the proton vertex, for the first time observed in a hard regime. HERA offers a unique possibility to study the nature of diffraction and to determine the Pomeron structure in terms of QCD.

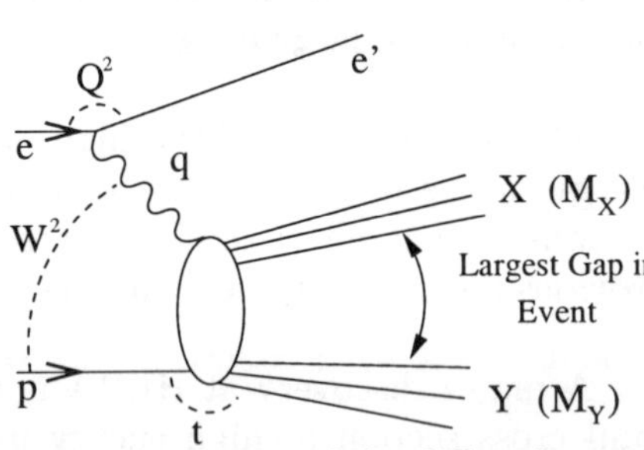

Figure 2. Sketch of the diffractive $e-p$ interaction

Inclusive DIS cross section and partonic structure of the Pomeron

The hard diffractive cross section can be defined by four kinematic variables conveniently chosen as Q^2, $x_{I\!P}$, β and t, where t is the squared four-momentum transfer to the proton, and $x_{I\!P}$ and β are defined as

$$x_{I\!P} \simeq \frac{Q^2 + M_X^2}{Q^2 + W^2} \qquad \beta \simeq \frac{Q^2}{Q^2 + M_X^2}; \tag{1}$$

$x_{I\!P}$ can be interpreted as the fraction of the proton momentum carried by the exchanged Pomeron and β is the fraction of the exchanged momentum carried by the quark struck by the photon. These variables are related to the Bjorken x scaling variable by the relation $x = \beta \cdot x_{I\!P}$.

Experimentally, the t variable is usually not measured or is integrated over. In analogy with non-diffractive DIS scattering, the measured cross section is expressed in the form of a three-fold diffractive structure function $F_2^{D(3)}(Q^2, x_{I\!P}, \beta)$:

$$\frac{\mathrm{d}^3\sigma\ (ep \to eXY)}{\mathrm{d}Q^2\ \mathrm{d}x_{I\!P}\ \mathrm{d}\beta} = \frac{4\pi\alpha^2}{\beta Q^4}\ (1 - y + \frac{y^2}{2})\ F_2^{D(3)}(Q^2, x_{I\!P}, \beta), \tag{2}$$

where y is the usual scaling variable, with $y \simeq W^2/s$. Conveniently, $F_2^{D(3)}$ is factorised in the form $F_2^{D(3)}(Q^2, x_{I\!P}, \beta) = f_{I\!P/p}(x_{I\!P}) \cdot F_2^D(Q^2, \beta)$, assuming that the $I\!P$ flux $f_{I\!P/p}(x_{I\!P})$ is independent of the $I\!P$ structure $F_2^D(Q^2, \beta)$, by analogy with the hadron structure functions, β playing the role of Bjorken x. The $I\!P$ flux is parameterized in a Regge inspired form. Including the reggeon ($I\!R$) trajectory in addition to the Pomeron, a good description of the data is obtained throughout the measured kinematic domain ($6.5 < Q^2 < 800$ GeV^2 $x_{I\!P} < 0.05$ and $0.01 < \beta < 0.9$)[3].

The reggeon contribution gets larger for increasing values of $x_{I\!P}$, which correspond to smaller energy (for given Q^2 and β values). It also gets larger for smaller values of β, which is consistent with the expected decrease with β of the reggeon structure function, following the meson example, whereas the Pomeron structure function is observed to be approximately flat in β. The comparison of $F_2^{D(2)}(Q^2, \beta)$ with the inclusive proton structure function, $F_2(Q^2, x)$, shows that the dynamic is different in the Pomeron and in the proton (see Fig. 3) for large β values. By analogy to the QCD evolution of the proton structure function, one can attempt to extract the partonic structure of the Pomeron from the Q^2 evolution of $F_2^{D(2)}(Q^2, \beta)$. The extracted distributions[3] are shown in Fig. 4 separately for the gluon and the singlet quark components as a function of z, the Pomeron momentum fraction carried by the parton entering the hard interaction. This distribution shows the dominance of gluons up to the high z values in the Pomeron partonic structure.

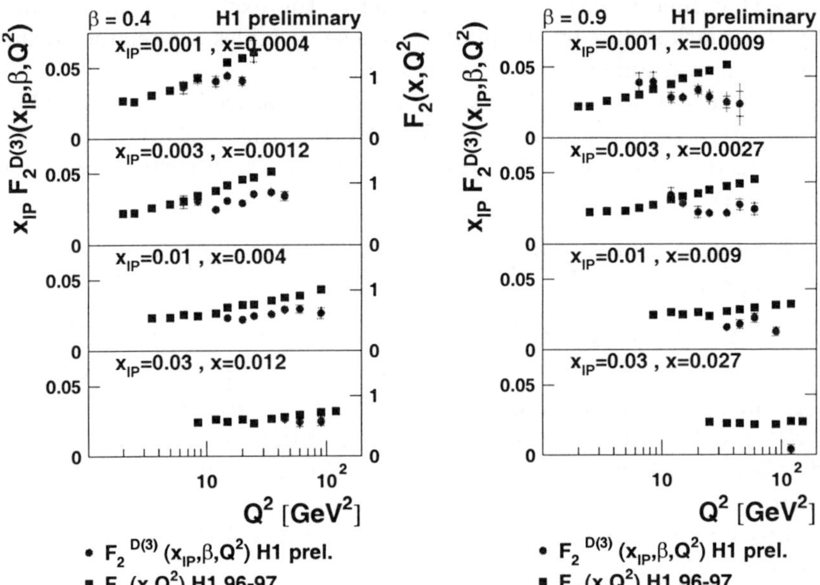

Figure 3. The $F_2^{D(3)}$diffractive structure function measurement as a function of Q^2 for different values of $x_{I\!P}$ in two bins in β. The inclusive proton structure function, $F_2(Q^2, x)$ is overlayed for the same Q^2 values and $x = \beta \cdot x_{I\!P}$.

The dominance of hard gluons in the Pomeron is confirmed by various analysis of the diffractive hadronic final state (jet production, energy flow, particle spectra and multiplicities, and event shape) providing a global consistent picture of diffraction[4,5,6].

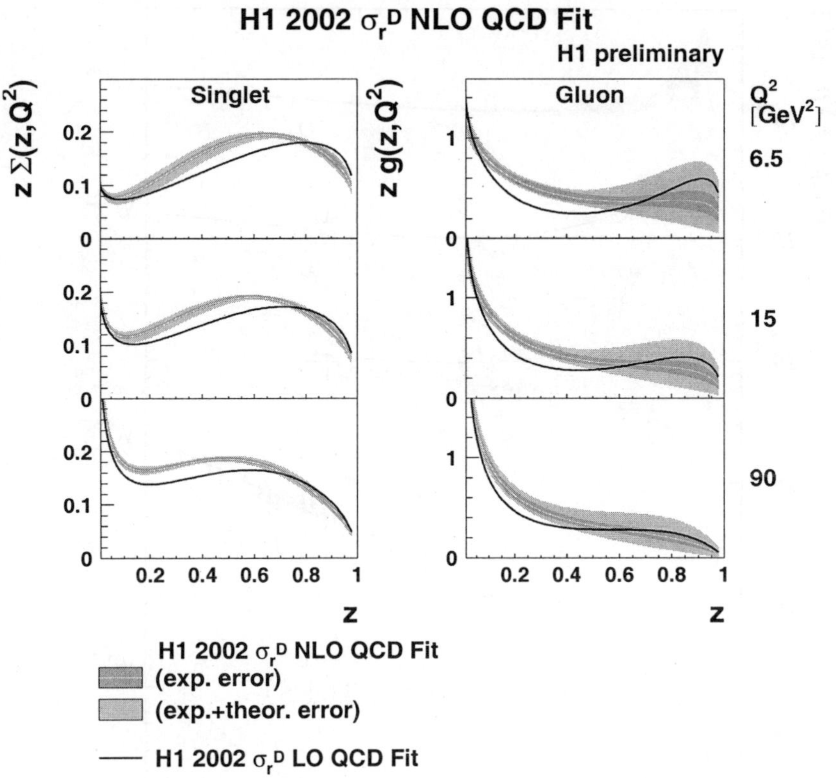

Figure 4. Quark (singlet) and gluon densities in the $I\!P$ extracted from the QCD fit of $F_2^{D(3)}$ as a function of z.

Exclusive diffractive final states

During these last decades, our knowledge of the hadronic structure has progressed a lot, mostly with the study of the inclusive inelastic reactions. The study of exclusive reactions will probably be dominant in the coming years, in this field, as they allow us to go one step deeper in our understanding of the hadronic structure. In particular they give access to momenta and helicity correlations of the hadron structure.

Furthermore, the exclusive elastic vector meson production and the deeply virtual Compton scattering (DVCS) studies: $\gamma^* p \to Xp$, where the

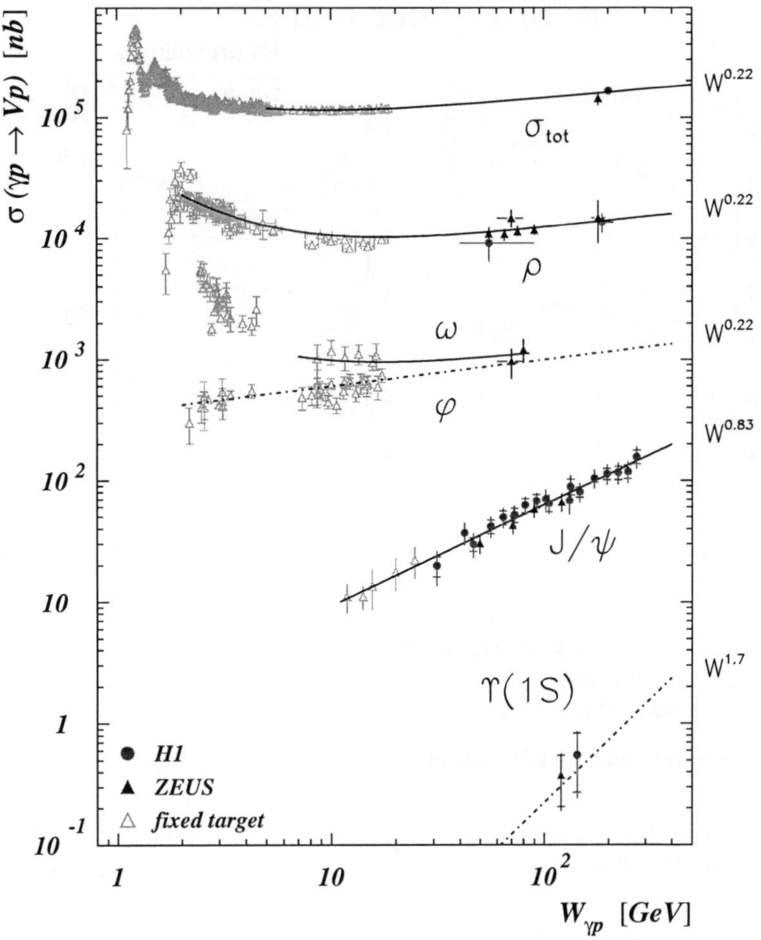

Figure 5. Diffractive cross sections as a function of the $\gamma^* - p$ system energy for various vector mesons in photo-production together with the total photo-production cross section.

hadronic system X consists only in a vector meson $(\rho, \omega, ...)$ or a photon (DVCS), provide a very interesting way to test the mechanism of diffraction and our understanding of the Pomeron structure.

Fig. 5 summarizes the W dependence of various elastic exclusive vector meson photo-production[7], together with the total photo-production cross section[8]. The light mesons (ρ, ω and ϕ) show a soft dependence in W, equivalent to that of the total cross section dependence, while this energy dependence is much steeper for J/ψ production. This regime change is interpreted as being due to the presence of a hard scale in the process, the (high) charm quark mass, making the J/ψ meson smaller than the confinement scale ($\sim 1fm$). In this case, it is natural to attempt a perturbative QCD description of the process, where the photon fluctuates into a quark-antiquark pair and the exchanged Pomeron is modeled by a pair of gluons. This leads, in a first approximation, to a cross section proportional to the gluon density squared, which is in good agreement (full line) with the data (points) shown on Fig. 6.

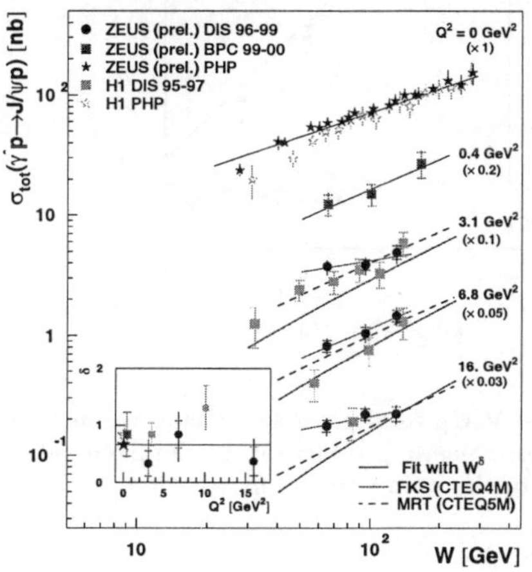

Figure 6. Exclusive J/ψ cross section in photoproduction and DIS Diffractive cross sections as a function of W.

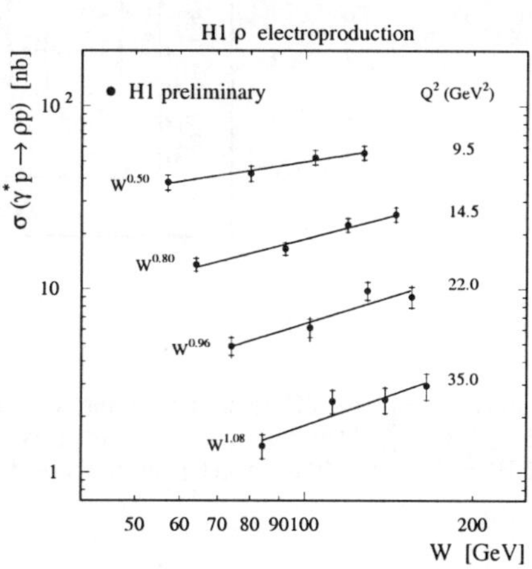

Figure 7. Exclusive ρ in DIS for different Q^2 values.

This figure also shows the agreement of the 2 gluon exchange model with measurement of exclusive J/ψ production in the DIS regime, where a second hard scale, Q^2 is present. As illustrated on Fig. 7, a modification of the W dependence also occurs for the elastic ρ production when the Q^2 increases, which highlights the appearance of Q^2 as a hard scale for a light vector meson production.

It is remarkable to see that the different vector meson cross sections present a universal behaviour as a function of the variable $Q^2 + M_V^2$, where M_V is the vector meson mass, as shown in Fig. 8 for $W = 75$ GeV. The cross sections were scaled by SU(4) factors, according to the quark content of the

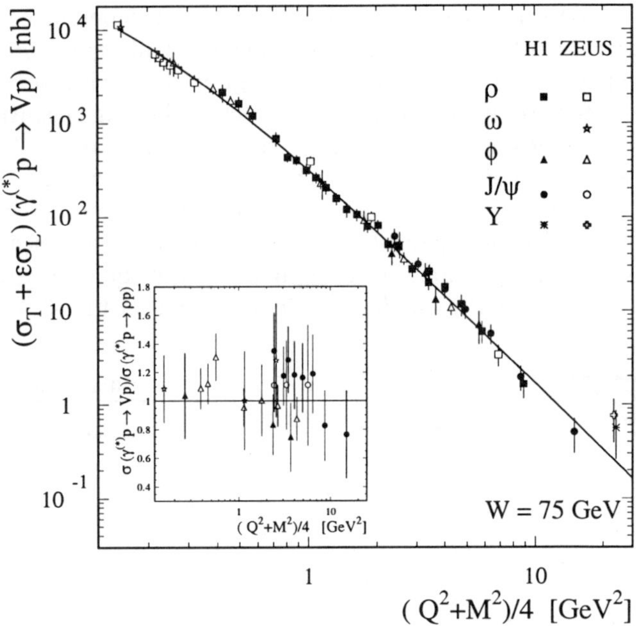

Figure 8. H1 and ZEUS measurements of the $\gamma p \to VMp$ total cross sections as a function of $(Q^2 + M_V^2)/4$ for $\rho, \omega, \phi, J/\psi$ and Υ vector meson elastic production, at the fixed value of $W = 75$ GeV. The insert presents the distance of the points to the fitted curve.

vector mesons[7]. This indicates that this variable choice seems to be a good scale to describe the vector meson production in a universal way, and that the cross section is at the first order not sensitive to the quark flavor (which would probably not be valid at lower energy).

However, the exclusive Υ cross section seems to be too high for this universal behaviour. In fact, the cross section measurement does not agree with simple QCD predictions. This is due to the fact that the transition from a virtual photon coming from an electron to an on shell particle forces the fractional momenta of the two gluons involved to be unequal. Therefore, in the cross section expression, the gluon density cannot just be squared, but two different x values have to be considered. When such skewing effects are included the agreement with QCD can be found again. This effect becomes important at high Q^2 values or high vector meson masses. To include such effects in the cross section calculations, the very powerful formalism of Generalised Parton Distributions (GPD) has been introduced[9].

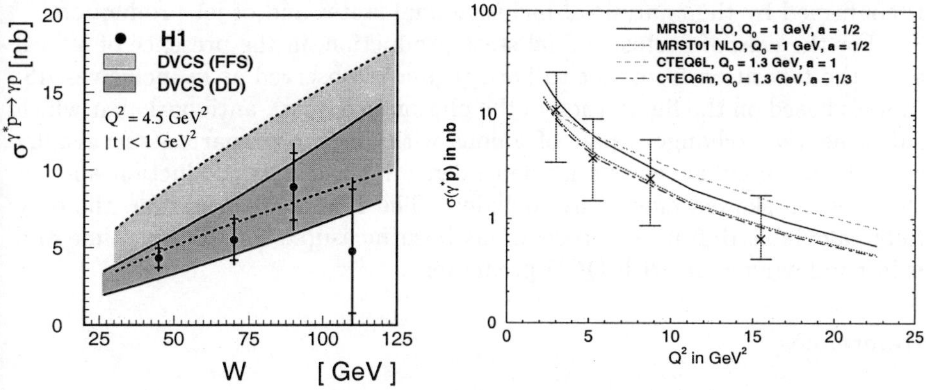

Figure 9. The H1 DVCS cross section measurement as a function of W (left) and of Q^2 (right). The measurement is compared to dipole model (DD) and QCD predictions (FFS for LO and right plot at NLO with different GPD inputs).

For vector meson production GPD appear quadratically in the cross section expression. A unique feature of DVCS is that they appear linearly in the interference term with the purely QED Bethe-Heitler process. Compared to vector meson production, DVCS is theoretically simpler (fully calculable) because the composite meson in the final state is replaced by the photon, thus avoiding large uncertainties due to the unknown meson wave functions. Therefore the DVCS measurement offers a particularly suitable channel to extract GPD. The first cross section measurement[10], shown in Fig. 9 as a function of W and Q^2, is in good agreement with the perturbative QCD prediction[11] and provides the first constrains on GPDs[12]

Conclusion

HERA experiments have produced a large amount of results in diffraction, which allow confrontations with QCD predictions, when one of the hard scales Q^2, the quark mass or t (not reported in this summary) is present in the process.

The QCD analysis of the total diffractive cross section, assuming factorisation into a Pomeron flux in the proton, shows that the corresponding parton distributions favor the dominance of hard gluons in the Pomeron. This is confirmed by the analysis of inclusive final states and of jet production.

For the case of exclusive final state production, in the presence of a hard scale the transition from soft to hard regime is observed as in inclusive DIS. Models based on the fluctuation of the photon in a quark-antiquark pair which subsequently exchange a pair of gluons with the proton-parton successfully reproduce the enhanced energy dependence for the J/ψ production and for the Υ when skewed effects are included. The DVCS cross section, the only fully calculable diffractive process, has been measured for the first time and is in good agreement with QCD predictions.

References

1. H1 Collab., C. Adloff et al., *Eur. Phys. J.* **C 21** (2001) 33-61, ZEUS Collab., J.Breitweg et al. *Eur. Phys. J.* **C 7** (1999) 609-630.
2. H1 Collab., T. Ahmed et al., *Nucl. Phys.* **B 429** (1994) 477, ZEUS Collab., M. Derrick et al., *Phys. Lett.* **B 315** (1993) 481.
3. P. Laycock, Proceedings of the DIS2002 Workshop, to be published in Acta Physica Polonica B,
   F.-P. Schilling, Proceedings of the DIS2002 Workshop, to be published in Acta Physica Polonica B,
   H1 Collab., C. Adloff et al., *Z. Phys.* **C 76** (1997) 613-629,
   ZEUS Collab., J. Breitweg et al. *Eur. Phys. J.* **C 1** (1998) 81-96.
4. H1 Collab., C. Adloff et al., *Phys. Lett.* **B B428** (1998) 206. ZEUS Collab., S. Chekanov et al. *Phys. Rev.* **D 65** (2002) 052001.
5. H1 Collab., C. Adloff et al. *Eur. Phys. J.* **C 5** (1998) 439-452.
6. H1 Collab., C. Adloff et al. *Eur. Phys. J.* **C 6** (1999) 421-436 ZEUS Collab., J. Breitweg et al. *Eur. Phys. J.* **C 5** (1998) 41-56.
7. H1 Collab., C. Adloff et al., *Phys. Lett.* **B 483** (2000) 360-372 and ref. therein,
   ZEUS Collab., J.Breitweg et al. *Eur. Phys. J.* **C 12** (2000) 393-410.
8. H1 Collab., S. Aid et al., *Z. Phys.* **C 69** (1995) 27-38, ZEUS Collab., M. Derrick et al., *Z. Phys.* **C 63** (1994) 391-408.
9. X. Ji, Phys. Rev. Lett. **78** (1997) 610, [hep-ph/9603249]. A. V. Radyushkin, Phys. Rev. **D 56** (1997) 5524, [hep-ph/9704207].
10. H1 Collab., C. Adloff et al., *Phys. Lett.* **B 517** (2001) 47.
11. A. Freund and M. Mc Dermott, *Eur. Phys. J.* **C 23** (2002) 651-674 and [hep-ph/0111472].
12. L. Favart, Proceedings of the QCD-N'02 Workshop, to be published in Nucl. Phys. A and [hep-ex/0106067].

RADIAL EXCITED STATES OF THE NUCLEON IN QUARK MODELS WITH DYNAMICAL CONFINEMENT

M. FIOLHAIS AND P. ALBERTO

Departamento de Física and Centro de Física Computacional, Universidade de Coimbra, P-3004-516 Coimbra, Portugal
tmanuel@teor.fis.uc.pt

B. GOLLI AND S. ŠIRCA

University of Ljubljana and J. Stefan Institute, Ljubljana, Slovenia

The Roper state is described in the framework of the chiral chromodielectric model as a single quark excitation in self-consistent meson potentials. The interplay between quark and meson excitations is discussed and we emphasize the role of non-quark degrees of freedom in the electro-production of nucleon excitations.

1 Introduction

The $P_{11}(1440)$ (Roper) resonance [1] is the lowest positive-parity N* state, most clearly visible in partial-wave decompositions of πN $\rightarrow$ πN and πN $\rightarrow$ $\pi\pi$N scattering, as well as in pion photoproduction. Although this four-star resonance is within the energy reach of many modern research facilities, the experimental analyses so far have not ventured beyond the determination of its mass, width, and electromagnetic couplings [2]. The only experimental information on the helicity amplitudes for Roper excitation with virtual photons comes from a very scarce data set obtained in older electroproduction experiments at DESY and NINA [3,4].

The electroexcitation of the Roper is also receiving considerable theoretical interest. In the spherically-symmetric SU(6) quark model, the Roper can be understood as a radial excitation ("breathing mode") of the proton, with one quark occupying the 2s state, yielding a $(1s)^2(2s)^1$ configuration. This physical picture implies a sizable Coulomb monopole contribution. Newer developments [5,6] have indicated a possible description of the Roper as a gluonic partner of the proton, represented as a (q^3g) hybrid baryon. In this approach, the monopole strength is expected to be highly suppressed, in contrast to the concept of "breathing". These two opposing concepts result in very different predictions for the Q^2-dependence of the transverse ($A_{1/2}$) and scalar ($S_{1/2}$) electro-production helicity amplitudes, and they both fail to reproduce the electromagnetic couplings in the real-photon limit. The Roper has been investigated as a hybrid baryon by using QCD sum rules [7], and in terms of vibrating flux-tubes between quarks [8,9]. Extensive studies in constituent, non-relativistic, relativised, and relativistic quark models also exist [10–15]. They generally result in good descriptions of the mass and width, but fail to reproduce the photo-couplings, with an exception of a non-relativistic quark model extended with vector-meson exchange [16]. An important addition to quark models are the mesonic degrees of freedom. A fair understanding of the photo-couplings was obtained by introducing meson-exchange currents between quarks [17], and in a relativistic quark model with a three-quark core and an admixture of a pion-baryon configuration [18].

In this work we use the framework of the chiral chromodielectric model to describe the nucleon and the Roper resonance as clusters of three valence quarks, coupled to clouds of σ and π mesons and to a chromodielectric field which dynamically confines the quarks[19,20]. The quarks are in the $(1\text{s})^3$ configuration for the nucleon and $(1\text{s})^2(2\text{s})^1$ for the Roper. The physical Roper emerges after the radial fields are determined in an iterative self-consistent variation using angular momentum and isospin projections.

In section 2 we introduce the model and in section 3 we explain how model states representing the nucleon and the Roper are obtained. The role of non-quark degrees of freedom in the Roper is discussed in section 4. In section 5 we present the electromagnetic form factors for the nucleon–Roper transition.

2 The model

The CDM contains quark and chiral meson degrees of freedom, in addition to a scalar-isoscalar chiral singlet chromodielectric field. The coupling of this field to fermions leads to quark confinement, and is an important feature of the model. The Lagrangian of the model reads [21]

$$\begin{aligned}\mathcal{L} = \mathrm{i}\bar{q}\gamma^\mu\partial_\mu q + \frac{g}{\chi}\bar{q}\left(\sigma_0 + \mathrm{i}\vec{\tau}\cdot\vec{\pi}\gamma_5\right)q + \frac{1}{2}\left(\partial_\mu\sigma_0\right)^2 + \frac{1}{2}\left(\partial_\mu\vec{\pi}\right)^2 + \frac{1}{2}(\partial_\mu\chi)^2 \\ -W(\sigma_0,\vec{\pi}) - U(\chi)\,. \qquad (1)\end{aligned}$$

Here $q(x)$ stands for the quark field operator, $\vec{\pi}(x)$ and $\sigma_0(x)$ are the chiral meson fields, pion and sigma, respectively (the arrow denotes isovector), and $\chi(x)$ is the chromodielectric field. There are two potential terms in Eq. (1): $W(\sigma,\vec{\pi})$, which is the Mexican hat potential for the chiral fields [21,22], and $U(\chi)$, the χ field potential. The Mexican hat potential is

$$W = \frac{\lambda}{4}\left(\sigma_0^2 + \vec{\pi}^2 - \nu^2\right)^2 + c\sigma_0 + W_0\,, \qquad (2)$$

where the parameters λ, ν, c and W_0 are related to the chiral meson masses and to the pion decay constant:

$$\lambda = \frac{m_\sigma^2 - m_\pi^2}{2f_\pi^2}, \quad \nu = f_\pi^2 - \frac{m_\pi^2}{\lambda}, \quad c = -f_\pi m_\pi^2 . \tag{3}$$

The constant W_0 in Eq.(2) ensures that $\min(W) = 0$. For the χ field potential we take

$$U(\chi) = \frac{1}{2} M^2 \chi^2 \tag{4}$$

where M is the mass of the field.

The vector current is conserved for the Lagrangian (1), which is almost chiral SU(2)×SU(2) symmetric. It is the (small) linear sigma term in (2) that explicitly breaks the chiral symmetry of (1), thus enforcing PCAC. The Mexican hat potential induces spontaneous chiral symmetry breaking, the vacuum expectation values of the chiral meson being $\langle 0|\vec{\pi}|0\rangle = 0$ and $\langle 0|\sigma_0|0\rangle = -f_\pi$, and a (dynamical) mass is acquired by the fermions. The parameters in the Mexican hat potential are the pion mass ($m_\pi = 0.14$ GeV), the pion decay constant ($f_\pi = 0.093$ GeV), and the sigma mass ($m_\sigma = 0.85$ GeV). In addition, there is one more parameter in model (1), the coupling constant g. For quadratic potentials such as (4), it turns out that the evaluated properties of the model states depend on $G = \sqrt{gM}$ [23], but not on g or M separately.

The CDM describes a system of interacting quark and meson fields. In the baryon sector of the model, one finds soliton solutions with three confined quarks due to the peculiar way the χ field couples to the fermions, which leads to an $r-$dependent quark mass, $m_q(r) = gf_\pi/\chi(r)$. The soliton requires $\chi(r) \to 0$ as $r \to \infty$, and hence $m_q(r) \to \infty$. This mechanism of increasing mass prevents the quarks to move too far away from the origin. The χ field can be regarded as a 0^{++} glueball [22]. It is an effective field mimicking the confining mechanism of QCD.

3 Baryons in the CDM model

Coherent states are assumed for all mesons and the starting point to describe a baryon is the hedgehog coherent state around a cluster of three valence quarks, which we write in the form

$$|Hh\rangle = N \exp\left\{\sum_{tm} (-1)^{1-m} \delta_{t,-m} \int_0^\infty \mathrm{d}k \sqrt{\frac{2\pi\omega_\pi(k)}{3}} \xi(k) a_{tm}^\dagger(k)\right\} \times$$
$$\exp\left\{\int_0^\infty \mathrm{d}k \sqrt{2\pi\omega_\sigma(k)} \eta(k) \tilde{a}^\dagger(k)\right\} \times$$

$$\exp\left\{\int_0^\infty \mathrm{d}k\sqrt{2\pi\omega_\chi(k)}\zeta(k)b^\dagger(k)\right\} \times \prod_{i=1,3} c_h^\dagger(i)|0\rangle\,, \tag{5}$$

where N is a normalization constant. The pion, the sigma and the chi amplitudes, $\xi(k)$, $\eta(k)$ and $\zeta(k)$, are related to the pion, sigma and chi radial profiles through

$$\phi(r) = \sqrt{\frac{2}{\pi}}\int_0^\infty k^2\mathrm{d}k j_1(kr)\xi(k)\,, \qquad \begin{Bmatrix}\sigma(r)\\ \chi(r)\end{Bmatrix} = \sqrt{\frac{2}{\pi}}\int_0^\infty k^2\mathrm{d}k j_0(kr)\begin{Bmatrix}\eta(k)\\ \zeta(k)\end{Bmatrix}\,. \tag{6}$$

The expectation values of the meson field operators in the hedgehog state are

$$\langle\pi_t(\boldsymbol{r})\rangle = \sqrt{\frac{4\pi}{3}}Y_{1t}^*(\hat{\boldsymbol{r}})\,\phi(r)\,, \langle\sigma(\boldsymbol{r})\rangle = \sigma(r)\,, \langle\chi(\boldsymbol{r})\rangle = \chi(r) \tag{7}$$

($t = 0, \pm1$ is the third component of isospin). In Eq. (5), $a_{tm}^\dagger$ is the creation operator for a p-wave pion with third components of isospin and angular momentum t and m, respectively, orbital wave function $\xi(k)$, and frequency $\omega_\pi = \sqrt{k^2 + m_\pi^2}$. Similarly, $\tilde{a}^\dagger(k)$ and $b^\dagger(k)$ create sigma and chi quanta in s-wave, with orbital wave functions $\eta(k)$ and $\zeta(k)$, respectively, and frequencies ω_σ and ω_χ. Finally, the operator $c_h^\dagger(i)$ creates an s-wave valence quark in a spin-isospin hedgehog state:

$$\langle\boldsymbol{r}|c_h^\dagger(i)|0\rangle = q_i(\boldsymbol{r}) = \frac{1}{\sqrt{4\pi}}\begin{pmatrix}u_i(r)\\ \mathrm{i}v_i(r)\,\boldsymbol{\sigma}\cdot\hat{\boldsymbol{r}}\end{pmatrix}|h\rangle\,, \quad |h\rangle = \frac{1}{\sqrt{2}}\left(|u\downarrow\rangle - |d\uparrow\rangle\right)\,. \tag{8}$$

The index i stands for all quantum numbers and distinguishes between different radial states. Quarks are always assumed to be in an s state ($j = 1/2$). Hence, only p-wave pions, and s-wave sigmas and chis may take part in the interactions.

Because (5) is not an eigenstate of angular momentum or isospin, the representative model states of the nucleon and Roper are obtained through a standard Peierls-Yoccoz projection on angular momentum (and isospin) [24],

$$|J, M; T = J, M_T\rangle = \frac{1}{\sqrt{F_J}}(-1)^{J+M}P^J_{M,-M_T}|Hh\rangle \tag{9}$$

where

$$P^J_{M,K} = \frac{2J+1}{8\pi^2} \int \mathrm{d}\Omega \, \mathcal{D}^{J^*}_{M,K}(\Omega) \, R(\Omega) \, . \tag{10}$$

Here $R(\Omega)$ is the rotation operator and $\mathcal{D}$ are the Wigner functions. In (9) F_J is the normalization factor given by

$$F_J = \langle Hh | P^J_{KK} | Hh \rangle \, , \tag{11}$$

which does not depend on K (so we choose $K = J$). For both the nucleon and the Roper, $J = T = \frac{1}{2}$. Due to the symmetry properties of the hedgehog, only a single projection, either in angular momentum or in isospin, is required since one operation in either space automatically selects the subspace $T = J$ in the other space. In Eq. (9) we project onto angular momentum.

The radial profiles $\phi(r)$, $\sigma(r)$, $\chi(r)$, $u_i(r)$, and $v_i(r)$ in Eqs. (7) and (8) are determined using (9) as a trial state, applying a variational principle to the angular momentum–isospin "projected energy", $E_J = \langle JM, TM_T | H_{\mathrm{CDM}} | JM, TM_T \rangle$, which is given by

$$E_J = \sum_{i=1}^{3} \epsilon_i + 4\pi \int_0^\infty r^2 \mathrm{d}r \left\{ \left[\left(\frac{\partial \sigma}{\partial r} \right)^2 + m_\sigma^2 \sigma^2 \right] + \left[\left(\frac{\partial \phi}{\partial r} \right)^2 + \frac{2\phi^2}{r^2} + m_\pi^2 \phi^2 \right] C_0^J \right.$$
$$\left. + \left[\left(\frac{\partial \chi}{\partial r} \right)^2 + M^2 \chi^2 \right] + \frac{\lambda}{4} \left[\phi^4 C_4^J + 2\phi^2 (\sigma^2 - 2 f_\pi \sigma) C_2^J + \sigma^4 - 4 f_\pi \sigma^3 \right] \right\} . \tag{12}$$

The "projection coefficients" $\mathcal{C}^J$ are expressed in terms of the normalization functions (11) which, in turn, depend on the intrinsic number of pions

$$N_\pi = 4 \int_0^\infty \mathrm{d}k k^2 \omega_\pi(k) \left[\int_0^\infty \mathrm{d}r r^2 \phi(r) j_1(kr) \right]^2 . \tag{13}$$

In (12), ϵ_i are quark energy eigenvalues. They are all equal for the nucleon. For the Roper two of them are equal (corresponding to the lowest state, 1s), while the third one corresponds to the 2s state.

The nucleon and the Roper resonance are therefore described as clusters of three quarks in radial-orbital configurations $(1\mathrm{s})^3$ and $(1\mathrm{s})^2$ $(2\mathrm{s})^1$ respectively, each surrounded by pion and σ-meson clouds and by a chromodielectric field, projected onto subspace with good angular momentum and isospin:

$$|\mathrm{N}_{\frac{1}{2},M_T}\rangle = \mathcal{N} \, P^{\frac{1}{2}}_{\frac{1}{2},-M_T} |Hh\rangle \, , \quad |\mathrm{R}'_{\frac{1}{2},M_T}\rangle = \mathcal{N}' \, P^{\frac{1}{2}}_{\frac{1}{2},-M_T} |Hh^*\rangle \tag{14}$$

($\mathcal{N}$ and $\mathcal{N}'$ are normalization factors). The Roper and the nucleon have different boson fields profiles since they can adapt to the corresponding quark configuration. The proper orthogonalization of states (14) is ensured by writing:

$$|\tilde{\mathrm{R}}\rangle = c_R^R|\mathrm{R}'\rangle + c_N^R|\mathrm{N}\rangle\,, \quad |\tilde{\mathrm{N}}\rangle = c_R^N|\mathrm{R}'\rangle + c_N^N|\mathrm{N}\rangle \tag{15}$$

where the coefficients result from the diagonalization of the Hamiltonian in the subspace spanned by $|\mathrm{R}'\rangle$ and $|\mathrm{N}\rangle$. However, we verified that the simpler orthogonalization procedure

$$|\mathrm{R}\rangle = \frac{1}{\sqrt{1-c^2}}(|\mathrm{R}'\rangle - c|\mathrm{N}\rangle)\,, \qquad c = \langle\mathrm{N}|\mathrm{R}'\rangle \tag{16}$$

is good enough.

A central point in our treatment of the Roper is the freedom of the chromodielectric profile, as well as of the chiral meson profiles, to adapt to the $(1s)^2(2s)^1$ configuration. Therefore, quarks in the Roper experience mean fields which are different from the mean boson fields felt by the quarks in the nucleon. In other words, the 'potential' breathes together with the quarks as illustrated in Figures 1 and 2.

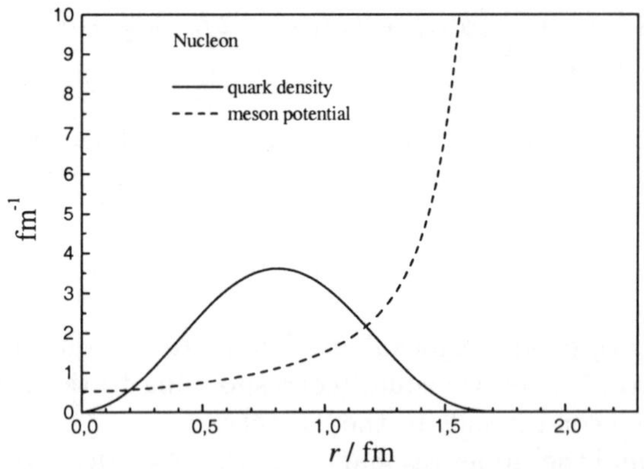

Figure 1. The baryon density (solid line) and the effective potential (dashed line) generated by the self-consistently determined π, σ and χ fields in the nucleon.

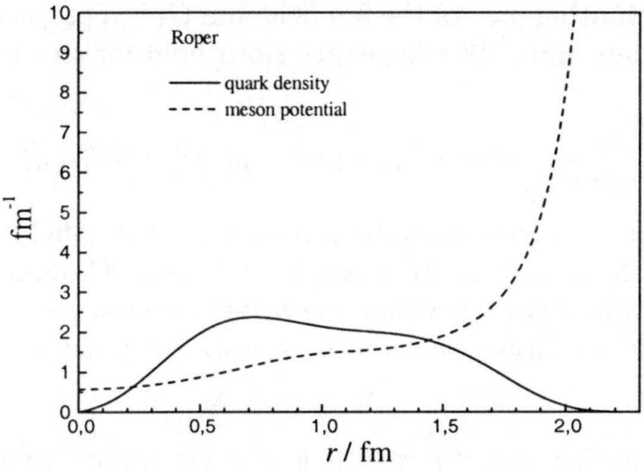

Figure 2. The baryon density (solid line) and the effective potential (dashed line) generated by the self-consistently determined π, σ and χ fields in the Roper.

4 Non-quark degrees of freedom

We have also investigated another possible type of excitation in which the quarks remain in the ground state configuration $(1s)^3$ while the chromodielectric field and the σ-field oscillate. Such oscillations can be described by expanding the boson fields as small oscillations around their ground-state values. For the σ-field we write:

$$\hat{\sigma}(\vec{r}) = \sum_n \frac{1}{\sqrt{2\varepsilon_n}} \varphi_n(r) \frac{1}{\sqrt{4\pi}} \left[\tilde{a}_n + \tilde{a}_n^\dagger\right] + \sigma(r) ,$$

where φ_n and ε_n satisfy the Klein-Gordon equation

$$\left(-\nabla^2 + m^2 + \frac{\mathrm{d}^2 V(\sigma(r))}{\mathrm{d}\sigma(r)^2}\right) \varphi_n(r) = \varepsilon_n^2 \varphi_n(r) . \tag{17}$$

A simple ansatz for the annihilation (creation) operator of the n-th mode is given by[a]

$$\tilde{a}_n = \int \mathrm{d}k\, \tilde{\varphi}_n(k) \left(\tilde{a}(k) - \sqrt{2\pi\omega_\sigma(k)}\, \eta(k)\right) , \qquad \tilde{a}_n|N\rangle = 0 , \tag{18}$$

where $\eta(k)$ and $\tilde{\varphi}_n(k)$ are the Fourier transforms of $\sigma(r)$ and $\varphi_n(r)$, respectively, and $\omega_\sigma(k) = \sqrt{k^2 + m_\sigma^2}$. The effective potential in (17) is given by

$$V_{\sigma\sigma}(r) = \lambda\left[C_2\phi(r)^2 + 3\sigma(r)(\sigma(r) + 2\sigma_v)\right] , \tag{19}$$

[a] This operator is not to be confused with the free field operator $\tilde{a}$ introduced in (5).

where σ is the fluctuating part of the full field and C_2 is a projection coefficient slightly smaller than unity. Similar expressions hold for the χ field with the effective potential

$$V_{\chi\chi}(r) = -\frac{3}{2\pi}\frac{g}{\chi(r)^3}\left[(\sigma(r)+\sigma_v)(u(r)^2-v(r)^2)+2\phi(r)u(r)v(r)\right] . \quad (20)$$

The effective potential turns out to be repulsive for the χ-field and attractive for the σ-field; in the latter case there exists at least one bound state with the energy ε_1 of typically 100 MeV below the σ-meson mass.

The ansatz for the Roper can now be simply extended as

$$|\mathrm{R}^*\rangle = c_1|\mathrm{R}\rangle + c_2\tilde{a}_\sigma^\dagger|\mathrm{N}\rangle , \quad (21)$$

where $\tilde{a}_\sigma^\dagger$ is the creation operator for the lowest vibrational mode. The coefficients c_i and the energy are determined by solving the generalized eigenvalue problem in the 2×2 subspace. The solution with the lowest energy corresponds to the Roper, while the orthogonal combination to one of the higher excited states with nucleon quantum numbers, e.g., the N(1710), provided the σ-meson mass is sufficiently small. The energy of the Roper is reduced (see Table 1), though the effect is small due to weak coupling between the state (16) and the lowest vibrational state with the energy ε_1. The reduction becomes more important if the mass of the σ-meson is decreased. The energy of the combination orthogonal to the ground state is close to $E_\mathrm{N}+\varepsilon_1$ with σ-meson vibrational mode as the dominant component.

m_σ	E_N	$2s--1s$	ΔE_R	$\Delta E_{\mathrm{R}*}$	c_2	ε_1
1200	1269	446	354	353	0.05	1090
700	1249	477	367	364	0.12	590

Table 1. For two σ-masses we show the nucleon energy (E_N), the Roper-nucleon energy splittings calculated from the single particle energy difference (2s–1s), the state (16) (ΔE_R) and the state (21) ($\Delta E_{\mathrm{R}*}$). All energies are given in MeV.

5 Nucleon–Roper electromagnetic transitions

The electromagnetic nucleon-Roper transition amplitudes as well as the transition amplitudes to higher excitations with nucleon quantum numbers represent an important test which may help distinguish between the models listed at the beginning. The transverse helicity amplitude is defined as

$$A_{1/2} = -\zeta\sqrt{\frac{2\pi\alpha}{k_W}}\int \mathrm{d}^3\boldsymbol{r}\ \langle\tilde{\mathrm{R}}_{+\frac{1}{2},M_T}|\boldsymbol{J}_{\mathrm{em}}(\boldsymbol{r})\cdot\boldsymbol{\epsilon}_{+1}\,\mathrm{e}^{\mathrm{i}\boldsymbol{k}\cdot\boldsymbol{r}}|\tilde{\mathrm{N}}_{-\frac{1}{2},M_T}\rangle \tag{22}$$

where k_W is the photon momentum at the photon point, and the scalar helicity amplitude as

$$S_{1/2} = \zeta\sqrt{\frac{2\pi\alpha}{k_W}}\int \mathrm{d}\boldsymbol{r}\,\langle\tilde{\mathrm{R}}_{+\frac{1}{2},M_T}|J^0_{\mathrm{em}}(\boldsymbol{r})\,\mathrm{e}^{\mathrm{i}\boldsymbol{k}\cdot\boldsymbol{r}}|\tilde{\mathrm{N}}_{+\frac{1}{2},M_T}\rangle\,. \tag{23}$$

Here J^μ_{em} is the EM current derived from the Lagrangian density (1):

$$J^\mu_{\mathrm{em}} = \sum_{i=1}^{3}\overline{q}_i\gamma^\mu(i)\left(\frac{1}{6}+\frac{1}{2}\tau_0(i)\right)q_i + (\vec{\pi}\times\partial^\mu\vec{\pi})_0\ . \tag{24}$$

The amplitudes (22) and (23) contain a phase factor ζ determined by the sign of the decay amplitude into the nucleon and the pion.

The new term in (21) *does not contribute* to the nucleon-Roper transition amplitudes. Namely, for an arbitrary EM transition operator $\hat{O}$ involving only quarks and pions we can write $\langle N|\tilde{a}_1\hat{O}|N\rangle = \langle N|[\tilde{a}_1,\hat{O}]|N\rangle = 0$, because of (18) and since the operators $\tilde{a}_n$ commute with $\hat{O}$. The amplitudes (22)–(23) obtained in the framework of the present calculation can be found in Ref. [19]. We found a relatively large discrepancy at the photon point which can be attributed to a too weak pion field in the model. Such small contribution was already noticed in the calculation of nucleon magnetic moments [23] and of the electroproduction of the Δ [25]. The pion contribution to the charged states only accounts for a few percent of the total amplitude.

Though the chromodielectric model gives only a qualitative picture of the lowest radially excited states of the nucleon and their electroproduction amplitudes, it yields some interesting features, in particular the possibility of σ-meson vibrations. Its main advantage over other approaches is that all properties, including the electromagnetic amplitudes and the resonance decay, can be calculated from a single Lagrangian without additional assumptions. It also allows us to treat exactly the orthogonalization of states which is particularly important in the description of nucleon radial excitations.

Acknowledgements

This work was supported by FCT (POCTI/FEDER), Portugal, and by The Ministry of Science and Education of Slovenia. MF acknowledges a travel grant from Fundação Calouste Gulbenkian (Lisbon), which made possible his participation in Hadrons 2002. He also would like to thank the organizers for the fine hospitality he enjoyed in Bento Gonçalves.

References

1. L. D. Roper, *Phys. Rev. Lett.* **12** (1964) 340.
2. D. E. Groom et al. (Particle Data Group), *Eur. Phys. J.* **C15** (2000) 1, and references therein.
3. C. Gerhardt, *Z. Phys. C* **4** (1980) 311.
4. B. Boden, G. Krösen, Research Program at CEBAF, Report of the 1986 Summer Study Group, V. Burkert (ed.), p. 121.
5. Z. Li, *Phys. Rev. D* **44** (1991) 2841,
6. Z. Li, V. Burkert, Z. Li, *Phys. Rev. D* **46** (1992) 70.
7. L. S. Kisslinger, Z. Li, *Phys. Rev. D* **51** (1995) R5986.
8. S. Capstick, P. R. Page, *Phys. Rev. D* **60** (1999) 111501.
9. G. L. Strobel, *Int. J. Theor. Phys.* **39** (2000) 115.
10. Fl. Stancu, P. Stassart, *Phys. Rev. D* **41** (1990) 916.
11. Z. Li, F. E. Close, *Phys. Rev. D* **42** (1990) 2207.
12. S. Capstick, *Phys. Rev. D* **46** (1992) 1965; *Phys. Rev. D* **46** (1992) 2864.
13. H. J. Weber, *Phys. Rev. C* **41** (1990) 2783.
14. S. Capstick, B. D. Keister, *Phys. Rev. D* **51** (1995) 3598.
15. F. Cardarelli, E. Pace, G. Salmè, S. Simula, *Phys. Lett. B* **397** (1997) 13.
16. F. Cano, P. González, *Phys. Lett. B* **431** (1998) 270.
17. U. Meyer, A. J. Buchmann, A. Faessler, *Phys. Lett. B* **408** (1997) 19.
18. Y. B. Dong, K. Shimizu, A. Faessler, A. J. Buchmann, *Phys. Rev. C* **60** (1999) 035203.
19. P. Alberto, M. Fiolhais, B. Golli, J. Marques, *Phys. Lett. B* **523** (2001) 273.
20. W. Broniowski, T. D. Cohen, M. K. Banerjee, *Phys. Lett. B* **187** (1987) 229.
21. T. Neuber, M. Fiolhais, K. Goeke, J. N. Urbano, *Nucl. Phys. A* **560** (1993) 909.
22. M. C. Birse, *Prog. Part. Nucl. Phys.* **25** (1990) 1.
23. A. Drago, M. Fiolhais, U. Tambini, *Nucl. Phys. A* **609** (1996) 488.
24. B. Golli and M. Rosina, *Phys. Lett. B* **165** (1985) 347; M. C. Birse, *Phys. Rev. D* **33**(1986) 1934.
25. M. Fiolhais, B. Golli, S. Širca, *Phys. Lett. B* **373** (1996) 229 ; L. Amoreira, P. Alberto and M. Fiolhais, *Phys. Rev. C* **62** (2000) 045202.

CONTRIBUTIONS

1. Lattice, Light-Front and Effective Theories

FREEZING OF THE QCD COUPLING CONSTANT AND THE PION FORM FACTOR

A. C. AGUILAR, A. MIHARA AND A. A. NATALE

Instituto de Física Teórica, Universidade Estadual Paulista, Rua Pamplona 145, 01405-900, São Paulo, SP, Brasil

aguilar@ift.unesp.br

The possibility that the QCD coupling constant (α_s) has an infrared finite behavior (freezing) has been extensively studied in recent years. We compare phenomenological values of the "frozen" QCD running coupling between different classes of solutions obtained through non-perturbative Schwinger-Dyson Equations. With these solutions were computed QCD predictions for the asymptotic pion form factor which, in turn, were compared with experiment.

1 SDE solutions and IR behavior of QCD

A non-perturbative method to investigate the infrared behavior of gluon and ghost propagators, and of the running coupling constant at low energies is through the solution of the Schwinger-Dyson equations (SDE)[1]. Early studies of the SDE for the gluon propagator in the Landau gauge concluded that the gluon propagator is highly singular in the infrared (IR)[2]. However, these results are discarded by simulations of QCD on the lattice at 95% confidence level[3], where it is shown that the gluon propagator probably is infrared finite. The lattice result is in agreement with two classes of SDE solutions. One proposed by Cornwall many years ago where the gluon acquires a dynamical mass[4], and another that has been extensively discussed by Alkofer and von Smekal[5] where the gluon propagator goes to zero when the momentum $q^2 \rightarrow 0$. The solutions differ due to the different approximations performed to solve the SDE, but in both cases there is a freezing of the coupling constant in the IR.

2 The non-perturbative coupling constants

We present here the coupling constants obtained through the SDE solutions. The first non-perturbative running coupling constant that we shall discuss was obtained by Cornwall[4], who used the pinch technique to derive a gauge invariant SDE. This non-perturbative coupling is equal to

$$\alpha_{sC}(q^2) = \frac{4\pi}{\beta_0 \ln\left[(q^2 + \frac{4M_g^2(q^2)}{\Lambda^2}\right]}, \; M_g^2(q^2) = m_g^2 \left[\frac{\ln\left(\frac{q^2+4m_g{}^2}{\Lambda^2}\right)}{\ln\left(\frac{4m_g{}^2}{\Lambda^2}\right)}\right]^{-12/11} \tag{1}$$

where $M_g(q^2)$ is a dynamical gluon mass. $\Lambda(\equiv \Lambda_{QCD})$ is the QCD scale parameter, $\beta_0 = 11 - \frac{2}{3}n_f$, where n_f is the number of flavors. In the above expression we are neglecting the effect of dynamical or bare fermion masses[4]. Some typical values are $m_g = 500 \pm 200$MeV and $\Lambda = 300$ MeV [4]

The other possibility for the IR finite running coupling was studied by Alkofer et al.[5]. We present below a fit for the numerical data of this solution

$$\alpha_{sA} = \begin{cases} 0.2161 + 9.2621 \exp\left(-2\frac{(q^2-0.0297)^2}{(0.6846)^2}\right) & : \; q^2 < 0.31 \text{ GeV}^2 \\ 1.4741 + 8.6072 \exp\left(-\frac{q^2-0.1626}{0.3197}\right) & : \; 0.31 < q^2 < 1.3 \text{ GeV}^2 \\ \frac{1.4978}{\ln(1.8488q^2)} & : \; q^2 > 1.3 \text{ GeV}^2 \end{cases} \tag{2}$$

3 The non-perturbative coupling and the pion form factor

It is known that the pion form factor, $F_\pi(Q^2)$, is quite dependent on the behavior of α_s at small momentum[6]. The asymptotic form factor is predicted by perturbative QCD[6,7]. It depends on the internal pion dynamics that is parameterized by the quark distribution amplitude of the pion. The QCD expression for the pion form factor is[7]

$$F_\pi(Q^2) = \int_0^1 dx \int_0^1 dy \, \phi^*(y, \tilde{Q}_y) T_H(x, y, Q^2) \, \phi(x, \tilde{Q}_x), \tag{3}$$

where $\tilde{Q}_x = \min(x, 1-x)Q$ and Q is the 4-momentum in Euclidean space transferred by the photon and $\phi(x, \tilde{Q}_x)$ is the pion wave function. In this work we use the model for the pion distribution amplitude proposed by Chernyak and Zhitnitsky[8]. This wave function was derived from QCD sum rules and it is written as

$$\phi(x, Q) = \frac{f_\pi}{2\sqrt{3}} x(1-x) \left\{6 + \left[30(2x-1)^2 - 6\right] \left(\frac{\alpha_s(Q)}{\alpha_s(\mu)}\right)^{\gamma_2}\right\}, \tag{4}$$

with $\mu = 500$ MeV and $\gamma_2 = 50/(99 - 6n_f)$.

The other function, $T_H(x, y, Q^2)$, is the hard-scattering amplitude . The lowest-order expression of $T_H(x, y, Q^2)$ is given by (see Braaten *et al.*[9])

$$T_H(x, y, Q) = \frac{64\pi}{3Q^2} \left\{\frac{2}{3} \frac{\alpha_s[(1-x)(1-y)Q^2]}{(1-x)(1-y)} + \frac{1}{3} \frac{\alpha_s(xyQ^2)}{xy}\right\}. \tag{5}$$

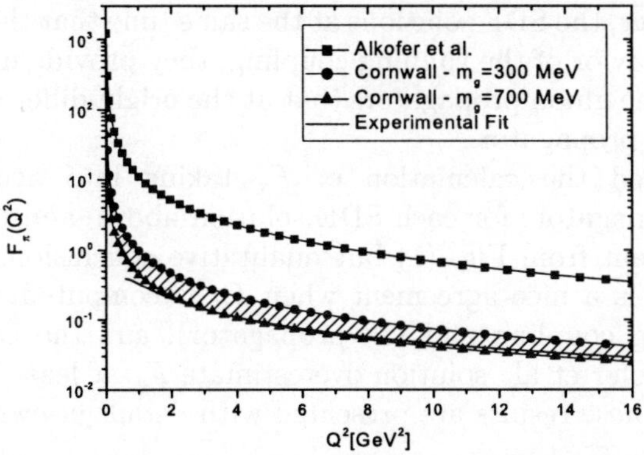

Figure 1. Pion form factor computed with the different non-perturbative running coupling constants. There is a nice agreement when F_π is computed with Cornwall's running coupling.

To compute the pion form factor using the non-perturbative running couplings proposed by Alkofer *et al.* Eq. (2) and Cornwall Eq. (1), we solved the integrals given by Eq.(3) substituting the quark distribution amplitude written in Eq.(4) and the expression of $T_H(x, y, Q^2)$ (Eq.(5)).

The pion form factor result for the different forms of the QCD coupling in the low momentum regime is shown in Fig.(1). We used for Eq.(1), the lower (300 MeV) and the upper (700 MeV) gluon mass values for a fixed $\Lambda =$ 300 MeV. These values defined the shaded area representing the expected range for the pion form factor, F_π.

In this same figure, we also compare our results with the experimental data (solid line) that was described by the least χ^2 fit ($\chi^2_{min} = 7.96742$) determined in Ref.[10]

$$F_\pi^{fit} = \frac{0.46895}{Q^2}\left(1 - \frac{0.3009}{Q^2}\right). \tag{6}$$

The results, using the running coupling of the Eq.(1), agree very nicely with the experimental data for a gluon mass value close to 700 MeV. On the other hand, the calculations with Alkofer et al. constant overestimate F_π at least by one order of magnitude.

4 Comments

We computed F_π using two distinct forms of the non-perturbative running coupling. However, the SDE solutions at the same time that they give the non-perturbative behavior of the running coupling, they provide non-perturbative expressions for the gluon propagators that at the origin differ drastically from the perturbative propagator.

We performed the calculation of F_π taking into account the non-perturbative propagators for each SDE solution above, and the quantitative results are different from Fig. (1) but qualitative conclusions are essentially the same: there is a nice agreement when F_π is computed with Cornwall's solution (running coupling and full propagator), and the calculations performed with Alkofer et al. solution overestimate F_π at least by one order of magnitude. All these results are presented with details elsewhere [11].

Acknowledgments

This research was supported by CNPq (AAN), FAPESP (ACA,AAN), CAPES (AM) and by Programa de Apoio a Núcleos de Excelência (PRONEX), Brazil.

References

1. C. D. Roberts and A. G. Williams, Prog. Part. Nucl. Phys., **33** 477 (1994).
2. S. Mandelstam, Phys. Rev. **D20** (1979) 3223; N. Brown and M. R. Pennington, Phys. Rev. **D38** (1988) 2266; **D39** (1989) 2723.
3. A. G. Williams *et al.*, hep-ph/0107029; F. D. R. Bonnet *et al.*, hep-lat/0101013; F. D. R. Bonnet *et al.*, Phys. Rev. **D62** (2000) 051501.
4. J. M. Cornwall, Phys. Rev. **D26** (1982) 1453; J. M. Cornwall and J. Papavassiliou, Phys. Rev. **D40** (1989) 3474, **D44** (1991) 1285.
5. R. Alkofer and L. von Smekal, Phys. Rept. (2001, in press), hep-ph/0007355; L. von Smekal, A. Hauck and R. Alkofer, Ann. Phys. **267** (1998) 1.
6. C.-R. Ji and F. Amiri, Phys. Rev. **D42** (1990) 3764; S. J. Brodsky, C.-R. Ji, A. Pang and D. G. Robertson, Phys. Rev. **D57** (1998) 245.
7. G. P. Lepage and S. J. Brodsky, Phys. Lett. **B87** (1979) 359; Phys. Rev. **D22** (1980) 2157.
8. V. L. Chernyak and I.R. Zhitnitsky, Phys. Rep. **112** (1984) 173.
9. E. Braaten and S.-M. Tse, Phys. Rev. **D35** (1987) 2255.
10. Tsung-Wen Yeh, hep-ph/0107192.
11. A. C. Aguilar, A. Mihara and A. A. Natale, Phys. Rev. **D65** (2002) 054011.

GAUGE FIELD DIVERGENCES IN THE LIGHT-FRONT

B.M. PIMENTEL, J.H.O. SALES
Instituto de Física Teórica
01405-900 São Paulo, Brazil
pimentel@ift.unesp.br

T. FREDERICO
Instituto Tecnológico de Aeronaútica, CTA
12228-900 São José dos Campos, Brazil
tobias@fis.ita.cta.br

The Bethe-Salpeter equation for the ground state of two fermions exchanging a gauge boson presents divergences in the momentum transverse, even in the ladder approximation projected in light-front. Gauge theories with light-front gauge also present the difficulty associated to the instantaneous term of the propagator of a system composed by fermions with boson-exchange interactions. We used a prescription that allowed an appropriate description of the singularity in the propagator of the gauge boson in the light-front.

1 Light-Front Dynamics: Definition

According to Dirac [1] "...the three-dimensional surface in space-time formed by a plane wave front advancing with the velocity of light. Such a surface will be called front for brevity ". An example of a light-front is given by the equation $x^+ = x^0 + x^3$.

A dynamical system is characterized by ten fundamental quantities: energy, momentum, angular momentum, and boost. In the conventional Hamiltonian form of dynamics one works with dynamical variables referring to physical conditions at some instant of time, the simple instant being given by $x^0 = 0$. Dirac found that other forms of relativistic dynamics variables refer to physical conditions on a front $x^+ = 0$. The resulting dynamics is called light-front dynamics, which Dirac called front-form for brevity.

The variables $x^+ = x^0 + x^3$ and $x^- = x^0 - x^3$ are called light-front time and longitudinal space variables respectively. Transverse variable $x^\perp = (x^1, x^2)$. Beware of many different conventions used in the literature. Here, we follow the conventions, notations and some useful relations from Ref.[2].

Instead of light-front field theory one will also find in the field theory literature in the infinite momentum frame, null-plane field theory and light-cone field theory. We prefer the word light-front since the quantization surface is a light-front (tangential to light cone).

By analogy with the light-front space-time variables, we define the longitudinal momentum $k^+ = k^0 + k^3$ and light-front energy $k^- = k^0 - k^3$.

For a free massive particle $k^2 = m^2$ leads to $k^+ \geq 0$ and the dispersion relation $k^- = \frac{(k^\perp)^2+m^2}{k^+}$.

The above dispersion relation is quite remarkable for the following reasons: (1) Even though we have a relativistic dispersion relation, there is no square root factor. (2) The dependence of the energy k^- on the transverse momentum $k^\perp$ is just like in the non-relativistic relation. (3) For k^+ positive (negative), k^- is positive (negative). This fact has several interesting consequences. (4) The dependence of energy on $k^\perp$ and k^+ is multiplicative and large energy can result from large $k^\perp$ and/or small k^+. This simple observation has drastic consequences for renormalization aspects [3] .

2 The Scalar Field Propagator

The Feynman propagator for the scalar field is

$$S(x^\mu) = \int \frac{d^4k}{(2\pi)^4} \frac{ie^{-ik^\mu x_\mu}}{k^2 - m^2 + i\varepsilon} . \tag{1}$$

and in terms of light-front variables [4], we have

$$S(x^+) = \frac{1}{2} \int \frac{dk^- dk^+ dk^\perp}{(2\pi)} \frac{ie^{\frac{-i}{2}k^- x^+}}{k^+ \left(k^- - \frac{k_\perp^2 + m^2 - i\varepsilon}{k^+}\right)} . \tag{2}$$

The Fourier transform of the single boson state propagator in the light-front time is given by:

$$\widetilde{S}(k^-) = \int dk^+ dk^\perp \frac{i}{k^+ \left(k^- - \frac{k_\perp^2 + m^2 - i\varepsilon}{k^+}\right)} . \tag{3}$$

3 Fermion Field

Let S_F denote the fermion field propagator in covariant theory

$$S_\mathrm{F}(x^\mu) = \int \frac{d^4k}{(2\pi)^4} \frac{i(\not{k}_\mathrm{on} + m)}{k^2 - m^2 + i\varepsilon} e^{-ik^\mu x_\mu} , \tag{4}$$

where $\not{k}_{\rm on} = \frac{1}{2}\gamma^+ \frac{(k^\perp)^2+m^2}{k^{+2}} + \frac{1}{2}\gamma^- k^+ - \gamma^\perp k^\perp$. Using light-front variables in Eq.(4), we have

$$S_{\rm F}(x^+) = \frac{i}{2}\int \frac{dk^- dk^+ dk^\perp}{(2\pi)} \left[\frac{\not{k}_{on} + m}{k^+ \left(k^- - k^-_{on} + \frac{i\varepsilon}{k^+}\right)} + \frac{\gamma^+}{2k^+} \right] e^{\frac{-i}{2}k^- x^+}. \quad (5)$$

We note that for the fermion field, the light-front propagator differs from the Feynman one by an instantaneous propagator.

4 Gauge Boson Propagator

Let $S^{\mu\nu}$ represent the gauge propagator,

$$S^{\mu\nu}(x^\mu) = \int \frac{d^4k}{(2\pi)^4} \frac{ie^{-ik^\mu x^\mu}}{k^2 + i\varepsilon} \left[\frac{-nkg^{\mu\nu} + n^\mu k^\nu + n^\nu k^\mu}{nk} \right], \quad (6)$$

where we choose the light-front gauge $A^+ = 0$, $n^\mu = (1, 0, 0, -1)$ and the metric tensor is given in reference[2].

The light-front components (6) can be written as $S^{+-} = S^{-+} = S^{++} = S^{+\perp} = 0$ and

$$S^{--} = 4\frac{ik^-}{k^+(k^2 + i\varepsilon)}, \quad S^{-\perp} = S^{\perp -} = 2\frac{ik^\perp}{k^+(k^2 + i\varepsilon)}, \quad S^{\perp\perp} = -1\frac{i}{k^2 + i\varepsilon}. \quad (7)$$

5 Interaction in First Order

We consider the fermion—anti-fermion system in the light-front with one-gauge boson exchange ($A^+ = 0$), for which the interaction Lagrangian density is given by

$$\mathcal{L}_I = g\overline{\Psi}_1 \gamma_\mu A^\mu \Psi_1 + g\overline{\Psi}_2 \gamma_\nu A^\nu \Psi_2. \quad (8)$$

The fermion corresponds to the field Ψ with rest masses m and the exchanged gauge boson to the field A^μ with mass $\mu = 0$. The coupling constant is g. The perturbative correction to the two-body propagator which comes from the exchange of one intermediate virtual boson, is

$$\Delta S_{g^2}(x^+) = (ig)^2 \int d\overline{x}_1^+ d\overline{x}_2^+ S_{k'}(x^+ - \overline{x}_1^+)(\gamma_\mu) S_k(\overline{x}_1^+) \quad (9)$$
$$S^{\mu\nu}(\overline{x}_2^+ - \overline{x}_1^+) S_p(x^+ - \overline{x}_2)(\gamma_\nu) S_{p'}(\overline{x}_2^+).$$

The intermediate boson propagates between the time interval $\overline{x}_2^+ - \overline{x}_1^+$. The labels in the particle propagators k and p indicate initial and k' and p' final states. We perform the Fourier transform from x^+ to P^- and for the total kinematical momentum P^+, which we choose positive, and $P^\perp$. The double integration in k^- is performed analytically in Eq.(10),

$$\Delta S_{g^2}(P^-) = \frac{-(ig)^2 i}{(4\pi)^2} \int \frac{dk^- dk'^-}{k^+ k'^+ (P^+ - k'^+)(P^+ - k^+)} \left\{ \frac{\not{k}'_{on} + m}{(k'^- - k'^-_{on} + \frac{i\varepsilon}{k'^+})} \right.$$

$$\times \gamma_- \frac{\not{k}_{on} + m}{(k^- - k^-_{on} + \frac{i\varepsilon}{k^+})} \frac{4(k^- - k'^-)}{(q^+)^2 \left(k^- - k'^- - q^-_{on} + \frac{i\varepsilon}{q^+}\right)} \frac{\not{p}'_{on} + m}{\left(p'^- - p'^-_{on} + \frac{i\varepsilon}{p'^+}\right)}$$

$$\times \gamma_- \frac{\not{p}_{on} + m}{\left(p^- - p^-_{on} + \frac{i\varepsilon}{p^+}\right)} + \frac{\not{k}'_{on} + m}{(k'^- - k'^-_{on} + \frac{i\varepsilon}{k'^+})} + \gamma_- \frac{\not{k}_{on} + m}{(k^- - k^-_{on} + \frac{i\varepsilon}{k^+})}$$

$$\times \frac{2(k^\perp - k'^\perp)}{(q^+)^2 \left(k^- - k'^- - q^-_{on} + \frac{i\varepsilon}{q^+}\right)} \frac{\not{p}'_{on} + m}{\left(p'^- - p'^-_{on} + \frac{i\varepsilon}{p'^+}\right)} \gamma_\perp \frac{\not{p}_{on} + m}{\left(p^- - p^-_{on} + \frac{i\varepsilon}{p^+}\right)}$$

$$+ [\gamma_\perp \to \gamma_-] + \frac{\not{k}'_{on} + m}{\left(k'^- - k'^-_{on} + \frac{i\varepsilon}{p'^+}\right)} \gamma_\perp \frac{\not{k}_{on} + m}{\left(k^- - k^-_{on} + \frac{i\varepsilon}{p^+}\right)}$$

$$\left. \frac{(-1)}{(q^+)^2 \left(k^- - k'^- - q^-_{on} + \frac{i\varepsilon}{q^+}\right)} \frac{\not{p}'_{on} + m}{\left(p'^- - p'^-_{on} + \frac{i\varepsilon}{p'^+}\right)} \gamma_\perp \frac{\not{p}_{on} + m}{\left(p^- - p^-_{on} + \frac{i\varepsilon}{p^+}\right)} \right\}.$$

6 Conclusion

From equation (10), we verified the existence of singularities in the components $(--)$ and $(-\perp)$ in the coordinate $q^+ = k^+ - k'^+$. We hoped to remove those singularities using the technique of displacement (δ^+) of the pole of the phase space in q^+ [4].

Acknowledgments

T.F and B.M. Pimentel thank to CNPq for partial support. J.H.O. Sales is supported by FAPESP/Brazil pos-doctoral fellowship. We acknowledge discussions with J.F.Libonati and J.Messias.

References

1. P.A.M. Dirac, Rev. Mod. Phys. **21**, 392 (1949).
2. J.B.Kogut and D.E.Soper, Phys.Rev.**D1** (1970) 2901.
3. S.D.Glasek and K.G.Wilson, Phys.Rev.**D49** (1994) 4214.
4. J.H.O. Sales, T.Frederico, B.V. Carlson and P.U.Sauer, Phys. Rev. **C61**:044003(2000); Phys. Rev. **C61**:064003(2001); Nucl. Phys.**A631** (1998) 574c.

LIGHT-FRONT MODEL OF THE KAON ELECTROMAGNETIC CURRENT

J.P.B.C. DE MELO AND L. TOMIO
Instituto de Física Teórica, Universidade Estadual Paulista, 01405-900, São Paulo, SP, Brazil
pacheco@ift.unesp.br

T. FREDERICO
Departamento de Física, ITA, Centro Técnico Aeroespacial, 12228-900, São José dos Campos, Brazil
tobias@fis.ita.br

The kaon electromagnetic form factor is extracted from both components of the current: J^+ and J^- with a pseudo-scalar coupling of the quarks to the meson. The pair production process does not contribute to J^+ in the Drell-Yan frame ($q^+ = 0$). However, the pair production process contribution is different from zero in J^- and this contribution is necessary to keep the rotational symmetry properties of the current.

1 Kaon Electromagnetic Current

Light-cone wave-functions models have been used in the literature [1,2] to describe pseudoscalar mesons [3–10]. In [8], the pion electromagnetic form factor in a pseudo-scalar coupling model are extracted using the components $J^+ = J^0 + J^3$ and $J^- = J^0 - J^3$ of the current, in the Drell-Yan frame ($q^+ = 0$). The pair term contribution was present in J^-, but not in J^+ where the contribution of the pair term vanishes. The pion J^+ current, with only light-cone valence wave function results to be equal to the covariant calculation, because the pair term vanishes. In the case of vector particles, even the J^+ current receives contribution from pair terms in order to respect the full covariance [11,12,13]. Therefore, in order to satisfy the angular condition for spin one particles, it is necessary to consider pair terms in the electromagnetic current J^+ (see [11] and references therein). Besides the valence contribution to the J^- current, the pair term is necessary for both, pseudoscalar and vector particles to keep the rotational symmetry properties of the current when using light-front coordinates. In this work, in order to calculate the electromagnetic form factor of the kaon, the components J^+ and J^- of the electromagnetic current are used.

The $J^{(\mu=\pm)}$ components of the electromagnetic current for the kaon receive contribution from the quark (q) and the antiquark ($\bar{q}$), given by

$$J_q^\mu(q^2) = ie_q g^2 N_c \int \frac{d^4k}{(2\pi)^4} Tr[S(k-m_{\bar{q}})\gamma^5 S(P'-k-m_q) \times \gamma^\mu S(P-k-m_q)\gamma^5]\Lambda(k,P')\Lambda(k,P) ,$$
$$J_{\bar{q}}^\mu(q^2) = [\, q \leftrightarrow \bar{q}\,] \text{ in } J_q^\mu(q^2) , \tag{1}$$

where the number of colors is $N_c = 3$, g is the coupling constant and e_q ($e_{\bar{q}}$) is the quark (antiquark) electric charge. The Breit frame is utilized, where the momentum transfer is $q^2 = -(\vec{q}_\perp)^2$, $P^0 = P'^0$ and $\vec{P}'_\perp = -\vec{P}_\perp = \vec{q}_\perp/2$. In order to regulate the divergent integral, we used $\Lambda(k,p) = N/((p-k)^2 - m_R^2 + i\epsilon)$, where m_R is the regulator mass and $m_{q(\bar{q})}$ is the quark (antiquark) mass. $S(P)$ is the fermion propagator. The light-front coordinates are defined as $k^+ = k^0 + k^3$, $k^- = k^0 - k^3$, $k_\perp = (k^1, k^2)$.

In the following, for the calculation of the pair term in the kaon electromagnetic current, we consider the method developed in refs.[12,14]. The contribution of the pair terms to J^+ and J^- are due to the matrix elements proportional to k^- in both cases (antiquark and quark on-shell). In order to calculate the electromagnetic form factor of the kaon, $F_{K^+}(q^2)$, we consider both J^+ and J^- components of the current.

Considering the expressions of the traces for the (+) component of the current, using light-front momenta, and identifying the functions $\Phi_{\bar{q}}^i(x,k_\perp)$, $\Phi_{\bar{q}}^f(x,k_\perp)$, one can obtain the electromagnetic form factor

$$F_{\bar{q}}^+(q^2) = e_q \frac{N^2 g^2 N_c}{P^+} \int \frac{d^2k_\perp dx}{2(2\pi)^3 x} \Big[-4\Big(f_1 x P^+ - x P^+ k_\perp^2 - 2 f_1 P^+ + 2k_\perp^2 P^+ + \frac{f_1 P^+}{x} - \frac{x P^+ q^2}{4} \Big) - \frac{4 f_1 P^+}{x} + 8P^+(x-1) m_q m_{\bar{q}} - 4x P^+ m_q^2 \Big] \times\ \theta(x)\theta(1-x)\ \Phi_{\bar{q}}^{*f}(x,k_\perp)\Phi_{\bar{q}}^i(x,k_\perp) ,$$
$$F_q^+(q^2) = [\, q \ \leftrightarrow\ \bar{q} \text{ in } F_{\bar{q}}^+(q^2)\,] \ . \tag{2}$$

The final expression for the electromagnetic form factor obtained with J^+ is the sum of two contributions from the quark and the antiquark currents:

$$F_{K^+}^+(q^2) = F_q^+(q^2) + F_{\bar{q}}^+(q^2) , \tag{3}$$

where the normalization is given by $F_{K^+}^+(0) = 1$. This expression is free of pair term contributions [8]. Therefore, the calculation of the kaon electromagnetic form factor with J^+, without light-front pair terms, gives the same result as the covariant one (see fig. (1)). The electromagnetic form factor obtained with J^- has contributions from the interval (I) $(0 < k^+ < P^+)$ and interval (II)

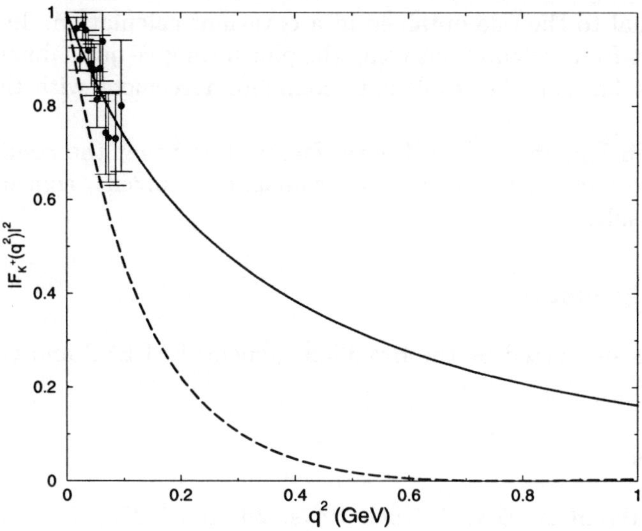

Figure 1. The kaon (K^+) form factor is given within the covariant and the light-front formalisms. The dashed line shows the result considering J^- without light-front pair terms. By adding the pair term to it, we obtain the result shown with solid line, which coincides with the covariant calculation and with the light-cone calculation with J^+. The experimental data (from ref. [15]) are also shown.

($P^+ < k^+ < P'^+$), where the pair term contributes in the integration (see [8,9]). The final expression for the electromagnetic form factor for the kaon, extracted from J^- is

$$F^-_{K^+}(q^2) = \left[F_q^{-(I)}(q^2) + F_{\bar{q}}^{-(I)}(q^2) + F^{-(II)}(q^2)\right] \ , \tag{4}$$

which is normalized by the charge, $F^-_{K^+}(0) = 1$.

2 Results

The parameters of the model are the constituent quark masses $m_q = m_u = 0.220$ GeV, $m_{\bar{q}} = m_{\bar{s}} = 0.419$ GeV, and the regulator mass $m_R = 0.946$ GeV, which are adjusted to fit the experimental value of the mean square charge radius $< r^2_{k+} >= 0.340\ fm^2$ [15]. With the above parameters, we obtain $< r^2_{k+} >= 0.354\ fm^2$, which is very close to the experimental value. The electromagnetic form factor is presented in fig.1. Due to the fact that J^+ does not have the light-front pair term contribution, the electromagnetic form factor is equal to the one obtained in a covariant calculation. In the case of J^-, the light-front calculations using the pair term give quite different results. As shown in fig. (1), the result is in complete agreement with the covariant calculation.

In conclusion, the kaon K^+ form-factor shows that the results obtained by using the components of the electromagnetic current, coincide with the covariant result.

Acknowledgements

This work is supported by the Brazilian agencies FAPESP and CNPq.

References

1. M. V. Terent'ev, Sov. J. Nucl. Phys. **24**, 106 (1976); L. A. Kondratyuk and M.V.Terent'ev , Sov. J. Nucl. Phys. **31**, 561 (1980).
2. J. P. B. C. de Melo and T. Frederico, Phys. Rev. **C55**, 2043 (1997), and references therein.
3. P.L.Chung, F. Coester and W.N. Polizou, Phys. Lett. **B205**, 545 (1988).
4. C. R. Ji and S. Cotanch, Phys. Rev. **D21**, 2319 (1990).
5. T. Frederico and G. Miller, Phys. Rev. **D45**, 4207 (1992); **50**, 210 (1994).
6. F. Cardarelli, I.L.Grach, I.M. Narodetskii, G. Salmé and S. Simula, Phys. Rev. **D53**, 6682 (1996).
7. F.Cardarelli et al., Phys. Lett. **B332**, 1 (1994).
8. J. P. B. C. de Melo, H. W. L. Naus and T. Frederico, Phys. Rev. **C 59**, 2278 (1999).
9. J. P. B. C. de Melo, T. Frederico, E. Pace and G. Salmé, Nucl. Phys. **A 707**, 399 (2002).
10. Bernard L. G. Bakker, Ho-Meoyng and Chueng-Ryong Ji, Phys. Rev. **D63**, 074014 (2001).
11. J. P. B. C. de Melo, T. Frederico, H. W. L. Naus and P. U. Sauer, Nucl. Phys. **A660**, 219 (1999).
12. J.P.B.C. de Melo, J.H.O. Sales, T. Frederico, and P.U. Sauer, Nucl. Phys. **A631**, 574 (1998).
13. Wolfgang Jaus, Phys. Rev. **D60**, 054026 (1999).
14. H. W. L. Naus, J.P.B.C. de Melo and T. Frederico, Few-Body Syst. **24**, 24 (1998).
15. S.R. Amendolia et al., Phys. Lett. **B178**, 435 (1986); Phys. Lett. **B277**, 168 (1986).

UNITARIZED CHIRAL PION-NUCLEON SCATTERING AMPLITUDE

J. SÁ BORGES

Instituto de Física, Universidade do Estado do Rio de Janeiro (UERJ), Rua São Francisco Xavier, 524, Maracanã, Rio de Janeiro, CEP 20550-900, RJ, Brazil
saborges@uerj.br

ISABELA P. CAVALCANTE

Departamento de Física, CCET, Universidade Federal de Mato Grosso do Sul (UFMS), Cidade Universitária, Campo Grande, CEP 79070-900, Brazil
ipc@dfi.ufms.br

In this work we apply the Inverse Amplitude Method to an alternative pion-nucleon amplitude and obtain a good description of pion-nucleon phase-shifts.

1 Introduction

Even though Quantum Chromodynamics (QCD) has achieved a great success in describing strong interactions, low energy hadron physics must still be modeled phenomenologically. A great theoretical improvement was made by means of the method of Chiral Perturbation Theory (ChPT) [1], which is an effective theory derived from the basis of QCD. The method consists of writing down chiral Lagrangians for the physical processes and uses the conventional technique of the field theory for the calculations.

In order to deal with baryons the theory faced problems related to fixing the scale for momenta and quark mass expansion. This led eventually to a method known as Heavy Baryon Chiral Perturbation Theory (HBChPT), which has been applied for instance to describe pion-nucleon scattering. The method yields a total amplitude and the corresponding partial waves satisfy only approximate elastic unitarity relation, since the leading amplitude is a real function of energy in the physical region. An interesting approach to implement unitarity in HBChPT amplitude is the Inverse Amplitude Method (IAM) [2]. IAM amplitude for pion-nucleon partial waves has been constructed and fitted to the experimental phase shifts by fixing nine free parameters [3].

On the other hand, two years ago [4] we developed a program to unitarize a pion-nucleon scattering amplitude obtained in a different framework, namely the hard-meson method of current algebra. All we need, for running this unitarization procedure is an approximate amplitude reproducing Weinberg prediction for S-wave scattering lengths. We imposed exact crossing relations and constructed auxiliary functions in order to respect approximate unitarity. Considering known the imaginary parts of partial wave amplitudes, we used the dispersion relation technique to arrive at a quasi-unitarized amplitude written in terms of known functions and a *two parameter* polynomial part. The strategy was to fit the isospin 3/2 P-wave phase shifts to experimental data, by fixing these *two parameters*, which were then used as inputs to evaluate phase shifts for $S_{11}-$, $S_{31}-$, $P_{11}-$, $P_{13}-$, $P_{31}-$, $D_{13}-$ and $D_{33}-$ waves. Some phase shifts of the model showed little similarity to the experimental results. In fact, the violation of unitarity of the resulting amplitudes seems to be very important mainly in the resonance region.

In the present work, we use the unitarity corrected amplitudes described above to construct IAM modified amplitudes in order to access the resonance region. We show that in IAM procedure each partial wave requires a set of parameters, while in our previous work we were able to fit the model amplitudes to the experimental data with only two parameters [4]. This is a consequence of the lack of crossing symmetry in the IAM unitarization scheme.

2 Formalism

We consider the reaction $\pi^a(k)+N(p) \to \pi^b(k')+N(p')$, which is described by the T-matrix amplitude

$$T^{ab}(p',k';p,k) = -A^{ab}(s,t,u) + \frac{i}{2}(\not{k}+\not{k}')\ B^{ab}(s,t,u),$$

with $s=(p+k)^2$, $t=(k'-k)^2$ and $u=(p-k')^2$. The invariant amplitudes A and B are decomposed as

$$A^{ab} = \delta^{ab}A^{+} + \frac{1}{2}[\tau^b,\tau^a]A^{-} \quad \text{and similarly for} \quad B^{ab}.$$

They exhibit the following symmetry properties under crossing:

$$A^{\pm}(s,t,u) = \pm\ A^{\pm}(u,t,s) \quad \text{and} \quad B^{\pm} = \mp\ B^{\pm}(u,t,s);$$

the amplitudes corresponding to definite isospins 1/2 and 3/2 are given by $A_{1/2} = A^{+} + 2\,A^{-}$, $A_{3/2} = A^{+} - A^{-}$ and similarly for $B_{1/2}$, $B_{3/2}$. We work in the center of mass system, so that $k=(\vec{k},w)$, $k'=(\vec{k}',w)$, $p=(-\vec{k},E)$. The total energy and scattering angle are given by $W=E+w$ and $|\vec{k}|^2\cos\theta = \vec{k}.\vec{k}'$, thus, in terms of these quantities, one has $s=W^2$, $t=-2|\vec{k}|^2(1-\cos\theta)$. In what follows, the nucleon mass is denoted by M.

For each isospin I the Pauli amplitudes are

$$F_{1I}(s,\cos\theta) = \frac{E+M}{8\pi W}\left[A_I(s,\cos\theta) + (W-M)B_I(s,\cos\theta)\right],$$

$$F_{2I}(s,\cos\theta) = \frac{E-M}{8\pi W}\left[-A_I(s,\cos\theta) + (W+M)B_I(s,\cos\theta)\right].$$

Partial wave amplitudes $f^{\pm}_{I\ell}$ are defined as $f^{\pm}_{I\ell}(s) = F_{1I\ell}(s) + F_{2I\ell\pm1}(s)$. For elastic scattering we have $\mathrm{Im}\, f^{\pm}_{I\ell}(s) = |\vec{k}|\ |f^{\pm}_{I\ell}(s)|^2$, which may be solved yielding $f^{\pm}_{I\ell}(s) = 1/|\vec{k}|\, e^{i\delta_{I\ell}(s)}\ \sin\delta_{I\ell}(s)$, where $\delta_{I\ell}(s)$ are real phase shifts.

For the low energy amplitudes, we consider the pion-nucleon scattering amplitudes obtained by Osypowsky [5] using the Ward identity technique, named current algebra amplitudes. Up to first-order in pion momenta, they are

$$A^{ca\,-} = \frac{\mu_V}{8Mf^2}(u-s) \quad \text{and} \quad B^{ca\,-} = \frac{1+\mu_V}{2f^2},$$

where $f = 94$ MeV is the pion decay constant and $\mu_V \simeq 3.7$. These amplitudes lead to well known Weinberg prediction for $S-$ wave scattering lengths.

We impose approximate unitarity on $S1$, $P1$, $P3$ and $D3$ partial waves by requiring that the first order corrected amplitudes $f^{(1)}_{i\ell}$ on the physical cut have imaginary parts related to the leading low energy amplitudes $f^{ca}_{i\ell}(s)$. Expressing functions A and B in terms of Pauli amplitudes we construct the auxiliary functions whose discontinuities are

$$\mathrm{Im}\ \mathcal{S}(s) = \frac{2|\vec{k}|}{W}\ A^{ca}_0(s),\ \mathrm{Im}\ \mathcal{D}(s) = \frac{2|\vec{k}|}{W}\ B^{ca}_0,\ \mathrm{Im}\ \mathcal{P}(s) = \frac{2|\vec{k}|}{W}\ A^{ca}_1(s).$$

We write subtracted dispersion relations for $\mathcal{S}$, $\mathcal{D}$ and $\mathcal{P}$ by introducing one pair of free parameters λ_1, λ_2 for *each* partial wave in the expressions

$$\mathcal{S}[\mathcal{P}](s) = s^2\lambda_1[\lambda_2] + A^{ca}_0[A^{ca}_1](s)\ G(s), \quad \mathcal{D}(s) = B^{ca}_0\ G(s),$$

where $G(s)$ is the two particle phase space integral. Crossing properties and partial wave total isospin dependence are then imposed to the first order corrected amplitudes. As in the procedure we used before, they loose unitarity in this final step. In order to restore it, we apply IAM to the corrected partial waves $t_{\ell I} = t^{ca}_{\ell I} + t^{(1)}_{\ell I}$, by writing

$$\tilde{t}_{\ell I}(s) = \frac{t^{ca}_{\ell I}(s)}{1 - t^{(1)}_{\ell I}(s)/t^{ca}_{\ell I}(s)}.$$

	S11	P11	P13	S31	P31	P33
λ_1	-8.3	-107	-49.0	1.1	18.0	-3.8
λ_2	12.	9.7	-10	-9.1	-94	-8.3

Table 1. $S_{11}, S_{31}, P_{11}, P_{31}, P_{13}, P_{33}$ partial waves.

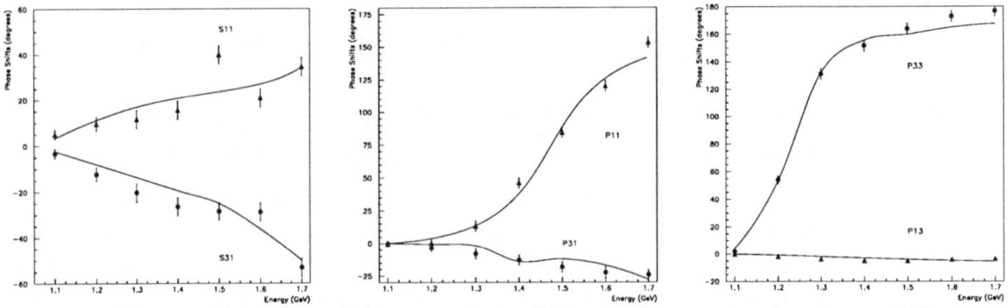

Figure 1. Pion-nucleon S- and P-wave phase shifts, in degrees, as functions of the *cms* energy, in GeV.

3 Preliminary Results

We performed the fits of $S_{1\,1}, S_{3\,1}, P_{1\,1}, P_{3\,1}, P_{1\,3}, P_{3\,3}$ partial waves in the form $\tilde{t}_{\ell\, I}$ above to experimental data, thus fixing the free parameters λ_1 and λ_2 independently for each wave. Their values are presented in table 1. The phase shifts thus obtained are shown in Figure 1, as functions of the *cms* energy, in GeV. Experimental data are from Ref. [6]. We observe that it is possible to fit the phase shifts of our model to the experimental data but, as crossing symmetry is badly violated in IAM approach, one cannot adjust two waves with the same values of parameters. This work is in progress. We will optimize the fitting procedure in order to get the same reasonable fits with a smaller number of parameters. We will also include D-wave predictions.

References

1. J. Gasser and H. Leutwyler, Nucl. Phys. B **250**, 1985 (465).
2. T. N. Truong, Phys. Lett. **61** (1988) 2526; **67**, (1991) 2260; A. Dobado and J. R. Peláez Phys. Rev. D **56** (1997) 3057.
3. A. Gómez Nicola, J. R. Peláez, Phys. Rev. D **62**, 2000 (017502); A. Gómez Nicola, J. Nieves, J. R. Peláez, E. Ruiz Arriola, Phys. Lett. B **486**, 2000 (77).
4. J. Sá Borges, Nucl. Phys. B **51**, 2000 (189).
5. H. J. Schnitzer, Phys. Rev. D **2**, 1970 (1621); P. Ford, Nucl. Phys. B**57** (1975) 25; E. T. Osypowski, Nucl. Phys. B 21 (1970) 615.
6. SAID on-line program, R. A. Arndt, R. L. Workmann *et al.*, *see* http://said.phys.vt.edu.

STRONG COUPLING QCD AND EFFECTIVE HADRONIC INTERACTIONS

S.M. ANTUNES, P.S. MORALES AND G. KREIN
Instituto de Física Teórica, Universidade Estadual Paulista,
Rua Pamplona 145, 01405-900 São Paulo, SP, Brazil
antunes@ift.unesp.br

The aim of this work is to implement the mechanism of *link rearrangement* - predicted in the strong coupling limit of Hamiltonian lattice QCD - in a constituent quark model in which constituent quarks, links and junctions are the dominant degrees of freedom. The implications of *link rearrangement* for the meson-meson interaction are investigated.

The mechanism of *link rearrangement* is predicted in the strong coupling limit of Hamiltonian lattice QCD. In hadron-hadron interactions, *link rearrangement* leads to effects similar to the traditional *quark exchange* with one gluon exchange. However, while the first is related to the confinement regime, the latter is related to the asymptotic freedom regime of QCD. Our aim in this contribution is to investigate possible means to differentiate both mechanisms.

The starting point is the model of Greenberg and Hietarinta (GH) [1]. This model provides a generalization of the traditional nonrelativistic quark model by introducing *links* and *junctions*. These are considered to be the result of a condensation of collective gluonic degrees of freedom. In this model, mesons are bound states of a quark and an antiquark and a link, and baryons are bound states of three quarks, three links and one junction.

The Hamiltonian is Color SU(3) local gauge invariant:

$$H = H_0 + V, \tag{1}$$

where H_0 is the part describing confinement, and V contains residual interactions. H_0 is of the form

$$H_0 = -\int d^3x\,[T_x q^{\dagger\alpha}(x)]\,N^{\beta}_{\alpha}(x,x)\,q_{\beta}(x) - \int d^3y\,[T_y \bar{q}^{\dagger}_{\alpha}(y)]\,\bar{N}^{\alpha}_{\beta}(y,y)\,\bar{q}^{\beta}(y)$$
$$+ \int d^3x d^3y\, q^{\dagger\alpha}(x) L^{\dagger\beta}_{\alpha}(x,y)\bar{q}^{\dagger}_{\beta}(y) V(x-y)\,\bar{q}^{\gamma}(y) L^{\delta}_{\gamma}(x,y) q_{\delta}(x). \tag{2}$$

Here, $\alpha, \beta, \cdots$ are SU(3) color indices and $T_z = \nabla^2_z/2m$ is the nonrelativistic kinetic energy operator. Also, $q^{\alpha}, q^{\dagger\alpha}, \bar{q}^{\alpha}, \bar{q}^{\dagger\alpha}$ are fermion operators that satisfy canonical anticommutation relations. $L^{\beta}_{\alpha}, L^{\dagger\beta}_{\alpha}$ are link operators, and $N^{\beta}_{\alpha}(x,x')$, $\bar{N}^{\beta}_{\alpha}(y,y')$ are link-number operators, that have Bose-type commutation relations [1].

Hadron states are eigenstates of H_0. A single-meson state is written as $|a\rangle = M_a^\dagger|0\rangle$, where $M_a^\dagger$ is given by

$$M_a^\dagger = \int d^3x d^3y \, \frac{e^{ip\cdot(x+y)/2}}{(2\pi)^{3/2}} \, \Phi_a^{ij} \, \psi_a(x-y) \, q_i^{\dagger\alpha}(x) L_\alpha^{\dagger\beta}(x,y) \bar{q}_{\beta j}^\dagger(y). \tag{3}$$

Here $|0\rangle$ is the vacuum state, annihilated by L_β^α and the quark and antiquark annihilation operators. Although we will not deal with baryons here, we mention that a single-baryon creation operator in the model includes a junction creation operator that obeys a bosonic commutation relation. While the link operator of the GH model has been shown to have justification in the strong coupling limit of Hamiltonian lattice QCD, the junction operator of the model seems to be an effective degree of freedom that possibly simulates excitations of a triple-junction of links [2].

The residual interactions in V in Eq. (1) include *link rearrangement*, *pair creation*, *short-distance elastic scattering* etc. Here we are interested in *link rearrangement* (LR). This interaction leads to effects similar to the traditional *quark exchange with one gluon exchange* (QEx+OGE) in hadron-hadron interactions. However, while LR is related to the confinement regime, QEx+OGE is related to the asymptotic freedom regime of QCD.

Our motivation for studying LR in hadron-hadron interactions is that qualitative differences between LR and QEx+OGE might provide insight towards the understanding of the confinement regime of QCD. With this motivation in mind, we consider meson-meson scattering. The interaction that simulates LR is written as (suppressing spin-flavor dependence)

$$\begin{aligned} V_{LR} = \int & d^3x_1 ... d^3y_2 \, q^{\dagger\mu_2}(x_2) L_{\mu_2}^{\dagger\nu_1}(x_2,y_1) \bar{q}_{\nu_1}^\dagger(y_1) q^{\dagger\mu_1}(x_1) L_{\mu_1}^{\dagger\nu_2}(x_1,y_2) \\ & \times \bar{q}_{\nu_2}^\dagger(y_2) V_{LR}(x_1,y_1;x_2,y_2) \, C_{\sigma_1\rho_1\sigma_2\rho_2}^{\mu_2\nu_1\mu_1\nu_2} \bar{q}^{\rho_1}(y_1) L_{\rho_1}^{\sigma_1}(x_1,y_1) q_{\sigma_1}(x_1) \\ & \times \bar{q}^{\rho_2}(y_2) L_{\rho_2}^{\sigma_2}(x_2,y_2) q_{\sigma_2}(x_2). \end{aligned} \tag{4}$$

The effect of this interaction on a two-meson state is pictorially represented in Fig. 1 below.

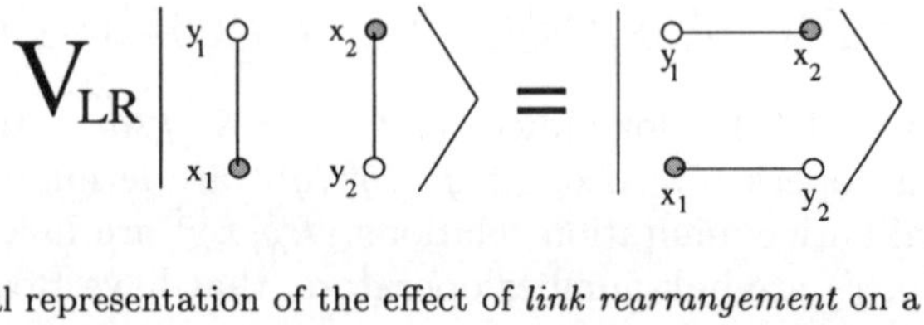

Figure 1. Pictorial representation of the effect of *link rearrangement* on a two-meson state.

In Born approximation, the meson-meson scattering amplitude is given by

$$A = \langle a'_1\, a'_2 | V_{LR} | a_1\, a_2 \rangle, \qquad |a_1\, a_2\rangle = M^\dagger_{a_1} M^\dagger_{a_2} |0\rangle. \tag{5}$$

Using for $M^\dagger$ the expression given in Eq. (3), one obtains that A can be written as $A = \sum_{i=1}^{4} A_i$, where

$$\begin{aligned}
A_1 &= \Phi^{*\mu_1\nu_2}_{a'_1}(x_1,y_2)\, \Phi^{*\mu_2\nu_1}_{a'_2}(x_2,y_1)\, \Phi^{\mu_2\nu_2}_{a_2}(x_2,y_2)\, \Phi^{\mu_1\nu_1}_{a_1}(x_1,y_1)\, V(x_1,y_1;x_2,y_2),\\
A_2 &= \Phi^{*\mu_2\nu_1}_{a'_1}(x_2,y_1)\, \Phi^{*\mu_1\nu_2}_{a'_2}(x_1,y_2)\, \Phi^{\mu_2\nu_2}_{a_2}(x_2,y_2)\, \Phi^{\mu_1\nu_1}_{a_1}(x_1,y_1)\, V(x_1,y_1;x_2,y_2),\\
A_3 &= A_2(a'_2 \leftrightarrow b'_2), \qquad\qquad A_4 = A_2(a'_2 \leftrightarrow b'_2).
\end{aligned} \tag{6}$$

The first thing one notices from this expression is that when the LR potential is a sum of *local 2-body* interactions as

$$V_{LR}(x_1,y_1;x_2,y_2) \equiv \frac{1}{2}\left[V_{qq}(x_1-x_2) + V_{\bar{q}\bar{q}}(y_1-y_2)\right] + V_{q\bar{q}}(x_1-y_2), \tag{7}$$

the scattering amplitude is *identical* to the QEx+OGE [3]. Now, nonlocal potentials are expected in QCD for long-distance interactions, since LR is generated by a plaquette in lattice QCD [2] - which involves four quarks. An extreme example of a nonlocal interaction is

$$V_{LR}(x_1,y_1;x_2,y_2) = \delta^{(3)}(x_1-x_2)\, \delta^{(3)}(y_1-y_2).$$

For this case, the meson-meson amplitude is a constant, independent of energy. This clearly shows that LR and QEx+OGE are qualitatively different for nonlocal LR potentials. In general, for a nonlocal LR interaction, LE and QEx+OGE are qualitatively different.

Next, we present an explicit numerical example. We consider a nonlocal LR potential of the form

$$V_{LR}(x_1,y_1;x_2,y_2) = V_q(x_1-x_2)\, V_{\bar{q}}(y_1-y_2). \tag{8}$$

For obtaining an analytical, closed form, V_q and $V_{\bar{q}}$, and the meson internal wave function are taken of a Gaussian form of size $a = 0.337$ GeV^{-2}. To be definite, we consider total isospin $I = 2$ scattering of two mesons of mass $M = 0.138$ GeV. The masses of the quarks are taken to be $m_q = m_u = m_d = 0.33$ GeV. In Fig. 2 we plot the cross sections for LR and QEx+OGE. In this figure we use the OGE coupling $\alpha_s = 0.6$, and normalize the LR potential to the OGE result. Clearly seen in Fig. 2 is the effect that, as expected, QEx+OGE is the dominant mechanism at higher energies, since local effects dominate at short-distances.

In conclusion, in this communication we have sketched a calculation in which the mechanism of *link rearrangement* - predicted in the strong coupling

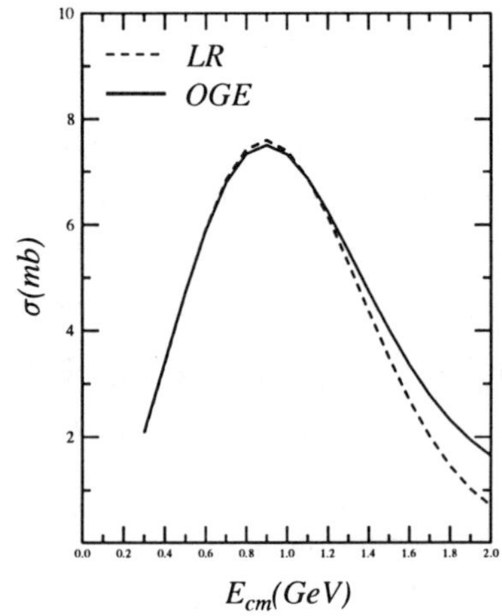

Figure 2. Total cross-section for total isospin $I = 2$ meson-meson scattering. Solid line: QEx+OGE, Dashed line: LR.

limit of Hamiltonian lattice QCD - was implemented in a constituent quark model and its consequences for meson-meson scattering examined. *Link rearrangement* leads to effects similar to the traditional *quark exchange* with one gluon exchange in hadron-hadron interactions. However while the first is related to the confinement regime, the latter is related to the asymptotic freedom regime of QCD. Both mechanisms lead to identical qualitative results when the rearrangement potential can be written as a sum of local quark-quark, antiquark-antiquark e quark-antiquark interactions. For general long-range nonlocal interactions, *link rearrangement* is qualitative and quantitatively different from *quark exchange* plus one-gluon exchange.

Acknowledgments

Work partially financed by the Brazilian agencies CAPES, CNPq and FAPESP (grant 00/05756 - 6), Brazil.

References

1. O.W. Greenberg and J. Hietarinta, *Phys. Rev. D* **22**, 993 (1980).
2. D. Robson, *Phys. Rev. D* **35**, 1018 (1987).
3. D. Hadjimichef, G. Krein, S. Szpigel, and J.S. da Veiga, *Ann. Phys. (N.Y.)* **268**, 105 (1998).

STUDY OF SCALING AT THE CHIRAL PHASE TRANSITION USING 2-FLAVOR LATTICE QCD DATA

TEREZA MENDES

IFSC, São Paulo University, Caixa Postal 369
13560-970 São Carlos SP, Brazil
mendes@if.sc.usp.br

The predicted $O(4)$ scaling at the chiral phase transition in the case of two degenerate quark flavors is investigated by an analysis of lattice QCD data. The data are unambiguosly normalized using the observed scaling along the so-called pseudocritical line, and compared to the universal scaling function obtained from simulations of the $3d$ $O(4)$ model. We find significantly better agreement for *larger* quark masses.

1 Introduction

The chiral transition for two-flavor QCD is predicted to be in the same universality class as the $3d$ $O(4)$ spin model [1]. This prediction has been confirmed for lattice QCD data in the Wilson-fermion case [2], but not for the staggered-fermion formulation. The comparison is usually done up to constants in the scaling functions and assuming infinite-volume behavior. Here we present an analysis of existing staggered-fermion data using additional scaling predictions from simulations of the $3d$ $O(4)$ model [3,4]. The data are unambiguosly normalized — using the observed scaling along the so-called pseudocritical line — and then compared to the $O(4)$ scaling function. In contrast with previous analyses, we find reasonable agreement in some cases, especially for the *larger* quark masses.

In order to check if two systems belong to the same universality class, one looks at universal quantities, such as the critical exponents

$$M_{h=0,\,t\to 0^-} \to |t|^\beta, \quad \chi_{h=0,\,t\to 0} \to |t|^{-\gamma}, \quad M_{t=0,\,h\to 0} \to h^{1/\delta}, \tag{1}$$

where M is the order parameter (e.g. the magnetization), χ is the corresponding susceptibility and

$$t = (T - T_c)/T_0, \quad h = H/H_0 \tag{2}$$

are the reduced temperature and magnetic field, respectively. More generally, one compares *scaling functions* for both systems. This comparison allows a more conclusive test, and is especially important for cases where the critical exponents cannot be determined with great accuracy.

2 Scaling functions and pseudo-critical line

The scaling ansatz is written for the free energy F_s in the critical region as

$$F_s(t,h) \;=\; b^{-d}\,F_s(b^{y_t}\,t, b^{y_h}\,h)\,, \tag{3}$$

where b is a rescaling factor, d is the dimension and y_t, y_h are related to the usual critical exponents: β, γ, δ. This implies that the magnetization be described by a universal function

$$M/h^{1/\delta} = f_G(t/h^{1/\beta\delta})\,. \tag{4}$$

Thus once the non-universal normalization constants T_0 and H_0 are determined for a given system in the universality class, its order parameter M scales according to the same scaling function f_G for all systems in this class. The susceptibility scales as

$$\chi \;=\; \partial M/\partial H \;=\; (1/H_0)\,h^{1/\delta-1}\,g(t/h^{1/\beta})\,, \tag{5}$$

where

$$g(z) \;=\; 1/\delta\,[f_G(z) - z/\beta f'_G(z)]\,. \tag{6}$$

For systems where a study at $H = 0$ is not possible, and the critical value T_c is not known with accuracy, an important region is the pseudo-critical line, defined by the points where the susceptibility χ shows a (finite) peak. This corresponds to the rounding of the divergence observed for $H = 0$, $T = T_c$. At each fixed h the peak in χ is given by $t_p \;=\; z_p\,h^{1/\beta\delta}$, and we have $M_p \;=\; h^{1/\delta}\,f_G(z_p)$, $H_0\,\chi_p \;=\; h^{1/\delta-1}\,g(z_p)$. Thus, the behavior along the pseudo-critical line is determined by the universal constants z_p, $f_G(z_p)$, $g(z_p)$. Critical exponents and the scaling function f_G are well-known for the $3d$ $O(4)$ model (see e.g.[3]) and the pseudo-critical line has been studied in [4]: we find that the peak susceptibilities are given by $z_p = 1.33(5)$.

Note that one may also consider the comparison for finite-size-scaling functions, since they are also universal and have the advantage of being valid for finite values of L, the linear size of the system. Such functions were studied for the $3d$ $O(4)$ model in [4].

3 Comparison with QCD data

For QCD the order parameter is given by the chiral condensate $<\overline{\psi}\,\psi>$, the analogue of the magnetic field is the quark mass m_q, and the reduced temperature is proportional to $6/g^2 - 6/g_c^2(0)$. Therefore, the chiral susceptibility peaks at $t_p \sim {m_q}^{1/\beta\delta}$.

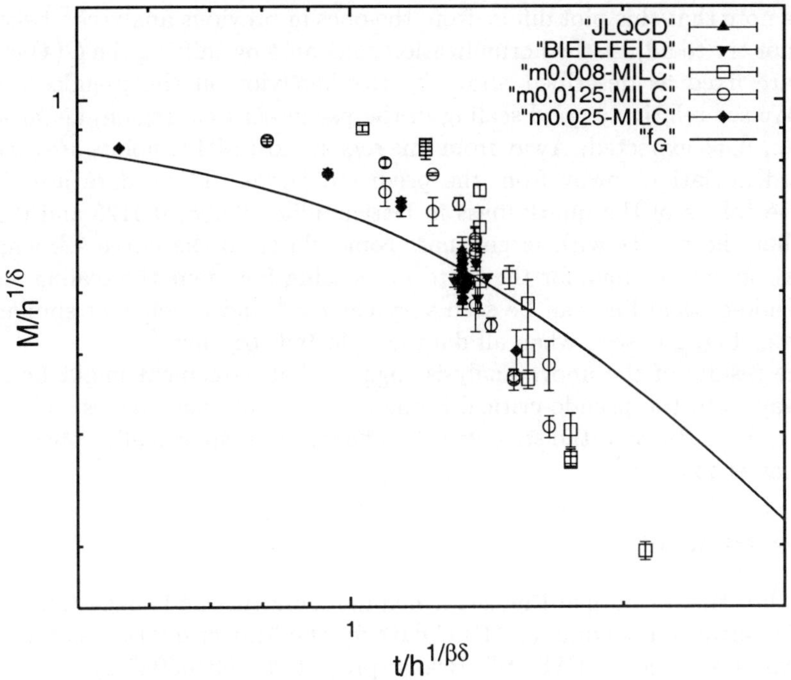

Figure 1. Comparison of QCD (staggered) data to the $O(4)$ scaling function. The data have been unambiguously normalized.

Extensive studies at $N_\tau = 4$ have been done by the Bielefeld[5], JLQCD[6] and MILC[7] groups. It was found that the susceptibility peaks scale reasonably well with the predicted exponents, but no agreement is seen in a comparison with the $O(N)$ scaling function [7]. This comparison was done up to multiplicative constants in the fields t and h. We redo this comparison here, performing a normalization of the QCD data.

Let $h \equiv N_\tau m_q/H_0$, $t \equiv (6/g^2 - 6/g_c^2(0))/T_0$. From the observed scaling along the pseudo-critical line (JLQCD and Bielefeld data) and the universal quantities z_p, $f_G(z_p)$ from the $O(4)$ model, we determine the normalization constants H_0 and T_0. This allows an unambiguous comparison of the MILC data to the scaling function f_G. We show this comparison in Fig. 1 below. (We also show the Bielefeld and JLQCD data. These data would correspond to a single point in the plot if the scaling behavior along the pseudo-critical line were perfect.)

We note that this plot differs from the ones in previous analyses [7] because we do not try to adjust the normalization constants by shifting the $O(4)$ curve (they are fixed as described above by the behavior on the pseudo-critical line). We see relatively good scaling in the pseudo-critical region, i.e. around $[z_p, f_G(z_p)]$, as expected. Away from this region most MILC points are several standard deviations away from the predicted curve. These data are given for three values of the quark mass in lattice units: 0.008, 0.0125 and 0.025. Note that the points with larger mass come closer to the curve. A similar comparison can be done for the finite-size-scaling function: the scaling along the pseudo-critical line can be seen very clearly [4] and a behavior similar to the one in Fig. 1 is seen when all data are plotted together [8].

The results of the above analysis suggest that agreement might be seen also away from the pseudo-critical if one considers heavier masses. We plan to carry out a study of the staggered two-flavor case especifically considering such heavier masses [9].

Acknowledgments

The author thanks Jürgen Engels for helpful comments and suggestions, and Doug Toussaint for sending us MILC data for the heavier quark masses. This work was supported by FAPESP, Brazil (project No. 00/05047-5).

References

1. R. Pisarski and F. Wilczek, *Phys. Rev.* **D29**, 338 (1984).
2. Y. Iwasaki, K. Kanaya, S. Kaya and T. Yoshié, *Phys. Rev. Lett.* **78**, 179 (1997).
3. J. Engels and T. Mendes, *Nucl. Phys.* **B572**, 289 (2000).
4. J. Engels, S. Holtmann, T. Mendes and T. Schulze, *Phys. Lett.* **B514**, 299 (2001).
5. F. Karsch and E. Laermann, *Phys. Rev.* **D50**, 6954 (1994).
6. S. Aoki et al. (JLQCD Collaboration), *Phys. Rev.* **D57**, 3910 (1998).
7. C. Bernard et al. (MILC Collaboration), *Phys. Rev.* **D61**, 054503 (2000).
8. T. Mendes, `hep-lat/0111044`.
9. A. Cucchieri, T. Mendes, in preparation.

VARIATIONAL FITS TO THE LATTICE ENERGY OF HEAVY-LIGHT MESONS

V. VIZCARRA AND G. KREIN
Instituto de Física Teórica, Universidade Estadual Paulista,
Rua Pamplona 145, 01405-900 São Paulo, SP
indhy@ift.unesp.br

We perform variational calculations of heavy-light meson masses using a fitted formula to a lattice two-quark potential. We examine the light quark mass dependence of the meson mass using the Schrödinger equation and the Dirac equation. For the Dirac equation, a saddle-point variational principle is employed, since the Dirac Hamiltonian is not bound from below.

Recently, Green, Koponen and Pennanen (GKP)[1] have adjusted lattice data for the energy of heavy-light ($Q\bar{q}$) mesons using a variational calculation within the context of the Schödinger equation, in which the heavy quark is considered static. For the potential they employ a fitted formula to a lattice two-quark potential

$$aV(r) = -\frac{0.309}{r/a} + 0.1649\, r/a + 0.629, \tag{1}$$

with lattice spacing $a = 0.18$ fm. Using a sum of three gaussians for the ansatz wave-function, a good fit to the lattice energy is obtained for a light quark mass of $m_q = 400\,\mathrm{MeV}$. In this communication we examine the question of how the variational meson energy changes as a function of m_q, comparing Schrödinger results with two relativistic calculations. Our motivation is that a mass of 400 MeV might not be small enough to warrant a Schrödinger approach of the kinetic energy. The question is particularly important in view that GKP[1] use the single-meson energy and the extracted quark mass to draw conclusions on the nature of the potentials acting in the two heavy-light system $Q^2\bar{q}^2$ when comparing to corresponding lattice data. The present work focus on the $Q\bar{q}$ system in the ground state.

The Hamiltonians used in this work are of the form

$$H_{Sch} = \frac{p^2}{2m_q} + V(r) \quad ; \quad H_{SR} = \sqrt{p^2 + m_q^2} + V(r)\,, \tag{2}$$

which are used in a Schrödinger equation, and

$$H_D = \vec{\alpha}\cdot\vec{p} + \beta\, m_q + \frac{1}{2}(1+\beta)\, V(r), \tag{3}$$

which is used in the Dirac equation. The factor $(1+\beta)/2$ of the potential in the Dirac Hamiltonian has the effect that only the upper component of the Dirac spinor contributes to the energy. The variational energy is the expectation value of the Hamiltonian,

$$\varepsilon[\phi] = \langle \phi(r)|H|\phi(r)\rangle, \tag{4}$$

where ϕ is the ansatz normalized wave function. For the Schrödinger equation, ϕ is taken of the form

$$\phi(r) = \left(\frac{\beta^2}{\pi}\right)^{3/4} \exp(-\beta^2 r^2/2), \tag{5}$$

where β is the variational parameter. For Dirac equation, since there is no lower bound to the energy, we employ the minimax variational principle[2]. This principle asserts that an appropriate variational principle for the Dirac equation is

$$\varepsilon[u,v] = \min_{\{u\}} \left[\max_{\{v\}} \langle \phi(r)|H|\phi(r)\rangle\right], \tag{6}$$

with

$$\phi(r) = \begin{pmatrix} i\,u(r) \\ \vec{\sigma}\cdot\hat{r}\,v(r) \end{pmatrix} \chi, \tag{7}$$

where χ is the spin 1/2 Pauli spinor. The ansatz for u is also taken to be Gaussian

$$u(r) = N \exp(-\beta^2 r^2/2). \tag{8}$$

For v, it has been shown[2] that the best variational ansatz is

$$v(r) = \gamma \left[\frac{\kappa+1}{\beta r} - \beta r\right] u(r), \tag{9}$$

where κ is the Dirac quantum number; $\kappa = -1$ for ground state. The variational parameters are β and γ. The normalization factor N is

$$N = \left(\frac{2}{2+3\gamma^2}\right)^{1/2} \left(\frac{\beta^2}{\pi}\right)^{3/4}. \tag{10}$$

The minimax principle of Eq. (6) means that the energy is a minimum with respect to β and a maximum with respect to γ.

Next, we write the explicit expressions for the energy eigenvalues. For the non-relativistic case the energy eigenvalue is given by

$$\varepsilon_{Sch}(\beta) = \frac{3}{2}\frac{\hbar^2}{2m_q}\beta^2 + \frac{4\beta^3}{\sqrt{\pi}} \int_0^\infty dr\, r^2\, V(r)\, e^{-\beta^2 r^2}, \tag{11}$$

while for the semi-relativistic case it is given by

$$\varepsilon_{SR}(\beta) = \frac{4}{\sqrt{\pi}\beta^4} \int_0^\infty dp\, p^2 \sqrt{p^2 + m_q^2} \left(-3 + 2\,\frac{p^2}{\beta^2} \right) e^{-\beta^2 r^2}$$
$$+ \frac{4\beta^3}{\sqrt{\pi}} \int_0^\infty dr\, r^2\, V(r)\, e^{-\beta^2 r^2}. \tag{12}$$

The Dirac energy is given by the expression

$$\varepsilon_D(\beta, \gamma) = \frac{3\gamma\beta}{2\pi(2+3\gamma^2)} + \frac{m_q}{4\pi} \frac{(2-3\gamma^2)}{(2+3\gamma^2)} \tag{13}$$
$$+ \frac{2}{2+3\gamma^2} \left(\frac{\beta^2}{\pi} \right)^{3/2} \int_0^\infty dr\, r^2\, V(r)\, e^{-\beta^2 r^2}. \tag{14}$$

The variational parameters, β for the Schrödinger case, and β and γ for the Dirac case, are determined from the nonlinear equations resulting from

$$\frac{\partial \varepsilon_{Sch}}{\partial \beta} = 0\,, \qquad \frac{\partial \varepsilon_{SR}}{\partial \beta} = 0\,,$$
$$\frac{\partial \varepsilon_D}{\partial \beta} = 0\,, \qquad \frac{\partial \varepsilon_D}{\partial \gamma} = 0\,. \tag{15}$$

After determination of the variational parameters from the resulting equations, one can examine the m_q dependence of the variational meson energy. The results for the different Hamiltonians using the fitted lattice potential of Eq. (1) are shown in the Table. Clearly seen in Table is that, as expected, for low values of $m_q \leq 800$ MeV the non-relativistic energy is very different from the relativistic ones. Moreover, one sees also that there is not much difference

Table 1. The variational meson energies for different values of the light quark mass. The non-relativistic energy ε_{Sch} includes the rest mass m_q.

m_q (MeV)	ε_{Sch} (MeV)	ε_{SR} (MeV)	ε_D (MeV)
100	1376	845	784
200	1171	872	833
400	1117	964	949
600	1185	1086	1083
800	1298	1226	1229
1000	1433	1377	1383
1500	1818	1773	1792
2000	2238	2200	2222

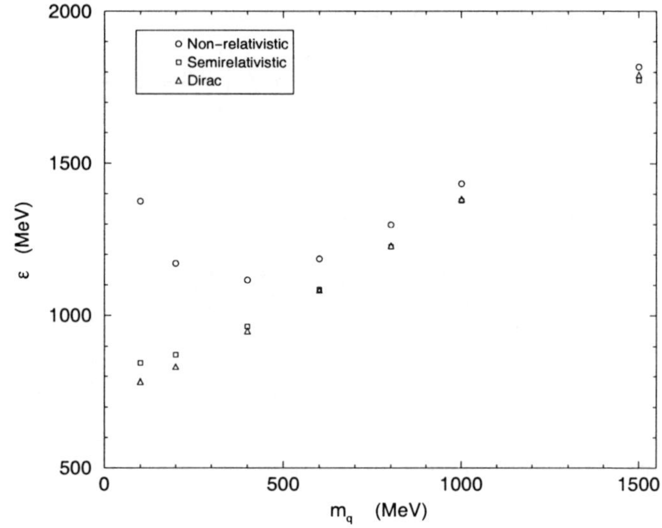

Figure 1. The variational meson energies for different values of the light quark mass. The non-relativistic energy ε_{Sch} includes the rest mass m_q.

between the semi-relativistic and Dirac energies for $m_q \geq 400$ MeV. The same results of the Table are represented graphically in Figure 1.

The two main conclusions of our work are: (1) For the fitted $Q\bar{q}$ lattice potential the Schrödinger energies are systematically larger than the corresponding semi-relativistic and Dirac energies. The differences start to become small for $m_q \geq 1500$ MeV. (2) The semi-relativistic and Dirac energies are not very different for a wide range of quark masses. For $m_q \geq 1500$ MeV the differences become small.

Work is in progress[3] to extend the work of GKP[1] on the $Q^2\bar{q}^2$ using relativistic kinetic energies

Acknowledgments

Work partially financed by CNPq and FAPESP (grant 00/11160-9), Brazil.

References

1. A.M. Green, J. Koponen and P. Pennanen, *Phys. Rev.* D **61**, 014014 (1999).
2. J.D.Talman, *Phys. Rev. Lett.* **57**, 1091 (1986).
3. V. Vizcarra and G. Krein, in progress.

NUCLEON ELECTROMAGNETIC OBSERVABLES IN A RELATIVISTIC QUARK MODEL

W.R.B. DE ARAÚJO, T. FREDERICO
Dep. de Física, Instituto Tecnológico de Aeronáutica, Centro Técnico Aeroespacial, 12.228-900 São José dos Campos, São Paulo, Brazil
wbaraujo@axpfep1.if.usp.br

M. BEYER
Fachbereich Physik, Universität Rostock, 18051 Rostock, Germany
beyer@darss.mpg.uni-rostock.de

H.J. WEBER
Dept. of Physics, University of Virginia, Charlottesville, VA 22904, U.S.A.
hjw@theory.phys.virginia.edu

In this work we show correlations between the electromagnetic observables for the nucleon in a constituent quark model in the Light-Front. In this model, the coupling between the nucleon and quark fields is expressed by an effective Lagrangean. We use confined and non-confined wave functions. The spin invariants for the isoscalar quark pair, given by a scalar, vector and mixed scalar-vector forms were tested in these observables.

1 Introduction

In this work, we focus our attention in two observables of the nucleon: the neutron charge mean square radius (r_n^2) and the proton electric to magnetic form factors ratio. Within a light-front relativistic model[1], we study the dependence of r_n^2 on the nucleon size parameterized by the proton charge radius (r_p). For artificially large protons, or small quark average momentum, we observe quite generally, that the contributions from F_1 and from the Foldy term to the neutron charge square radius tend to cancel as suggested by Isgur[2]. Concerning the proton electric form factor, we observe that at high momentum transfers, the $N-3q$ model presents numerical results below the experimental data. We conclude that this is due to the change of sign of the electric form factor of the model near $2(GeV/c)^2$, for the models we have tested.

2 Nucleon electromagnetic current and numerical results

The electromagnetic form factors are extracted from the plus component

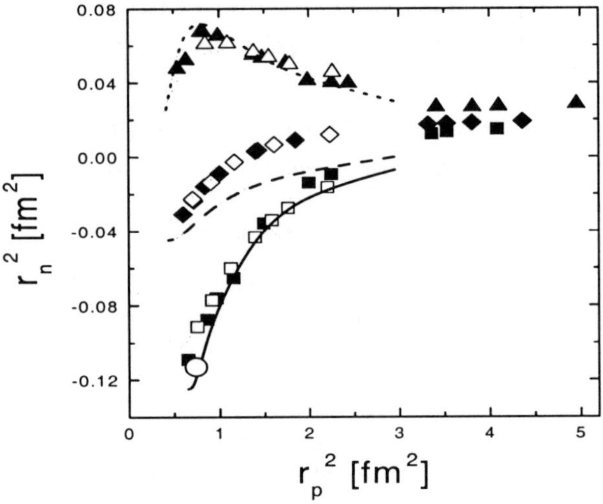

Figure 1. Neutron mean square charge radius as a function of the proton mean square charge square radius. Results for modified Gaussian wave function with α equal to 1 (full square), 1/2 (full diamond) and 0 (full triangle). Results for modified power-law wave function with α equal to 1 (open square), 1/2 (open diamond) and 0 (open triangle). The experimental values are within the empty circle.

of the current for momentum transfers satisfying the Drell-Yan condition $q^+ = q^0 + q^3 = 0$. The microscopic operator of the nucleon electromagnetic current is a sum of four terms: $J_N^+(q^2) = J_{aN}^+(q^2) + 4J_{bN}^+(q^2) + 2J_{cN}^+(q^2) + 2J_{dN}^+(q^2)$. In each term of the nucleon plus component of the current, from J_{aN}^+ to J_{dN}^+, the quark momenta are on-k^--shell. To construct the nucleon current, we use an effective Lagrangian for the N-q coupling which is written as[1]

$$\mathcal{L}_{\mathrm{N-3q}} = \alpha m_N \epsilon^{lmn} \overline{\Psi}_{(l)} i\tau_2 \gamma_5 \Psi^C_{(m)} \overline{\Psi}_{(n)} \Psi_N + (1-\alpha)\epsilon^{lmn} \overline{\Psi}_{(l)} i\tau_2 \gamma_\mu \gamma_5 \Psi^C_{(m)} \overline{\Psi}_{(n)} i\partial^\mu \Psi_N + H.C. \quad (1)$$

where τ_2 is the isospin matrix, the color indices are $\{l, m, n\}$ and ϵ^{lmn} is the totally antisymmetric symbol. The conjugate quark field is $\Psi^C = C\overline{\Psi}^\top$, where $C = i\gamma^2\gamma^0$ is the charge conjugation matrix; α is a parameter to dial the spin coupling parameterization. The mass of the constituent quark is $m = 0.22 GeV$[1]. The different models of the momentum component of the wave function correspond to the choices of the harmonic and power-law forms[3]: $\Psi_{\mathrm{HO}} = N_{\mathrm{HO}} \exp(-M_0^2/2\beta^2)$, $\Psi_{\mathrm{Power}} = N_{\mathrm{Power}}(1 + M_0^2/\beta^2)^{-p}$, and modified

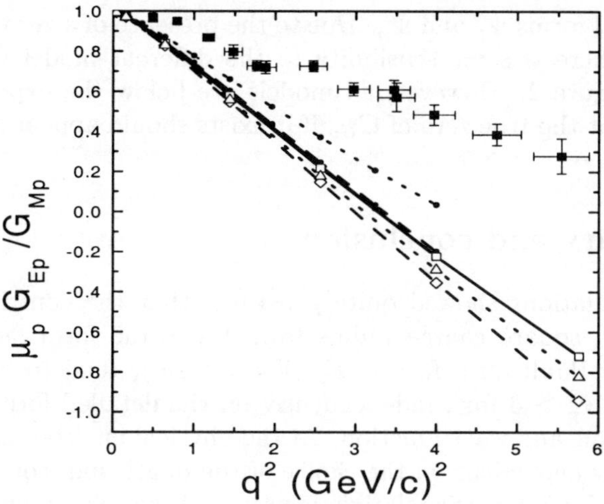

Figure 2. Proton form factor ratio $\mu_p G_{Ep}/G_{Mp}$ as a function of momentum transfer q^2. Results for modified power-law wave function with α equal 1 (solid line with dots), 1/2 (dashed line with dots), 0 (short-dashed line with dots). Results for power-law wave function with α equal 1 (open square with solid line), 1/2 (open diamond wih dashed line) and 0 (open triangle with dot line). Experimental data from M.K. Jones et al.[Jefferson Lab Hall A collaboration], Phys. Rev. Lett. **84**, 1398 (2000) and O.Gayou et al, nucl-ex/0111010.

harmonic and power-law wave functions, $\Psi'_{\rm HO} = N'_{\rm HO} \frac{\exp(-M_0^2/2\beta^2)}{\beta_1^2 - M_0^2}$, $\Psi'_{\rm Power} = N'_{\rm Power} \frac{(1+M_0^2/\beta^2)^{-p}}{\beta_1^2 - M_0^2}$, where M_0^2 is the free three-quark squared mass and the normalization is determined by the proton charge. The width parameters are β and β_1, with the constraint $\beta_1 < 3m$.

The functional dependence of the neutron mean square charge radius, r_n^2, with proton mean square charge radius is shown in Figure 1. The relativistic regime of the model for $r_p^2 < 3$ fm^2 is characterized by the strong sensitivity to different spin coupling schemes. For large proton sizes, the average momenta of the quarks are small compared to its constituent masses, and the sensitivity to different spin coupling schemes is strongly suppressed for $r_p^2 > 3$ fm^2. Also in this limit, we observe that r_n^2 tends to small values, due to the cancellation of the dominant Foldy term by the leading term in a q/m expansion of ${F_1}^2$. We show in Figure 2 the results for the ratio of $\mu_p G_{Ep}/G_{Mp}$ for different spin coupling schemes and momentum components of the wave function, using also the polinomial forms Ψ_p and Ψ'_p. Due to the presence of a zero in G_{Ep} around $2(GeV/c)^2$, there is some sensibility to the different model wave functions, as seen in Figure 2. However, all models are below the experimental data, indicating that the true zero of G_{Ep} if it exists should appear at larger values of q^2.

3 Summary and conclusion

Our calculations showed quite generally, that the contributions to the neutron mean square charge radius from the Dirac form factor, F_1 (r^2_{1n}), and from the Pauli form factor, F_2 (Foldy term), tend to cancel for large nucleon sizes, $r^2_p > 3$ fm^2, independently on the detailed form of quark spin coupling scheme and wave function. At the physical nucleon scale the value of r^2_{1n} is strongly dependent on the choice of the quark spin coupling scheme, a consequence of the intrinsic relativistic nature of the constituent quark motion inside the nucleon.

The new data on the ratio $\mu_p G_{Ep}/G_{Mp}$ indicate the necessity to improve the wave function models, include other (e.g. axial-vector quark pair) spin coupling forms, and/or a description of constituents quarks beyond the models discussed in the present work.

Acknowledgments

WRBA thanks FAPESP (Fundação de Amparo à Pesquisa do Estado de São Paulo) for financial support and LCCA/USP for providing computational facilities. TF thanks CNPq (Conselho Nacional de Pesquisas) and FAPESP, Brazil.

References

1. W.R.B. de Araújo, E.F. Suisso, T.Frederico, M.Beyer and H.J. Weber, Phys. Lett. **B**478 (2000) 86; E.F. Suisso, W.R.B. de Araújo, T.Frederico, M.Beyer and H.J. Weber, Nucl. Phys. **A694** (2001) 351.
2. N. Isgur, Phys. Rev. Lett. **83** (1999) 272.
3. S.J. Brodsky, H.-C. Pauli and S.S.Pinsky, Phys. Rep. **301** (1998)299.

2. Hadron Models

NON-LOCAL CHIRAL QUARK MODELS AT FINITE T AND μ

D. GOMEZ DUMM[*,**] AND N. N. SCOCCOLA[**,†,††]
[*] *IFLP, Physics Department, Universidad Nacional de La Plata, C.C. 67, (1900) La Plata, Argentina*
[**] *CONICET, Argentina*
[†] *Physics Department, Comisión Nacional de Energía Atómica, Av.Libertador 8250, (1429) Buenos Aires, Argentina*
[††] *Universidad Favaloro, Solís 453, (1078) Buenos Aires, Argentina*
dumm@venus.fisica.unlp.edu.ar

Chiral quark models with non-local covariant separable interactions at finite temperature and chemical potential are investigated. In particular we study the chiral phase transition for some definite non-local regulators. We find that in all cases the transition is of first order for low values of T, turning into a smooth cross-over at a certain "end point". Predictions for the position of this point are given.

1 Introduction

Within QCD, asymptotic freedom implies that when either the temperature T or the chemical potential μ are high the effective couplings become small. Thus, for a given hadron system one expects that at a certain point a phase transition (or cross-over) takes place, and the system changes to a new phase in which color is screened rather than confined and chiral symmetry is restored. Unfortunately, so far, it has not been possible to obtain detailed information about the corresponding $T - \mu$ phase diagram directly from QCD. In this situation, different effective models have been proposed. One of the most popular is the Nambu-Jona-Lasinio model [1], in which the quark fields interact via local four point vertices which satisfy chiral symmetry. Some covariant nonlocal extensions of the NJL model have been studied in the last few years [2]. Non-locality arises naturally in the context of relevant approaches to low-energy quark dynamics, such as the instanton liquid model [3] and the Schwinger-Dyson re-summation techniques [4]. In addition, non-local models have several advantages over the local scheme: the effective interaction is finite to all orders in the loop expansion, soft regulators lead to small NLO corrections, etc. It has been also shown [5] that a proper choice of the non-local regulator and the model parameters can lead to some form of quark confinement, in the sense of a quark propagator without poles at real energies. In this note we discuss the extension of this type of models to finite temperature and chemical potential.

2 Formulation

We consider the following Euclidean action

$$S_E = \int d^4x \left[\bar{\psi}(x)\left(-i\partial\!\!\!/ + m_c\right)\psi(x) - \frac{G}{2}\, j_a(x) j_a(x) \right] , \tag{1}$$

where m_c is the current quark mass and the current $j_a(x)$ is given by

$$j_a(x) = \int d^4y\; d^4z\; r(y-x)\; r(x-z)\; \bar{\psi}(y)\Gamma_a \psi(z) \,. \tag{2}$$

Here $\Gamma_a = (1, i\gamma_5\vec{\tau})$ and $r(x-y)$ is a non-local regulator function. However, this regulator is local in momentum space. To proceed we perform a standard bosonization of the theory, introducing meson fields $\sigma(x)$ and $\vec{\pi}(x)$. Then we carry out the mean field approximation by expanding these fields around their translational invariant vacuum expectation values $\bar{\sigma}$ and $\bar{\pi}_i$, and neglecting the fluctuations $\delta\sigma(x)$ and $\delta\pi_i(x)$. The mean values of the pion fields $\bar{\pi}_i$ vanish due to symmetry reasons.

The partition function in the grand canonical ensemble for finite temperature T and chemical potential μ can be obtained through the replacement

$$\int \frac{d^4p}{(2\pi)^4}\, F(p_4,\vec{p}) \quad \rightarrow \quad \sum\!\!\!\!\!\!\int_p F(p_4,\vec{p}) \equiv T \sum_{n=-\infty}^{n=\infty} \int \frac{d^3p}{(2\pi)^3}\, F(\omega_n - i\mu, \vec{p}) \;, \tag{3}$$

where ω_n are the Matsubara frequencies corresponding to fermionic modes, $\omega_n = (2n+1)\,\pi T$. The grand canonical thermodynamic potential in the mean field approximation is then given by

$$\Omega_{MF}(T,\mu) = -4N_c \left(\sum\!\!\!\!\!\!\int_p \log\left[p^2 + \Sigma^2(p^2)\right] \right) + \frac{\bar{\sigma}^2}{2G} \tag{4}$$

with the quark self-energy $\Sigma(p^2)$ defined as $\Sigma(p^2) = m_c + \bar{\sigma}\; r^2(p^2)$. The minima of the thermodynamical potential are obtained from the solutions of $\partial\Omega_{MF}/\partial\bar{\sigma} = 0$. This leads to the gap equation. In addition, given the thermodynamical potential the expressions for all other relevant quantities like the quark vacuum expectation value $\langle\bar{q}q\rangle$ and the quark density ρ_q can be easily derived [6]. In local models the sums over the Matsubara frequencies can be easily carried out and one obtains rather simple expressions in terms of the usual occupation numbers. However, for the covariant regulators we are considering here the situation is more complicated, since the analytic structure of the quark propagator in the complex plane can be much richer in this case: there might be a rather complicate structure of poles and cuts. Depending on the shape of the regulator and the parameters of the model, it might exist a situation in which there are no cuts, and the propagator has no poles along the real axis. Such situation can be interpreted as a sort of realization of confinement [5]. This is e.g. the case of the Gaussian regulator, $r(q^2) = \exp\left(-q^2/2\Lambda^2\right)$.

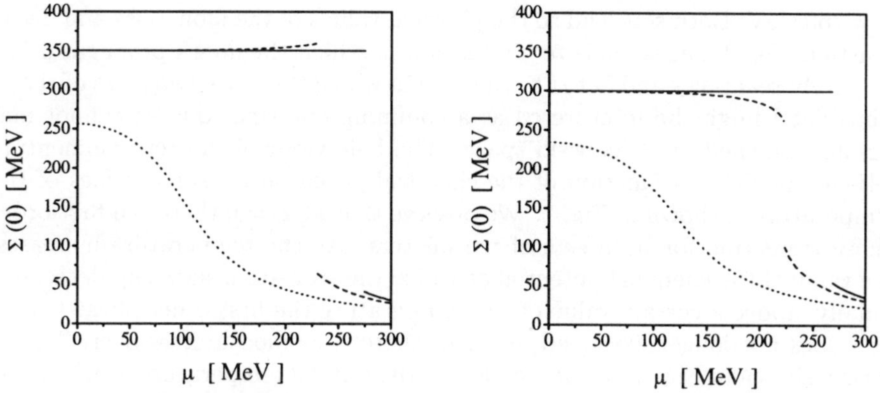

Figure 1. Self-energy as function of the chemical potential for some values of the temperature. Full, dashed and dotted lines correspond to $T = 0$, $T = 50$ MeV and $T = 100$ MeV respectively. The left (right) panel displays the results for Set I (Set II).

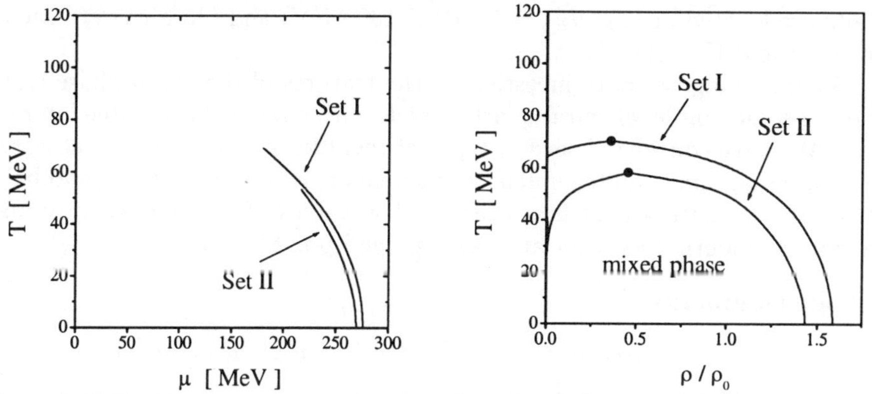

Figure 2. Critical temperatures as a function of the chemical potential (left) and the quark density (right). The fat dots in the right panel indicate the position of the end points.

3 Results and conclusions

We have found that the expressions for the relevant quantities can be given in terms of the poles and cuts of the quark propagator, and generalized (complex) occupation numbers can be naturally defined. We skip here a description of the analytical procedure (details can be found in Ref. 6), and just quote some numerical results obtained for the Gaussian regulator. Results for the Lorentzian and the instanton model regulators are found to be qualitatively similar [6]. We consider two sets of values for the parameters of the model: Set I corresponds to $G = 50\text{GeV}^{-2}$, $m_c = 10.5\text{MeV}$ and $\Lambda = 627\text{MeV}$, while for Set II the respective values are $G = 30\text{GeV}^{-2}$, $m_c = 7.7\text{MeV}$ and $\Lambda = 760\text{MeV}$. Both sets lead to the physical values of the pion mass and decay constant. Set I corresponds to a situation in which the quark propagator has no purely real poles and Set II to the case in which there are two pairs of them. Thus Set I might be interpreted as a confining one since quarks cannot materialize on-shell in Minkowski space. The behaviour of the zero-momentum self-energy $\Sigma(0)$ as function of the chemical potential for some values of the temperature is shown in Fig. 1. We observe that at $T = 0$ there is a first order phase transition for both sets of parameters. As the temperature increases, the value of the chemical potential at which the transition shows up decreases. Finally, above a certain value of the temperature the first order phase transition does not longer exist and, instead, there is a smooth cross-over. This is clearly shown in Fig. 2, where we display the critical temperature at which the phase transition occurs as a function of the chemical potential (left panel) and as a function of the density (right panel). The point at which the first order phase transition ceases to exist is usually called "end point". In our case these points are located at (T_E, μ_E) =(70 MeV, 180 MeV) and (55 MeV, 210 MeV) for Sets I and II, respectively.

Summarizing, we have investigated the features of the chiral phase transition in some non-local chiral quark models with covariant separable interactions. We have considered several types of regulators finding that in all cases the phase diagram is quite similar. In particular, we obtain that for two light flavors the transition is of first order at low values of the temperature and becomes a smooth cross-over at a certain "end point".

Acknowledgments

This work was supported in part by Fundación Antorchas, Argentina.

References

1. See e.g. S. Klevansky, *Rev. Mod. Phys.* **64**, 649 (1992).
2. G. Ripka, *Quarks Bound by Chiral Fields*, Oxford University Press, 1997.
3. T. Schaefer and E. Schuryak, *Rev. Mod. Phys.* **70**, 323 (1998).
4. C.D. Roberts and S.M. Schmidt, *Prog. Part. Nucl. Phys.* **45S1**, 1 (2000).
5. R.S. Plant and M.C. Birse, *Nucl. Phys.* A **628**, 607 (1998).
6. D. Gómez Dumm and N.N. Scoccola, *Phys. Rev.* D **65**, 074021 (2002).

COMPARISON BETWEEN $p\bar{p}$ ANNIHILATION INTO TWO PIONS AND THREE PIONS

E. A. VEIT AND M. BETZ
Instituto de Física, Universidade Federal do Rio Grande do Sul, Cx. P. 15051, CEP 91501-970, Porto Alegre, Brazil
eav@if.ufrgs.br and betz@if.ufrgs.br

J. HAIDENBAUER
Institut für Kernphysik, Forschungszentrum Jülich GmbH, 52425 Jülich, Germany
j.haidenbauer@fz-juelich.de

A comparison between $p\bar{p}$ annihilation into two pions and three pions is made in a baryon-exchange model with initial-state interactions. Annihilation diagrams involving nucleon and Δ-resonance exchanges are included consistently in both channels. The influence of the Δ exchange turns out to be strongly reduced for three-pion annihilation as compared to two-pion annihilation. Nevertheless, cross sections for annihilation into three uncorrelated pions can be similar in magnitude to those for annihilation into two pions.

1 Introduction

In this paper we focus on results of a combined study of proton-antiproton annihilation into two and three uncorrelated pions in a baryon-exchange model.[1] By "uncorrelated", we mean that these pions are not the decay products of an intermediate heavy-meson resonance. Our investigation was done in a distorted-wave Born approximation, employing the most recent $N\bar{N}$ model of the Jülich group[2] for the initial-state interaction, and guided by the same principles applied in that reference for the calculation of annihilation into two-meson channels.

The nucleon-antinucleon interaction developed by the Jülich group is derived in a time-ordered formalism and specifically also the amplitude for annihilation into two pions is calculated in time-ordered perturbation theory. In contrast, in order to avoid the proliferation of diagrams, the three-pion annihilation amplitude is calculated in Feynman-type perturbation theory. This procedure implies that the treatment of the off-shell behaviour of the three-pion annihilation amplitude is different from that used in the Jülich model of annihilation into two pions.[2] In order to treat the two-pion annihilation amplitude in the same fashion as the three-pion one, i.e. in Feynman-type perturbation theory, we refit the $NN\pi$ and $N\Delta\pi$ vertex parameters used in the model.

We start out from the two-pion annihilation channel by using available experimental data on $p\bar{p} \to \pi^+\pi^-$ to determine the free parameters of our model, i.e. the cutoff masses at the $NN\pi$ and $N\Delta\pi$ vertices, for several choices of form factors. We then turn to annihilation into three uncorrelated pions and discuss the relative importance of the NN, $N\Delta$ and $\Delta\Delta$ exchanges, as well as the relative sizes of the cross sections, compared to those for annihilation into two pions.

2 The Model

The basis of the present work is the Jülich model for $N\bar{N}$ scattering and annihilation.[2] In this model, heavier mesons are treated as stable particles in the description of $N\bar{N}$ annihilation into two mesons. In accordance with this approach, the uncorrelated three-pion channel is here considered separately from the channels in which a pion and a heavy meson, which can decay into two pions, are formed. Neglecting interferences between the various channels seems acceptable for the aim of this work.

The Born transition amplitudes $V^{N\overline{N}\to n\pi}$ $(n = 2, 3)$ are given by the sums of the Feynman tree diagrams involving N and Δ exchanges.[1] The complete amplitudes are obtained by summing such diagrams over all permutations of the final-state pions. The vertices in these diagrams correspond to the standard coupling terms in the Lagrangian density, with pseudovector pion-nucleon coupling. The values of the $NN\pi$ and $N\Delta\pi$ coupling constants are taken from the Bonn NN model.[3] We assume that the $\Delta\Delta\pi$ coupling constant is related to the $NN\pi$ one by the $SU(2) \times SU(2)$ quark–model relation.[4]

The vertices include form factors to regularize the calculation and to take into account the extended hadron structure. In order to investigate the sensitivity of our model to the details of these form factors, we employ three different parameterizations in the present study, given by the functions

$$\mathcal{F}_M(p) = \frac{\Lambda^2_{\alpha X\pi} - M^2_X}{\Lambda^2_{\alpha X\pi} - p^2} \quad \text{(monopole type)}\,, \tag{1}$$

$$\mathcal{F}_G(p) = exp\Big[-\frac{(p^2 - M^2_X)^2}{\Lambda^4_{\alpha X\pi}}\Big] \quad \text{(gaussian type)}\,, \tag{2}$$

and

$$\mathcal{F}_P(p) = \frac{\Lambda^4_X}{\Lambda^4_X + (p^2 - M^2_X)^2} \quad \text{(Pearce type}\ ^{5,6})\,, \tag{3}$$

with Λ_X the cutoff mass associated with a baryon line of type X (N or Δ), of mass M_X and four-momentum p. In Eq. (1) and (2), α denotes the type of the other baryon present at the vertex and $\Lambda_{\alpha X\pi}$ represents the cutoff mass corresponding to the $\alpha X\pi$ vertex. In order to reduce the number of free parameters, we assume $\Lambda_{N\Delta\pi} = \Lambda_{\Delta N\pi} = \Lambda_{\Delta\Delta\pi} \equiv \Lambda_\Delta$.

In order to combine the Feynman annihilation amplitudes with the initial-state distorted wave, it is necessary to specify a prescription for the energy components of the four-momenta of their external N and $\bar{N}$ legs. We set both equal to their on-energy-shell value (i.e., half the total available energy in the center-of-mass frame).

3 Results and Conclusions

In this contribution, we concentrate on the cross sections for $p\bar{p}$ annihilation into two and three uncorrelated charged pions. Results for neutral pions, as well as for the various charged and uncharged branching ratios, will appear elsewhere.[1]

To get insight into the influence of the vertex form factors on the results for $p\bar{p} \to 3\pi$, we accomplished fits to the reaction $p\bar{p} \to \pi^+\pi^-$ using each of the three parameterizations with two sets of models, one with Λ_N larger than Λ_Δ and the other with Λ_N smaller than Λ_Δ. Thus, we have in total six parameter sets, all determined in such a way as to provide an adequate description of the total and differential $p\bar{p} \to \pi^+\pi^-$ cross sections. The parameter set A corresponds to form factors which suppress Δ exchange compared to nucleon exchange. The parameter set B corresponds to a quite strong Δ exchange, hopefully providing some sort of upper bound for the Δ contribution when used to describe annihilation into three uncorrelated pions.

The relative importance of just N exchange for the two- and three-pion annihilation cross sections is illustrated in Table 1 for the two parameter sets. The quoted ranges are based on results for the three different form factors, calculated for two sample laboratory kinetic energies (65 MeV and 220 MeV). For the parameter set A, the contribution from Δ exchange is indeed not very important for two pions and is completely negligible for three pions. For the parameter set B, the Δ-exchange contribution dominates over the N contribution in two-pion annihilation, although, because of the constructive interference between them, both contributions are important in building up the cross section. In the three-pion channel, for parameter set B, there is only a moderate effect from the Δ exchange.

As a consequence of these features, the total annihilation cross section into three uncorrelated pions is generally significantly larger if we assume that N

Table 1. Relative sizes of total cross sections calculated with different parameter sets (first column): i) N exchange only compared to whole $p\bar{p} \to \pi^+\pi^-$ cross section (second column); ii) NN exchange only compared to whole $p\bar{p} \to \pi^+\pi^-\pi^0$ cross section (third columnn); whole $p\bar{p} \to \pi^+\pi^-\pi^0$ cross section compared to whole $p\bar{p} \to \pi^+\pi^-$ cross section.

Set	$\sigma(N)/\sigma(N{+}D)$	$\sigma(NN)/\sigma(NN{+}ND{+}DD)$	$\sigma(\pi^+\pi^-\pi^0)/\sigma(\pi^+\pi^-)$
A	73% - 98%	97% - 101%	45% - 128%
B	9% - 26%	57% - 80%	3% - 39%

exchange dominates the two-pion decay (parameter set A). This is illustrated by the last column of Table 1, which shows the ratio of the total cross sections for annihilation into $\pi^+\pi^-\pi^0$ and $\pi^+\pi^-$. While this ratio is of order unity for set A, it can be quite small for set B.

Not unexpectedly, the actual magnitude of the three-pion annihilation cross section depends rather severely on the choice of the vertex form factors. This dependence is reduced by requiring consistency between the treatments of the two- and three-pion decay channels, but ultimately it cannot be avoided since the two-pion data can be fitted about equally well with models that differ quite significantly in the details of their dynamics. With due concession for these uncertainties, our calculations show that annihilation into three uncorrelated pions should not be disregarded.

Acknowledgments

"Centro Nacional de Supercomputação da Universidade Federal do Rio Grande do Sul (UFRGS)", DLR (Germany; BRA W0B 2F) - CNPq (Brazil; 910133/94-8), and FAPERGS (Fundação de Amparo à Pesquisa do Estado do Rio Grande do Sul), Brazil.

References

1. M . Betz, E. A. Veit and J. Haidenbauer, Eur. Phys. J. A **14**, 113-124 (2002).
2. V. Mull and K. Holinde, Phys. Rev. C **51**, 2360 (1995).
3. R. Machleidt, K. Holinde, and Ch. Elster, Phys. Rep. **149** (1987) 1; version given in Table 9 therein.
4. J. J. De Swart, Rev. Mod. Phys. **35**, 916 (1963).
5. B. C. Pearce, and B. K. Jennings, Nucl. Phys. **A528**, 655 (1991).
6. C. Schütz, J. W. Durso, K. Holinde, and J. Speth, Phys. Rev. C **49**, 2671 (1994).

THE MODIFIED FUZZY BAG MODEL: FITTING THE MASSES OF THE LOW-LYING BARYONS

H. T. COELHO
Departamento de Fsica, Universidade Federal de Pernambuco,
50670-901 Recife, Brazil
heliotcoelho@terra.com.br

M. DILLIG
Institut für Theoretische Physik III, der Universität Erlangen-Nürnberg,
Erlangen, Germany
mdillig@theorie3.physik.uni-erlangen.de

F. G. PILOTTO, C.A.Z. VASCONCELLOS
Instituto de Física, Universidade Federal do Rio Grande do Sul,
91501-970 Porto Alegre, RS, Brazil
pilotto@if.ufrgs.br, cesarzen@if.ufrgs.br

In this work we revisit the fuzzy bag model, by introducing as a new feature explicitly the requirement of energy-momentum conservation over a finite smooth bag surface (as opposed to the MIT bag or relativistic potential models). Including in addition center-of-mass and one-gluon exchange corrections, the model parameters are determined from a consistent fit of the baryon masses in the lowest baryon octet and decuplet. As one immediate and crucial consequence, the bag constant B acquires a radial dependence, whose functional form is completely determined from the internal consistency of the model. Together with chiral symmetry respected from the coupling to the external pion cloud, our approach is a first step in modeling, in a consistent way, the transition of effective hadronic degrees of freedom towards a deconfined quark gluon plasma at very high densities.

1 Introduction

In this work we develop a new version of the fuzzy bag model. As the main element, we incorporate and explore explicitly the conservation of energy and momentum over a smooth, finite bag surface. As one immediate consequence, the bag pressure B acquires a radial dependence, $B = B(r)$, whose functional form is completely determined within the consistency of the model. This property is of paramount importance for the correct interpretation of the model as a bag model, with the outward pressure of the quarks inside the bag locally balanced by the inward pressure of the non-perturbative vacuum outside the bag. Clearly this feature is absolutely missing in relativistic potential models; it is also absent in the formulation of the original fuzzy bag model[1] and its predecessor, the MIT bag model[2]: in both versions the non-perturbative QCD vacuum is parameterized by a constant B in the Lagrangian density. As in previous approaches[3,4], we consider a bag with an interior region with free quarks, a surface region, where the quark wave function is gradually suppressed, and an exterior region, where the quark wave function vanishes. The physical picture underlying such a modeling is intuitively very appealing: it combines the interior region of the bag as the perturbative vacuum of QCD with the exterior region of the non-perturbative vacuum of QCD. The finite surface region with a typical scale of about half a fermi models the smooth transition region between the two vacua, as expected from a gradual outside-inside restoration of chiral symmetry.

In our model, the Lagrangian density is[3]

$$\mathcal{L}_{FBM} = \frac{i}{2}[\bar{\psi}\gamma^{\mu}\partial_{\mu}\psi - \partial_{\mu}\bar{\psi}\gamma^{\mu}\psi] - B(r)F(r) - \bar{\psi}[m_q + (1+\gamma^0)V(r)]\psi, \quad (1)$$

where, for a given spherically symmetric mean field potential for a single quark given as $V(r) = \frac{1}{2}V_0 + V_c(r)$, the fuzzyness of the bag surface is parameterized in the function $F(r)$ (details involving notation can be found in references[3,4]). The divergence of the energy-momentum tensor $\partial_\mu T^{\mu\nu} = 0$ implies

$$\partial_r\,[B(r)\,F(r)] + \sum_q \bar{\psi}\,(1+\gamma^0)\,\psi\,\partial_r V_c(r) = 0\,. \quad (2)$$

We remark that, in contrast to the MIT bag, in our model the energy-momentum conservation is imposed, rather than derived from requiring the action to be stationary under deformations of the bag surface. Moreover, it is important to point out that the above equation cannot be satisfied for a constant B: the structure of the model imposes that both the quark wave functions and $F(r)$ are self-consistently determined from a set of coupled equations for a given $V_c(r)$, which then unavoidably enforces a radial dependence of the bag pressure. Focusing below on the baryons in their ground state (i. e. with purely s-wave quarks), $\bar{\psi}\,(1+\gamma^0)\,\psi$ shows only a radial dependence, reflecting the underlying spherical symmetry; for excited states the admixture of angular momenta quite naturally leads to an angular dependence in B. From the equation above it is possible to determine completely the form of $B(r)$ for a given choice of the suppression function and confining potential. This internal consistency avoids the high degree of arbitrariness, as introduced in various approaches, and allows a very natural extrapolation of the model from free baryons into dense hadronic matter, such as encountered in the investigation of neutron star properties.

By demanding the suppression function and the confining potential to be continuous for all values of r, we finally find[5] that, in the surface region, characterized by two radial parameters ($R_0 < r < R_1$), $B(r)$ satisfies

$$B(r) = \frac{1}{\mathcal{F}(r)} \left(B_0 - \frac{1}{2\pi} \sum_q \int_{R_0}^{r} \mathrm{d}r \, \frac{u(r)^2}{r^2} \frac{d\mathcal{V}_c(r)}{dr} \right) . \tag{3}$$

Due to the continuity of $B(r)$, then $B_0 = B(R_0)$. Requiring the product $B(r)\,F(r)$ to be finite for all r, from $\mathcal{F}(R_1) = 0$ the expression inside the parenthesis in Eq. (3) has to vanish for $r = R_1$. Thus, as a consequence of energy-momentum conservation, $B(r)$ is completely determined for all values of r. The next step of our calculation is to consider the Euler-Lagrange equations and its solutions. After normalization of the solutions, we determine the baryon masses, given by the sum of the quark energies and the volume contribution,

$$M = \sum_q E_q + \frac{2}{3} \sum_q \int_{R_0}^{R_1} \mathrm{d}r \, r \, u(r)^2 \, \frac{d\mathcal{V}_c(r)}{dr} . \tag{4}$$

This expression is to be compared with the corresponding one in the MIT bag model: in our model, the volume contribution reduces in its final form to an integration within the range of the two radial parameters R_0 to R_1, which involves thus only the surface of the bag. For a more realistic estimate of the masses, we improve this result by including center-of-mass and one-gluon exchange corrections as well as the pion field to respect chiral symmetry. As a result, the pion-quark interaction induces renormalization effects on the hadronic wave functions and observables. Expecting only moderate corrections from the pion cloud, the corresponding calculations are readily performed in a framework similar to the non-relativistic Chew-Low model of the nucleon-pion interaction[4,5]. Here we neglect both baryon recoil and the influence from the vacuum due to the presence of anti baryons; we stress, however, that in spite of complex technicalities, a relativistic extension of such calculations is readily feasible.

2 Results

As a first test of the consistency of the model, we provide in Table 1 our results for fitting the masses of the low-lying baryons (for the details see Refs.[4,5]).

Table 1. Results for the masses of the low-lying baryons (in MeV).

baryon	N	Δ	Λ	Σ	Σ^*	Ξ	Ξ^*	Ω^-
experim.	940	1232	1116	1193	1392	1318	1533	1672
theory	961	1232	1138	1195	1381	1312	1497	1671

Table 2. Separate contributions to the total mass of each baryon (in MeV).

baryon	N	Δ	Λ	Σ	Σ*	Ξ	Ξ*	Ω^-
E_B	960	960	1125	1125	1125	1289	1289	1453
E_{BF}	300	300	274	274	274	249	249	224
$(\Delta E_B)_{cm}$	-126	-126	-104	-104	-104	-88	-88	-77
$(\Delta E_B)_g^M$	-126	126	-126	-82	104	-100	85	71
$(\Delta E_B)_\pi$	-47	-28	-31	-18	-18	-38	-38	0

In Table 2 we have listed the separate contributions which build up each baryon mass: E_B is the sum of the energy eigenvalues of each quark, E_{BF} is the volume energy due to the term $B(r)\,F(r)$ in the Lagrangian density, $(\Delta E_B)_{cm}$ is the center of mass correction, $(\Delta E_B)_g^M$ is the color magnetic energy, and $(\Delta E_B)_\pi$ is the self-energy due to pion-quark interaction.

3 Conclusion and outlook

The approach presented is a first step in current attempts to improve the modeling of quark dynamics and confinement beyond standard bag models. Promising extensions and applications are for example the investigation of baryon form factors at large momentum transfers, or the immersion of the quark bag into hadronic matter towards a more realistic investigation of the transition of effective (mesons and baryons) to constituent (quark and gluons) degrees of freedom in the high density domain of neutron star matter.

Acknowledgments

Supported in part by the Forschungszentrum Jülich (FZJ), Germany, under contract 41445386 (COSY-62).

References

1. Y. Nogami, A. Suzuki and N. Yamanishi, *Can. J. Phys.* **62**, 554 (1984).
2. A. Chodos, R. L. Jaffe, C. B. Thorn and V. Weisskopf, *Phys. Rev.* **D9**, 3471 (1974).
3. C. A. Z. Vasconcellos, H. T. Coelho, F. G. Pilotto, B. E. J. Bodmann, M. Dillig and M. Razeira, *Eur. Phys. J.* **C4**, 115 (1998).
4. F. G. Pilotto, C. A. Z. Vasconcellos and H. T. Coelho, *Mod. Phys. Lett.* **A17**, 543 (2002).
5. F. G. Pilotto and C. A. Z. Vasconcellos, *submitted for publication.*

THREE-PION-EXCHANGE POTENTIAL

JOEL C. PUPIN
Instituto de Física Teórica, Universidade Estadual Paulista
01405-900, Rua Pamplona 145, São Paulo, SP, Brazil
pupin@fermanet.com.br

Using the chiral symmetry, we calculated the dominant contribution to the nucleon - nucleon potential due to the exchange of three non-correlated pions. This contribution is isovetor with pseudoscalar and axial components. The pseudoscalar component is dominant, it has a range of 1.0 fm and it contributes in the pion channel.

1 Introduction

The nucleon-nucleon interaction due to exchange of three uncorrelated pions can be represented by the coupling of two processes $\pi N \to \pi\pi N$ at tree level, as shown in Figure 1. This coupling corresponds to the amplitude

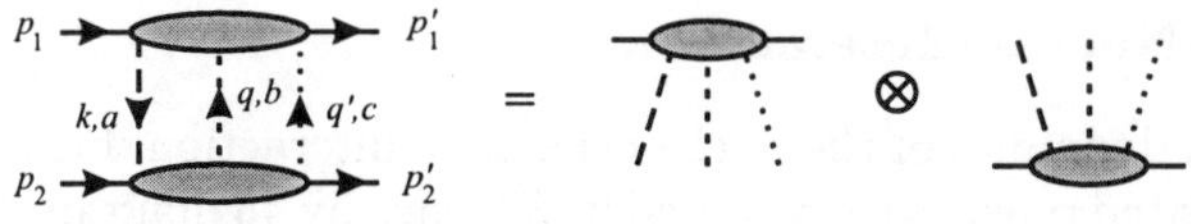

Figure 1. Decomposition of $NN \to NN$ in two intermediate processes $\pi N \to \pi\pi N$.

$$T_{3\pi} \;=\; \frac{1}{3\,!}\int \frac{d^4Q}{(2\pi)^4}\int \frac{d^4Q'}{(2\pi)^4}\;\Delta(k)\;\Delta(q)\;\Delta(q')\;T^{(1)}_{abc}\;T^{(2)}_{cba}\;, \tag{1}$$

where $Q = (q' + q + k)\,/2$ and $Q' = (q' - q)\,/2$ are integration variables, $\Delta(k)$, $\Delta(q)$ and $\Delta(q')$ are propagators of the pions a, b and c, and $T^{(1)}_{abc}$ and $T^{(2)}_{cba}$ are pion production amplitudes for nucleon 1 and 2. The factor $1/3!$ is due to the symmetry of the intermediate three-pion state.

There are many alternative ways of implementing chiral symmetry. In particular, the subset of diagrams in the pure pion-nucleon sector, corresponding to the minimal realization of chiral symmetry in this problem, may be evaluated by means of non-linear Lagrangians with πN couplings that may be either pseudovector (PV) or pseudoscalar (PS). It is important to stress that these Lagrangians, in spite of their different aspects, have the same dynamical content and the physical results do not depend on the particular version one adopts.

Using PV Lagrangian, pion production is represented by the diagrams in Figure 2. These diagrams correspond to the sum $T_{cba} = T^{\pi}_{cba} + T^{ct}_{cba} + \boxed{T^{res}_{cba}}$, where T^{π}_{cba} is the amplitude from pion pole diagram in the t-channel, T^{ct}_{cba} is the amplitude from contact diagram, and denotes a rest, T^{res}_{cba}, whose dynamical content is indicated by the frame in Figure 2.

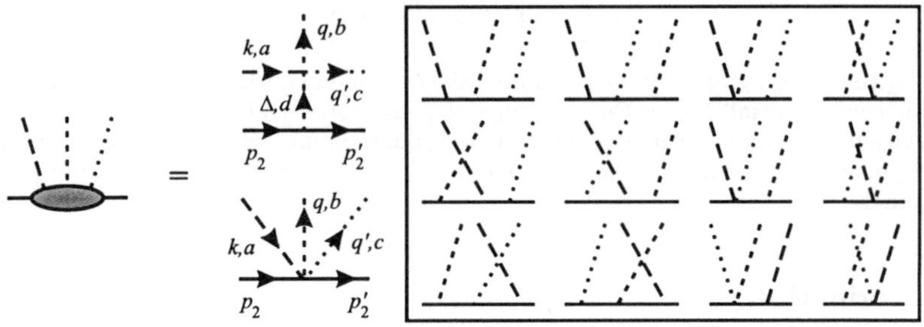

Figure 2. Feynman diagrams generated by PV Lagrangian for the process $\pi N \to \pi\pi N$.

At threshold, the approximation $T_{cba} \cong T^{\pi}_{cba} + T^{ct}_{cba}$ reproduces correctly the leading contribution at threshold[1].

2 Nucleon-Nucleon Interaction

The chiral dynamics of the nucleon-nucleon interaction due to exchange of three uncorrelated pions is represented in PV case by 45 diagrams with 2 loops, involving just pions and nucleons. As the exact calculation of this interaction is too extensive, we use the dominant contribution of pion production at threshold of the previous section and obtain the class of diagrams [a] in Figure 3.

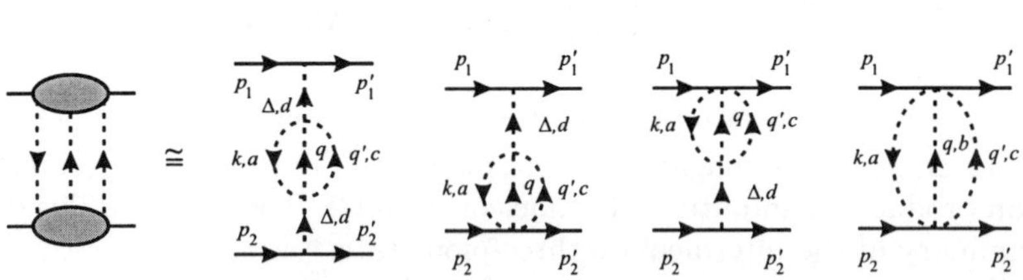

Figure 3. PV diagrams for the NN interaction due to exchange of three uncorrelated pions.

This approximation yields the isovectorial amplitude

$$T = \left(\frac{g_A}{8f_\pi^3}\right)^2 \boldsymbol{\tau}^{(1)}\cdot\boldsymbol{\tau}^{(2)}\; [\bar{u}\,\gamma_5\,\gamma^\mu u]^{(1)}\, [\bar{u}\,\gamma_5\,\gamma^\nu u]^{(2)}\; \frac{1}{3!}\int \frac{d^4Q}{(2\pi)^4}\frac{1}{(Q-\Delta/2)^2-\mu^2}$$

[a]The last diagram was calculated in reference[2].

$$\times \int \frac{d^4Q'}{(2\pi)^4} \frac{1}{\left[(Q/2+\Delta/4+Q')^2-\mu^2\right]\left[(Q/2+\Delta/4-Q')^2-\mu^2\right]}$$

$$\times \left\{ 32\Delta_\mu\Delta_\nu \, \frac{6\Delta\cdot Q\,\Delta\cdot Q - 2\mu^2\Delta\cdot Q + \mu^4 + 8\Delta_\rho\Delta_\sigma Q'^\rho Q'^\sigma}{(\Delta^2-\mu^2)^2} \right.$$

$$- 8\left(\Delta_\mu g_{\nu\rho}+\Delta_\nu g_{\mu\rho}\right) \frac{6\Delta\cdot Q\,Q^\rho + 3\Delta\cdot Q\,\Delta^\rho - \mu^2 Q^\rho - \mu^2\Delta^\rho/2 + 8\Delta_\sigma Q'^\rho Q'^\sigma}{\Delta^2-\mu^2}$$

$$\left. + 12Q_\mu Q_\nu + 6\Delta_\nu Q_\mu + 6\Delta_\mu Q_\nu + 3\Delta_\mu\Delta_\nu + 16 g_{\mu\rho}g_{\nu\sigma}Q'^\rho Q'^\sigma \right\} . \tag{2}$$

Some terms contribute to one-pion exchange too, renormalizing pion mass and πN coupling constant. As we are working with the physical pion mass μ and the physical coupling constant $g_{\pi N}$, these effects are already taken into account in one-pion exchange. Therefore, these terms will be neglected here.

Going to center of mass and non-relativistic limit, we obtain the potential in configuration space [b]

$$V_{\rm SS}^{\rm P} = \Pi \left(1 + \frac{7}{ax} + \frac{27}{a^2x^2} + \frac{60}{a^3x^3} + \frac{60}{a^4x^4}\right)\left\{\gamma - 3\gamma^2 + 3\gamma^3\left[1-\alpha(1-\alpha)\right]\right\} \tag{3}$$

$$V_{\rm T}^{\rm P} = \Pi \left(1 + \frac{10}{ax} + \frac{45}{a^2x^2} + \frac{105}{a^3x^3} + \frac{105}{a^4x^4}\right)\left\{\gamma - 3\gamma^2 + 3\gamma^3\left[1-\alpha(1-\alpha)\right]\right\} \tag{4}$$

$$V_{\rm SS}^{\rm A} = -\frac{\Pi}{a\,x}\left(1 + \frac{6}{ax} + \frac{15}{a^2x^2} + \frac{15}{a^3x^3}\right)\left\{\gamma^2 - \gamma^3\left[1-\alpha(1-\alpha)\right]\right\} , \tag{5}$$

where

$$\Pi = \frac{32}{3}\left(\frac{g_A}{8f_\pi^3}\right)^2 \frac{\mu^7}{(4\pi)^5}\; \boldsymbol{\tau}^{(1)}\cdot\boldsymbol{\tau}^{(2)} \int_0^1 d\alpha \int_0^1 d\gamma \; u^4 \; \frac{e^{-a\,x}}{x^3} , \tag{6}$$

$$x = \mu r , \qquad a = \sqrt{\frac{1}{1-\gamma} + \frac{1}{\alpha(1-\alpha)\,\gamma}} . \tag{7}$$

3 Conclusions

Inspecting the graphics for the spin-spin channel in Figure 4, one learns that the contribution of the axial component is quite small and hence the

[b] Superscripts P and A correspond to exchanges of pseudoscalar and axial systems and subscripts SS and T, to spin-spin and tensor channels.

3π-exchange potential is dominated by the pseudoscalar component. In the spin-spin and tensor channels, three-pion exchange tends to add to the one-pion exchange and is visible up to 1.0 fm only for the spin-spin channel. In both channels, its contribution is smaller than that obtained with the exchange of one meson a_1 as in reference[3].

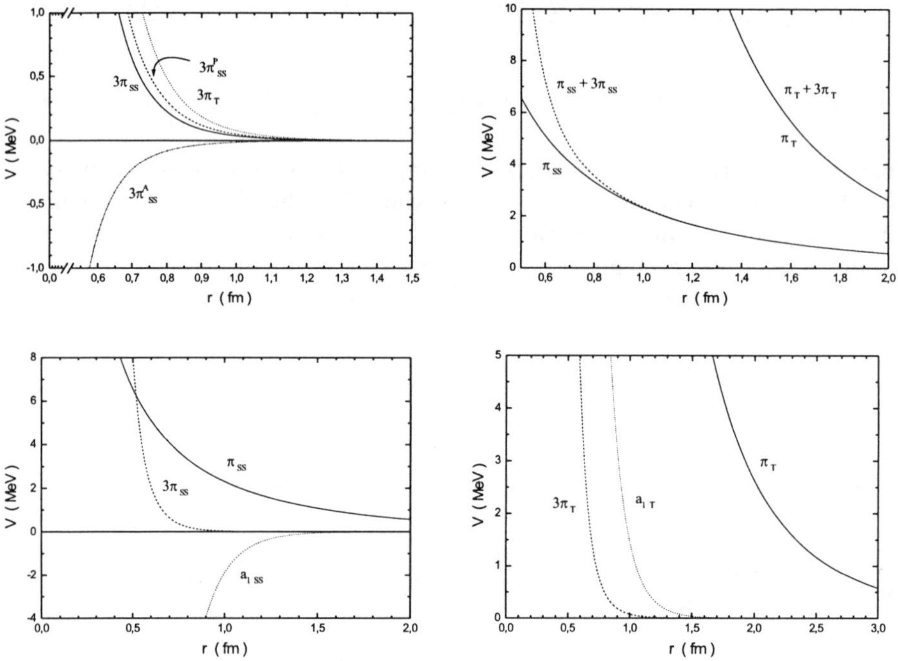

Figure 4. Profile functions of π, a_1, and 3π exchanges.

Acknowledgments

We acknowledge the support of FINPE Group (Instituto de Física, Universidade de São Paulo) and of FAPESP (Brazilian Agency, Grant No. 94/03469-7).

References

1. M.G. Olsson and L. Turner, Phys. Rev. Lett. **20** (1968) 1127.
2. J.C. Pupin and M.R. Robilotta, Phys. Rev. C **60** (1999) 014003.
3. N. Kaiser, Phys. Rev. C **61** (2000) 014003.

DYNAMICAL NATURE OF THE NUCLEAR PSEUDOSPIN AND ITS ISOSPIN ASYMMETRY

R. LISBOA, M. MALHEIRO, A. DELFINO
Instituto de Física, Universidade Federal Fluminense, Niterói, Brazil
mane@if.uff.br

P. ALBERTO, M. FIOLHAIS
Departamento de Física and Centro de Física Computacional, Universidade de Coimbra, P-3004-516 Coimbra, Portugal
tmanuel@teor.fis.uc.pt

M. CHIAPPARINI
Departamento de Física Teórica, Universidade do Estado do Rio de Janeiro, Brazil
chiappa@uerj.br

Pseudospin symmetry in nuclei is investigated by solving the Dirac equation with Woods-Saxon scalar and vector radial potentials. We relate the pseudospin interaction with a pseudospin-orbit term in a Schroedinger-like equation for the lower component of the Dirac spinor. We show that this term gives a large contribution to the energy splittings of pseudospin partners, so that the near pseudospin degeneracy arises from a significant cancellation among the different terms in that equation. This is a manifestation of the dynamical character of this symmetry in the nucleus. We analyze the isospin dependence of the pseudospin symmetry in a nuclear isotope chain by including a vector-isovector potential V_ρ and a Coulomb potential and conclude that V_ρ gives the main contribution to the observed pseudospin isospin asymmetry.

1 Introduction

Pseudospin symmetry was introduced in the late 60's to account for a degeneracy in single-particle nucleon levels of certain nuclei, with quantum numbers $(n, \ell, j = \ell + 1/2)$ and $(n - 1, \ell + 2, j = \ell + 3/2)$ [1,2]. These levels have the same quantum number $\tilde{\ell} = \ell + 1$, called "pseudo" orbital angular momentum, and a "pseudo" spin quantum number, $\tilde{s} = 1/2$. In general, for a $n\ell j$ state, we have $\tilde{\ell} = \ell \pm 1$ if $j = \ell \pm 1/2$. Pseudospin symmetry is exact when doublets with $j = \tilde{\ell} \pm \tilde{s}$ are degenerate. Since then the origin of pseudospin symmetry in nuclei has been a subject of intense theoretical research.

Recently, Ginocchio [3] uncovered the relativistic character of the symmetry, noting that the pseudo-orbital angular momentum is the orbital angular momentum of the lower component of the Dirac spinor. This is an exact SU(2) symmetry for the Dirac Hamiltonian [4] with an attractive scalar potential, S, and a repulsive vector potential V, when they have the same magnitude, i.e., $\Sigma = S+V = 0$. This would explain why there are quasi-degenerate pseudospin doublets in nuclei, since in relativistic mean field (RMF) models nuclear saturation arises from an extensive cancellation between a large attractive scalar potential and a large repulsive vector potential. However, since Σ acts as a binding potential, one cannot set $\Sigma = 0$ in the nucleus, because then there would not exist any bound states. Other authors have studied the role of pseudospin-orbit coupling in pseudospin symmetry [5,6].

2 Dynamical pseudospin symmetry and isospin asymmetry

Pseudospin symmetry is related to an invariance under a SU(2) transformation of the Dirac Hamiltonian with central scalar S and vector V potentials

$$H = \boldsymbol{\alpha} \cdot \boldsymbol{p} + \beta(m + S) + V \ , \tag{1}$$

whose generators are [4] $S_i = s_i \, \frac{1}{2}(1 - \beta) + \tilde{s}_i \, \frac{1}{2}(1 + \beta)$ where $\tilde{s}_i = \frac{\boldsymbol{\sigma}\cdot\boldsymbol{p}}{p} s_i \frac{\boldsymbol{\sigma}\cdot\boldsymbol{p}}{p}$, $s_i = \sigma_i/2$. The commutator of S_i with (1) is zero if $\Sigma = 0$ or $\frac{d\Sigma}{dr} = 0$ [7].

The Dirac equation $H\Psi = \epsilon\Psi$ can be written as a pair of second-order differential equations for the upper and lower components. Setting $\Delta = V - S$, V and S being radial functions, and defining $\epsilon = E + m$, we have

$$p^2\Psi_+ = \frac{\Delta'}{E + 2m - \Delta}\left(\frac{\partial}{\partial r} - \frac{1}{r}\boldsymbol{\sigma} \cdot \boldsymbol{L}\right)\Psi_+ + (E + 2m - \Delta)(E - \Sigma)\Psi_+ \tag{2}$$

$$p^2\Psi_- = \frac{\Sigma'}{E - \Sigma}\left(\frac{\partial}{\partial r} - \frac{1}{r}\boldsymbol{\sigma} \cdot \boldsymbol{L}\right)\Psi_- + (E + 2m - \Delta)(E - \Sigma)\Psi_- \ , \tag{3}$$

where $\Psi_\pm = [(1 \pm \beta)/2]\Psi$ are the upper and lower components and the primes denote derivatives with respect to r. The $\boldsymbol{\sigma} \cdot \boldsymbol{L}$ term in Eq. (3) is the pseudospin-orbit term.

We solved the Dirac equation with the Hamiltonian (1) with scalar and vector potentials with Woods-Saxon shape: $U(r) = U_0/\big(1 + \exp[(r - R)/a]\big)$. We fitted the parameters of this potential to the neutron spectrum of ^{208}Pb, obtaining the values $R = 7$ fm, $\Delta_0 = 650$ MeV, $\Sigma_0 = -66$ MeV and $a = 0.6$ fm. We then varied the diffuseness, radius and Σ_0 separately, keeping all the other parameters fixed and showed that the splittings of the pseudospin doublets vary notably, namely they decrease as $|\Sigma_0|$ decreases and diffuseness increases, sometimes reversing sign. The details can be found in ref. [7].

The effect of the pseudospin-orbit term on pseudospin energy splittings can be assessed by writing (3) as a Schroedinger-like equation, dividing it by an energy- and r-dependent effective mass $m^* = (E + 2m - \Delta)/2$ and taking

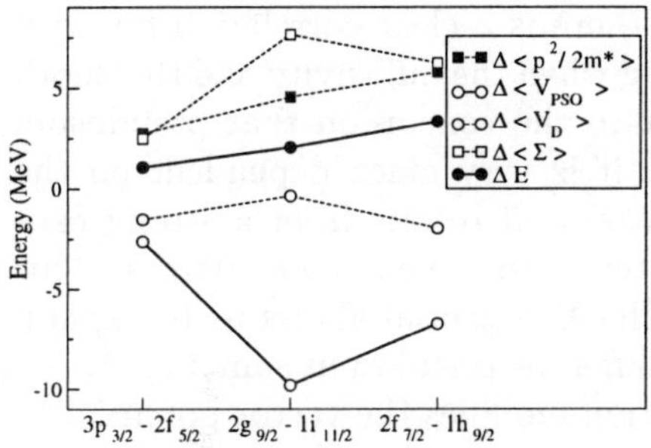

Figure 1. Differences for the energy terms in (4) for ($1i_{11/2}$, $2g_{9/2}$), ($2f_{5/2}$, $3p_{3/2}$) and ($1h_{9/2}$, $2f_{7/2}$) pseudospin partners.

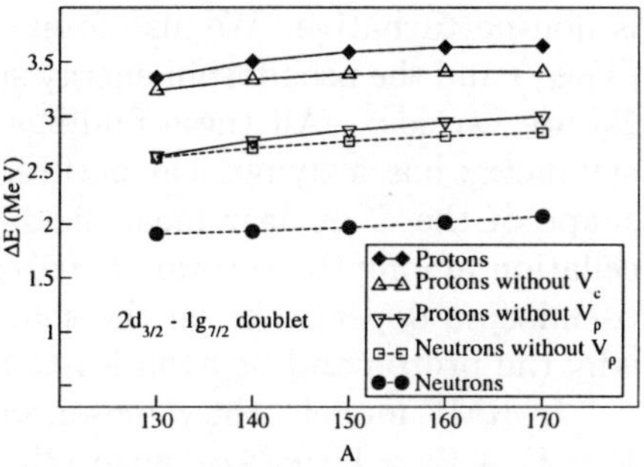

Figure 2. Effect of Coulomb and ρ potentials V_c and V_ρ in the splitting of the ($2d_{5/2}$, $1g_{7/2}$) pseudospin partners in a Sn isotopic chain.

the expectation value of each term relative to Ψ_-:

$$\left\langle \frac{p^2}{2m^*} \right\rangle + \langle V_{\rm PSO} \rangle + \langle V_{\rm D} \rangle + \langle \Sigma \rangle = E \ , \tag{4}$$

where

$$\left\langle \frac{p^2}{2m^*} \right\rangle = \int \Psi_-^\dagger \frac{p^2}{2m^*} \Psi_- \, {\rm d}^3 \boldsymbol{r} \Big/ \int \Psi_-^\dagger \Psi_- \, {\rm d}^3 \boldsymbol{r} \ , \tag{5}$$

$$\langle V_{\rm PSO} \rangle = {\rm P} \int \Psi_-^\dagger \frac{1}{2m^*} \frac{\Sigma'}{E - \Sigma} \frac{1}{r} \boldsymbol{\sigma} \cdot \boldsymbol{L} \Psi_- {\rm d}^3 \boldsymbol{r} \Big/ \int \Psi_-^\dagger \Psi_- \, {\rm d}^3 \boldsymbol{r} \tag{6}$$

$$\langle V_{\rm D} \rangle = -{\rm P} \int \Psi_-^\dagger \frac{1}{2m^*} \frac{\Sigma'}{E - \Sigma} \frac{\partial \Psi_-}{\partial r} \, {\rm d}^3 \boldsymbol{r} \Big/ \int \Psi_-^\dagger \Psi_- \, {\rm d}^3 \boldsymbol{r} \ , \tag{7}$$

$$\langle \Sigma \rangle = \int \Psi_-^\dagger \Sigma \Psi_- \, {\rm d}^3 \boldsymbol{r} \Big/ \int \Psi_-^\dagger \Psi_- \, {\rm d}^3 \boldsymbol{r} \ , \tag{8}$$

where 'P' denotes the principal value of the integral. These terms can be identified, respectively, as a kinetic term, a pseudospin-orbit term, a potential term related to what is sometimes called Darwin term and the mean value of the Σ potential with respect to the lower component, Ψ_-.

In Fig. 1 are plotted the differences of the terms in Eq. (4) between each member of the three topmost pseudospin partners in ^{208}Pb, ($1i_{11/2}$, $2g_{9/2}$), ($2f_{5/2}$, $3p_{3/2}$) and ($1h_{9/2}$, $2f_{7/2}$). One sees that the contribution of $V_{\rm PSO}$ for the pseudospin energy splittings is larger than the splittings themselves and has the opposite sign of the kinetic and $\langle \Sigma \rangle$ terms. For all doublets there is a significant cancellation of $\langle V_{\rm PSO} \rangle$ with the kinetic and $\langle \Sigma \rangle$ terms. A similar analysis was made in ref. [6], concluding that the pseudospin-orbit potential is non-perturbative. We also found that there is a clear correlation between $\langle V_{\rm PSO} \rangle$ and the pseudospin energy splitting when the diffusivity and the depth Σ_0 are varied [7]. All these findings point to the conclusion that pseudospin symmetry has a dynamical nature, since it is very much dependent on the shape of the Σ nuclear mean field potential and results from a strong cancellation among the several contributing terms to the energy splittings. This pseudospin dependence on the shape of the Σ potential allows us to explain why the proton and neutron levels have different pseudospin splittings [8].

In RMF models, the ρ meson interaction modifies the vector potential to $V = V_\omega + V_\rho = V_\omega \pm \frac{g_\rho}{2}\rho_0$, where the plus sign refers to protons and the minus sign refers to neutrons, and ρ_0 is the time component of the ρ field which is proportional to $Z - N$. The magnitude of the ρ potential is around 4-8 MeV. For heavy nuclei, $\rho_0 < 0$ and therefore $V_p < V_n$ which implies $|\Sigma_n| < |\Sigma_p|$. This means that, for neutron-rich nuclei, the pseudospin symmetry is favored for neutrons. Notice, however, that the Coulomb potential, which has the opposite sign of V_ρ for protons, also affects proton spectra. Both of these effects can be seen in Fig. 2, where is plotted the effect of V_ρ and the Coulomb potential on the energy splitting of the $(2d_{5/2}, 1g_{7/2})$ pseudospin partners along the Sn isotopic chain from $A = 130$ to $A = 170$. We see that the potential V_ρ gives the main contribution for the observed isospin asymmetry.

Acknowledgments

We acknowledge financial support from FCT (POCTI), Portugal, and from CNPq/ICCTI Brazilian-Portuguese scientific exchange program.

References

1. K. T. Hecht and A. Adler, *Nucl. Phys.* **A137**, 129 (1969).
2. A. Arima, M. Harvey, and K. Shimizu, *Phys. Lett.* **B30**, 517 (1969).
3. J. N. Ginocchio, *Phys. Rev. Lett.* **78**, 436 (1997); *i*bid, *Phys. Rept.* **315**, 231 (1999).
4. J. N. Ginocchio and A. Leviatan, *Phys. Lett.* **B245**, 1 (1998). *i*bid, *Phys. Rev. Lett.* **82**, 4599 (1999).
5. J. Meng, K. Sugawara-Tanabe, S. Yamaji, P. Ring, and A. Arima, *Phys. Rev.* **C58**, R628 (1998).
6. S. Marcos, M. López-Quelle, R. Niembro, L. N. Savushkin, and P. Bernardos, *Phys. Lett.* **B513**, 30 (2001).
7. P. Alberto, M. Fiolhais, M. Malheiro, A. Delfino, and M. Chiapparini, *Phys. Rev. Lett.* **86**, 5015 (2001); *i*bid, *Phys. Rev.* **C65**, 034307 (2002).
8. J. Meng and I. Tanihata, *Nucl. Phys.* **A650**, 176 (1999).

CENTER-OF-MASS CORRECTIONS IN RELATIVISTIC MEAN FIELD DESCRIPTIONS OF LIGHT NUCLEI

S. S. AVANCINI

Departamento de Física, Universidade Federal de Santa Catarina, Trindade, C.P. 476, CEP: 88040-900, Florianópolis, Brasil
fsc1sav@fsc.ufsc.br

P. ALBERTO AND M. FIOLHAIS

Departamento de Física and Centro de Física Computacional, Universidade de Coimbra, Coimbra, Portugal
tmanuel@teor.fis.uc.pt

The non-linear sigma-omega model is used to describe light nuclei, such as the alpha particle, in a relativistic mean field calculation. It turns out that, in systems with few particles, the spurious center-of-mass energy is relatively large. In our calculation we remove that energy by means of the Peierls-Yoccoz projection method, assuming a coherent state description for the mesons. We compare this correction to the total energy with the usual harmonic oscillator correction to the COM motion, showing that the former is not appropriate for light nuclei. Finally, we also compare the Hartree and Hartree-Fock approximations, stressing the relevance of the exchange term in few body systems.

1 Non-linear Walecka Model

The non-linear Walecka model[1] consists in a relativistic quantum field theory which describes the nucleons as Dirac spinors interacting by the exchange of scalars and vector mesons. We start with the Lagrangian density

$$\begin{aligned} L \; = \; & \overline{\psi}(x)(i\gamma^\mu\partial_\mu - M)\psi(x) \; + \; \tfrac{1}{2}\partial^\mu\phi(x)\partial_\mu\phi(x) \; - \; U(\phi) \; + \; g_\sigma\overline{\psi}(x)\psi(x)\phi(x) \\ & -\tfrac{1}{4}\Omega^{\mu\nu}(x)\Omega_{\mu\nu}(x) + \tfrac{1}{2}m_\omega^2\omega^\mu(x)\omega_\mu(x) - g_\omega\overline{\psi}(x)\gamma^\mu\psi(x)\omega_\mu(x) \\ & -\tfrac{1}{4}\vec{R}^{\mu\nu}(x)\vec{R}_{\mu\nu}(x) + \tfrac{1}{2}m_\rho^2\vec{\rho}^\mu(x)\vec{\rho}_\mu(x) \; - \; g_\omega\overline{\psi}(x)\gamma^\mu\vec{\tau}\psi(x)\vec{\rho}_\mu \end{aligned} \tag{1}$$

where ψ, ϕ, ω and ρ are the nucleon, the scalar sigma meson, the vector ω meson and the vector ρ meson fields respectively. The scalar field couples to the nucleons through the Yukawa term $\overline{\psi}\psi\phi$, the vector meson ω through the conserved nucleon current $\overline{\psi}\gamma^\mu\psi$ and finally the vector meson ρ couples to the isovector current $\overline{\psi}\gamma^\mu\vec{\tau}\psi$. The sigma meson moves in a nonlinear potential:

$$U(\phi) = \frac{1}{2}m_\sigma^2\phi^2(x) \; + \; \frac{\kappa}{3!}\phi^3(x) \; + \; \frac{\lambda}{4!}\phi^4(x) \,.$$

The field tensors for the vector mesons are given by $\Omega^{\mu\nu} = \partial^\mu\omega^\nu - \partial^\nu\omega^\mu$ and $\vec{R}^{\mu\nu} = \partial^\mu\vec{\rho}^\nu - \partial^\nu\vec{\rho}^\mu$. From the Euler-Lagrange equations one obtains the equations of motion for the fields. Here the solution of these equations will be considered in the mean field approximation(MFA) where: i) the meson fields are substituted by their expectation values, i.e., they are treated as classical fields, ii) the Dirac sea of negative energy states is neglected (no-sea approximation). In the nuclear context the sources of the meson fields are obtained by summation over all the positive energy occupied orbits in the Slater determinant of the baryons.

In what follows, we will focus our study on the alpha particle. The lowest energy alpha particle configuration is spherical with 4 nucleons occupying the lowest s-wave state

$$\psi(r) = \frac{1}{\sqrt{4\pi}} \begin{pmatrix} u(r) \\ iv(r)\vec{\sigma}\cdot\hat{r} \end{pmatrix} \chi_\lambda^\tau \chi_\eta^s \,, \tag{2}$$

where χ_η^s and χ_λ^τ describe the spin and isospin quantum numbers. It is easy to show that, for the alpha particle, the isovector current density vanishes and only the zero component of the nucleon current density survives. Therefore, since the source for the ρ meson is zero there is no contribution from this meson. Our intention here is to study the dynamics of the nuclear many-body problem, so we need a full quantum wave function. It is well known the difficulties or impossibility of considering the complete solution of the quantum field equations. The closest quantum analog of the mean field is the coherent state. These states, if conveniently chosen, result in the same MFA equations for the fields, however, they contain fluctuations of the fields which have important consequences for center-of-mass corrections. We begin with the ϕ meson field

$$\hat{\phi}(\vec{r}) = \frac{1}{(2\pi)^{\frac{3}{2}}} \int \frac{d^3k}{\sqrt{2\omega_\phi}} \left[b(\vec{k})e^{i\vec{k}\cdot\vec{r}} + b^\dagger(\vec{k})e^{-i\vec{k}\cdot\vec{r}} \right] \tag{3}$$

where $\omega_\phi^2 = m_\sigma^2 + \vec{k}^2$. The coherent state is defined as

$$|\phi\rangle = \sqrt{N_\phi}\exp\left(\int d^3k\ \eta(\vec{k})b^\dagger(\vec{k})\right)|0\rangle \,,$$

where $b(\vec{k})|\phi\rangle = \eta(\vec{k})|\phi\rangle$ and N_ϕ is a normalization factor. The $\eta(\vec{k})$ coefficients are determined by imposing the condition $\phi_0(r) = \langle\phi|\hat{\phi}(\vec{r})|\phi\rangle$, where ϕ_0 is the solution for the ϕ-field in the MFA. Analogously to the ϕ-field one can conveniently define the coherent state, $|\omega\rangle$, for the vector field. Therefore the state vector representing the alpha particle is given by $|\Psi\rangle = |\psi\rangle|\phi\rangle|\omega\rangle$. The Hamiltonian H is obtained from the T^{00} component of the energy-momentum tensor. Using the coherent states just defined one can calculate the expectation value of the normal ordered Hamiltonian in the state $|\Psi\rangle$,

$$E = \langle\Psi| : H : |\Psi\rangle \;=\; \langle\omega|\langle\phi|\langle\psi|H|\psi\rangle|\phi\rangle|\omega\rangle \;=\; E_{MFA} \;+\; E_{Fock} \qquad (4)$$

where

$$E_{Fock} = 4\int dr\; r^2 \frac{g_\omega^2}{8\pi m_\omega^2}(u^2+v^2)^2 \;. \qquad (5)$$

With the coherent states chosen conveniently, the expression of the energy functional will contain one term which is exactly the same that would be obtained in the MFA and a second one which is the Fock term. The state $|\Psi\rangle$ contains fluctuations of the meson fields which will be taken into account in the center-of-mass corrections. The effects of the Fock term which usually are not considered in the literature will be treated in this work.

2 Peierls-Yoccoz center-of-mass projection

The physical quantum state of the alpha particle should be a momentum eigenstate since the model Hamiltonian is translationally invariant. Nevertheless our MFA solution corresponds to a localized wave function which violates the translational symmetry. Localized states, in general, contain spurious center-of-mass energy which should be removed. In order to remove the spurious center-of-mass motion, the non-relativistic Peierls-Yoccoz(PY) projection[2] can be used. In this technique momentum eigenstates are produced through a linear combination of localized states

$$|\vec{p}\rangle \;=\; N_p\int d^3a e^{-i(\hat{\vec{p}}-\vec{p})\cdot\vec{a}}|\Psi\rangle \;=\; N_{\vec{p}}\int d^3a e^{i\vec{p}\cdot\vec{a}}U(\vec{a})|\Psi\rangle \equiv N_{\vec{p}}P_{\vec{p}}|\Psi\rangle \;, \qquad (6)$$

where the translation operator $U(\vec{a}) = \exp(-i\hat{\vec{p}}\cdot\vec{a})$. The overlap between displaced states is given by $\langle\Psi|U(\vec{a})|\Psi\rangle$. Using the PY procedure we obtain the energy of the alpha particle with momentum $\vec{p}=0$

$$E_{\vec{p}=0} \;=\; \frac{\int d^3\, a\langle\Psi| : H_N \;+\; H_\sigma \;+\; H_\omega : U(\vec{a})|\Psi\rangle}{\int d^3a\langle\Psi|U(\vec{a})|\Psi\rangle} \;. \qquad (7)$$

The last expression contains a Fock contribution that is usually neglected in the MFA.

3 Results and Conclusions

There are two center-of-mass corrections often employed in the literature[5]. The first consists in the harmonic oscillator approximation $E_{CM}^{(HO)} = -\frac{3}{4}\hbar\omega$

	NL1	NL2	NL3	NL2+Fock
E_{MFA}	15.99	14.48	16.57	144.76
$E_{MFA}+E_{CM}$	27.50	27.76	29.30	179.24
$E_{MFA}+E_{CM}(HO)$	35.36	33.84	35.90	164.13
$E_{\vec{p}=0}$	66.76	69.60	74.80	280.80
Exp.En.	28.30	28.30	28.30	28.30
$\langle r^2\rangle^{1/2}_{MFA}$	2.04	1.90	1.94	1.20
$\langle r^2\rangle^{1/2}_{\vec{p}=0}$	1.83	1.72	1.76	1.09
$\langle r^2\rangle^{1/2}(Exp.)$	1.60	1.60	1.60	1.60

Table 1. MFA results for the alpha particle. Compared are the results for the center-of-mass corrected binding energy using $E^{(HO)}_{CM} = 41A^{-1/3} = -19.37MeV$, $E_{CM} = -\langle P^2_{CM}\rangle/2AM$ and the PY projection. The projected and non-projected radius are also shown. In the last column we show results including the Fock term for the NL2 parameter set. Energies in MeV and radius in fm.

with $\hbar\omega = 41A^{-1/3}$. The second consists in calculating directly from the nucleons wave functions $E_{CM} = -\frac{\langle P^2_{CM}\rangle}{2AM}$ where $P_{CM} = -i\sum_\nu \nabla_\nu$.

The operator P_{CM} is the classical center-of-mass momentum which can be evaluated in terms of the nucleon wave functions u and v. These procedures are not completely satisfactory because in both cases the meson motion is neglected and we also do not obtain a center-of-mass corrected wave function. In this work we employ the PY projection technique, which removes the center-of-mass motion including the meson fields motion. In our calculations the non-linear Walecka model parameter sets[3,4] NL1, NL2 and NL3 have been used. In Table 1 we compare the alpha particle binding energy in the MFA using the COM corrections discussed above. In the last column of Table 1 we show for the parameter set NL2 the alpha particle binding energy including the Fock term . From Table 1 we see that the results using the procedures often employed in the literature are apparently better. Nevertheless these techniques do not describe the wave function and give radius larger than the experimental value. We believe that a more appropriate parameterization may result in better results. A detailed investigation of the relationship between the CM projection results and the parameter sets used in the calculations will be published elsewhere.

References

1. B.D. Serot and J.D. Walecka, *Adv. Nucl. Phys.* **16**(1986) 1.
2. P. Ring and P. Schuck, The Nuclear Many-Body Problem, Springer-Verlag, 1980.
3. P. G. Reinhard, *Rep. Prog. Phys.* **55** (1989) 439-514.
4. G. A. Lalazissis, J. Konig and P. Ring, *Phys. Rev.* **C** (1997) 540-543.
5. S. Typel and H.H. Wolter, *Nucl. Phys.* **A** 656 (1999) 331-364.

3. High-Energy Collisions

INITIAL CONDITIONS FROM HIGH DENSITY QCD FOR ULTRA-RELATIVISTIC HEAVY ION COLLISIONS

A. L. AYALA FILHO AND V. P. GONÇALVES

Instituto de Física e Matemática, Univ. Federal de Pelotas
Caixa Postal 354, 96010-090 Pelotas, RS, Brazil
ayala@ufpel.tche.br, barros@ufpel.tche.br

In this contribution we propose an improvement of the EKS nuclear parton distributions for the small x region of high energy processes, where the perturbative high parton density effects cannot be disregarded. We analyze the behavior of the ratios xG_A/xG_N and F_2^A/F_2^D and verify that at small x they are strongly modified when compared to the EKS predictions.

The theoretical understanding of the nuclear structure function F_2^A in the full kinematic region has progressed in recent years, with several models which describe the experimental data with quite success [1]. Here we will restrict ourselves to the descriptions which use the DGLAP evolution equations to describe the behavior of the nuclear parton distributions. Recently, Eskola, Kolhinen and Salgado [2] have shown that the experimental results presenting nuclear shadowing effects can be described using the DGLAP evolution equations with adjusted initial parton distributions. The basic idea of this framework is the same as in the global analysis of parton distributions in the free proton: they determine the nuclear parton densities at a wide range of x and $Q^2 \geq Q_0^2 = 2.25$ GeV^2 through their perturbative DGLAP evolution by using the available experimental data from lA DIS and Drell-Yan (DY) measurements in pA collisions as constraint. EKS have expressed the results in terms of the nuclear ratios $R_f^A(x, Q^2)$ for each parton flavor f in a nucleus with A nucleons ($A > 2$), at $10^{-6} \leq x \leq 1$ and $2.25\,GeV^2 \leq Q^2 \leq 10^4\,GeV^2$. The results of EKS seems to show that, in the kinematic region of the present data, the high density dynamical effects are small enough to be described by the DGLAP evolution equation with a suitable set of non-perturbative initial conditions. The main shortcoming of the EKS parameterizations in the calculations of eA, pA or AA processes is associated to the small x predictions, where the solution of the DGLAP equations reduces to the well known limit of the double logarithm approximation (DLA). This limit is characterized by a strong growth of the gluon distribution, with a similar behavior for the F_2 structure function, which implies a high parton density in this kinematic region. However, when the density of gluons and quarks becomes very high the physical processes of interaction and recombination of partons, not considered in the DGLAP equations, become important in the parton cascade and these effects should be expressed in a new evolution equation. Therefore, the EKS description should be improved to include the high parton density effects when smaller values of x are considered. This is the case, for instance, in the calculation of the minijet cross section at LHC. Here we will consider the Glauber-Mueller approach for the high density effects, which is a common limit of the current high density approaches in the kinematic region which we are interested (see Ref. [3] for details). Thus we intend to obtain no model dependent predictions.

The deep inelastic scattering off a nucleus is usually interpreted in a frame where the nucleus is going very fast. In this case the nuclear shadowing is a result of an overlap in the longitudinal direction of the parton clouds originated from different bound nucleons. Thus low x partons from different nucleons overlap spatially creating much larger parton densities than in the free nucleon case. This leads to a large amplification of the nonlinear effects expected in QCD at small x. In the target rest frame, the electron-nucleus scattering can be visualized in terms of the propagation of a small $q\bar{q}$ pair in high density gluon fields through much larger distances than it is possible with free nucleons. Following Gribov [4] and considering the Glauber-Mueller approach to estimate the high density effects, we may write (See Ref. [3] for details)

$$F_2^A(x,Q^2) = \frac{R_A^2}{2\pi^2}\sum_{u,d,s}\epsilon_i^2\int_{\frac{1}{Q^2}}^{\frac{1}{Q_0^2}}\frac{d^2r_t}{\pi r_t^4}\{C + ln(\kappa_q(x,r_t^2)) + E_1(\kappa_q(x,r_t^2))\}\ , \quad (1)$$

where $\kappa_q = (2\alpha_s A/3R^2)\,\pi\, r_t^2\, xG_N(x,\frac{1}{r_t^2})$. The main shortcoming of this approach is that the large x effects (the antishadowing and the EMC effect) were disregarded, which implies predictions only for the asymptotic behavior (large s/small x) of the observables.

As discussed before the EKS parameterization, although describes the current experimental fixed target data quite well, is not a good approximation for the small x region in the perturbative regime, where the DGLAP equations predicts a strong growth of the parton distributions (the DLA limit). To improve the EKS description we propose the substitution of DLA limit of the DGLAP evolution, present in the EKS framework, by the Glauber-Mueller evolution. This procedure implies to calculate the nuclear structure function using the following expression [3] $F_2^A(x,Q^2) = F_2^A(x,Q^2)[\mathrm{EKS}] - (1/A)\,F_2^A(x,Q^2)[\mathrm{DLA}] + (1/A)\,F_2^A(x,Q^2)[\mathrm{eq.~(1)}]$, with $F_2^A(x,Q^2)[\mathrm{EKS}] = R_{F_2}^A \times F_2^N(x,Q^2)$, where $R_{F_2}^A$ is obtained in terms of a combination of nuclear parton ratios $R_f^A(x,Q^2)$ [2]. Moreover, $F_2^A(x,Q^2)[\mathrm{DLA}]$

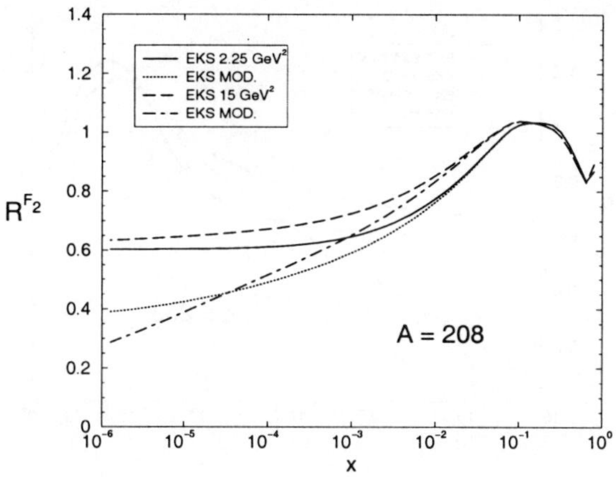

Figure 1. Our predictions for the behavior of the ratio $R^{F_2} = F_2^A/F_2^D$ (denoted as EKS MOD) for two values of virtualities ($Q^2 = 2.25$ and $15\,GeV^2$).

is calculated using the parton distribution from DLA limit of DGLAP evolution. The above procedure implies the inclusion of: (a) the full DGLAP evolution equation in all kinematic region; (b) the non-perturbative nuclear corrections in the nuclear parton distributions, that describe the experimental fixed target data, and (c) the high density effects present in the parton evolution at small x in the perturbative regime. Therefore, we are able to describe the parton evolution in all kinematic region. A similar description can be used to estimate the nuclear gluon distribution xG_A.

Using the above procedure we can calculate the ratios $R^{F_2} = F_2^A/F_2^D$ and $R_G = xG_A/xG_N$, where F_2^D is the deuterium structure function. In Fig. 1 we present our predictions for the behavior of the ratio $R^{F_2} = F_2^A/F_2^D$ (denoted as EKS MOD) for two values of virtualities ($Q^2 = 2.25$ and $15\,GeV^2$). For comparison the EKS predictions are also shown. We verify that while for the region of large values of x ($\geq 10^{-2}$) our predictions are almost identical to the EKS results, at small x the differences are large. The results shown in Fig. 1 predict a sizeable modification in the quark distribution at low x ($x \leq 10^{-3}$). Thus, we expect that the lA cross section will present a strong reduction when compared to the lp in this kinematic region. Also the Drell-Yan production in heavy ion collisions should be strongly modified when compared to the pp one at LHC energy ($x \approx 10^{-4}$).

In Fig.2 we present our predictions for the behavior of the ratio

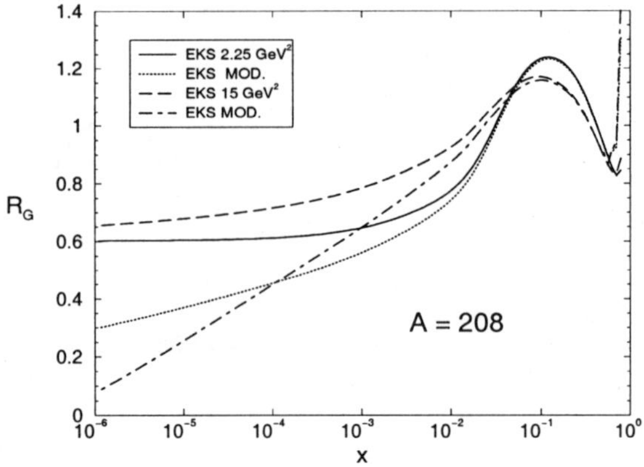

Figure 2. Our predictions for the behavior of the ratio $R_G(x, Q^2) = xG_A/xG_N$ (denoted as EKS MOD) for two values of virtualities ($Q^2 = 2.25$ and $15\,GeV^2$).

$R_G(x, Q^2) = xG_A/xG_N$ (denoted as EKS MOD) for two values of virtualities ($Q^2 = 2.25$ and $15\,GeV^2$). For comparison the EKS predictions are also shown. We verify that while for the region of large values of x ($\geq 10^{-2}$) our predictions are almost identical to the EKS results, at small x the differences are large. We can see from both figures that the EKS assumption that $R_G = R^{F_2}$ at small values of x is strongly modified by the high density effects and its Q^2 evolution, since $R_G \ll R^{F_2}$ for $Q^2 = 15\,GeV^2$. Therefore, we predict a large modification in the quarkonium production at LHC, as well as in its bound states (e.g. J/Ψ, Ψ', Υ, ...). Furthermore, as the high density effects are important already to $Q^2 \approx 2.25\,GeV^2$, also the minijet production will be modified at LHC. At RHIC kinematical region the high density effects do not significantly modify the gluon distribution and we expect a similar prediction of the quarkonium production from both models.

References

1. See, *e.g.*, M. Arneodo. *Phys. Rep.* **240** (1994) 301; G. Piller, W. Weise. *Phys. Rep.* **330** (2000) 1.
2. K. J. Eskola, V. J. Kolhinen, C. A. Salgado. *Eur. Phys. J.* **C9** (1999) 61; K.J. Eskola, V. J. Kolhinen, P. V. Ruuskanen. *Nucl. Phys.* **B535** (1998) 351.
3. A. L. Ayala Filho, V. P. Gonçalves. *Eur. Phys. J.* **C20** (2001) 343.
4. V. N. Gribov. *Sov. Phys. JETP* **29** (1969) 483.

PARTON SATURATION IN PROTON-NUCLEUS COLLISIONS

A. L. AYALA FILHO AND V. P. GONÇALVES
Instituto de Física e Matemática, Univ. Federal de Pelotas
Caixa Postal 354, 96010-090 Pelotas, RS, Brazil
ayala@ufpel.tche.br, barros@ufpel.tche.br

In this contribution we investigate the role of the high density effects in the heavy quark production cross section in pA processes at RHIC and LHC. We use, as initial condition, a gluon distribution consistent with fixed target nuclear data and the Glauber-Mueller approach to describe the high density effects. We demonstrate that this process can be used as a probe of the presence of the high density effects. Moreover, we include these effects in the calculation of the heavy quark production in AA collisions, verifying that they cannot be disregarded both in the estimates of quarkonium suppression and in the initial conditions of the quark-gluon plasma.

The relativistic collider facilities RHIC and LHC will for the first time provide the opportunity to systematically study the physics of hot and ultra-dense matter in hadron-nucleus (pA) and nucleus-nucleus (AB) collisions at energies that are orders of magnitude larger than the energies of the current accelerators. The systematic study of pA collisions at the same energies is essential to gain insight into the structure of the dense medium effects. Such effects, as the energy loss and shadowing, are absent or small in pp collisions, but become increasingly prominent in pA collisions, and are of major importance in AA reactions. By comparing pA and AA reactions involving very heavy nuclei, one may be able to distinguish basic hadronic effects that dominate the dynamics in pA collisions, from a quark-gluon formation predicted to occur in heavy ion AA collisions.

One of the nuclear medium effects is the nuclear shadowing, which is the modification of the target parton distributions so that $xq^A(x,Q^2) < Axq^N(x,Q^2)$, as expected from a superposition of pp interactions. The current experimental data presenting nuclear shadowing can be described reasonably using the DGLAP evolution equations with adjusted initial parton distributions [1]. However, this parameterization does not include the dynamical saturation effects of the parton distributions predicted at small x and large A. In this kinematical regime, the density of quarks and gluons becomes very high and the processes of interaction and re-combination between partons, not present in the DGLAP evolution, should be considered. Recently, we have proposed a procedure to improve the nuclear parton distributions and include the perturbative high density effects in the predictions of the inclusive observables in eA processes [2] (see also Ref. [4]). We have verified that the ratios $R^{F_2} = F_2^A/F_2^D$ and $R_G = xG_A/xG_N$ are strongly modified in the region of small x if the high density effects are included in the calculations. Here we analyze in detail the heavy quark production in pA processes in order to investigate the presence and magnitude of the high density effects. Our analysis is motivated by the fact that the heavy quark production at RHIC and LHC energies is dominated by initial state gluons. Thus, the pA process becomes a valuable source of information about the gluon distribution in the nucleus. Since the probability for making a heavy quark pair is proportional to square of gluon distribution, any depletion in number of gluons will make a significant difference in the number of heavy quark pairs produced.

For a review of the high density scenario see e.g. Refs.[2,3]. Let us consider the heavy quark production. For high energies and at leading order (LO), the production is dominated by gluon fusion, with charm and bottom quarks produced basically in the process $gg \to Q\overline{Q}$ with $Q = c$ and b. The LO cross section for a proton p colliding with a nucleus A is then:

$$\sigma_{pA} = \int d\tau \int dx_F \;\; \frac{1}{s} \frac{1}{\sqrt{x_F^2 + 4\tau}} \, xG_p(x_1, \mu^2) \, xG_A(x_2, \mu^2) \; \times$$
$$\hat{\sigma}_{gg \to Q\overline{Q}}(x_1, x_2, m_Q, \mu^2) \; , \qquad (1)$$

where $x_F = x_1 - x_2$, $\tau = M^2/s$ and M^2 is the invariant mass of the virtual gluon in the sub-process and xG_i the gluon distributions evaluated at momentum fraction x and momentum scale μ^2. To estimate the modification of heavy quark production cross section due to the high density effects, we calculate the effective exponent α, which is given by $\alpha = 1 + \left(ln \frac{\sigma_{pA}}{A\sigma_{pN}} \Big/ lnA \right)$. If the particle production is not modified by the presence of nuclear matter, then $\alpha = 1$. We calculate the above expression considering, as an input in the nuclear cross sections, the parameterization of the nuclear effects in the gluon distributions, which is given by $xG_A(x_i, \mu^2) = R_G(x_i, \mu^2) \, xG_N(x_i, \mu^2)$, with R_G calculated considering the high density effects [2].

In Fig. 1 we present the effective exponent α as a function of the c.m. energy $s^{1/2}$ for $A = 208$. We verify that the high density effects are sizeable at high values of energy.The general behavior of the exponent can be understood as follows. When the integrations in Eq. (1) are taken, the nuclear gluon distribution is evaluated in the x_2 interval given by $\tau < x_2 < \sqrt{\tau}$. Thus, when the energy grows, the x_2 interval goes to the small x region. For charm production, for example, the anti-shadowing region in xG_A dominates the integration for energies smaller than 80 GeV. For bigger energies, the shadowing region dominates and the exponent is smaller than 1. In bottom production, the x_2 interval is dislocated to larger values of x, which implies

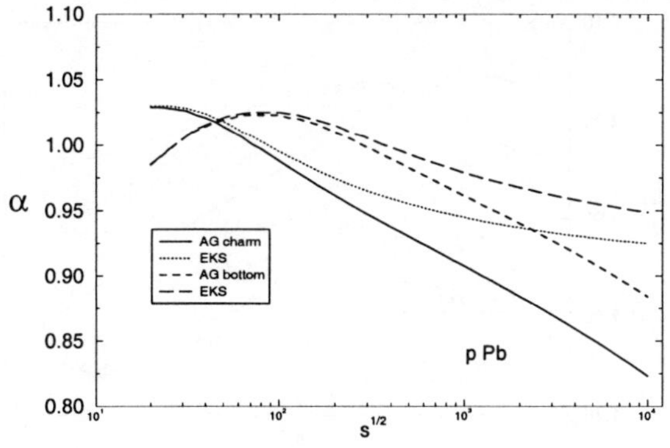

Figure 1. The effective exponent α as a function of the c.m. energy $s^{1/2}$ for $A = 208$.

that the anti-shadowing region in xG_A dominates the production at energies in the interval 100-200 GeV and α is close to one even for RHIC energies. For bigger values of energy, the suppression is sizeable and the effective exponent is smaller than 1. In general grounds, we can see that the high density effects, described by the Glauber-Mueller formula, predicts a reduction of the pA cross section when compared with DGLAP-EKS description of nuclear effects. For LHC energies, the difference is very important, since the EKS evolution predicts an almost constant α with energy, while the generalized approach predicts a decrease in α, almost proportional to $\ln s^{1/2}$.

The strong modification of the heavy quark production in pA process indicates that high density effects on the gluon distribution will play an important role in the calculation of the initial conditions and signatures of the quark-gluon plasma in heavy ion collisions. To illustrate this point, we present in Fig. 2 the exponent β calculated by the expression $\beta = 2 + \left(ln \frac{\sigma_{AA}}{A^2 \sigma_{pN}} \Big/ lnA \right)$ for the heavy quark production in Pb-Pb collisions as a function of $\sqrt{s}$. The exponent presents a similar feature to the exponent α shown in Fig. 1. For energies close and smaller than 100 GeV, the most important contribution comes from the anti-shadowing region of xG_A, and the exponent is bigger than 2. For larger energies, the small x region gives the more important contribution. When compared to the EKS prediction, the high density effects strongly reduces the production cross sections. The absolute values of the exponent β are 1.98 and 2.03 for $c\bar{c}$ and $b\bar{b}$, respectively, at RHIC energy. For the LHC energy, both exponents are reduced to 1.81 and 1.90, respectively. As we can see, the high density effects will strongly reduce the production

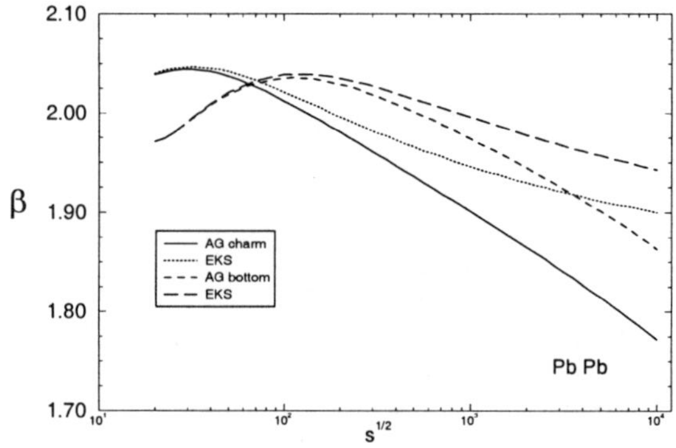

Figure 2. The exponent β calculated by the expression $\beta = 2 + \left(ln \frac{\sigma_{AA}}{A^2 \sigma_{pN}} \Big/ lnA \right)$ for the heavy quark production in Pb-Pb collisions as a function of $\sqrt{s}$.

cross sections in AA processes and should be taken into account to calculate the cross section of the hard QCD process in the first stage of the heavy ion collisions. These results have important implications in the signatures of the QCD phase transition, specially in the suppression of quarkonium production since our results indicate that the production rate for this process should be strongly modified by the presence of the high density effects. Consequently, if the analysis of the pA process demonstrate the presence of the these effects in the kinematic regions of RHIC and LHC, we have that the current estimates of quarkonium suppression in AA processes should be completely re-analyzed.

References

1. K. J. Eskola, V. J. Kolhinen, C. A. Salgado. *Eur. Phys. J.* **C9** (1999) 61.
2. A.L. Ayala Filho, V. P. Gonçalves. *Eur. Phys. J.* **C20** (2001) 343.
3. A.L. Ayala Filho, V. P. Gonçalves. *Phys. Lett.* **B534** (2002) 76.
4. A.L. Ayala Filho, V. P. Gonçalves, *in these proceedings.*

PROMPT J/ψ PRODUCTION FROM TEVATRON TO LHC

C. BRENNER MARIOTTO[1,2], J. DAMET[2,3], G. INGELMAN[2,4]

[1] *Institute of Physics, Univ. Fed. do Rio Grande do Sul, Box 15051, CEP 91501-960 Porto Alegre, Brasil*

[2] *High Energy Physics, Uppsala University, Box 535, S-75121 Uppsala, Sweden*

[3] *Laboratory for High Energy Physics, University of Bern, Bern, Switzerland*

[4] *Deutsches Elektronen-Synchrotron DESY, Hamburg, Germany*

mariotto@if.ufrgs.br

Models with essential non-perturbative QCD dynamics and describing Tevatron data on high-$p_\perp$ charmonium are extrapolated to give predictions of prompt J/ψ production at the LHC. Differences of up to an order of magnitude occur. An important point is here the treatment of higher order perturbative QCD effects.

The interplay of hard and soft QCD effects has been demonstrated in Tevatron data [1] for high-$p_\perp$ J/ψ, ψ' and Υ, which are up to factors of 50 above the pQCD prediction in the Colour Singlet Model (CSM) [2], where a colour singlet $c\bar{c}$ pair is produced at the parton level. This deficit can be explained by letting a fraction of the more abundant colour octet $c\bar{c}$ pairs be transformed into colour singlet through some soft QCD dynamics. This has been described in different models: the Colour Octet (COM) [3], Colour Evaporation (CEM) [4], Soft Colour Interaction (SCI) [5] and Generalized Area Law (GAL) [6] models, which can be made to fit these Tevatron data. In this work we study the extrapolation of CEM, SCI and GAL models to the LHC energy and examine its theoretical uncertainty [7,8]. These models are based on a similar phenomenological approach, where soft colour interactions can change the colour state of a $c\bar{c}$ pair from octet to singlet. They employ the same hard processes to produce a $c\bar{c}$ pair regardless of its spin state. Of importance for high-$p_\perp$ J/ψ production are NLO tree diagrams with a third hard parton that balances the $p_\perp$ of the $c\bar{c}$ pair. The most important one is the process $gg \to c\bar{c}g$ with a gluon exchange in the t channel, since the basic $2 \to 2$ process $gg \to gg$ (followed by a gluon splitting $g \to c\bar{c}$) has a much larger $\mathcal{O}(\alpha_s^2)$ cross section than the LO $gg \to c\bar{c}$ and $q\bar{q} \to c\bar{c}$ production processes. Still higher orders than NLO can be important at the Tevatron and LHC energies, since many gluons can be emitted and their virtuality need not be very large to allow a split into a $c\bar{c}$ pair. Higher order processes can be approximately described by the parton shower approach available in the PYTHIA [9] Monte Carlo, where in all basic QCD $2 \to 2$ processes the incoming and outgoing partons may branch as described by the DGLAP equations. In CEM [4,10] the

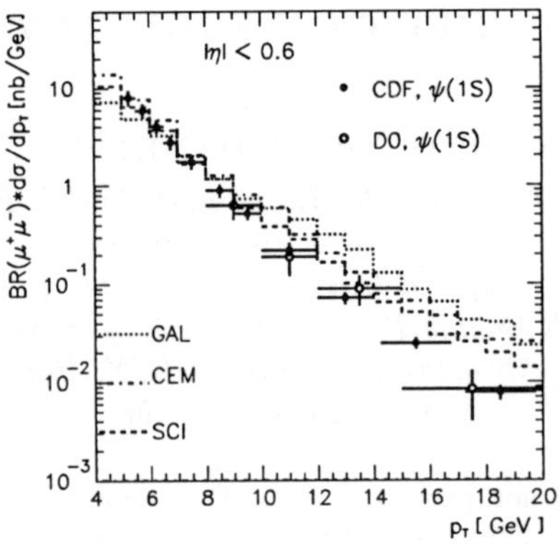

Figure 1. Distribution in transverse momentum of prompt J/ψ as observed by CDF and D0 [1] in $p\bar{p}$ interactions at the Tevatron and obtained in the CEM, SCI and GAL models.

exchange of soft gluons is assumed to randomize the colour state, implying a probability 1/9 that a $c\bar{c}$ pair is colour singlet and produces charmonium if its mass is below the threshold for open charm production, $m_{c\bar{c}} < 2m_D$. The fraction of a specific charmonium state i, relative to all charmonia, is given by a non-perturbative parameter ρ_i ($\rho_{J/\psi} = 0.4 - 0.5$). In SCI [5,11,12] it is assumed that colour-anticolour, corresponding to non-perturbative gluons, can be exchanged between partons emerging from a hard scattering and hadron remnants, leading to different topologies of the confining colour string-fields and thereby to different hadronic final states. The probability to exchange a soft gluon between parton pairs is given by a phenomenological parameter R. The mapping of $c\bar{c}$ pairs below the threshold for open charm production is based on spin statistics resulting in a fraction of a specific quarkonium state i with total angular momentum J_i and main quantum number n_i given by $f_i = \Gamma_i / \sum_k \Gamma_k$, where $\Gamma = (2J_i + 1)/n_i$. In GAL [6], a generalization of the area law suppression e^{-bA} (with A the area swept out by the string in energy-momentum space) gives a dynamic probability $R = R_0(1 - e^{-b\Delta A})$ for two string pieces to interact depending on the area difference ΔA resulting from the changed string topology. This favours shorter strings and thereby quarkonium production. The parameters R_0 and b are obtained from a fit to both HERA and LEP data. The comparison of the CEM, SCI and GAL models with the Tevatron data is shown in Fig. 1. As can be seen, all models give a quite decent description of the data. Although the shape is not perfect in the tail of the distribution, it is quite acceptable given the simplicity of the models. The overall normalization is correctly given by the models. For the CEM

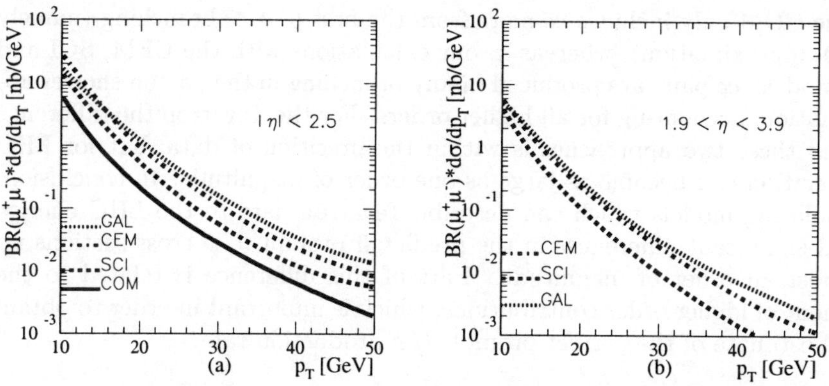

Figure 2. Differential cross sections in transverse momentum for J/ψ in pp collisions at $\sqrt{s} = 14$ TeV based on the CEM, SCI and GAL models. In all cases the J/ψ is required to have $p_\perp > 10$ GeV and decay into $\mu^+\mu^-$, required to be within the indicated rapidity coverage of the ATLAS (a) and LHC-b (b) detectors. For comparison, the COM results from ref.[8] are included in (a).

this is obtained by setting $\rho_{J/\psi} = 0.43$ and the charm quark mass to 1.5 GeV. The SCI and GAL models have not been tuned to these data, but the result is also sensitive to the charm quark mass (taken as default $m_c = 1.35$ GeV in PYTHIA 5.7). The parton densities used are the default CTEQ3L [13] for SCI and CTEQ4L [13] for CEM and GAL. One should note that an arbitrary K factor is not needed in any of the models, since higher order pQCD processes were included through the parton showers. Based on the ability of these models to correctly reproduce the Tevatron data and keeping all parameters fixed from this comparison, we can now extrapolate them to the LHC energy. Results of these models for LHC energies, *i.e.* $\sqrt{s} = 14$ TeV, are shown in Fig. 2 for the acceptance regions of the future LHC experiments ATLAS and LHC-b. For ATLAS , the $p_\perp$ distributions in Fig. 2a are quite similar for the three models, although they differ somewhat in the high-$p_\perp$ tail. These predictions should, however, not be taken as very precise in view of the simplicity of these models that attempt to describe unknown non-perturbative QCD phenomena. The overall normalization, which *e.g.* is sensitive to the value of the charm quark mass, should not be considered to be better than within a factor two. For the LHC-b pseudo-rapidity cuts, SCI is suppressed relative to CEM and GAL, showing that the models have a somewhat different rapidity behaviour. We can also see in Fig. 2a that the COM result [8] is significantly lower than the other model results. As we have found recently [7], part of this difference is because COM only includes $c\bar{c}$ pairs from the first $g \to c\bar{c}$ branching (namely a NLO approximation), whereas in our calculations with the CEM, SCI and GAL models, $c\bar{c}$ pairs are produced in any branching in the parton shower approximation, accounting for all higher orders. For the Tevatron the difference between these two approaches is within the precision of data, but for LHC the deviation can become as large as one order of magnitude. In conclusion, extrapolating models which can describe Tevatron data to the LHC energy, we find significant differences in the predicted prompt J/ψ cross sections up to almost an order of magnitude. Part of this difference is related to the treatment of higher order contributions, which is important in order to obtain a good estimate of the correct prompt J/ψ production rate.

Acknowledgments

This work was supported by the Swedish Research Council and by CAPES and CNPq, Brazil.

References

1. F. Abe *et al.*, CDF collaboration, *Phys. Rev. Lett.* **75**, 4358 (1995); *ibid.* **79**, 572 (1997); *ibid.* **79**, 578 (1997); S. Abachi *et al.*, D0 collaboration, Fermilab-CONF-96/249E; *Phys. Lett.* B **370**, 239 (1996).
2. R. Baier, R. Rückl, *Z. Phys.* C **19**, 251 (1983).
3. E. Braaten, T.C. Yuan, *Phys. Rev.* D **50**, 3176 (1994) ; **52**, 6627 (1995); G.T. Bodwin *et al.*, *Phys. Rev.* D **51**, 1125 (1995); **55**, 5853 (1997).
4. J.F. Amundson *et al.*, *Phys. Lett.* B **390**, 323 (1997).
5. A. Edin, G. Ingelman, J. Rathsman, *Phys. Lett.* B **366**, 371 (1996); *Z. Phys.* C **75**, 57 (1997).
6. J. Rathsman, *Phys. Lett.* B **452**, 364 (1999).
7. J. Damet, G. Ingelman, C. B. Mariotto, hep-ph/0111463.
8. M.A. Sanchis-Lozano *et al.*, *Phys. Lett.* B **406**, 232 (1997) ; *Nucl. Phys.* B (Proc. Suppl.) **86**, 543 (2000).
9. T. Sjöstrand *et al.*, *Comput. Phys. Commun.* **135**, 238 (2001).
10. C. Brenner Mariotto, M.B. Gay Ducati, G. Ingelman, *Eur. Phys. J.* C **23**, 527 (2002).
11. R. Enberg, G. Ingelman, N. Timneanu, *J. Phys.* G: *Nucl. Part. Phys.* **26**, 712 (2000); *Phys. Rev.* D **64**, 114015 (2001).
12. A. Edin, G. Ingelman, J. Rathsman, *Phys. Rev.* D **56**, 7317 (1997).
13. H.L. Lai *et al.*, *Phys. Rev.* D **55**, 1280 (1997) ; D **51**, 4763 (1995).

IMPROVED MAPPING OF $c\bar{c}$ ONTO CHARMONIUM STATES

C. BRENNER MARIOTTO[*,†], M.B. GAY DUCATI[*], G. INGELMAN[†,††]

[*] *Institute of Physics, Univ. Fed. do Rio Grande do Sul, Box 15051, CEP 91501-960 Porto Alegre, Brasil*

[†] *High Energy Physics, Uppsala University, Box 535, S-75121 Uppsala, Sweden*

[††] *Deutsches Elektronen-Synchrotron DESY, Hamburg, Germany*

mariotto@if.ufrgs.br, gay@if.ufrgs.br, gunnar.ingelman@tsl.uu.se

We discuss the relative rates of different charmonium states and introduce an improved model for mapping the continuous $c\bar{c}$ mass spectrum on the physical charmonium resonances.

The theoretical description of charmonium production separates the hard and soft parts of the process based on the factorisation theorem in QCD. Perturbative QCD (pQCD) should be applicable for $c\bar{c}$ production, since the charm quark mass m_c is large enough to make $\alpha_s(m_c^2)$ a small expansion parameter. The formation of bound hadron states occurs through processes with small momentum transfers making α_s large and preventing the use of perturbation theory. This forces us to use phenomenological models to describe the formation of charmonium states from perturbatively produced $c\bar{c}$ pairs. The Colour Evaporation (CEM) [1] and Soft Colour Interaction (SCI) [2,3] models are based on a similar phenomenological approach, where soft colour interactions can change the colour state of a $c\bar{c}$ pair from octet to singlet. They employ the same hard pQCD processes to produce a $c\bar{c}$ pair regardless of its spin state. A colour singlet $c\bar{c}$ pair with an invariant mass below the threshold for open charm ($m_{c\bar{c}} < 2m_D$) will then form a charmonium state. Mapping $c\bar{c}$ pairs onto charmonium states is based on phenomenological parameters taken from comparison to data (CEM), or from spin statistics (SCI), the latter resulting in a fraction of a specific quarkonium state i with total angular momentum J_i and main quantum number n_i given by $f_i = \Gamma_i / \sum_k \Gamma_k$, where $\Gamma_i = (2J_i + 1)/n_i$. The CEM and SCI models give in general a good description of charmonium data both from fixed target and Tevatron energies [3,4,5]. However, the overall energy-independent factors which describe the mapping of $c\bar{c}$ pairs onto different charmonium states are not satisfactorily understood. In particular, spin statistics gives an acceptable ratio ψ'/ψ at the Tevatron [3], but fails at fixed target energies where an extra suppression by a factor four is required[4]. In other words, this ratio has an energy dependence which has not been accounted for. In this work we try to understand this problem by developing a model which introduces a correlation between the invariant mass of the $c\bar{c}$ pair and the mass of the final charmonium state.

The $c\bar{c}$ pair is produced in a pQCD process with a continuous distribution of its invariant mass $m_{c\bar{c}}$ and must be mapped onto the discrete spectrum of charmonium states. The soft interactions that turn the pair into a colour singlet may very well change its mass by a few hundred MeV, which is the typical scale of the soft interactions, but larger mass shifts should be suppressed. The different charmonium states (η_c, J/ψ, χ_c, ψ') are, however, separated in mass over a region of about 1 GeV. It therefore seems likely that the probability to form a particular charmonium state will depend on the original value of $m_{c\bar{c}}$ and not just on an overall factor, such as spin statistics. Thus, there should be a relatively larger probability to form states that are nearby in mass, than those further away on the mass scale. Based on these considerations we have constructed the following model.

The smearing of the $c\bar{c}$ mass due to soft interactions is considered as

$$G_{sme}(m_{c\bar{c}}, m) = \exp\left(-\frac{(m_{c\bar{c}} - m)^2}{2\sigma_{sme}^2}\right), \tag{1}$$

where the gaussian width σ_{sme} should be a few hundred MeV. The different charmonium resonances i are represented by functions $F_i(m_i, m)$ around their physical masses m_i. For a $c\bar{c}$ pair with a given mass $m_{c\bar{c}}$, the probability to form a certain charmonium state is then

$$\mathcal{P}_i(m_{c\bar{c}}) = \frac{\int G_{sme}(m_{c\bar{c}}, m) F_i(m_i, m) dm}{\sum_j \int G_{sme}(m_{c\bar{c}}, m) F_j(m_j, m) dm} \approx \frac{s_i G_{sme}(m_{c\bar{c}}, m_i)}{\sum_j s_j G_{sme}(m_{c\bar{c}}, m_j)}, \tag{2}$$

where the sum is over all charmonium resonances. The approximate result, which is used in our practical calculations, is obtained by letting $F_i(m_i, m) = s_i \delta(m - m_i)$, *i.e.* neglecting the width of the very narrow charmonium states and including the previously used spin statistical factor $s_i = (2J_i + 1)/n_i$.

The smearing of $m_{c\bar{c}}$ across the threshold $2m_D$ for open charm, implies non-zero contributions for charmonium also above the $D\overline{D}$ threshold as well as some open charm production for $m_{c\bar{c}}$ originally below this threshold. For a given original $m_{c\bar{c}}$ value, the probability to end up in the region above $D\overline{D}$ threshold is given by the area of the smearing gaussian in that region, *i.e.*

$$A(m_{c\bar{c}}) = \frac{1}{\sqrt{2\pi}\sigma_{sme}} \int_{2m_D}^{\infty} \exp\left(-\frac{(m_{c\bar{c}} - m)^2}{2\sigma_{sme}^2}\right) dm\,. \tag{3}$$

The probability to end up in a specific resonance is then given by $(1 - A)\mathcal{P}_i$ and is shown in Fig. 1 for the different charmonium states. For a given original $c\bar{c}$ mass $m_{c\bar{c}}$ one can here see the fractional production of different states, as

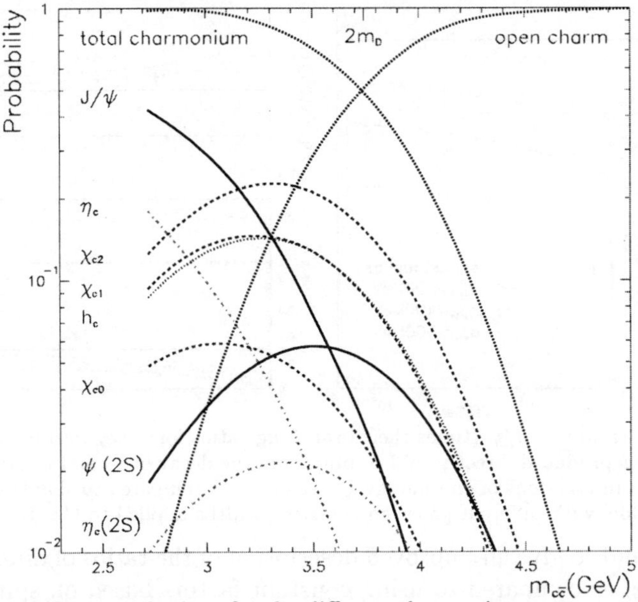

Figure 1. Probability distributions for the different charmonium states as obtained in the model with gaussian smearing ($\sigma_{sme} = 400\,MeV$). The resulting total probability for charmonium production and the remainder as open charm production are also shown.

well as the sum of all charmonium states together with the remainder giving open charm. By folding these charmonium probability functions with the distribution in $m_{c\bar{c}}$ obtained from pQCD, *i.e.* $d\sigma_{c\bar{c}}/dm_{c\bar{c}}$ (LO processes and $m_c = 1.35$ GeV), the cross section for a given charmonium state is

$$\sigma_i = \int_{2m_c}^{\sqrt{s}} dm_{c\bar{c}} \frac{d\sigma_{c\bar{c}}}{dm_{c\bar{c}}} (1 - A(m_{c\bar{c}}))\mathcal{P}_i(m_{c\bar{c}})\,. \tag{4}$$

Applying this mapping procedure to the CEM model we obtain the results in Fig. 2. As opposed to the simple spin statistics factor, this model gives a reasonable description of the observed ratio of ψ' to J/ψ production and fractions of J/ψ produced directly, coming from decays of χ_c states and from ψ'. In particular, the model gives a characteristic energy dependence of the kind indicated by the data. Considering all observables from Fig. 2, the preferred value for the σ_{sme} is 400 MeV or lower. This fits well with the expectation for the non-perturbative dynamics that this model should describe. The essential new ingredient of this model, namely introducing a correlation between the invariant mass of the $c\bar{c}$ pair and the masses of the different charmonium

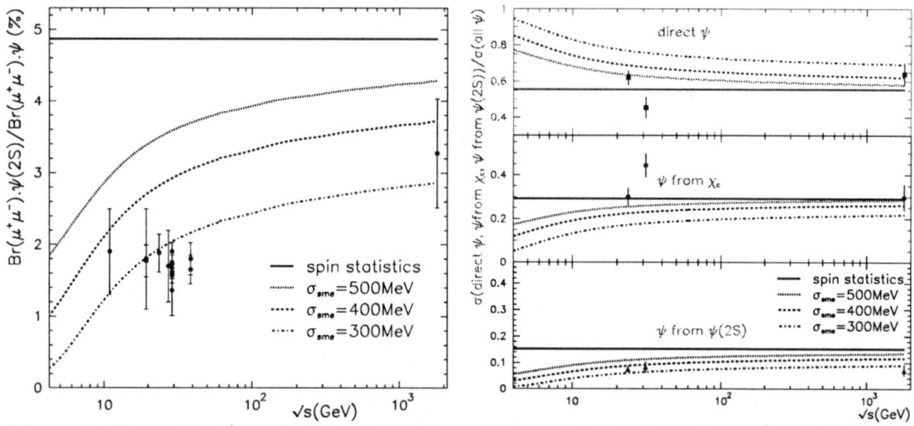

Figure 2. The ratio ψ' / J/ψ (times their branching ratios for decay into $\mu^+\mu^-$) (left) and fractions of J/ψ produced directly, and coming from the decay of χ_c and ψ' states (right) in hadron-hadron interactions of cms energy $\sqrt{s}$. Data [6,7,8] compared to simple spin statistics and to our model with different gaussian smearing widths applied to CEM.

states, is found to give an improved description of the ratios of different charm onium states as compared to using constant factors based on spin statistics. In particular, these ratios acquire an energy dependence which is particularly strong at low energies where threshold effects are more pronounced.

Acknowledgments

This work was partially financed by CAPES, Brazil, and by the Swedish Research Council.

References

1. J.F. Amundson *et al.*, *Phys. Lett.* B **390**, 323 (1997); O.J.P. Éboli, E.M. Gregores, F. Halzen, *Phys. Rev.* D **60**, 117501 (1999).
2. A. Edin, G. Ingelman, J. Rathsman, *Phys. Lett.* B **366**, 371 (1996); *Z. Phys.* C **75**, 57 (1997).
3. A. Edin, G. Ingelman, J. Rathsman, *Phys. Rev.* D **56**, 7317 (1997).
4. C. Brenner Mariotto, M.B. Gay Ducati, G. Ingelman, *Eur. Phys. J.* C **23**, 527 (2002).
5. J. Damet, G. Ingelman, C.B. Mariotto, JHEP **0209**, 014 (2002).
6. M.C. Abreu *et al*, *Phys. Lett.* B **444**, 516 (1998), and references therein.
7. F. Abe *et al*, *Phys. Rev. Lett.* **75**, 4358 (1995); **79**, 572 , **79**, 578 (1997).
8. L. Antoniazzi *et al*, *Phys. Rev. Lett.* **70**, 383 (1993); V. Koreshev *et al*, *Phys. Rev. Lett.* **77**, 4294 (1996).

ISOBARIC THERMAL MODEL FOR HADRON PRODUCTION

C.E. AGUIAR, A.G. GRUNFELD, T. KODAMA AND L.L.S. PORTUGAL
Instituto de Física, Univ. Federal do Rio de Janeiro Cidade Universitária, Caixa Postal 68528, 21945-970 Rio de Janeiro, RJ, Brazil
tkodama@if.ufrj.br

In this work, we study this point in detail and propose the use of isobaric equation of state instead of iso-volumetric one. In our opinion, this framework seems to be more realistic for the description of the hadronic scenario in the final state of the collision since there is no predefined volume of the fireball. Our purpose is to calculate, within this thermal model, the ratios of particle abundances and compare them with most recent data from BNL (RHIC) experiment and also with the previous calculations.

The observation of the quark-gluon plasma (QGP) phase is one of the main goals of the ultra-relativistic heavy ion collision physics. Some signatures, such as strangeness enhancement, J/Ψ suppression etc., are extensively analyzed for this purpose. As these signatures are of thermodynamical nature, it is fundamental to verify whether thermal equilibrium is reached in these processes. Some authors claim that there are evidences that both thermal and chemical equilibrium are reached in some stage of the collision [1]. This is because the final hadronic spectra are well described by a simple thermal model. Usually grand canonical ensembles for relativistic gas of hadronic resonances are used for these analysis. In the frame of the grand canonical ensemble, the system is supposed to hadronize, after reaching a chemical equilibrium, from a unique, homogeneous massive fireball[2]. It is pointed out that the effect of excluded volume in the equation of state is important for better fit of the experimental data [3].

On the other hand, it is not clear at all whether the use of grand canonical ensemble is the most appropriate description of the scenario. In particular, when the system is finite and excluded volume of constituent hadrons is not negligible compared to the size of the system, its effect on the equation of state becomes much more enhanced. Furthermore, it is hard to believe that the system is in statistical equilibrium as a whole. The apparent agreement of the observed particle ratio with the statistical equilibrium does not necessarily mean that the whole system should be in equilibrium. When we consider the hydrodynamical scenario for the space-time evolution of the system, a local statistical equilibrium would be more reasonable. That is, the apparent global statistical equilibrium can be understood in terms of a collection of smaller hadronic clusters or fireballs, each of which is described in terms of equilibrated hadronic gas. One possible approach is the use of isobaric ensemble [4] instead of iso-volumetric one [3]. These two descriptions are equivalent for infinite systems (or zero excluded volume) but give different equation of states for a finite system. An isobaric ensemble describes the statistical equilibrium of a finite system under a given pressure, so that it is more adequate to be applied for the description of local equilibrium, dividing the whole system into finite clusters or fireballs. In this work, we study whether the use of isobaric ensemble is more adequate for the description of the hadronic scenario in the final state of the collision. For this purpose we analyze several hadronic observables. We have first studied particle multiplicities distribution, which has different properties in GC and Isobaric ensembles. As a second step, we are currently analyzing how that differences manifest in particle ratio event by event fluctuations at chemical freeze out.

The isobaric partition function for a system composed by α specie is given by

$$Z_P(\beta, \mu_1, ..., \mu_\alpha, \beta P) = \frac{1}{\beta P - \sum_j^\alpha \Phi_j(\beta, m_j) e^{\beta\mu_j - \beta P v}} \tag{1}$$

where the chemical potential μ_j for each particle is composed by

$$\mu'_\alpha = \mu_B B_\alpha + \mu_S S_\alpha + \mu_{I_3} I_{3\alpha}, \tag{2}$$

here B_α, S_α and $I_{3\alpha}$ are the baryon, strangeness and the three-component of isospin quantum numbers for the specie α (see ref.[5] for details of calculation). The average volume of the system and the mean value for the abundance of the particles of species i are respectively given by

$$\langle V \rangle = \frac{1 + \sum_i^\alpha v \Phi_i e^{\beta\mu_i - \beta v P}}{\beta P - \sum_i^\alpha \Phi_i e^{\beta\mu_i - \beta P v}} \tag{3}$$

and

$$\langle N_i \rangle = \frac{\Phi_i e^{\beta\mu_i - \beta v P}}{\beta P - \sum_j^\alpha \Phi_j e^{\beta\mu_j - \beta P v}}. \tag{4}$$

As the particle ratio data cannot distinguish the two ensembles, it is interesting to look for another observables which manifest the difference among the two schemes. As pointed out in ref.[3], the multiplicity distribution for the isobaric ensemble is much different from that of the grand canonical ensemble. Therefore, it will be interesting to investigate the event-by-event fluctuations of the particle ratio. For this purpose, let us study in more detail the probability distribution of particles in our scheme.

The probability of finding N particles in the system described by the GC ensemble is:

$$P_N^{GC} = \frac{e^{\mu\beta N} Z(\beta, N, V)}{\mathcal{Z}_{GC}(\beta, \mu, V)}. \tag{5}$$

In normal equilibrium states, this distribution is narrow and leads to a dispersion $D = (\langle N^2 \rangle - \langle N \rangle^2)^{1/2} \sim (\langle N \rangle)^{1/2}$. For systems that are described by ideal gases, the above P_N is the Poisson distribution. We will show, that in the isobaric ensemble, the corresponding distribution is geometric, leading to a broad multiplicity distribution.

Now, we will study the abundancies multiplicity distribution in the isobaric ensemble. The probability of finding N particles in that system is:

$$P_N^{(P)} = \frac{e^{\mu\beta N} Z_P(\beta, N, V)}{\mathcal{Z}_P(\beta, \mu, P)} = \frac{\sum_{V_a} e^{-P\beta V_a} e^{\mu\beta N} Z(\beta, N, V_a)}{\mathcal{Z}_P(\beta, \mu, P)}. \tag{6}$$

Considering the excluded volume, this probability becomes

$$P_N^{(P)} = \frac{\sum_{V_a} e^{\mu\beta N - P\beta V_a} (1/N!) \phi^N(\beta, m)(V_a - vN)^N}{\sum_{V_a} e^{-P\beta V_a} \sum_N^M e^{\mu\beta N} (1/N!) \phi^N(\beta, m)(V_a - vN)^N}. \tag{7}$$

Integrating in volume

$$P_N^{(P)} = (1-\eta)\eta^N = \frac{1}{\langle N \rangle + 1} exp\left[-Nln\left(1 + \frac{1}{\langle N \rangle}\right)\right] \tag{8}$$

where

$$\eta = \frac{\phi e^{\beta\mu - \beta P v}}{\beta P}; \tag{9}$$

it is easy to see that

$$\eta = \frac{\langle N \rangle}{\langle N \rangle + 1}. \tag{10}$$

The variance is

$$D \equiv (\langle N^2 \rangle - \langle N \rangle^2)^{1/2} = (\langle N^2 \rangle + \langle N \rangle^2)^{1/2} \sim \langle N \rangle. \tag{11}$$

Under the condition $\langle N \rangle \gg 1$ eq. (8) yields

$$P_N^{(P)} = \frac{1}{\langle N \rangle} exp\left[\frac{-N}{\langle N \rangle}\right]. \tag{12}$$

Considering α specie and taking into account the excluded volume v (the same for all species), the distribution P_N^P becomes

$$P^{(P)}(N_1...N_\alpha) = (1 - \sum_\alpha \eta_\alpha)\ N!\,\Pi_i\ \frac{{\eta_i}^{N_i}}{N_i!}. \tag{13}$$

Let us suppose now that we have k independent clusters (or fireballs) containing $N_1, ..., N_k$ particles (for one specie) with probabilities $P_{N_1}^{(P)}, ..., P_{N_k}^{(P)}$. The probability of appearing N particles from k clusters is

$$\Pi_N^k = \sum_{N_1,...,N_k \geq 0} P_{N_1}^P, ..., P_{N_k}^P \delta(N_1 + ... + N_k - N). \tag{14}$$

The average multiplicity is:

$$\langle N \rangle = \sum_{N \geq 0} N \Pi_N^k = K \langle n \rangle \tag{15}$$

and the dispersion is $D \sim \langle N \rangle$. If $N \gg k$ and $\langle N \rangle \gg k$ then

$$\Pi_N^k \sim \frac{1}{\langle N \rangle} \left(\frac{N}{\langle N \rangle} \right)^{k-1} exp\left[-k \frac{N}{\langle N \rangle} \right]. \tag{16}$$

Summing up, the dispersion for the isobaric and grand canonical ensemble are different. It is straightforward to see this effect when considering a vanishing excluded volume. Consequently, the fluctuation of the particles ratios will depend on the ensemble. We are currently analyzing the particles ratios fluctuation (considering excluded volume) and its dependence on the number of clusters.

References

1. P. Braun-Munzinger, I. Heppe and J. Stachel, *Phys. Lett.* **B 465**, 15 (1999).
2. P. Braun-Munzinger, D. Magestro, K. Redlich and J. Stachel, *Phys. Lett.* **B 518**, 41 (2001).
3. G. D. Yen, M. Gorenstein, W. Greiner and S. N. Yang, *Phys. Rev.* **C 56**, 2210 (1997).
4. M. Gorenstein, *Sov. J. Nucl. Phys.* **31**, 845 (1980).
5. C. E. Aguiar, A. G. Grunfeld, T. Kodama, L. L. Portugal, Proceedings of the XXII Brazilian Meeting on Particles and Fields, São Lourenço, Minas Gerais, Brazil.

PRELIMINAR POLARIZATION STUDIES OF Ξ^-- Ξ^+

C.J. SOLANO SALINAS

Centro de Investigaciones y Estudios Avanzados - CINVESTAV, México
Universidade Federal de Itajubá - UNIFEI, Itajubá, MG, Brazil
javier@mda.cinvestav.mx

F.R.A. SIMÃO

Centro Brasileiro de Pesquisas Físicas, Rio de Janeiro, Brazil
simao@cbpf.br

Using data from Fermilab fixed-target experiment E791, we have measured polarization for Ξ^- and Ξ^+ hyperons in π^- - nucleon interactions at 500 GeV/c. The polarization is measured as function of the angular distribution, transversal to the production plane, and over the ranges of Feynman (x_F) and p_T^2 $-0.12 \leq x_F \leq 0.12$ and $0 \leq p_T^2 \leq 4(GeV/c)^2$. We did not find a convinced evidence for polarization non-zero. Our preliminary results only have statistical errors.

1 Introduction

Since the unexpected measurement of substantial polarization for inclusively produced lambda particles by 300 GeV/c protons on Berylium [1], it has been determined that hyperons produced at high energy have non-zero polarization. The measurements all [2] seem to be consistent with a polarization given by:

$$\vec{P}_{a+A\to b+X} = f_{a,b}(x_F, p_T^2, A)\hat{n}\,, \tag{1}$$

A being the atomic weight of the target and $\hat{n}$ the normal to production plane. Several theoretical ideas have been proposed to explain these phenomena. Anderson [3] proposed that polarization from the soft, semi-classical process of quark-antiquark pair production by tunneling in a confined color field. DeGran and Miettinen [4] used the parton recombination model to relate the polarization to a Thomas precession like term in the recombination process.

There are a few results of hyperon polarization in π^--nucleus interactions. For Λ^0 in $\pi^- - Cu$ interactions at 230 GeV/c, Barlag [5] found, for $x_F > 0$ and $p_T^2 > 1(GeV/c)^2$, a polarization of -0.28±0.09±0.02. For Ξ^- and Ξ^+ there not studies with a π^- beam, but just a few results with a proton and Σ^- beam finding a polarization of 10%.

2 The Experiment and the Sample

As a byproduct of our charm program in Fermilab experiment E791, we collected a large sample of $\Lambda^0/\bar{\Lambda}^0$, Ξ^-/Ξ^+, and Ω^-/Ω^+ hyperons. Initially, we used the Ξ^- and Ξ^+ samples to measure the polarization production reported here, but we expect to continue the studies with the other hyperons.

Experiment E791 recorded data from 500 GeV/c π^- interactions in five thin foils (one platinum and four diamond) separated by gaps of 1.38 to 1.39 cm. Each foil had a thickness of approximately 0.4% of a pion interaction length (0.5 mm for the upstream platinum target, and 1.6 mm for each of the carbon targets). The E791 spectrometer [6] in the Fermilab Tagged Photon Laboratory was a large-acceptance, two-magnet spectrometer augmented by eight planes of multiwire proportional chambers (MWPC) and six planes of silicon microstrip detectors (SMD) for beam tracking. The magnets provided a total transverse momentum kick of 512 MeV/c. Downstream of the target there were 17 planes of SMD's for track and vertex reconstruction, 35 drift chamber planes, two MWPC's, two multicell threshold Čerenkov counters, electromagnetic and hadronic calorimeters (with apertures about 70 by 140 mr), and a muon detector. An important element of the experiment was its extremely fast data acquisition system [7] which was combined with a very open transverse-energy trigger to record a data sample of 2×10^{10} events. The trigger required that the total "transverse energy" (i.e., sum of the products of energy observed times the tangent of the angle from the target to each calorimeter segment) be at least 3 GeV.

The $\Lambda^0/\bar{\Lambda}^0$ sample was from approximately 7% of the overall sample recorded for the experiment. The total signal, taken as the sum of background subtracted signal in each bin, was $2\,587\,870 \pm 1\,780$ Λ^0's and $1\,690\,030 \pm 1\,500$ $\bar{\Lambda}^0$'s. Ξ^- were selected via the decay mode $\Xi^- \to \Lambda^0\pi^-$ and Ω^- via the decay mode $\Omega^- \to \Lambda^0 K^-$. Beginning with a Λ^0 candidate, we added a third, distinct track as a possible pion or kaon daughter. All three tracks were required to have hits in the drift chamber region only. For these samples, we removed the requirement on the Λ^0 impact parameter. The invariant mass of the candidate hyperon (calculated from the known mass and measured momentum of the Λ^0, as determined by its two decay tracks, together with the third track) was required to be between 1.290 and 1.350 GeV/c^2 for the Ξ^- and between 1.642 and 1.702 GeV/c^2 for the Ω^-. As with the Λ^0 sample, the Ξ^-/Ξ^+ and Ω^-/Ω^+ invariant mass plots were fit to a Gaussian signal plus linear background for each interval of x_F and p_T^2. With the final selection criteria and the full E791 data set, we found $996\,180 \pm 1\,200$ Ξ^-, $706\,620 \pm 1\,020$ Ξ^+, $8\,750 \pm 110$ Ω^-, and $7\,460 \pm 100$ Ω^+ after background subtraction. Again, these numbers and their errors come from the sum of signals in all bins. We checked that the Ξ^- contamination for the Ω^-'s, after all cuts, was negligible. Selection criteria for the particle and antiparticle samples were identical. However, geometrical acceptances and reconstruction efficiencies were not necessarily the same, mostly due to an inefficient region in the drift chambers produced by the negative pion beam. To evaluate this effect, a large sample of Monte Carlo (MC) events was created using the PYTHIA/JETSET event generators [8]. These were projected through a detailed simulation of the E791 spectrometer to simulate "data" in digitized format which was then processed through the same computer reconstruction code as that used for data from the detector. Candidate events were then subjected to the same selection criteria as that used for data. To account for correlations between x_F and p_T^2, efficiencies were determined in bins of the two parameters.

3 Procedure

To study the polarization of the Ξ one looks at the angular distribution of the lambda in the decay $\Xi^- \rightarrow \Lambda^0\pi^-$ with $\Lambda^0 \rightarrow p^+\pi^-$ (and the corresponding conjugate decay), with respect to the normal to the production plane. This plane is defined as the one formed by the beam direction and the momentum of the Ξ. The angular distribution of Λ particles produced is not isotropic but depends of cosine of 'the angle between the momentum of Λ (at the Ξ rest frame) and the quantization direction of spin ($\alpha_{\Xi^-} = -\alpha_{\Xi^+} = -0.456$)

$$\frac{1}{N_o}\frac{dN_\Xi}{dcos(\theta_{\Lambda\hat{n}})} = \frac{1}{2}(1 + \alpha_\Xi\, P_\Xi\, cos(\theta_{\Lambda\hat{n}}))\ ,\ \text{with}\ cos(\theta_{\Lambda\hat{n}}) = \frac{\vec{p}_{\Lambda_{CM_\Xi}}.\hat{n}}{\left|p_{\Lambda_{CM_\Xi}}\right|}. \quad (2)$$

For the actual state of our analysis in hyperon polarization, we corrected the data by the efficiency of MC. After that, we fitted the angular distributions of polarization for Ξ^- and Ξ^+ and the results you can see in the Tables.

4 Results and Conclusions

Most of the intervals we checked (see Tables) are compatible with no polarization. In few intervals we have non zero polarization but just with statistical errors. We need to continue the study to look for systematical effects. A future step is the study of other two hyperons, $\Lambda^0/\bar{\Lambda}^0$ and Ω^-/Ω^+

Polarization of Ξ^-				
P_{Ξ^-}	$0 \leq p_T^2 \leq 1$	$1 \leq p_T^2 \leq 2$	$2 \leq p_T^2 \leq 3$	$3 \leq p_T^2 \leq 4$
$-.12 \leq x_F \leq -.06$	0.058 ± 0.048	0.008 ± 0.136	0.472 ± 0.206	0.382 ± 0.284
$-.06 \leq x_F \leq 0.00$	0.120 ± 0.032	0.026 ± 0.090	0.350 ± 0.150	0.176 ± 0.198
$0.00 \leq x_F \leq 0.06$	0.008 ± 0.048	0.180 ± 0.126	0.248 ± 0.204	0.144 ± 0.282
$0.06 \leq x_F \leq 0.12$	0.068 ± 0.156	$-.174 \pm 0.312$	$-.436 \pm 0.518$	0.598 ± 0.606

Polarization of Ξ^+				
P_{Ξ^+}	$0 \leq p_T^2 \leq 1$	$1 \leq p_T^2 \leq 2$	$2 \leq p_T^2 \leq 3$	$3 \leq p_T^2 \leq 4$
$-.12 \leq x_F \leq -.06$	$-.106 \pm 0.056$	$-.236 \pm 0.138$	$-.148 \pm 0.222$	$-.170 \pm 0.310$
$-.06 \leq x_F \leq 0.00$	$-.096 \pm 0.034$	$-.128 \pm 0.092$	$-.294 \pm 0.158$	$-.224 \pm 0.238$
$0.00 \leq x_F \leq 0.06$	$-.082 \pm 0.048$	$-.148 \pm 0.122$	$-.198 \pm 0.218$	0.168 ± 0.290
$0.06 \leq x_F \leq 0.12$	0.058 ± 0.160	$-.524 \pm 0.326$	0.062 ± 0.360	$-.786 \pm 0.784$

Acknowledgements

This research was supported by the FAPEMIG (Brazil) and by the CONACYT (México).

References

1. G. Bunce *et al.*, Phys. Rev. Lett. **36** (1976) 1113.
2. B. Lundberg *et al.*, Phys. Rev. **D 40** (1989) 3557; A. Morelos *et al.*, Phys. Rev. Lett. **71** (1993) 2172; P.M. Ho *et al.*, Phys. Rev. **D 44** (1991) 3402.
3. B. Anderson *et al.*,Phys. Lett. **B 85** (1979) 417.
4. A. DeGran and H.I. Miettinen, Phys. Rev. **D 24** (1981) 2419.
5. S. Barlag *et al.*,Phys. Lett. **B 325** (1994) 531.
6. J.A. Appel, *Ann. Rev. Nucl. Part. Sci.* **42** (1992) 367; E.M. Aitala, *et al.* (E791 Collaboration), EPJdirect **C 4** (1999) 1-67.
7. S. Amato, *et al.*, Nucl. Instr. and Methods A 324 (1993) 535.
8. *PYTHIA* 5.7 and *JETSET* 7.4; H.U. Bengtsson and T. Sjöstrand, Computer Physics Commun **46** (1987) 43; T. Sjostrand, CERN-TH.7112/93.

PROTON-PROTON AND PROTON-ANTIPROTON CROSS-SECTIONS AT VERY HIGH ENERGIES

ERASMO FERREIRA
Instituto de Física, Universidade Federal do Rio de Janeiro, Rio de Janeiro 21945-970, RJ, Brazil.
Erasmo@if.ufrj.br

FLAVIO I. M. PEREIRA
MCT-Observatório Nacional, Rua General José Cristino, 77, Rio de Janeiro, 20921-400, RJ, Brazil.
flavio@on.br

Results obtained with the model of the stochastic vacuum for hadron-hadron scattering are extrapolated to the very high energies of 14 and 30 TeV, with predictions for values of total cross-section, slope parameter and hadronic sizes.

1 Introduction

The model of the stochastic vacuum (MSV) provides a framework for a simple and effective description of the high-energy pp and p$\bar{\text{p}}$ data, which leads to a determination of relevant parameters of non-perturbative QCD. The gluon correlator is the main ingredient of the QCD vacuum that determines the phenomenological consequences of the model [1]. This correlation function determined by the vacuum structure of QCD is characterized by three parameters, namely, the gluon condensate $\langle g^2FF\rangle$, the correlation length a, and the parameter κ ($0 \leq \kappa \leq 1$) that determines the ratio of the contributions of confining and non-confining parts of the correlator. Gauge invariance is an important property of the correlator.

Application of MSV in the eikonal framework to pp and p$\bar{\text{p}}$ scattering leads to a profile function $J(b, S/a, \kappa)$, where b is the impact parameter and S is the proton extension parameter. In therms of this profile, the total cross-section and the slope parameter are given by

$$\sigma^{\mathrm{T}}(s) = [\langle g^2FF\rangle a^4]^2 I_0 \quad \text{and} \quad B(s) = \frac{1}{2}\frac{I_2}{I_0}a^2 \equiv Ka^2 \; , \tag{1}$$

where I_0 and I_2 are two moments of the profile function $J(b, S/a, \kappa)$,

$$I_0(S/a, \kappa) = \int J(b, S/a, \kappa)d^2\mathbf{b} \quad \text{and} \quad I_2(S/a, \kappa) = \int b^2 J(b, S/a, \kappa)d^2\mathbf{b} \; . \tag{2}$$

Parameterizing conveniently the moment I_0 and the ratio $K = I_2/2I_0$ in terms of the hadronic radius S, through the forms

$$I_0 = \alpha(S/a)^{\beta} \quad \text{and} \quad K = \eta + \gamma(S/a)^{\delta} \,, \tag{3}$$

we find for a given κ a set of parameters α, β, γ, η, δ determined by fitting $\alpha(S/a)^{\beta}$ and $\eta + \gamma(S/a)^{\delta}$ to the values of I_0 and K (as functions of S/a) obtained by numerical integration of Eqs.(2) [2]. For $\kappa = 3/4$ suggested by lattice calculations the values of these parameters are $\alpha = 0.6532 \times 10^{-2}$, $\beta = 2.791$, $\eta = 2.030$, $\gamma = 0.3293$, and $\delta = 2.126$. By using the input (accelerator) data at 541 and 1800 GeV we determine the QCD parameters

$$a = 0.32\pm0.01 \text{ fm}, \ \langle g^2FF\rangle = 2.7\pm0.1 \text{ GeV}^4 \text{ and } \langle g^2FF\rangle a^4 = 18.7\pm0.4\,. \tag{4}$$

2 Extrapolations for 14 and 30 TeV

In the model the energy dependence of the observables is due to a slow variation of the hadronic sizes. Using the data at 541 and 1800 GeV, the relation between S/a and $\ln s$ obtained from Eqs.(1) and (2) can be parameterized in the form

$$\frac{S}{a} = 1.788 + 0.384(\ln\sqrt{s})^{0.72} \,, \tag{5}$$

where the exponent 0.72 was introduced to put $\sigma^{\mathrm{T}}(s)$ in the unitarity bound $\ln^2\sqrt{s}$. This parameterization of the proton size allows us to predict values of the pp cross-section at the energies of 14 TeV of LHC and 30 TeV of cosmic rays of Akeno [3] and Fly's Eyes [4] experiments, and Nikolaev [5]. Our values are shown in the Table 1.

3 Other Parameterizations and Predictions

Several authors have made parameterizations and predictions for pp

$\sqrt{s}$ (TeV)	σ^T (mb)	B (GeV^{-2})	S_p (fm)	S/a
14	93	19	$1.20 = 1.39\ \mathrm{R_p}$	3.74
30	101	21	$1.24 = 1.43\ \mathrm{R_p}$	3.85

Table 1. $R_p = 0.862 \pm 0.012$ fm is the proton electromagnetic radius.

and $p\bar{p}$ scattering at very high energies, and it is interesting to compare their values. Some of these predictions are Donnachie-Landshoff [6]:

$$\sigma^{\rm T}_{pp,p\bar{p}} = 21.70\ s^{0.0808}\ ({\rm mb})\,,$$

$$\sigma^{\rm T}_{pp,p\bar{p}}(14\ {\rm TeV}) = 101\ {\rm mb} \quad {\rm and} \quad \sigma^{\rm T}_{pp,p\bar{p}}(30\ {\rm TeV}) = 115\ {\rm mb}\ .$$

Honda et al. [3] (Akeno):

$$\sigma^{\rm T}_{pp} = 38.5 + 1.37\ \ln^2(\sqrt{s}/10\ {\rm GeV})\ ({\rm mb})\,,$$

$$\sigma^{\rm T}_{pp,p\bar{p}}(14\ {\rm TeV}) = 110\ {\rm mb} \quad {\rm and} \quad \sigma^{\rm T}_{pp,p\bar{p}}(30\ {\rm TeV}) = 126\ {\rm mb}\ .$$

Ávila, Luna and Menon (ALM)[7]:

$$\sigma^{\rm T}_{pp,p\bar{p}}(14\ {\rm TeV}) = 113 \pm 5\ -\ 140 \pm 7\ {\rm mb}\ .$$

Block et al.[8]:

$$\sigma^{\rm T}_{pp,p\bar{p}}(14\ {\rm TeV}) = 108.0 \pm 3.4\ \ {\rm mb}\ .$$

Avila and Benitez [9]:

$$\sigma^{\rm T}_{pp,p\bar{p}}(14\ {\rm TeV}) = 73\ -\ 76\ {\rm mb} \quad {\rm and} \quad \sigma_{pp,p\bar{p}}(30\ {\rm TeV}) = 116\ -\ 136\ {\rm mb}\ .$$

These values are shown in the Fig.1 for comparison with our extrapolation of $\sigma^{\rm T}(s)$ to high energies. The figure also includes the data of E-811 experiment[10].

4 Conclusions

We have studied the pp and $p\bar{p}$ systems, describing its high-energy data in terms of non-perturbative QCD parameters, provided by MSV, and related the energy dependence of the observables with the change of the proton size. Our extrapolations predict values of the total cross section, slope parameter, and hadronic radius at energy of 14 TeV of LHC. The extrapolation to cosmic rays energies at 30 TeV are comparable to the values given by Akeno [3] and Fly's Eyes [4] experiments, and to parametric extrapolations made by other authors.

Acknowledgments

We thank M. Menon, E. Luna, R. Ávila and A. Martini for fruitful discussions and helpful informations.

References

1. Dosch, H. G., Ferreira, E. and Krämer, A., *Phys. Lett.* **B289** (1992) 153; *ibid.* **B318** (1993) 197; *Phys. Rev.* **D50**(1994) 1992.
2. Ferreira, E. and Pereira, F., *Phys. Rev.* **D55**(1997) 130; *Phys. Lett.* **B399**(1997) 177.
3. Honda, M. et al., *Phys. Rev. Lett.* **70** (1993) 525 - (Akeno).
4. Baltrusaitis, R. M. et al., *Phys. Rev. Lett.* **52**(1984) 1380.
5. Nikolaev, N. N., *Phys. Rev.* **D48** (1993), R1904.

6. Donnachie, A. e Landshoff, P. V., *Phys. Lett.* **B296** (1992), 227.
7. Ávila, R. F., Luna, E.G. S. and Menon, M. J., *Braz. Jour. of Phys.* **31** (2001) 567.
8. Block, M. M. et al., *Phys. Rev.* **D60** (1999) 054024.
9. Avila, C. and Benitez, J., in LISHEP 2002.
10. Avila, C. et al., *Phys. Rev. Lett.* **B445** (1999) 419 - (E-811).

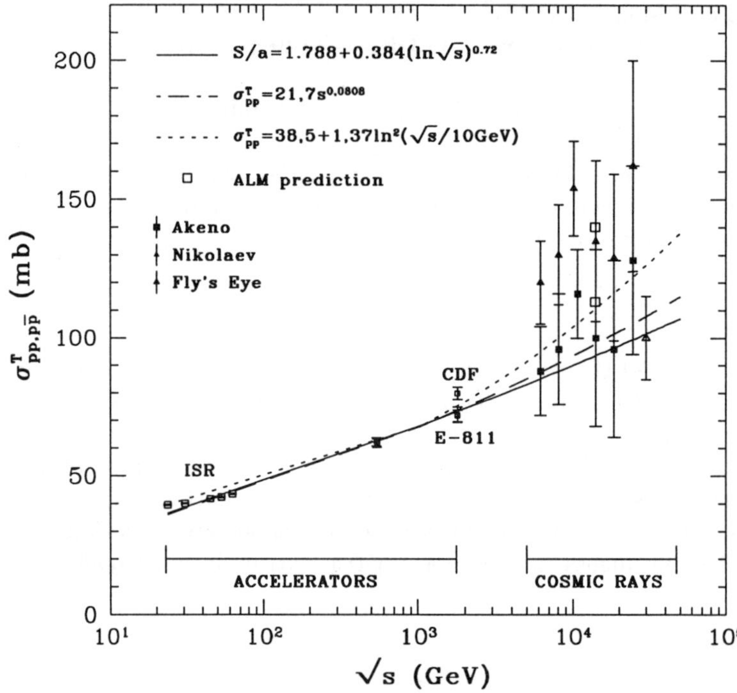

Figure 1. Predictions for pp and p$\bar{\text{p}}$ scattering with our extrapolation of $\sigma^{\mathrm{T}}(s)$ to high energies. The figure also includes the data of E-811 experiment[10].

COMPARATIVE STUDY BETWEEN A QCD INSPIRED MODEL AND A MULTIPLE DIFFRACTION MODEL

E.G.S. LUNA, A.F. MARTINI AND M.J. MENON

Instituto de Física 'Gleb Wataghin', UNICAMP
13083-970, Campinas, SP, Brazil
luna@ifi.unicamp.br, martini@ifi.unicamp.br, menon@ifi.unicamp.br

A comparative study between a QCD Inspired Model (QCDIM) and a Multiple Diffraction Model (MDM) is presented, with focus on the results for *pp* differential cross section at $\sqrt{s} = 52.8$ GeV. It is shown that the MDM predictions are in agreement with experimental data, except for the dip region and that the QCDIM describes only the diffraction peak region. Interpretations in terms of the corresponding eikonals are also discussed.

1 Introduction

Despite the great amount of model descriptions of the experimental data on high-energy elastic hadron scattering, a pure QCD treatment is still absent [1]. In order to look for insights that could provide further developments, and since in the *eikonal representation* unitarity may be automatically verified, we perform a comparative study between two eikonal approaches: a QCD Inspired Model (QCDIM) [2] and a Multiple Diffraction Model (MDM) [3]. In the Eikonal representation [4], the scattering amplitude is expressed as the Fourier transform of $1 - \exp\{i\chi(s,b)\}$, where b is the impact parameter and $\chi(s,b)$ the *complex* eikonal function. The MDM and the QCDIM are characterized by the following choices for the eikonal function.

2 Multiple diffraction model

In the Glauber multiple diffraction formalism the eikonal in the momentum transfer space is given by [4] $\chi(s,q) = G_A G_B f$, where $q = \sqrt{-t}$ is the momentum transfer, $G_{A,B}$ are the hadronic form factors and f the elementary (constituent-constituent) amplitude. In a recent version of this approach for *pp* elastic scattering, the following parameterizations have been used[3] $Im f(s,q) = C(s)h(q)$, $G_A = G_B = 1/(1 + q^2/\alpha^2)(1 + q^2/\beta^2)$, where $h(q) = (1 - q^2/a^2)/(1 + q^4/a^4)$. The real part of the elementary amplitude (and therefore that of the eikonal) has been calculated through a first order singly subtracted derivative dispersion relation. The model has 5 parameters and the experimental data on *pp* elastic scattering above 10 GeV are

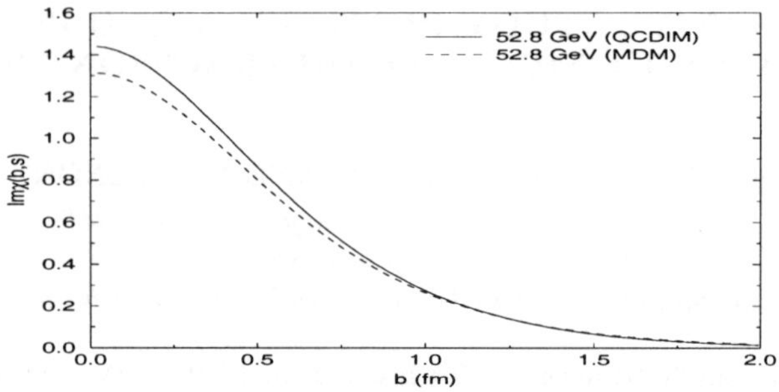

Figure 1. The imaginary parts of the eikonal at $\sqrt{s} = 52.8$ GeV calculated with the Multiple Diffraction Model (dashed line) and the QCD Inspired Model (solid line).

well described with the following values and parameterizations [3]: $\alpha^{-2}(s) = 2.57 - 0.217\ln(s) + 0.0243\ln^2(s)$, $C(s) = 14.3 - 1.65\ln(s) + 0.159\ln^2(s)$, $a^2 = 8.20$ GeV^2, $\beta^2 = 1.80$ GeV^2 and $K = -172$ (the subtraction constant).

3 QCD inspired model

We consider here a version [2] that treats both pp and $p\bar{p}$ scattering by means of crossing symmetry in the eikonal function $\chi_{pp}^{p\bar{p}}(b,s) = \chi^{even}(b,s) \pm \chi^{odd}(b,s)$, where the real and imaginary parts are generated by the substitution $s \to s\,e^{-i\pi/2}$. The even eikonal is given by the sum of three contributions, gluon-gluon, quark-gluon and quark-quark, which are individually factorizable into a product of a cross section $\sigma(s)$ by an impact parameter space distribution $W(b;\mu)$: $2\chi^{even}(b,s) = i[\sigma_{qq}(s)W(b;\mu_{qq}) + \sigma_{qg}(s)W(b;\mu_{qg}) + \sigma_{gg}(s)W(b;\mu_{gg})]$, where $\mu_{qg} \equiv \sqrt{\mu_{qq}\mu_{gg}}$ and $W(b;\mu) = \mu^2/96\pi(\mu b)^3 K_3(\mu b)$. The "parton-parton" contributions are parameterized by

$$\chi_{qq}(b,s) = i\,\Sigma_{gg}\left(a + b\,\frac{m_0}{\sqrt{s}}\right) W(b;\mu_{qq}), \quad \chi_{qg}(b,s) = i\,\Sigma_{gg}\,a'\ln\frac{s}{s_0}\,W(b;\mu_{qg}),$$

where $\Sigma_{gg} = 9\pi\alpha_s^2/m_0^2$ is a normalization constant and the gg cross section reads

$$\sigma_{gg}(s) = C'_{gg}\,\Sigma_{gg}\,N_g^2\sum_{l=0}^{5}\left\{\frac{a(l) - \frac{b(l)}{l-\epsilon}}{l-\epsilon} - \tau_0^{l-\epsilon}\left[\frac{a(l) - \frac{b(l)}{l-\epsilon}}{l-\epsilon} + \frac{b(l)}{l-\epsilon}\ln\tau_0\right]\right\}.$$

The odd eikonal is parameterized as $2\chi^{odd}(b,s) = C_{odd}\Sigma_{gg}\frac{m_0}{\sqrt{s}}\,e^{i\pi/4}\,W(b;\mu_{odd})$. The parameters were determined through the analysis of elastic pp and $p\bar{p}$ scattering [2].

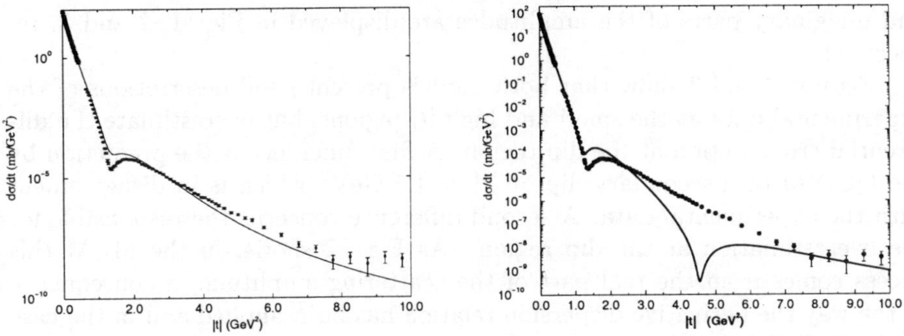

Figure 2. The differential cross section for pp scattering at $\sqrt{s} = 52.8$ GeV with the descriptions from the MDM (left) and QCDIM (right).

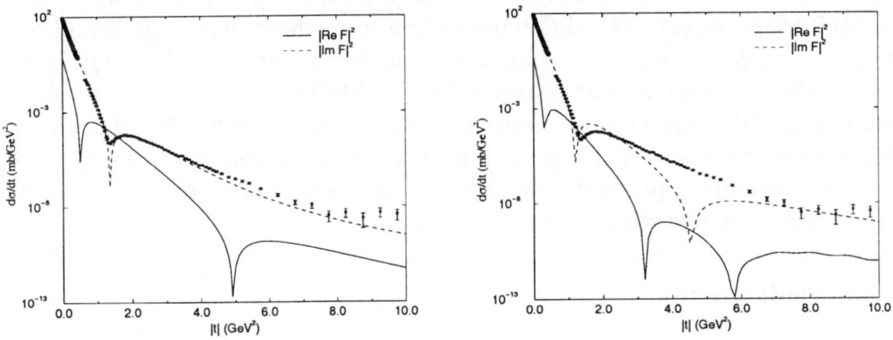

Figure 3. Contributions to the differential cross section from the real and imaginary parts of the scattering amplitude: MDM (left) and QCDIM (right).

4 Results and Discussion

The predictions from both models for total cross sections and the ρ parameter may be found in the corresponding references [2,3]. Here we want to focus on the results for the pp at 52.8 GeV, since the differential cross section corresponds to the highest momentum transfer interval with data available. The predictions from both models for the imaginary parts of the eikonal, the differential cross section and the corresponding contributions from the real and imaginary parts of the amplitudes are displayed in Figs 1, 2 and 3, respectively.

Figures 2 and 3 show that both models present good descriptions of the experimental data at the small and high $|t|$ regions, but overestimate the differential cross section at the dip region. A first difference is the prediction by the QCDIM of a secondary dip at $|t| \sim 4.5$ GeV^2 which is in disagreement with the experimental data. A second difference concerns the responsible for the overestimation at the dip region. As Fig. 2 shows, in the MDM this excess comes from the real part of the scattering amplitude, a consequence of the way the derivative dispersion relation has been applied and in the case of the QCDIM the excess comes from the imaginary part of the amplitude, which also predicts the position of the dip in the wrong place. These differences come from the corresponding imaginary parts of the eikonal, as shown in Fig.1. For large b-values (above ~ 1.0 fm) the two results are quite similar and this explains why both models present good descriptions at small-$|t|$ values (diffraction peak). The differences come from the central region and are associated with the overestimation of the opacity in the case of the QCDIM. Despite the naive ideas that characterize the MDM it was possible to show that the QCDIM predicts an hadron blacker than it should be at small-b. That means that some of the elementary contributions (qq, qg, gg) may have been overestimated, probably those associated with semi-hard interactions. We are presently investigating these aspects.

Acknowledgements

Work supported by FAPESP and CNPq, Brazil. We are thankful to the organizers of this excellent and stimulating Workshop, in particular to Cesar Vasconcellos and Victoria Herscovitz.

References

1. G. Matthiae, *Rep. Prog. Phys.* **57**,743 (1994).
2. M.M. Block et al., *Phys. Rev.* **D60**, 054024 (1999).
3. A.F. Martini and M.J. Menon, XXII Brazilian National Meeting on Particles and Fields (2001), www.sbf1.if.usp.br (to be published); A.F. Martini, et al., Phys. Rev. **D59**, 116006 (1999).
4. R.J. Glauber, in *Lectures in Theoretical Physics*, edited by W. E. Brittin *et al.* (Interscience, New York, 1959) Vol. I, p.315; V. Franco and G.K. Varma, Phys. Rev. C **18**, 349 (1978).

HADRON PRODUCTION OF MAJORANA NEUTRINOS AT VLHC ENERGIES

F.M.L. DE ALMEIDA JR., Y.A. COUTINHO, J.A. MARTINS SIMÕES
Instituto de Física, Universidade Federal do Rio de Janeiro, RJ, Brazil
marroqui@if.ufrj.br, yara@if.ufrj.br, simoes@if.ufrj.br

M.A.B. DO VALE
Natural Sciences Department, FUNREI, MG, Brazil
aline@funrei.br

The Very Large Hadron Collider (VLHC) is being proposed as a 50+50 TeV hadron collider to extend the energy frontier beyond the LHC. Since 1998-1999 the option of a ep collider operating with the 3 TeV proton booster has been considered. This design uses a 80 GeV electron beam to produce ep collisions with a luminosity of 2600 pb-1/yr with a center of mass energy of 1TeV. We study the discovery potential of this proposed ep collider for detecting new neutral heavy Majorana leptons suggested by different extensions of the Standard Model, using the channel $e^- p \to e^+$+jets.

1 Introduction

The recent experimental results on solar and atmospheric neutrino properties confirm that standard neutrinos have small masses and oscillate. However, the nature of those massive neutrinos - if they are Dirac or Majorana particles - is still an open question. This is connected with lepton number conservation and to the general symmetries of some extension of the standard model. The smallness of neutrino masses is generally understood as a consequence of the see-saw mechanism. This brings the question of the possibility of new heavy neutrino states. So far none of these new states was experimentally observed with masses up to $M_N \simeq 100$ GeV. For higher masses there are many suggestions of experimental possibilities in the next high energy hadron-hadron colliders [1], in electron-positron linear accelerators, in neutrinoless double-beta decay. Another experimental possibility is found on electron-proton colliders with a positron in the final state. This means a very clear experimental signature that violates the electron leptonic number by two units. In this paper we investigate the discovery potential for recently proposed electron- proton colliders for detecting new neutral heavy Majorana leptons as suggested by extensions of the Standard Model. Since 1998-1999 the option of an electron-proton collider for the Very Large Hadron Collider at Fermilab operating with the proton booster has been considered. We study the reaction $e^- p \longrightarrow e^+ +$ jets and present estimates for the signal and Standard Model background.

2 The Very Large Hadron Collider

There is a growing work on the options for a future Very Large Hadron Collider (VLHC). The low magnetic field (approximately 2 T) collider design in a tunnel can also be used for a future high magnetic field (approximately 10 T) collider. These designs could also be used for electron-proton colliders. In the low-field version electron-proton collisions would be produced by an 80 GeV electron ring and a 3 TeV proton booster ($\sqrt{s} = 980$ GeV), with a luminosity $\mathcal{L}=2600$ pb^{-1}/yr, while in the high-field option one could reach a center-of-mass energy $\sqrt{s} = 6320$ GeV and a luminosity $\mathcal{L} = 1400$ pb^{-1}/yr with the collision of a 200 GeV electron beam with a 50 TeV proton beam. These two options are considered in the present work.

3 The Model

New heavy leptons are present in many possible extensions of the standard model. In the superstring inspired scenario we have new doublet vector leptons. In mirror models there are new fermions with right handed doublets and left handed singlets. Some see-saw extensions of the standard model include new weak-singlet fermions.

The general mixing of light and heavy Majorana neutrinos has new mixing parameters as well as new possible CP violating phases. It is well known that in neutrinoless double beta decay the theoretical estimates depend strongly on mixing angles and CP phases. As a consequence, the comparison between the experimental limits and theory is model dependent. This is also the case for other processes. In this paper we are interested in the study of the discovery potential of new electron-proton colliders for new heavy Majorana neutrinos. From the experimental point of view, instead of start comparisons of very particular models with a great number of parameters and hypothesis one must start looking for the lightest of the new possible Majorana neutrinos, coupled to the electron family. This reduces the number of unknown parameters to two; the heavy neutrino mass and one new mixing angle, that should be determined experimentally.

In this paper we will adopt a general parameterization that applies to all the above models. We will suppose that mixing angles for heavy-to-light neutrinos and new heavy neutrino masses are independent parameters. The light neutrinos couplings to the neutral Z are given by $g_{V,A} = g_{V,A}^{SM} - sin^2\theta_{\text{mix}}/2$. Taking the experimental values for $g_{V,A}$ and the Standard Model predictions we obtain a small upper bound for θ_{mix}. A recent estimate gives $\sin^2\theta_{\text{mix}} < 0.0052$ with 95% C.L. This limit value is used throughout this paper for all curves and distributions. The decay of the Majorana neutrino into muons and taus are also possible. In ref. [2] it was found that for the muon sector the experimental mixing angle limit with 95% C.L. is much smaller than for the electron sector.

Most extended models predict new fermions and new gauge vector bosons. Since we presently have no signal for new interactions, we will make the hypothesis that the new heavy neutrino states behave as $SU_L(2)\otimes U_Y(1)$ basic representations and interact mostly with the standard model gauge bosons. We can resume the new particle interactions in the neutral and charged current lagrangians:

$$\mathcal{L}_{nc} = -\frac{g}{4c_W} sin\theta_{mix} Z_\mu \overline{\psi_N}\gamma^\mu (1-\gamma_5)\psi_\nu + h.c.. \tag{1}$$

and

$$\mathcal{L}_{cc} = -\frac{g}{2\sqrt{2}} sin\theta_{mix} W_\mu \overline{\psi_N}\gamma^\mu (1-\gamma_5)\psi_e + h.c. \tag{2}$$

and similar terms for the other leptonic families. In the Majorana neutrino case, each completely neutral heavy lepton is coupled to all charged leptons but we are considering only the lower mass state.

The decay modes for these leptons, in the Majorana case [1] must include both signatures $N \longrightarrow \ell^\mp W^\pm$ and $N \longrightarrow \nu_\ell\,(\bar{\nu}_\ell)\,Z$, with $\ell = e, \mu, \tau$. In this work we have considered the Majorana neutrino decay only to the first family since for the other families we must add more mixing parameters.

4 Results

In the final visible particles we have imposed the following criteria to select events: a) for $|\eta_{e^+}| < 3.5$ and $E_{e^+} > 5$ GeV we must have the number of $e^+ > 0$; b)for $|\eta_{h_i}| < 3.5$ and $E_{h_i} > 5$ GeV we must have the number of hadrons > 5. With these two conditions, more than 99 % of the background events were eliminated for both energies.

In order to reconstruct the Majorana neutrino invariant mass we have employed the algorithm cluster PYCELL from the PYTHIA package, with 60 azimuthal divisions and 26 divisions in the electron beam direction. For $\sqrt{s}=$ 980 GeV we have applied the condition $0.7 < \sqrt{(\Delta\eta)^2 + (\Delta\phi)^2}$ while for $\sqrt{s}=$ 6320 GeV this condition was changed to $0.3 < \sqrt{(\Delta\eta)^2 + (\Delta\phi)^2}$ with

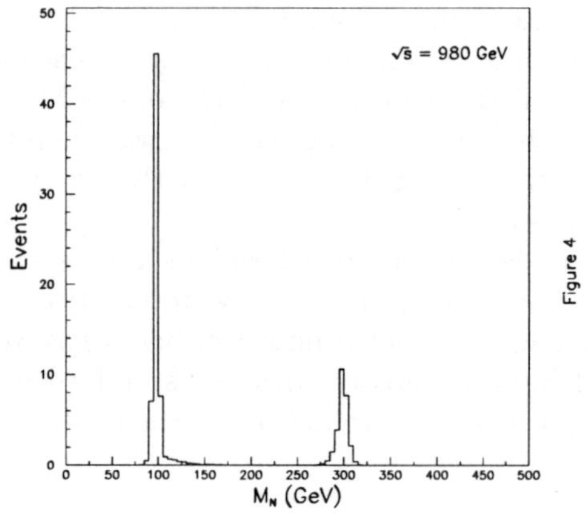

Figure 1. Majorana neutrino invariants.

$|\eta| < 3.5$ for both energies. After these conditions, we have further demanded that: c) number of clusters > 1 and Number of clusters with 1 track $= 1$; d) cluster invariant mass in the range (70, 90) GeV. If we choose the clusters obtained with the surviving events then we can reconstruct the Majorana neutrino invariant mass as shown in Figure 1. The same result is obtained for $\sqrt{s} = 6320$ GeV.

5 Conclusions

The present work shows that heavy Majorana neutrinos could be produced in the electron-proton option for the VLHC. The very clear signature of a positron in the final state allows to separate efficiently the signal events from the background and to reconstruct the Majorana neutrino invariant mass.

Acknowledgments

This work was partially supported by the following Brazilian agencies: CNPq and FAPERJ.

References

1. F.M.L. Almeida Jr., Y.A. Coutinho, J.A. Martins Simões, P.P. Queiroz Filho and C.M. Porto, Phys. Lett. B**400** (1997) 331; A. Ferrari et al., Phys. Rev. D**62** (2000) 013001-1. D**44** (1991) 2836 ; F.M.L. Almeida Jr., J.H. Lopes, J.A. Martins Simões, P. P. Queiroz Filho and A. J. Ramalho, Phys. Rev. D**51** (1994) 5990.
2. F.M.L. Almeida Jr., Y.A. Coutinho, J.A. Martins Simões and M.A.B. do Vale, Phys. Rev. D**62** (2000) 075004.

NONPERTURBATIVE QCD TREATMENT OF J/Ψ PHOTOPRODUCTION

H.G. DOSCH

Institut für Theoretische Physik, Universität Heidelberg, Philosophenweg 16, D-6900 Heidelberg, Germany
H.G.Dosch@Thphys.uni-heidelberg.de

ERASMO FERREIRA

Instituto de Física, Universidade Federal do Rio de Janeiro, Rio de Janeiro 21945-970, RJ, Brazil.
Erasmo@if.ufrj.br

We present results of a non-perturbative QCD calculation of elastic vector meson production in photon-proton collisions at high energies. We use the non-perturbative model of the stochastic QCD vacuum which has been used before to calculate hadron-hadron high energy scattering. In the applications of the model, fundamental parameters of non-perturbative QCD are obtained, and the energy dependence of observables are related to the influence of the energy on the parameters of the transverse hadron wave-functions. In the application of the stochastic vector model to photo-production processes here presented we use light cone wave functions for the photon and vector mesons to calculate photo-production cross-sections for the J/ψ meson. The energy dependence of the integrated cross-section is treated in the hard+soft pomeron schedule, and the t-dependence of the differential cross-section is obtained. The calculation is made with no free parameters, and the comparison with data is very satisfactory.

1 Calculation of Ψ photo-production in MSV

Photo-production of vector mesons presents characteristic features of soft hadronic interactions, virtual or real photons appearing in these processes as true hadronic structures of quark-antiquark pairs interacting with the proton. The QCD mechanisms governing these interactions combine perturbative contributions of gluon exchanges between individual quarks with ingredients of non-perturbative hadron-hadron interactions.

We here present the results of a non-perturbative QCD calculation of exclusive elastic photo-production of J/Ψ mesons. We build appropriate light-cone wave-functions for the photon and the vector meson and obtain the overlap that accounts for the transition responsible for the $\gamma\mathrm{p} \to J/\Psi\mathrm{p}$ process. We use the stochastic vacuum model to evaluate the eikonal amplitude of the interaction between proton and the photon-meson system. We adopt the same values of the QCD vacuum parameters that describe successfully purely

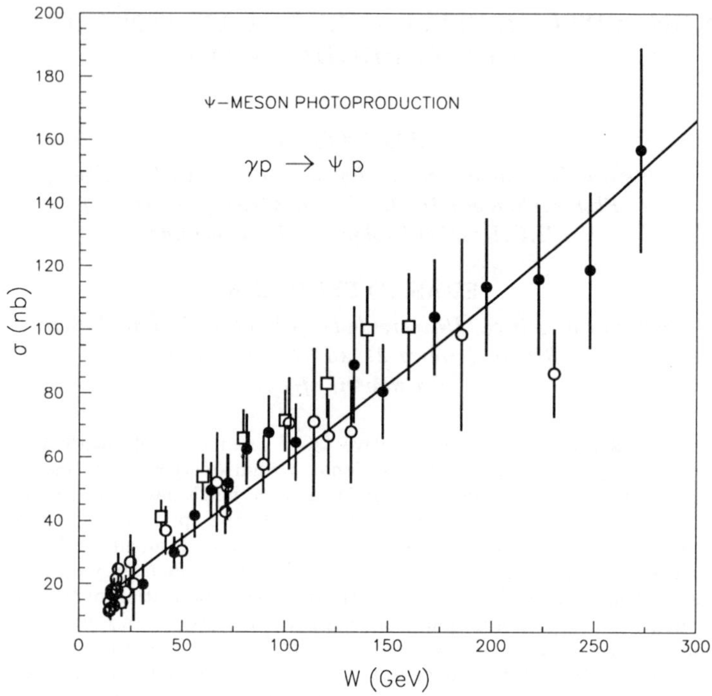

Figure 1. Energy dependence of the integrated elastic cross-section for Ψ photo-production.

hadronic processes (such as pp total and differential elastic cross-sections), so that our calculation has no arbitrary quantities. We obtain the forward amplitude and the dependence on the momentum transfer t, which we then compare with the experimental data.

We also investigate a phenomenological model based on a combination of soft and hard pomeron exchanges to describe the observed energy dependence of the photo-production cross-sections.

2 Conclusions

A pure QCD calculation without perturbative ingredients, and with no new arbitrary parameters in the non-perturbative stochastic vacuum model, accounts successfully for the energy dependence of the integrated elastic cross-section and for the low-t angular distribution in $J\Psi$ photo-production.

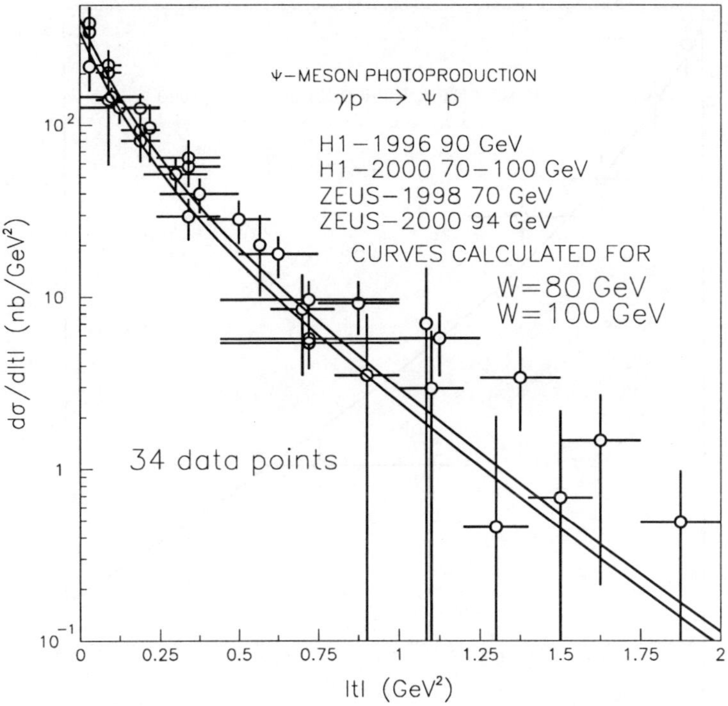

Figure 2. t dependence $d\sigma/dt$ in Ψ photo-production at 70 GeV.

Acknowledgments

We thank CNPq and DAAD for financial support of this collaboration.

References

1. H.G. Dosch, E. Ferreira and A. Krämer, Phys. Rev. D **50**, 1992 (1994).
2. A. Donnachie and P.V. Landshoff, Phys. Lett. B **185**, 403 (1987).
3. A. Donnachie and P.V. Landshoff, Phys. Lett. B **348**, 213 (1995).
4. M. Wirbel, B. Stech and M. Bauer, Z. Phys. C **29**, 637 (1985).
5. NMC, M. Arneodo *et al.*, Nucl. Phys. **B429**, 503 (1994).
6. ZEUS Collab., M. Derrick *et al*, Phys. Lett. B **380**, 220 (1996).

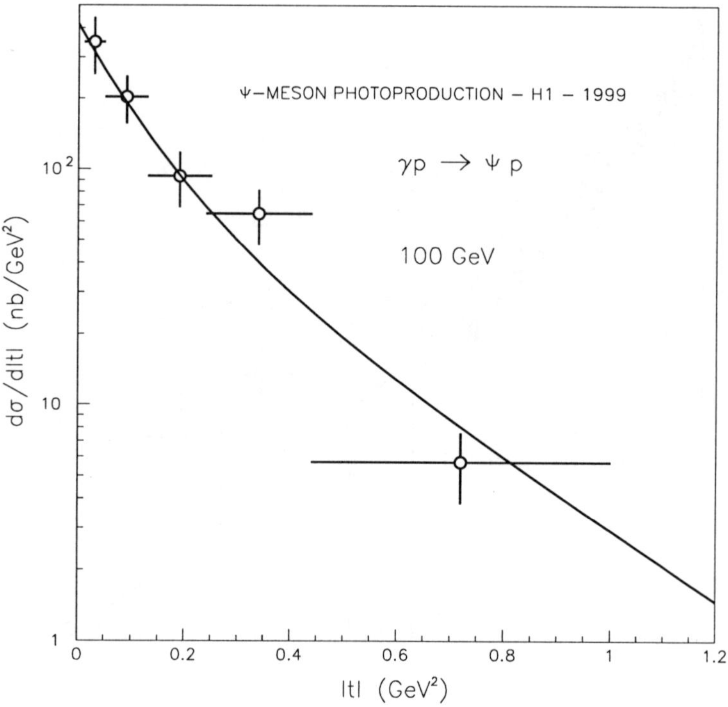

Figure 3. t dependence $d\sigma/dt$ in Ψ photo-production at 100 GeV.

7. EMC, J.J. Aubert *et al.*, Nucl. Phys. **B213**, 1 (1983).
8. M. Binkley *et al.*, Phys. Rev. Lett. **48**, 73 (1982).
9. NMC, M. Arneodo *et al.*, Phys. Lett. B **332**, 195 (1994).

ENTROPY BEHAVIOUR WITH NUCLEON-ANTINUCLEON ASYMMETRY

J.B. SILVA , A. DELFINO AND M. MALHEIRO

Instituto de Física, Universidade Federal Fluminense, 24210-340, Niterói, Brazil
jbsilva@if.uff.br

The recent STAR Collaborations at RHIC found anti-proton-to-proton ratio about 0.6. Constraining the baryonic density to satisfy this experiment data, a set of hadronic models is studied to investigate the baryonic entropy per particle behaviour. We also investigate the baryonic chemical potential as a function of the anti-nucleon-nucleon ratio. Our results show a systematic increasing(decreasing) of the entropy per particle (baryonic chemical potential) as this ratio increases.

1 Introduction

The ultimate goal of ultra-relativistic nucleus-nucleus collisions is to study the properties of interacting matter under extreme conditions of high density and temperature. Recently [1], we studied a class of different relativistic hadronic mean field models (RHMF) to investigate phase transition from hadron phase to Quark-Gluon Plasma (QGP). The QGP phase was described by a perturbative QCD equation of state [2]. In that study, only the infinite symmetric nuclear matter has been brought to represent the hadron phase.

Recent STAR Collaborations at RHIC which found anti-proton-to-proton ratio ($\frac{\rho_{\bar{N}}}{\rho_N}$) about 0.6 [3], motivate us to extend our previous study to a realistic case. Constraining $0 \leq \frac{\rho_{\bar{N}}}{\rho_N} \leq 1$, we calculate the baryonic entropy per particle S/A as well as the baryonic chemical potential μ. Our study shows a strong model dependence of S/A and μ as a function of $\frac{\rho_{\bar{N}}}{\rho_N}$. Even for the specific experimental data case $\frac{\rho_{\bar{N}}}{\rho_N} = 0.6$, this dependence is already reasonable. The hadronic models we use are described in Ref. [1]. In general they are generalizations of the Walecka model, containing cubic and quartic self mesonic interaction ($C1, Q1, NL1, NL2, NL3$ and $NLSH$ [4,5]), scalar derivative coupling (ZM and $ZM3$) [6] and self vector interactions ($Q2, G1$ and $G2$)[4].

2 RHMF Models Equation of State

We describe the hadronic (meson-baryon) phase (H) by the presented in Ref. [1,4,5,6], in the Hartree approximation extended to finite temperature.

FIGURE:1

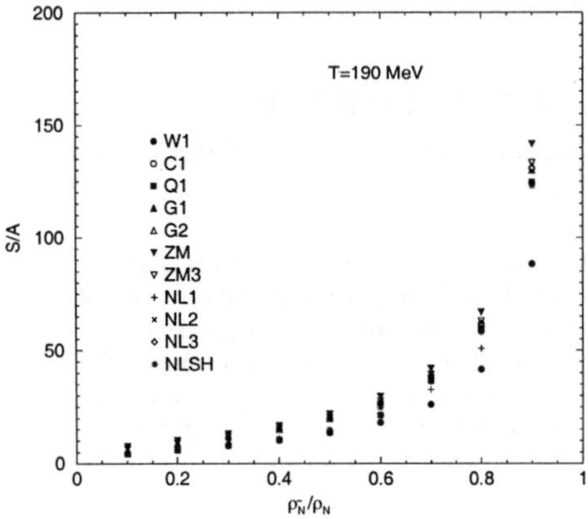

FIGURE:2

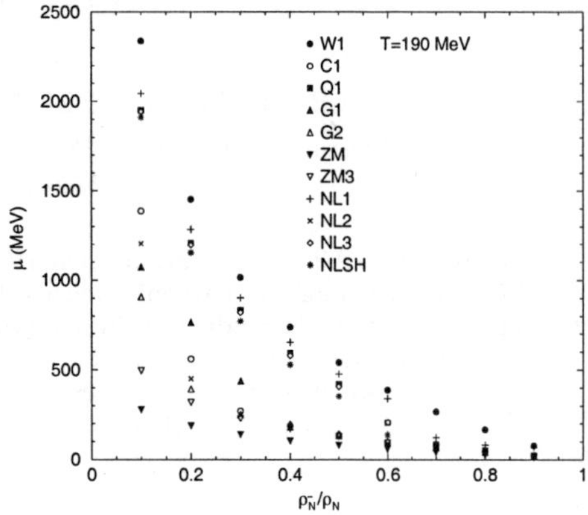

Except for ZM and $ZM3$ models [1], the pressure reads

$$
\begin{aligned}
p(M^*,\rho_B) = & -\frac{m_s^2}{g_s^2}\Phi^2\left(\frac{1}{2}+\frac{\kappa_3}{3!}\frac{\Phi}{M}+\frac{\kappa_4}{4!}\frac{\Phi^2}{M^2}\right) \\
& -\frac{1}{2g_v^2}\left(1+\eta_1\frac{\Phi}{M}+\frac{\eta_2}{2}\frac{\Phi^2}{M^2}+\frac{\eta_3}{3!}\frac{\Phi^3}{M^3}\right)m_v^2W^2 \\
& -\frac{1}{4!g_v^2}\left(\zeta_\circ+\zeta_1\frac{\Phi}{M}\right)W^4+W\rho_B \\
& +\frac{1}{3}\frac{\gamma}{(2\pi)^3}\int d^3k\frac{k^2}{E^*(k)}\left(f_+(k,T)+f_-(k,T)\right), \qquad (1)
\end{aligned}
$$

where Φ and W are the scalar and vector meson fields respectively, $E^*(k) = (\mathbf{k}^2 + M^{*2})^{1/2}$ and $f_\pm(k,T)$ stand for the Fermi-Dirac distribution for baryons (anti-baryons). The baryon density ρ_B is defined by $\rho_B = \frac{\gamma}{(2\pi)^3}\int d^3k\ [f_+(k,T) - f_-(k,T)]$. Thus, in the zero density regime one has the same number of baryons and anti-baryons.

3 Results and Conclusions

3.1 Behaviour of $\frac{S}{A}$ and μ as a function of the anti-nucleon-nucleon ratio

The entropy per particle $\frac{S}{A}$, vs. anti-nucleon-nucleon ratio $\frac{\rho_{\bar{N}}}{\rho_N}$ is given in Figure 1. We found, for all the hadronic models, a systematic increasing of $\frac{S}{A}$ as $\frac{\rho_{\bar{N}}}{\rho_N}$ increases. In Fig.2, we show the chemical potential μ as a function of

FIGURE:3

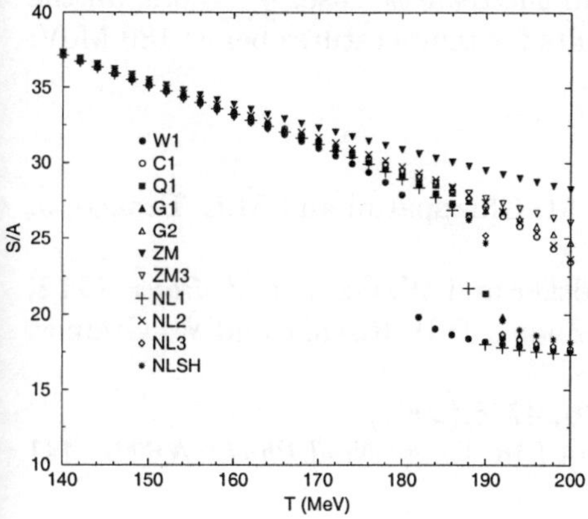

FIGURE:4

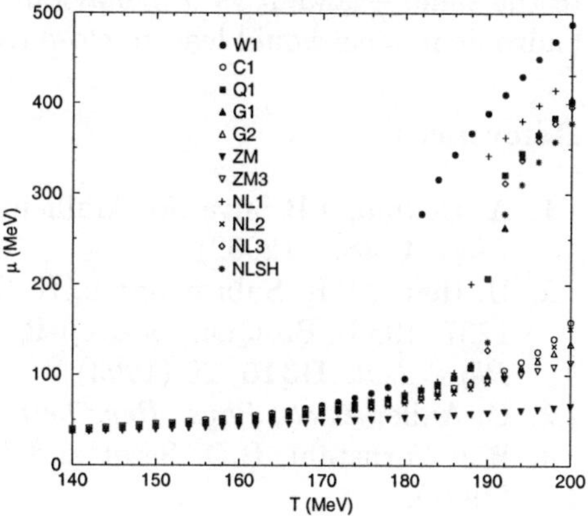

$\frac{\rho_{\bar{N}}}{\rho_N}$. Now, μ decreases as $\frac{\rho_{\bar{N}}}{\rho_N}$ increases.

In both figures, our calculation fixed $T = 190$ MeV for $\frac{S}{A}$ and μ versus $\frac{\rho_{\bar{N}}}{\rho_N}$ behavior. Unless the hadronic model itself presents a nucleon- anti-nucleon phase transition (see next section), Fig.1 and Fig.2 describe standard $\frac{S}{A}$ and μ as a function of $\frac{\rho_{\bar{N}}}{\rho_N}$. Let us remark that the increasing of $\frac{\rho_{\bar{N}}}{\rho_N}$ goes in the same direction of the nucleon-anti-nucleon symmetry. In the particular case in which the number of nucleons and anti-nucleons are equal, $\frac{\rho_{\bar{N}}}{\rho_N}$=1 and μ=0.

3.2 Behaviour of $\frac{S}{A}$ and μ as a function of the temperature for $\frac{\rho_{\bar{N}}}{\rho_N} = 0.6$

Fig. 3 shows $\frac{S}{A}$ as a function of T for a fixed $\frac{\rho_{\bar{N}}}{\rho_N} = 0.6$. We see that $\frac{S}{A}$ presents an abrupt decreasing for T around 180 MeV and 190 MeV for some models ($W1, NL1, NL3, NLSH, Q1$ and $G1$). These temperatures are associated which the critical temperatures to cause a hadronic phase transition to the anti-nucleon-nucleon plasma. In the models where such phase transitions do not appear ($ZM, ZM3, G2, NL2$ and $C1$) the curves are smooth.

Fig. 4 shows μ as a function of T for a fixed $\frac{\rho_{\bar{N}}}{\rho_N} = 0.6$. When the hadronic model itself presents hadronic phase transition to the anti-nucleon-nucleon plasma, an abrupt increasing of μ manifests. This behaviour has the same pattern of $\frac{S}{A}$ vs T above discussed. Again, in the models where such phase transitions do not appear the curves are smooth.

To conclude, our study shows a hadronic model dependence for specific cases in which the hadronic model itself presents a phase transition to the anti-nucleon-nucleon plasma. This dependence occurs only when we approach the critical temperature. Below this temperature, essentially all the models lead to the same $\frac{S}{A}$ and μ vs T behaviour. In the realistic case, $\frac{\rho_{\bar{N}}}{\rho_N} = 0.6$, all the hadronic models would lead to close results for temperatures below 180 MeV.

References

1. A. Delfino, J.B Silva, M. Malheiro, M. Chiapparini and M.E. Bracco, *J. Phys.* **G28**, 1 (2002).
2. U. Heinz, P.R. Subramanian, H. Stöcker and W. Greiner, *J. Phys.* **G12**, 1237 (1986); Bo-Qiang Ma, Qi-Ren Zhang, D.H. Rischke and W. Greiner, *Phys. Lett.* **B315**, 29 (1993).
3. C. Aldler *et al., Phys. Rev. Lett.* **86**, 4778 (2001).
4. R.J. Furnstahl, B.D. Serot and Hua-Bin Tang, *Nucl.Phys.* **A615**, 441 (1997).
5. P. G. Reinhard, *Rep. Prog. Phys.* **52**, 439 (1989);
   G. A. Lalazissis, J. Koening and P. Ring, Phys. Rev. **C55**, 540 (1997), M. M. Sharma, M. A. Nagarajan, and P. Ring, *Phys. Lett.* **B312**, 377 (1993).
6. J. Zimanyi, S. A. Moszkowski, *Phys. Rev.* **C42**, 1416 (1990),
   A. Delfino, C. T. Coelho, M. Malheiro, Phys. Rev. **C51**, 2188, (1995); *ibid Phys. Lett.* **B345**, 361, (1995).

PAIRS OF CHARGED HEAVY-LEPTONS FROM AN $SU(3)_L \otimes U(1)_N$ MODEL AT $e^- - e^+$ COLLIDERS

J.E. CIEZA MONTALVO

Instituto de Física, Universidade do Estado do Rio de Janeiro
CEP 20559 – 900 *Rio de Janeiro, Brazil*
cieza@uerj.br

M.D. TONASSE

Instituto Tecnológico de Aeronáutica, Centro Técnico Aeroespacial,
Praça Marechal do Ar Eduardo Gomes 50,
12228-900 São José dos Campos, SP, Brazil
tonasse@fis.ita.cta.br

We investigate the production and signatures of pairs of charged heavy-leptons using the $SU(3)_L \otimes U(1)_N$ electroweak model in e^-e^+ colliders (NLC and CLIC).

1 Introduction

In this work we address to a new class of heavy-fermions, *i.e.*, the heavy-leptons which can be produced by strong and electroweak forces based on the $SU(3)_C \otimes SU(3)_L \otimes U(1)_N$ (3-3-1 for short) semi simple symmetry group [1]. In this model the new leptons require not new generations, as occur in most of the heavy-lepton models [2]. It is a chiral electroweak model whose left-handed charged heavy-leptons, which we denote by $P_a = E$, M and T, together with the associated ordinary charged leptons and its respective neutrinos, are accommodated in $SU(3)_L$ triplets. So, we will study the production mechanism for heavy-exotic leptons in e^-e^+ colliders such as NLC ($\sqrt{s} = 500$ GeV) and CLIC ($\sqrt{s} = 1000$ GeV). Here we study the production of a pair of exotic leptons and summarize our results.

2 Cross-Section Production

Pair production of exotic particles is, to a very good approximation, a model independent process, since it proceeds through a well known electroweak interaction. This production mechanism can be studied through the analysis of the reactions $e^-e^+ \to L^-L^+$, provided that there is enough available energy ($\sqrt{s} \geq 2M_L$). We will analyze the following processes: $e^-e^+ \to L^-L^+$, $e^-e^+ \to \gamma\gamma \to L^-L^+$, $e^-e^+ \to Z\gamma \to L^-L^+$ and $e^-e^+ \to ZZ \to L^-L^+$. The first process takes place through the exchange of a photon, a boson Z^0 and $Z^{0'}$ in the s channel, while the other processes take place through the exchange of heavy lepton in the t and u channel. Using the interaction Lagrangians, it is easy to evaluate the cross section for the process $e^+e^- \to L^+L^-$, involving a neutral current, from which we obtain:

$$\begin{aligned}
\left(\frac{d\sigma}{d\cos\theta}\right)_{P^+P^-} &= \frac{\beta\alpha^2\pi}{s^3}\Bigg\{\left[2sM_P^2 + \left(M_P^2 - t\right)^2 + \left(M_P^2 - u\right)^2\right] \\
&+ \frac{1}{2\sin^2\theta_W\cos^2\theta_W\left(s - M_{Z,Z'}^2 + iM_{Z,Z'}\Gamma_{Z,Z'}\right)}\Big[2sM_P^2 g'^{PP}_V g^\ell_V \\
&+ g'^{PP}_V g^\ell_V\left[\left(M_P^2 - t\right)^2 + \left(M_P^2 - u\right)^2\right] + g'^{PP}_A g^\ell_A\Big(\left(M_P^2 - u\right)^2 \\
&- \left(M_P^2 - t\right)^2\Big)\Big]\Bigg\} + \frac{\beta\pi\alpha^2}{16\cos^4\theta_W\sin^4\theta_W}\frac{1}{s\left(s - M_{Z,Z'}^2 + iM_{Z,Z'}\Gamma_{Z,Z'}\right)^2} \\
&\times\Bigg\{\left[\left(g'^{PP}_V\right)^2 + \left(g'^{PP}_A\right)^2\right]\left[\left(g^\ell_V\right)^2 + \left(g^\ell_A\right)^2\right]\left[\left(M_P^2 - u\right)^2 + \left(M_P^2 - t\right)^2\right] \\
&+ 2sM_P^2\left[\left(g'^{PP}_V\right)^2 - \left(g'^{PP}_A\right)^2\right]\left[\left(g^\ell_V\right)^2 + \left(g^\ell_A\right)^2\right] \\
&+ 4g'^{PP}_V g'^{PP}_A g^\ell_V g^\ell_A\left[\left(M_P^2 - u\right)^2 - \left(M_P^2 - t\right)^2\right]\Bigg\} \\
&+ \frac{\beta\pi\alpha^2}{8\sin^4\theta_W\cos^4\theta_W s\left(s - M_Z^2 + iM_Z\Gamma_Z\right)\left(s - M_{Z'}^2 + iM_{Z'}\Gamma_{Z'}\right)} \\
&\times\Bigg\{2sM_P^2\left(g^\ell_V + g^\ell_A\right)\left(g^{PP}_V g'^{PP}_V - g^{PP}_A g'^{PP}_A\right) + \left(M_P^2 - t\right)^2\Big[\left(\left(g^\ell_V\right)^2 + \left(g^\ell_A\right)^2\right) \\
&g^{PP}_V g'^{PP}_V + g^{PP}_A g'^{PP}_A - 2g^\ell_V g^\ell_A g^{PP}_V g'^{PP}_A - 2g^\ell_V g^\ell_A g^{PP}_A g'^{PP}_V\Big] \\
&+ \left(M_P^2 - u\right)^2\Big[\left(\left(g^\ell_V\right)^2 + \left(g^\ell_A\right)^2\right)\left(g^{PP}_V g'^{PP}_V + g^{PP}_A g'^{PP}_A\right) \\
&+ 2g^\ell_V g^\ell_A g^{PP}_V g'^{PP}_A + 2g^\ell_V g^\ell_A g^{PP}_A g'^{PP}_V\Big]\Bigg\},
\end{aligned} \tag{1}$$

where

$$g^{PP}_{V,A} = \frac{a_L \pm a_R}{2}, \qquad g'^{PP}_{V,A} = \frac{a'_L \pm a'_R}{2}.$$

The primes (′) are for the case when we take a boson Z', $\Gamma_{Z,Z'}$ are the total widths of the boson Z and Z' [3], $\beta = \sqrt{1 - 4M_P^2/s}$ is the velocity of the heavy-lepton in the c. m. of the process, α is the fine structure constant which we take equal to $\alpha = 1/128$, $g_{V,A}^{\ell}$ are the standard coupling constants, M_Z is the mass of the Z boson, $\sqrt{s}$ is the c. m. energy of the e^-e^+ system, $t = M_P^2 - (1 - \beta\cos\theta)s/2$ and $u = M_P^2 - (1 + \beta\cos\theta)s/2$, where θ is the angle between the heavy-lepton and the incident electron, in the c. m. frame. For Z' boson we take $M_{Z'} = (0,6 - 3)$ TeV, since $M_{Z'}$ is proportional to VEV v_χ [4,5]. For the standard model parameters we assume PDG values, *i.e.* $M_Z = 91.02$ GeV, $\sin^2\theta_W = 0.2315$ and $M_W = 80.33$ GeV [6].

Another way to produce a pair of heavy exotic leptons is through the elastic reactions of the type $e^-e^+ \to \gamma\gamma \to L^-L^+$, $e^-e^+ \to ZZ \to L^-L^+$ and $e^-e^+ \to Z\gamma \to L^-L^+$. The three processes take place through the exchange of the exotic lepton in the t and u channels. So the cross section for the production of a pair L^-L^+ in the e^-e^+ collisions, we obtained by convoluting the cross section for the sub-process $\gamma\gamma \to L^-L^+$, $Z\gamma \to L^-L^+$ and $ZZ \to L^-L^+$, with the two photon, $Z\gamma$ and ZZ luminosities in these collisions, that is

$$\sigma = \int_{\tau_{min}}^{1} \frac{dL}{d\tau} d\tau\, \hat{\sigma}(\hat{s} = x_1 x_2 s)$$
$$= \int_{\tau_{min}}^{1} \int_{\ln\sqrt{(\tau)}}^{-\ln\sqrt{(\tau)}} \frac{dx_1}{x_1} f_{V/\ell}(x_1) f_{V/\ell}(x_2) \int \frac{d\hat{\sigma}}{dcos} dcos \qquad (2)$$

where $V = \gamma, Z$ and $\frac{d\hat{\sigma}}{dcos} dcos$ is the sub-process cross section.

3 Results and Conclusions

In the figures we present the cross section for the process $e^+e^- \to L^+L^-$, for the NLC and CLIC. In all calculations we take $\sin^2\theta_W = 0.2315$, $M_Z = 91.188$ GeV.

First in Fig. 1, in the left, considering that the only limitation for the discovery of charged heavy leptons is the total number of events [7,8], we take into account that the expected integrated luminosity for the NLC, will be of order of $6 \times 10^4 pb^{-1}/yr$ which gives a total of: $\simeq 2.5 \times 10^4$ lepton pairs produced per year, for the process $e^-e^+ \to L^-L^+$, where we considered the mass of the heavy lepton equal to 200 GeV and mass of the boson $Z' = 1200$ GeV, while considering the mass of the boson $Z' = 2000$ GeV the production of pair of heavy leptons will be of order of 2.2×10^4.

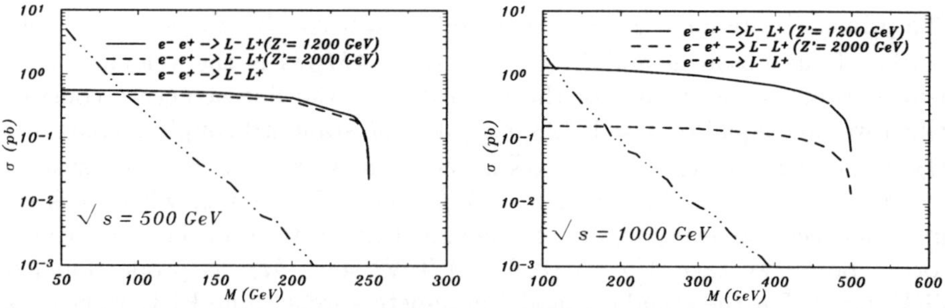

Figure 1. Cross section for the process $e^+e^- \rightarrow L^+L^-$ (left panel: NLC; right panel: CLIC).

In Fig. 1, in the right, taking into account that the integrated luminosity for the CLIC will be of order of $2 \times 10^5 pb^{-1}/yr$, then the statistics that we can expect for this collider is a little larger, so for the process $e^-e^+ \rightarrow L^-L^+$, considering the mass of the heavy lepton equal to 200 GeV and the mass of the boson Z' equal to 1200 GeV, we will have a total of $\simeq 2 \times 10^5$ lepton pairs produced per year, while for the boson $Z' = 2000$ GeV, we will have $\simeq 3 \times 10^4$ respectively.

References

1. V. Pleitez and M. D. Tonasse, *Phys. Rev.* **D48** (1993) 2353.
2. A review on the other types of heavy-leptons is given by P. H. Frampton, P. Q. Hung and M. Sher, *Phys. Rep.* **330** (2000) 263.
3. J. E. Cieza Montalvo and M. D. Tonasse, *Nucl. Phys.* **B623** (2002) 325.
4. F. Pisano and V. Pleitez, *Phys. Rev.* **D 46**, (1992) 410; R. Foot, O. F. Hernandez, F. Pisano and V. Pleitez, *ibid* **47** (1993) 4158.
5. P.H. Frampton, *Phys. Rev. Lett.* **69** (1992) 2889.
6. C. Caso *et al.* (Particle Data Group), *Eur. Phys. J.* **C3** (1998) 1.
7. V. Barger, T. Han and J. Ohnemus, *Phys. Rev.* **D37** (1988) 1174.
8. I. Hinchliffe, *Int. J. Mod. Phys.* **A4**(1989) 3867.

CHARGE ASYMMETRIES IN CHARM PHOTO-PRODUCTION

J. MAGNIN

Centro Brasileiro de Pesquisas Físicas, Dr. Xavier Sigaud 150 - CEP 22290-180, Rio de Janeiro - Brasil
jmagnin@cbpf.br

E. CUAUTLE

Instituto de Ciencias Nucleares, Universidad Autónoma de México, Apdo. Postal 70-543, 04510, D.F. México

G. HERRERA AND A. SÁNCHEZ-HERNÁNDEZ

Centro de Investigacion y Estudios Avanzados (CINVESTAV), Apdo. Postal 14-740, 07000, D.F. México

Charm quarks are expected to be produced mainly by the photon-gluon fusion mechanism in $\gamma - N$ interactions. However, a small part of the total charm cross section originates in similar processes to those appearing in charm hadro-production through the resolved (hadronic) component of the photon. Although the contribution of the resolved part of the photon is small at fixed target energies, it can help to understand the small but sizeable charge asymmetries measured in charm hadron photo-production experiments.

1 Introduction

The photon-gluon fusion mechanism followed by the fragmentation of the charm quark accounts for the main features of charm hadron production in $\gamma - N$ interactions [1].

However, small asymmetries in D^+/D^- and $D^0/\bar{D}^0$ production have been observed in experiments [1,2]. An asymmetry has been observed also in Λ_c^+/Λ_c^- photo-production [2] but with large errors bars. These asymmetries mean that a different number of particles than anti-particles are being produced in the final state, indicating that hadronization mechanisms are not so simple as the independent fragmentation of c-quarks. In fact, the usual parton model calculation based on the $\gamma g \to c\bar{c}$ diagram plus Peterson fragmentation of c quarks to charm hadrons cannot explain the observed asymmetries even including NLO (Next to Leading Order) effects.

The solution of the problem could be the *resolved* or *hadronic* component of the photon and the recombination mechanism together with a careful consideration of color string fragmentation, as we shall discuss in the following.

2 Charm photo-production in QCD

The total differential cross section for charm production as a function of $z = 2p_{l,c}/\sqrt{s}$, the momentum fraction of the charm quark, is given by [3]

$$\frac{d\sigma}{dz} = s\int dp_T^2 \frac{xg^N(x,Q^2)}{E(\sqrt{s}-m_T e^{y_c})}\frac{d\hat{\sigma}_{\gamma g}}{d\hat{t}} +$$
$$\Sigma_{i,j}\frac{\sqrt{s}}{2}\int dp_T^2\, dy_{\bar{c}} x_1 f_i^\gamma(x_1,Q^2) x_2 f_j^N(x_2,Q^2)\frac{1}{E}\frac{d\hat{\sigma}_{i,j}}{d\hat{t}}\,, \tag{1}$$

where x, x_i are the momentum fractions of the corresponding partons, p_t^2 is the transverse momentum squared, $y_{\bar{c}}$, E are the rapidity and energy of the $\bar{c}$-quark, s is the c.m. energy. $d\hat{\sigma}_{\gamma g}/d\hat{t}$ and $d\hat{\sigma}_{i,j}/d\hat{t}$ are the $\gamma\ g \to c\bar{c}$ and $parton_i\ parton_j \to c\bar{c}$ elementary cross sections. The sum is over light quark flavors and gluons.

The first term of Eq. (1) is known as the point-like part while the second one is called the resolved or hadronic contribution of the photon. This later part arises naturally when Next to Leading Order contribution to charm photo-production are taken into account.

While for the first term in Eq. (1) the only possible hadronization mechanism is the fragmentation of the charm quark to a charm hadron, for the second term, in addition to fragmentation, recombination of the charm quark with the debris of the hadron-like photon is possible. In order to be able to calculate this second hadronization process, the double differential cross section of having a charm quark accompanied by a spectator quark from the hadron-like photon structure is needed. This calculation has been carried out, in the context of charm hadro-production, in Ref. [4]. Its expression in the parton model is

$$\frac{d^2\sigma}{dz\, dx_v} = \frac{\sqrt{s}}{2} x_v f_v^\gamma(x_v,Q^2)\int dp_T^2\, dy_{\bar{c}} x_1 f_i^\gamma(x_1,Q^2) x_2 f_j^N(x_2,Q^2)\frac{1}{E}\frac{d\hat{\sigma}_{i,j}}{d\hat{t}}\,, \tag{2}$$

where x_v, f_v^γ are the momentum fraction and parton distribution function of the spectator quark.

3 Hadronization

Hadronization of charm quarks to D-mesons coming from the point-like contribution can only occur via the formation of two color strings among the $c - \bar{c}$ pair and the debris of the nucleon (See Fig. 1). The differential cross section can be well described by convoluting the charm differential cross

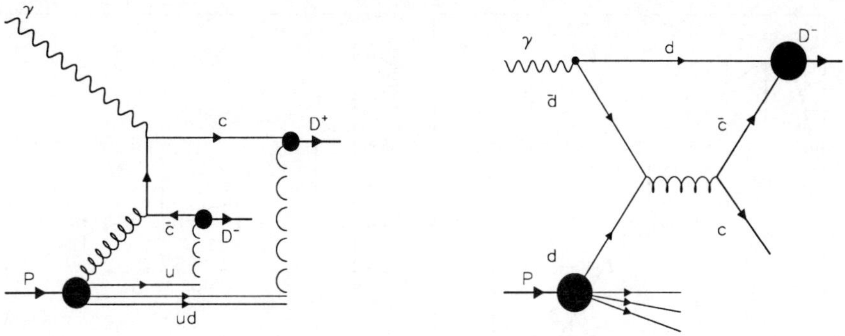

Figure 1. Typical hadronization processes in photo-production: fragmentation via color strings (left) and recombination with the debris of the photon (right).

section with a Peterson fragmentation function (PFF) as

$$\frac{d\sigma_H}{dx_F} = \int \frac{dy}{y}\frac{d\sigma}{dz} D_{H/c}(y),\; y = \frac{x_F}{z}\,, \tag{3}$$

where $D_{H/c}$ is a PFF, $d\sigma/dz$ is given by the first term of Eq. (1) and $x_F = 2p_{l,D}/\sqrt{s}$ is the momentum fraction of the D meson.

Contributions of the hadron-like component to hadronization are of two different types, namely, the fragmentation of the charm quark in a similar way it happens for the point-like component, and the recombination of the charm quark with the debris of the photon (see Fig. 1). The first contribution can be calculated by using Eq. (3) by just replacing the differential cross section $d\sigma/dz$ by the second term of Eq. (1). The recombination part of the total differential cross section is given, in turn, by

$$\frac{d\sigma_H}{dx_F} = \frac{\sqrt{s}}{2}\int x_v z^\star \frac{d^2\sigma}{dx_v dz}\; R(x_v, z, x_F)\frac{dx_v}{x_v}\frac{dz}{z}\,, \tag{4}$$

with $d^2\sigma/dx_v dz$ given by Eq. (2), $z^\star = 2E_c/\sqrt{s}$ and R a recombination function. The total cross section, which is constructed as the sum of the fragmentation and recombination parts, is compared to experimental data in Fig. 2.

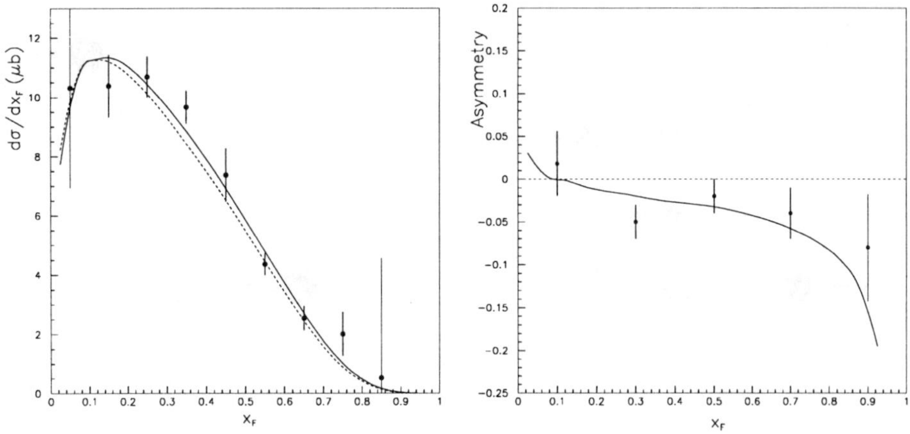

Figure 2. D-meson cross section (left) and D^+/D^- asymmetry (right). Experimental data were taken from Refs [1,2]. The solid curves show the results of calculations.

4 Conclusions

The model gives a reasonably good description of both, the differential cross section and asymmetry for D-meson photo-production. Moreover, the hadronic part of the photon plays a key role to give the production asymmetries observed in experimental data. For baryon production, the model predicts a positive asymmetry (more baryons than antibaryons), in agreement with experimental results.

Acknowledgments

J.M wishes to thank the warm hospitality at the Hadron 2002.

References

1. J.C. Anjos *et al.* (E691 Collaboration), *Phys. Rev. Lett.* **62** (1989) 513.
2. P.L. Frabetti *et al.* (E687 Collaboration), *Phys. Lett.* **B370** (1996) 222.
3. R.K. Ellis and P. Nason, *Nucl. Phys.* **B312** (1989) 551.
4. V.G. Kartvelishvili, A.K. Likhoded and S.R. Slabospitskii, *Sov. J. Nucl. Phys.* **28** (1978) 678; *ibid.***33** (1981) 434.

THE HIGH DENSITY EFFECTS IN THE DRELL-YAN PROCESS

M.A. BETEMPS [1], M.B. GAY DUCATI [1], A.L. AYALA FILHO [2]

[1] *Instituto de Física, Universidade Federal do Rio Grande do Sul, Caixa Postal 15051, CEP 91591-970 Porto Alegre, RS - Brazil*
mandrebe@if.ufrgs.br, gay@if.ufrgs.br

[2] *Instituto de Física e Matemática, Universidade Federal de Pelotas, Caixa Postal 354, CEP 96010-090 Pelotas, RS - Brazil*
ayala@ufpel.tche.br

The high density effects in the Drell-Yan process ($q\bar{q} \to \gamma^* \to l^+l^-$) are investigated for pA collisions at RHIC and LHC energies. In particular, we use a set of nuclear parton distributions that describes the present nuclear eA and pA data in the DGLAP approach including the high density effects introduced in the perturbative Glauber-Mueller approach.

1 Introduction

Drell-Yan production (DY) together with deep inelastic scattering (DIS) are powerful processes to investigate the parton distribution in the nucleus[1] and the nucleon. In this work, we investigate the high density effects in DY production at RHIC and LHC kinematical region, based on the Glauber-Mueller approach for DIS [2,3,4]. In the infinite momentum frame, in proton-nucleus collisions, the DY process corresponds to the annihilation of a quark (antiquark) from the proton with antiquark (quark) from the nucleus into a virtual photon (vector boson), which afterwards decays into a massive lepton pair [5]. These processes are sensitive to the sea antiquark content of the nucleus, which is important at high energy kinematical regime. The reactions at RHIC and LHC open the opportunity to study the physics of dense matter in hadron-nucleus (pA) and nucleus-nucleus (AA) scatterings at very high energies. Thus the high density effects are expected to play an important role in the description of those high energy processes [6].

The current experimental data for DY process, evidencing nuclear effects, can be reasonably described using DGLAP evolution equations with properly adjusted initial parton distributions and A-dependent parameterization (EKS parameterization) [7] . However, this parameterization does not include the dynamical saturation effects of the parton distributions. The DGLAP evolution only considers emission processes, but for the kinematical regime of RHIC and LHC, the parton interaction and the recombination processes should be important due to the high density of quarks and gluons. Recently, it was proposed in Ref. [8] a modification in the EKS nuclear parameterization to take into account the high density effects encoded in the AGL formulation [3,4] for the Glauber-Mueller approach (AG parameterization). In the next section, we obtain a nuclear quark distribution based on AG parameterization and compare it with the EKS one for the DY cross section[9].

2 The High Density Effects in the Drell-Yan Cross Section

We expect that the nuclear high density effects [2,3,4] will be important in the description of the parton evolution in the heavy ion collisions, since these effects can strongly modify the partons distribution in a nucleus[7,8]. At small x, the parton evolution may create much larger nuclear parton density than in a free nucleon [3,4]. Therefore, the high density effects probably will give rise to nonlinear dynamics expected in QCD at small x, which can reduce the further growth of the nuclear parton distribution[10].

To estimate the high density effects we consider the Glauber multiple scattering theory[2,11]. In this approach[3,4], extending the AG parameterization[8], considering a symmetrical sea, the nuclear sea quark distribution is[9]

$$\begin{aligned} xq^A_{u,d,s}(x,Q^2) &= xq^A_{u,d,s}(x,Q^2)[\mathrm{EKS}] - \frac{3}{4A}F_2^A(x,Q^2)[\mathrm{DLA}] \\ &+ \frac{3}{4A}F_2^A(x,Q^2)[\text{ Eq. 6 from Ref. } ^8]\,, \end{aligned} \tag{1}$$

where $F_2(x,Q^2)$ are the structure functions defined in the DIS process. The $F_2(x,Q^2)$[DLA] is given by the Born term of the Eq. 6 from Ref.[8]. This expression gives a nuclear quark distribution driven by the DGLAP evolution (EKS parameterization) corrected by unitarity effects calculated in DLA in a perturbative Glauber-Mueller approach. Only sea quarks evolutions are modified in this approach. The nuclear valence quark behavior is obtained from the EKS parameterization. The nuclear and nucleon sea quarks distribution ratio, as a function of Bjorken x, is shown in Fig. 1 to the u flavour. The AG results are compared with the EKS ones for the Lead nucleus. We can note, from the Fig. 1, that the high density effects encoded in the Glauber-Mueller approach grows as x tends to zero. The EKS parametrization shows the sea quark ratio scaling at this region. The high density effects start to be sizeable for x lower then 10^{-2}.

In order to investigate all assumptions presented here, we use the nuclear parton distribution (Eq. 1) to estimate the DY cross section, which at LO in

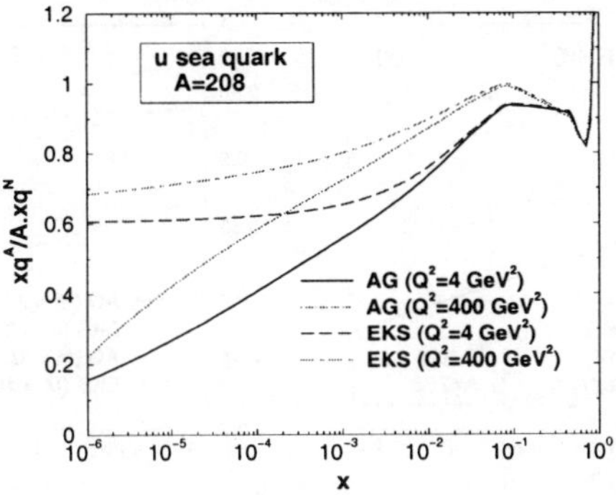

Figure 1. The sea quark distribution function ratio $R = \frac{xq^A}{A.xq^N}$ for flavour u.

pA collisions is given by

$$\frac{d\sigma_{DY}^{pA}}{dydM^2} = \frac{4\pi\alpha^2}{9sM^2} \sum_{i=u,d,s} e_i^2 [q_i^p(x_1, M^2)\bar{q}_i^A(x_2, M^2) + x_1 \leftrightarrow x_2], \tag{2}$$

where M^2 is the lepton pair invariant squared mass, s is the squared center-of-mass energy and x_1, x_2 are the momenta fraction carried by the quarks (antiquarks). To parameterize the modification of the DY cross section due to the high density effects, we use an effective exponent (α) defined in the relation $\sigma_{DY}^{pA} = A^\alpha \sigma_{DY}^{pN}$, where σ_{DY}^{pA} and σ_{DY}^{pN} are the proton-nucleus and proton-nucleon DY cross sections, respectively. If α is close to one the nuclear effects in the process are negligible. A crucial requirement for the use of the DY process for the study of the quark structure of nuclei is that the factorized form of tho cross section remains valid for nuclei.

In the Fig. 2 we show the α exponent as a function of the rapidity y ($x_{(1/2)} = \sqrt{M^2/s}\, e^{(+/-)y}$) in DY process at RHIC ($\sqrt{s} = 350$ GeV) and LHC ($\sqrt{s} = 8.8$ TeV) energies for two mass values. The figure shows the results for α obtained with the AG and EKS parameterizations and the rapidity region acceptance of the detectors. From the kinematics, we know that small values of y are related with $x_1 \approx x_2$. An increase in the y provides higher values of x_1 and smaller x_2. For RHIC and LHC energies the α exponent diminishes with y. It means that the high density effects increase with the rapidity, since with this increase the small x_2 region is reached. Such effects are more significant at small values of M, once the difference between the two parameterizations is bigger for the values of $M = 2$ GeV. Analyzing the results for RHIC and LHC, we can see that the high density effects are more evident at LHC energies. It

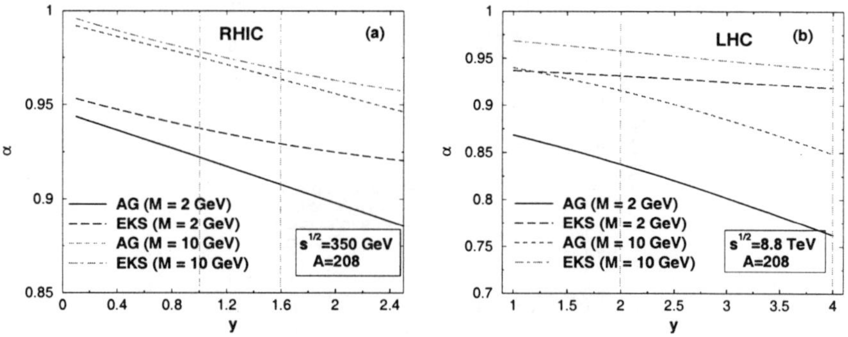

Figure 2. *The α exponent for DY process in pA collisions at (a) RHIC and (b) LHC.*

occurs because, as energy increases, the same y interval corresponds to smaller values of x, both for x_1 and x_2. The high density effects provide a reduction in the differential cross section of about 15% at RHIC energies and about of 55% at LHC energies. The strong modification of the DY production in pA process indicates that the high density effects on the quark distribution will play an important role in the DY production in AA collisions. This shows that the high density effects should be important in the description of parton evolution in heavy ion collisions at high energies.

Acknowledgments

This work was partially financed by CNPq and FAPERGS, Brazil.

References

1. A. L. Ayala F$^{\underline{o}}$, M. B. Gay Ducati, L. N. Epele and C. A. Garcia Canal, *Phys. Rev.***C49** (1994) 489.
2. A. H. Mueller, *Nucl. Phys.* B **335**, 115 (1990).
3. A. L. Ayala, M. B. Gay Ducati and E. M. Levin, *Nucl. Phys.* B **439**, 1997 (3)05.
4. A. L. Ayala, M. B. Gay Ducati and E. M. Levin, *Nucl. Phys.* B **511**, 1998 (3)55.
5. S. D. Drell and T. M. Yan, *Phys. Rev. Lett.* **25**, 1970 (3)16.
6. K. J. Eskola, et al. hep-ph/0108093.
7. K. J. Eskola, et al. *Eur. Phys. J.* **C9**(1999) 61.
8. A. L. Ayala Filho, V. P. Gonçalves, *Eur. Phys. J.* **C20** (2001) 343.
9. M. A. Betemps, MSc Dissertation, IF-UFRGS, (2002).
10. M. B. Gay Ducati, *Braz. J. Phys.* **31** (2001) 115.
11. R. J. Glauber, *Phys. Rev.* **100** (1955) 242.

PARTON SATURATION EFFECTS TO THE DRELL-YAN PROCESS IN THE COLOR DIPOLE PICTURE

M.A. BETEMPS, M.B. GAY DUCATI, M.V.T. MACHADO
Instituto de Fisica, Universidade Federal do Rio Grande do Sul, Caixa Postal 15051, 91501-970, Porto Alegre, Brasil
mandrebe@if.ufrgs.br, gay@if.ufrgs.br, magnus@if.ufrgs.br

We report on the results obtained in the study of the parton saturation effects, taken into account through the multi-scattering Glauber-Mueller approach, applied to the Drell-Yan (DY) process described in the color dipole picture. As a main result, one shows that those effects play an important role in the estimates of the DY differential cross section at RHIC energies.

The Drell-Yan (DY) process, i.e. the production of massive lepton pairs in hadronic collisions, in conjunction with deep inelastic scattering (DIS), has been one of the important processes probing strong interaction physics. Recently, in connection with the availability of high energy accelerators, a great attention has been focused on the small-x region of QCD. There, the parton densities become high and the limits of the perturbative methods are tested. Such a region presents the onset of the saturation phenomenon (at scale Q_s^2), i.e. the taming of the parton (mostly gluon) distribution due to nonlinear dynamics associated with unitarity effects[1] .

In this contribution, we study the high energy DY cross section in the target rest frame. In this case, the relevant degrees of freedom are the projectile wavefunction and the dipole-proton effective cross section. The underlying process is the scattering of a parton from the projectile structure function off the target color field. This parton radiates (bremsstrahlung) a massive photon, which subsequently decays into a lepton pair. The interaction with the target can occur before or after the photon emission. A remarkable feature emerging is that the $\gamma^* q$-proton (or hadron) interaction can be described by the same $q\bar{q}$ (color dipole)-proton cross section as in DIS. Although diagramatically no dipole is present, the interference among graphs results in a product of two quark amplitudes in the DY cross section testing the external gluonic field at two different transverse positions (impact parameters), in a similar way to DIS[2].

In such a representation, the photo absorbtion cross section on deep inelastic is described by the convolution of the wavefunctions, Ψ_{γ^*}, from the virtual photon and the interaction dipole cross section, σ_{dip}. The wavefunctions are considered taking into account the simplest photon Fock state configuration, i.e., a $q\bar{q}$-pair fluctuation (dipole), whose transverse separation $r_\perp$ remains fixed during the interaction since its lifetime is larger than the collision one. The σ_{dip} is modeled phenomenologically based on a matching between hard and soft domains, constrained by the DIS available data. Small dipole size configurations can be described through pQCD, whereas the large size ones belong to the non-perturbative domain. Hence, one can write the photo absorptions cross section as a function of the Bjorken scaling variable $\tilde{x} = x_{Bj}$, the photon virtuality scale $\tilde{Q}^2 = Q^2$ and the quark momentum fraction α in the quantum mechanical form[3],

$$\sigma_{T,L}(\gamma^* p \to q\bar{q}) = \int d^2 r_\perp \int_0^1 d\alpha \, |\Psi^{T,L}_{q\bar{q}}(\alpha, r_\perp, \tilde{Q}^2)|^2 \, \sigma_{dip}(\tilde{x}, r_\perp) \, , \quad (1)$$

where the simplest phenomenological realization of the dipole cross section is given by the GBW model[4], $\sigma_{dip} = \sigma_0[1 - \exp(-r_\perp^2 Q_s^2/4)]$, with the parameters determined by fitting small-x HERA data.

In a similar way, the cross section for radiation of a virtual photon from a quark after scattering on a proton, has the following factorized form in impact parameter representation [5],

$$\frac{d\,\sigma_{T,L}(qp \to \gamma^* X)}{d\ln\alpha} = \int d^2 r_\perp \, |\Psi^{T,L}_{\gamma^* q}(\alpha, r_\perp, \tilde{Q}^2)|^2 \, \sigma_{dip}(\tilde{x}, \alpha r_\perp), \quad (2)$$

where $\Psi_{\gamma^* q}$ is the quark wavefunction representing the $\gamma^* q$ fluctuation. Here, $r_\perp$ is the photon-quark transverse separation, $\alpha r_\perp$ is the $q\bar{q}$ separation and α is the fraction of the light-cone momentum of the initial quark taken away by the photon. Note the difference with the DIS case, where the dipole separation is just $r_\perp$ and now the dipole cross section is evaluated for $\alpha r_\perp$. The scaling variable $\tilde{x}$ is taken as $x_2 = x_1 - x_F$, which are the usual invariants in the fast proton system: $x_{1,2}$ are the longitudinal momentum fractions of the projectile (target) parton and x_F is the Feynman variable. We stress that such identification is far from clear in the dipole picture: for instance, one can also choose $\tilde{x} = \alpha x_2$. In the rest frame, the quark carries momentum fraction x_1/α of the parent hadron, where x_1 is the proton's momentum fraction carried by the photon. The hadronic differential cross section to Drell-Yan process is expressed in the factorized form [5],

$$\frac{d\,\sigma^{DY}}{dM^2\,dx_F} = \frac{\alpha_{\rm em}}{6\,\pi M^2}\,\frac{1}{(x_1 + x_2)} \int_{x_1}^1 \frac{d\alpha}{\alpha} F_2^p\left(\frac{x_1}{\alpha}, \tilde{Q}^2\right) \frac{d\sigma(qp \to \gamma^* X)}{d\ln\alpha} \quad (3)$$

where the factor $\alpha_{\rm em}/(6\,\pi M^2)$ is due to the photon decay into a lepton pair, coming from electrodynamics. The dilepton invariant mass is M^2. Our input for σ_{dip} is constrained by the standard gluon distribution from the target corrected by saturation effects in high energy limit, encoded in the Glauber-Mueller (GM) approach [6]. In Eq.(3), the structure of the projectile is described by the proton inclusive structure function $F_2^p(x, \tilde{Q}^2)$, in which was chosen $\tilde{Q}^2 = M^2$. Other possible identification relying on plausible arguments is $\tilde{Q}^2 = (1 - x_1)M^2$ (Ref. [7]). Our input for the dipole cross section is given by

$$\sigma_{dip}^{GM}(\tilde{x}, r_\perp) = \frac{\pi^2 \alpha_s}{3}\, r^2\, xG_N^{GM}(\tilde{x}, \frac{4}{r_\perp^2}), \tag{4}$$

$$xG^{GM}(\tilde{x}, Q^2) = \frac{2R^2}{\pi^2}\int_{\tilde{x}}^{1} \frac{dx'}{x'} \int_{1/\tilde{Q}^2}^{1/Q_0^2} \frac{dr_\perp^2}{r_\perp^4}\,(\gamma_E + \ln[\kappa_G(x', r_\perp^2)] + E_1[\kappa_G(x', r_\perp^2)]),$$

where γ_E and $E_1(x)$ are the Euler constant and the exponential integral. The packing factor is $\kappa_G = (3\pi\alpha_s r_\perp^2/2R^2)\, xG^{\mathrm{DGLAP}}$. The saturation scale Q_s^2 is defined through the relation $\kappa_G(x, Q_s^2) = 1$. The target transverse size is R, which was taken to be $R^2 = 5\ \mathrm{GeV}^{-2}$ (strong unitarity corrections) supported by previous studies[3]. The xG was parameterized by the GRV94 gluon density.

In a recent work[8], we have calculated the DY cross section for the available small-x_2 data, and found that the dipole picture in that kinematical region ($x_2 \geq 10^{-2}$) is being tested in its limit of applicability. As an improvement, we have introduced a reggeonic piece to account for the large x_2 range in the phenomenological form $\sigma_R(x, r_\perp) = N_R\, r_\perp^2 x^{0.4525}(1-x)^3$, which supplements Eq.(4). Towards smaller x_2, such contribution turns out to be negligible. In Ref.[7], the procedure was to correct the saturation scale by a threshold factor $Q_s^2 \to Q_s^2\,(1 - x)^5$, and the numeric equivalence between the dipole picture (using GBW model) and NLO DGLAP calculations is shown[7], relying on an educated guess for the scales $\tilde{x}$ and $\tilde{Q}^2$.

In order to address the color dipole picture at high energies[8] (small x_2), the DY differential cross section for RHIC energies, $\sqrt{s} = 500$ GeV, is shown in Fig. (1) for the same fixed x_F. In this case, x_2 reaches values of order 10^{-4} and unitarity effects are important. The solid lines are the GM estimates, Eq.(4), and the dot-dashed lines are the rest frame calculations with DGLAP gluon distribution (without parton saturation effects). The GM results underestimate those ones without parton saturation, showing important role of the unitarity effects. Moreover, we have obtained a result numerically equivalent to the LO Breit frame calculations at the RHIC energies, suggesting consistency between both approaches. From the plots one verifies that the deviations due to saturation effects are larger as M diminishes, corresponding to smaller x_2 values. In absolute values, the corrections at RHIC energies reach up to ≈ 20 %, showing that saturation effects are important in a reliable description of the observables.

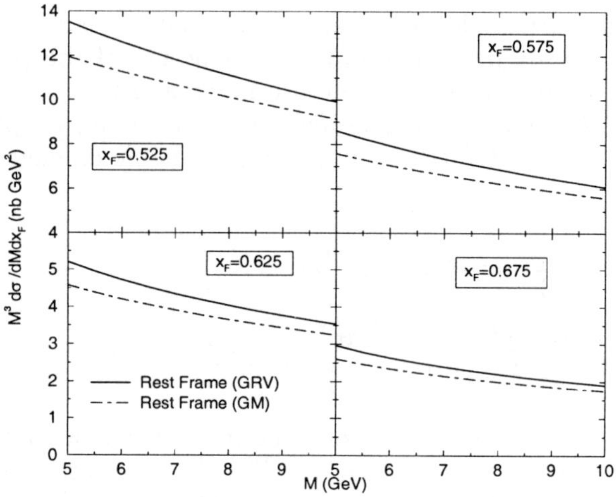

Figure 1. The DY differential cross section per nucleon versus M for the RHIC energies ($\sqrt{s}$ = 500 GeV) at fixed x_F in pd reaction. The solid lines are GM estimates, whereas dot-dashed ones are without gluon saturation calculation.

Acknowledgments

The authors enjoyed the lively scientific atmosphere of this meeting. This work is supported by CNPq, Brazil. MVTM thanks Jörg Raufeisen for discussions.

References

1. M.B. Gay Ducati. *Braz. J. Phys.* **31** (2001) 115.
2. S.J. Brodsky, A. Hebecker, E. Quack, *Phys. Rev.* D **55**, 1997 (2)584.
3. M.B. Gay Ducati, M.V.T. Machado, *Phys. Rev.* D **65**, 2002 (1)14019.
4. K. Golec-Biernat, M. Wüsthoff, *Phys. Rev.* D **59**, 1999 (0)14017.
5. B.Z. Kopeliovich, J. Raufeisen, A.V. Tarasov, *Phys. Lett.* B **503**, 2001 (9)1.
6. A.L. Ayala, M.B. Gay Ducati, E.M. Levin, *Nucl. Phys.* B **511**, 1998 (3)55.
7. J. Raufeisen, J.-C. Peng, G.C. Nayak, *Phys. Rev.* D **66**, 2002 (0)34024.
8. M.A. Betemps, M.B. Gay Ducati, M.V.T. Machado, *Phys. Rev.* D **66**, 2002 (0)14018.

HEAVY QUARK PRODUCTION AS A BASELINE FOR THE QUARKONIUM PRODUCTION

M.B. GAY DUCATI AND L.F. MACKEDANZ
Universidade Federal do Rio Grande do Sul, Porto Alegre, RS - Brazil
gay@if.ufrgs.br and thunder@if.ufrgs.br

V.P. GONÇALVES
Universidade Federal de Pelotas, Pelotas, RS - Brazil
barros@ufpel.tche.br

In this contribution, we analyze the energy dependence of the ratio $\sigma^{J/\Psi}/\sigma^{DY}$, as well as the ratio $(\sigma^{J/\Psi}/\sigma^{Q\bar{Q}})$, in pA and AA processes, considering the high density effects expected to be present in the high energy collisions. We verify that a large quarkonium suppression is predicted if these effects are included in the calculations. Moreover, we demonstrate that the pA process can be used to make it evident and to estimate the magnitude of the high density effects.

One of main predictions of QCD is the transition from the confined/chirally broken phase to the deconfined/chirally symmetric state of quasi-free quarks and gluons, the so-called quark-gluon plasma (QGP). Recently the heavy ion collisions have provided strong evidence for the formation of a QGP, with the first results of RHIC marking the beginning of a collider era in the experiments with relativistic heavy ions, as well as the era of detailed studies of the characteristics of the QGP. Theoretically, the early evolution of these nuclear collisions is governed by the dominant role of gluons, due to their large interaction probability and the large gluonic component in the initial nuclear wave functions. Such extreme condition is expected to significantly influence QGP signals and should modify the hard probes produced at early times of the heavy ion collision. One of the signatures of the formation of the Quark-Gluon Plasma (QGP) is the J/Ψ suppression [1], with the ratio between J/Ψ production and Drell-Yan production used as baseline to measure this suppression. However, as shown in Ref. [2], the high density effects will strongly reduce the heavy quark production cross sections in AA processes and should be taken into account to calculate the cross section of the hard QCD process in the first stage of the heavy ion collisions. Therefore, it is important to reanalyze the current estimates of quarkonium suppression in AA processes considering the high density effects. Furthermore, the systematic study of pA collisions at the same energies is essential to gain insight into the structure of the dense medium effects. Such effects, as the energy loss and shadowing[3], are absent or small in pp collisions, but become increasingly prominent in pA collisions, and are of major importance in AA reactions. By comparing pA and AA reactions involving very heavy nuclei, one may be able to distinguish basic hadronic effects that dominate the dynamics in pA collisions, from a quark-gluon formation predicted to occur in heavy ion AA collisions.

We will consider the quarkonium production using the Color Evaporation Model (CEM) [4]. Initially, the $q\overline{q}$ pair will in general be in a color octet state. It subsequently neutralizes its color and binds into a physical resonance. Color neutralization occurs by interaction with the surrounding color field. If we enforce that a colorless object is already present at the perturbative level then a hard interaction with the surrounding color field is assumed. However, it is possible that the colorless object is produced at the nonperturbative level. The CEM provides a simple and general phenomenological approach to color neutralization. In CEM, quarkonium production is treated identically to open heavy quark production with exception that in the case of quarkonium, the invariant mass of the heavy quark pair is restricted to be below the open meson threshold, which is twice the mass of the lowest meson mass that can be formed with the heavy quark. Consequently, the cross sections for J/Ψ production in this model is given by: $\sigma_{J/\Psi} = \frac{1}{9}\rho_{J/\Psi}\int_{2m_c}^{2m_D} dm \frac{d\sigma_{c\bar{c}}}{dm}$, where the factor 1/9 is the probability to have a color singlet pair and the $\rho_{J/\Psi}$ factor account for the fraction of the total charmonium carried by the J/Ψ state. The differential cross section present in the above is proportional to the convolution between the gluon distribution of the projectile and the target. Thus, the quarkonium production is a valuable search of information about the gluon distribution in the nucleus for pA and AA processes. Moreover, since the probability for making a heavy quark pair is proportional to square of gluon distribution[3], any depletion in number of gluons will make a significant difference in the number of heavy quark pairs produced. One of the nuclear medium effects present in the quarkonium production is the nuclear shadowing, which is the modification of the target parton distributions so that $xq^A(x,Q^2) < Axq^N(x,Q^2)$, as expected from a superposition of pp interactions. The current experimental data presenting nuclear shadowing can be described reasonably using the DGLAP evolution equations with adjusted initial parton distributions [5]. However, this parameterization does not include the dynamical saturation (high density) effects of the parton distributions predicted at small x and large A. Recently, a procedure to improve the nuclear parton distributions and include the perturbative high density effects has been proposed in Ref. [6], which has verified that the ratio $R_G = xG_A/xG_N$ is strongly modified in the region of small x if the high density effects are in-

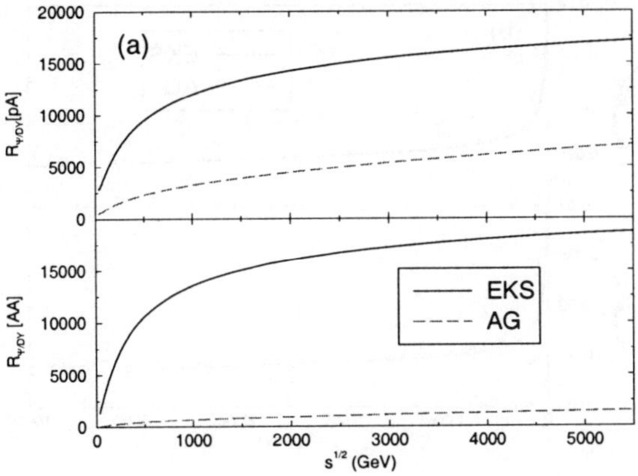

Figure 1. Our results for the energy dependence of the ratio $\sigma^{J/\Psi}/\sigma^{DY}$ for pA and AA processes.

cluded in the calculations. Here, we calculate $\sigma_{J/\Psi}$ using as input in the nuclear cross sections, the parameterization of the nuclear effects in the gluon distributions, which is given by $xG_A(x_i,\mu^2) = R_G(x_i,\mu^2)\,xG_N(x_i,\mu^2)$, with R_G calculated considering the high density effects [6]. In Fig. 1 we present our results for the energy dependence of the ratio $\sigma^{J/\Psi}/\sigma^{DY}$ for pA and AA processes. For the calculation of the Drell-Yan cross section we include the high density effects in the nuclear quark distributions (For details of this calculation see Ref. [7]) . Since these effects are larger in the gluon case, we have that the ratio will be strongly modified when compared to the calculations without the inclusion of the high density effects. We verify that this modification is large already in pA process and RHIC energy, which implies that this process could be used to make it evident the presence and estimate the magnitude of the high density effects. Moreover, we predict a very large suppression of the quarkonium production in AA processes, associated only with the presence of these effects, independently of the production or not of a quark-gluon plasma.

In Fig. 2 we present our results for the energy dependence of the ratio $\sigma^{J/\Psi}/\sigma^{Q\bar{Q}}$ between the bound and open states cross section. Since both processes depend similarly on the nuclear gluon distribution this ratio should be not strongly dependent on the inclusion of the HD effects. This expectation is verified in our results. On the other hand, the production of a QGP modify the quarkonium cross section and consequently the behavior of this ratio. We conclude that the $\sigma^{J/\Psi}/\sigma^{DY}$ ratio is not adequate as a signature of QGP,

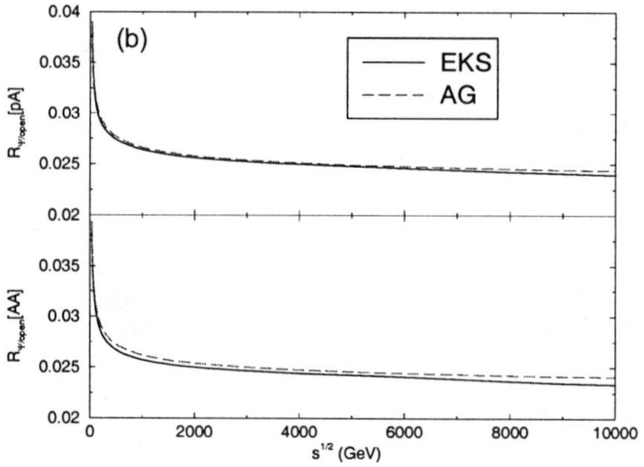

Figure 2. Our results for the energy dependence of the ratio $\sigma^{J/\Psi}/\sigma^{Q\bar{Q}}$ between the bound and open states cross section.

since it is strongly modified by the inclusion of the HD effects. However, the experimental analysis of the $\sigma^{J/\Psi}/\sigma^{Q\bar{Q}}$ ratio can be used as a baseline to make it evident the presence of a new state of matter.

Acknowledgements

This work was supported by CNPq, Brazil.

References

1. T. Matsui, H. Satz, *Phys. Lett.* B **178**, 1986 (4)16.
2. A.L. Ayala Filho, V. P. Gonçalves. *Phys. Lett.* **B534**(2002) 76.
3. L. N. Epele, C. A. Garcia Canal, M. B. Gay Ducati, *Phys. Lett.* B **226**, 1989 (1)67.
4. J. F. Amudson *et al.* *Phys. Lett.* B **390**, 1997 (3)23; M. B. Gay Ducati, C. B. Mariotto, *Phys. Lett.* B **464**, 286 (1999); M. B. Gay Ducati, V. P. Gonçalves, C. B. Mariotto, *Phys. Rev.* **D65** (2002)037503 .
5. K. J. Eskola, V. J. Kolhinen, C. A. Salgado, *Eur. Phys. J.* **C 9** (1999) 61.
6. A. L. Ayala Filho, V. P. Gonçalves, *Eur. Phys. J.* **C 20** (2001) 343.
7. M. A. Betemps, M. B. Gay Ducati, A. L. Ayala Filho, *in these proceedings*.

CHARMONIUM-HADRON CROSS SECTION IN THE DIPOLE-CONDENSER MODEL

M.S. KUGERATSKI AND F.S. NAVARRA
Instituto de Física, Universidade de São Paulo
C.P.66318, 05315-970, SP, Brazil
msks@if.usp.br

The purpose of this work is to evaluate the cross sections for the dissociation of J/Ψ in collision with a nucleon, using the dipole condenser model (DCM). The nucleon is represented by an external (condenser) color field and the charmonium is represented by a color dipole. Preliminary studies have shown that we are in line with the estimates obtained with short-distance QCD models.

Charmonium-hadron cross sections became recently a crucial ingredient for understanding the forthcoming RHIC data on charmonium production. More than two decades ago, Peskin and Bhanot [1] showed that one could apply perturbative QCD to calculate the interactions between bound states of heavy quarks and light hadrons. The justification of their approach comes basically from the fact that the heavy quark mass provides a large energy scale. Later their results have been rederived in the context of the operator product expansion (OPE) [2] and more recently in terms of the QCD factorization theorem [3]. In all these works there are some non-perturbative input, such as certain matrix elements involving gluon operators, the heavy quark wave function inside the bound state or moments of the gluon distribution function in the hadron.

In this work we calculate the charmonium-hadron cross section using all the basic ideas quoted above but with simplifying assumptions, so as to render the calculations simple, analytic and to be able to treat more complex interactions, like those involving the χ states or the $(\overline{Q}-Q)_8 - g$ pre-resonant (or color octet) state. We call it the dipole-condenser model (DCM). The starting point is the assumption that the charmonium (dipole) is small compared with the hadron (condenser). As a consequence, the $\overline{Q}-Q$ pair will interact mostly with the external color field but not with the (quark) sources. Moreover, the external color field is considered to have only low momentum components ("soft gluons") and thus is able to transfer only a small ammount of energy, which will be barely enough to dissociate the bound state. The typical binding energy is $\epsilon \simeq 0.6$ GeV. Therefore, in a first approximation

$$\epsilon << M \tag{1}$$

where M is the mass of the bound state ($M \simeq 3$ GeV). The binding energy is also small compared to the collision energy

$$\epsilon << \sqrt{s}\,. \tag{2}$$

Approximation (1) justifies the use of quantum mechanical perturbation theory and approximation (2) justifies the use of the eikonal approximation, which, in this case, implies that the condenser follows a straight line trajectory and remains essentially undisturbed during the interaction. The dipole is excited and dissociated into two mesons, 1 and 2, which, in the charmonium rest frame, have nearly the same momenta and are non-relativistic

$$|\vec{p_1}| \simeq |\vec{p_2}| \quad |\vec{p_i}|/m << 1\,. \tag{3}$$

where m is the charm quark mass. These last assumptions are not essential but allow for a purely analytical evaluation of the cross section. Due to all these approximations the validity of the model is restricted to energies far from reaction threshold. With these assumptions we can write the interaction Hamiltonian as

$$H_{int} = g(T_1^a \vec{E}_1^a \vec{r}_1 + \overline{T}_2^b \vec{E}_2^b \vec{r}_2) \tag{4}$$

where T^a ($\overline{T}^b$) are the generators of color group SU(3) in the fundamental (conjugate) representation. The indices 1 and 2 represent the quark and antiquark respectively; $\vec{r_1}$ and $\vec{r_2}$ are the quark and antiquark coordinates in the charmonium rest frame. $\vec{E}_1^a$ and $\vec{E}_2^b$ are the chromoeletric fields generated by the hadron in motion (condenser) and "felt" by quark and antiquark in the bound state. They have to be Lorentz transformed to the charmonium rest frame, bringing to our calculation a Lorentz gamma factor, which is the source of the energy dependence ($\sqrt{s}$) of our results. We shall for the moment neglect the magnetic component, since it does not do any work on the charges and thus is not effective in the energy transfer. We can represent this external field by:

$$\vec{E}(r_e,t) = \gamma \vec{E}_0 exp\left(-\frac{(X-x_e)^2}{d^2} - \frac{(Y-y_e)^2}{d^2}\right) exp\left(-\gamma^2 \frac{[vt-z_e]^2}{d^2}\right) \tag{5}$$

with $e = 1,2$. X, Y and Z are the hadron coordinates. $Z = vt$, because the hadron moves with velocity $\vec{v}$ along the z axis. We neglect the deflection of the hadron trajectory, because we are studying reactions in the non-perturbative regime, i.e., of the low momentum transfer. X and Y are related with the impact parameter b by $b^2 = X^2 + Y^2$. Notice that, by simplicity, we choose one preferential direction for the field, in this case, the x-axis. Neglecting the CM motion, (4) can be rewritten as

$$H_{int} = g(\frac{\lambda^a}{2}E_1^a + \frac{\lambda^{b^T}}{2}E_2^b)(\frac{x_1 - x_2}{2}). \tag{6}$$

Also for the sake of simplicity, from now on, in (6) we take $x_1 - x_2 \simeq a$, where a is the typical separation between quark and antiquark. Initially the quark-antiquark pair is in a localized region of the space. The initial wave function of system has one spatial part and one color part

$$\Psi_{in} = f(r_1, r_2)c_n d_n \tag{7}$$

where c_n and d_n , with $n = 1, 2, 3$, are the initial color vectors for quark and antiquark respectively, taken in a color singlet state. We choose

$$f(r_1, r_2) = N_i\, exp\left(-\frac{\vec{r_1}^{\,2}}{a^2}\right) exp\left(-\frac{\vec{r_2}^{\,2}}{a^2}\right) exp(-i\varepsilon_i t) \tag{8}$$

where ε_i is the charmonium initial energy and N_i is a normalization constant. The initial wave function Ψ_{in} describes the confinement of quarks and also asymptotic freedom, as it allows the quarks to be independent inside the bag. Under the action of the external field the initial wave function Ψ_{in} evolves to a final state Ψ_f:

$$\Psi_f = t(r_1, r_2)c_j d_k \tag{9}$$

where c_j and d_k, with $j, k = 1, 2, 3$, are the quark and antiquark final color vectors and

$$t(r_1, r_2) = N_f\; exp\,(i\vec{p_1}.\vec{r_1})\; exp\,(i\vec{p_2}.\vec{r_2})\; exp\,(-i\varepsilon_f t) \tag{10}$$

where $\vec{p_1}$ and $\vec{p_2}$ are the quark momenta, ε_f is the final energy of the $c - \bar{c}$ pair and N_f is a normalization constant. The Ψ_f function represents the wave function of free particles. In the model, the quarks become free, inside the condenser and then suffer hadronization, which always occurs with probability one and does not change the cross section, which is thus given by

$$d^6\sigma = 2\pi \int_0^\infty db\, b\, \overline{|T_{fi}|^2} \cdot \frac{V}{(2\pi^3)} d^3p_1 \cdot \frac{V}{(2\pi^3)} d^3p_2\,. \tag{11}$$

The transition amplitude squared can be easily computed from (6), (7) and (9):

$$|T_{fi}|^2 = |<|\Psi_f|H_{int}|\Psi_{in}>|^2 = T^*_{fi} T_{fi}\,. \tag{12}$$

Since color is not observed, we take the average of the all initial color states, and the sum of all final states:

$$|T_{fi}|^2 \to \overline{|T_{fi}|^2} \equiv \frac{1}{3}\sum_n \frac{1}{8}\sum_a \sum_j \sum_k |T_{fi}|^2 \,.$$

In performing the integrations in (11) we make use of the inequalities (3):

$$\varepsilon_f - \varepsilon_i \simeq \frac{p_1^2}{2m} + \frac{p_2^2}{2m} - \epsilon$$

where $\epsilon = -(m + m - M)$ is binding energy ($\epsilon < 0$). As it is clear from (6), we will need to know the matrix element $< gE >$. Following Kuzmenko and Simonov [4], we take it to be equal to $< gE > \simeq 1$ GeV/fm. Inserting the typical hadron size of $d = 0.8$ fm, a charmonium size of $a = 0.2$ fm, small binding energy $\epsilon \simeq 0$ and Lorentz factor of $\gamma = 10$ in (11) we obtain:

$$\sigma \simeq 2.15\,\mathrm{mb}\,.$$

For the moment our aim is just to estimate the order of magnitude. As it can be seen, our number is very close to previuos short-distance QCD estimates[3,2]. A detailed study will be published elsewhere.

Acknowledgments

This work has been supported by FAPESP and CNPq, Brazil.

References

1. M.E. Peskin, *Nucl. Phys.* **B156**, 365 (1979) ; G. Bhanot and M.E. Peskin, *Nucl. Phys.* **B156**, 391 (1979).
2. D.Kharzeev and H. Satz, *Phys. Lett.* **B366**, 316 (1996); **B356**, 365 (1995); **B334**, 155 (1994).
3. Y. Oh, S. Kim and S.H. Lee, *Phys. Rev.* **C65** 067901 (2002).
4. V.Kuzmenko and Y.A.Simonov, *Phys. Lett.* **B 494** 81 (2000).

ON THE ROLE OF THE SUBTRACTION CONSTANT IN ANALYTIC MODELS FOR HIGH ENERGY FORWARD SCATTERING

R.F. ÁVILA, E.G.S. LUNA AND M.J. MENON
UNICAMP - 13083-970 Campinas, SP, Brazil
rfa@ifi.unicamp.br, luna@ifi.unicamp.br, menon@ifi.unicamp.br

Analytic models for high energy forward scattering are characterized by the use of dispersion relations and parameterizations of the experimental data on total cross sections and the ρ parameter. In this communication we investigate the role played in the fit results by the subtraction constant, present in the dispersion relations. Making use of two analytic models with different asymptotic inputs and taking also into account the cosmic ray information on proton-proton cross sections at the highest energies, we show that the subtraction constant constrains the fits in both low and high energy regions.

Two fundamental quantities in the investigation of the high energy hadron-hadron scattering are the total cross section and the ρ parameter defined respectively as

$$\sigma_{tot}(s) = ImF(s,t=0)/s \; ; \; \rho(s) = ReF(s,t=0)/ImF(s,t=0)\,.$$

Analytic models are characterized by the use of dispersion relations, connecting real and imaginary parts of the amplitude. For hadron-hadron interactions Froissart-Martin bound demands at least one subtraction constant, here represented by K, and pp and $p\bar{p}$ scattering may be simultaneously investigated by combining even and odd amplitudes. In a recent work we analyzed forward pp and $\overline{p}p$ scattering through two analytic models with singly subtracted derivative dispersion relation and considering the subtraction constant as a free fit parameter [1]. In order to investigate the role of the subtraction constant, in this communication we perform the same analysis, but attributing the value zero to the subtraction constant. We present here only the results of a complete list of references and detailed calculations may be found in this previous paper [1]. As explained in this reference, we analyze $\overline{p}p$ and pp accelerator data on σ_{tot} and ρ above $\sqrt{s} = 10$ GeV and also include the discrepancies in the informations from cosmic ray experiments for σ_{tot}^{pp}: Akeno and Fly's Eye results and the Nikolaev and Gaisser, Sukhatme, Yodh (GSY) results. For this reason we consider two Ensembles of data: *Ensemble I* with $\overline{p}p$ and pp accelerator data + Akeno + Fly's Eye and *Ensemble II* with $\overline{p}p$ and pp accelerator data + Nikolaev + GSY. For pp and $\overline{p}p$ scattering, Crossing Symmetry and Analyticity allows to connect $\sigma_{tot}(s)$ and $\rho(s)$ through two compact and symmetric formulas, $\rho^{pp}\sigma_{tot}^{pp}(s) = E(\sigma_+) + O(\sigma_-)$ and $\rho^{p\bar{p}}\sigma_{tot}^{p\bar{p}}(s) = E(\sigma_+) - O(\sigma_-)$, where, $E(\sigma_+)$ and $O(\sigma_-)$ are analytic transforms relating real and imaginary parts of Even (+) and Odd (−) amplitudes, respectively. At high energies they can be expressed in a derivative form, given in the previous paper [1]. With this approach, $\rho(s)$ may be determined through suitable input parameterizations for the total cross sections. As in the previous work,we shall use two different choices for these inputs, namely, the Donnachie-Landshoff (DL) parameterization [2] $\sigma_{tot}^{pp}(s) = Xs^{\epsilon} + Ys^{-\eta}$, $\sigma_{tot}^{p\bar{p}}(s) = Xs^{\epsilon} + Zs^{-\eta}$ and the Kang-Nicolescu (KN) model with maximal Odderon contribution [3], $\sigma_{tot}^{pp}(s) = A_1 + B_1 \ln s + C \ln^2 s$, $\sigma_{tot}^{\bar{p}p}(s) = A_2 + B_2 \ln s + C \ln^2 s + 2Rs^{-\frac{1}{2}}$. For both parameterizations (DL and KN) we perform simultaneous fits to $\sigma_{tot}(s)$ and $\rho(s)$ in the cases of ensembles I and II, assuming the subtraction constant $K = 0$ and comparing with our previous results with K as free fitting parameter. The results with the DL parameterization are shown in Figs.1 and 2 and those with the KN parameterization in Figs.3, 4. Concerning the total cross section, Figs. 1 and 3 show that at the high energy region, the results with $K = 0$ and K free are different. In the case of the DL parameterization the values are higher for K free than for $K = 0$ (see for example the curves at 10^5 GeV); the contrary seems to happen in the case of the KN parameterization. The results for $\rho(s)$ in Figs. 2 and 4 indicate that at the lower energy region ($\sqrt{s} \sim 10 - 20$ GeV) different behaviors are also obtained with $K = 0$ and K free, mainly in the case of the DL parameterization. We conclude that taking $K = 0$ constraint the fits at both high and low energy regions, which means that the increase of the total cross section may be faster than predicted in analysis that assume $K = 0$.

Acknowledgements

Work supported by FAPESP and CNPq, Brazil. We are thankful to the organizers of this excellent and stimulating Workshop, in special to Cesar Vasconcellos and Victoria Herscovitz.

References

1. R.F. Ávila, E.G.S. Luna, M.J. Menon, it in LISHEP 2002 - Session C - Workshop on Diffractive Physics, February, 4 - 8, 2002, Rio de Janeiro, Brazil (to be published), hep-ph/0206068.
2. A. Donnachie and P. V. Landshoff, Z. Phys. C **2**, 55 (1979); Phys. Lett. B **387**, 637 (1996).
3. K. Kang and B. Nicolescu, Phys. Rev. D **11**, 2461 (1975).

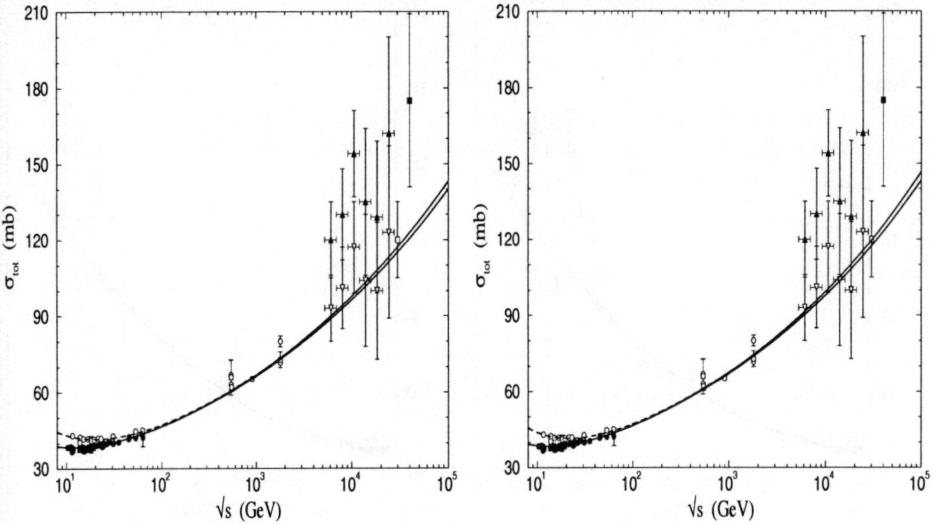

Figure 1. Total cross section. Simultaneous fits to σ_{tot} and ρ with the DL parameterization: pp (solid) and $\overline{p}p$ (dashed) and $K = 0$ (left) and K as free fit parameter (right). For each reaction, the lower curve corresponds to Ensemble I and the higher one to Ensemble II.

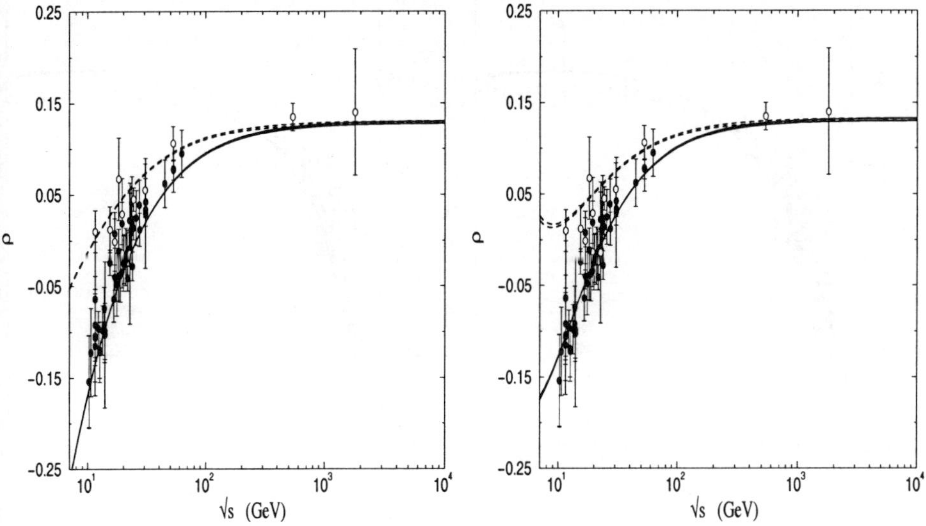

Figure 2. ρ parameter. Fits to σ_{tot} and ρ with the DL parameterization (same legend as Fig. 1).

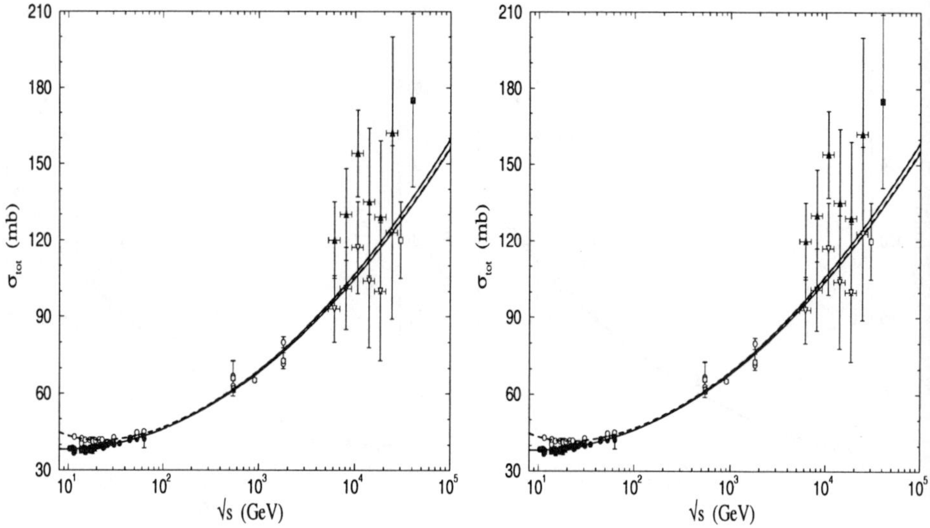

Figure 3. Total cross section. Fits to σ_{tot} and ρ with the KN parameterization (same legend as Fig. 1).

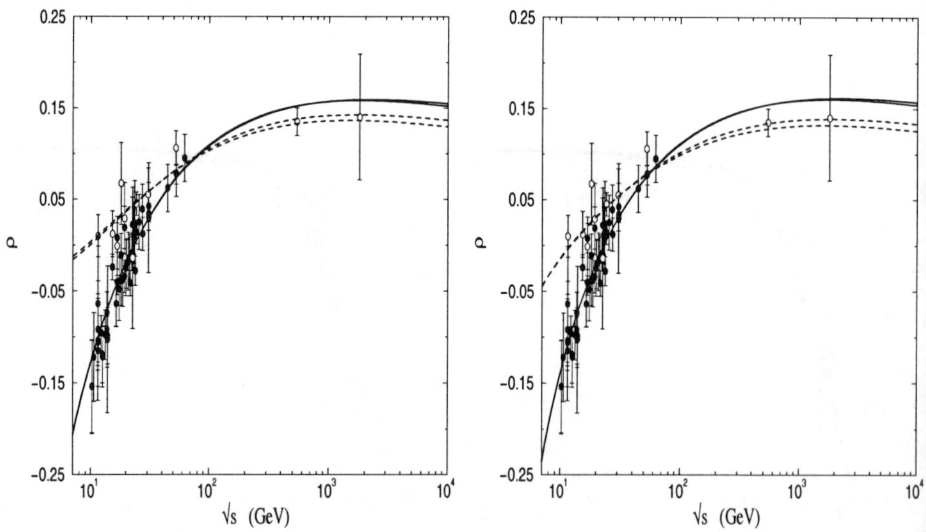

Figure 4. ρ parameter. Fits to σ_{tot} and ρ with the KN parameterization(same legend as Fig. 1) .

PROBING THE STRUCTURE OF THE $Z\gamma Z$ VERTEX AT VLHC

Y.A. COUTINHO, A.J. RAMALHO, R. WALSH AND S. WULCK
Instituto de Física, Universidade Federal do Rio de Janeiro,
Rio de Janeiro, RJ, Brazil
yara@if.ufrj.br, ramalho@if.ufrj.br, walsh@if.ufrj.br, steniow@if.ufrj.br

We analyze the prospects of probing anomalous $Z\gamma Z$ couplings at the proposed *ep* mode of the Very Large Hadron Collider (VLHC), by means of the process $e^- + p \longrightarrow e^- + \gamma + X$. The 95% C.L. limits for the form factors that describe the $Z\gamma Z$ vertex are calculated for the *ep* collider of the VLHC (in its low- and high-field options), and for the current energies and luminosities of the DESY HERA ep collider.

1 Introduction

The interactions among vector gauge bosons continue to attract the attention of both theorists and experimentalists. From the standard model point of view, these interactions are determined by an underlying non-Abelian gauge symmetry, and at the lowest order of perturbation theory are described by trilinear vertices. Possible deviations from nonstandard signals, if detected, would point to new physics. Anomalous vertices are an important ingredient in some extended models that allow for composite vector bosons or gauge vector bosons strongly interacting with each other [1]. The experimental studies of the vector gauge boson self-interactions are at an early stage, especially those in which only neutral vector bosons are involved [2].

The VLHC program [3] allows for two different *ep* collider designs: the low-field version with an 80 GeV electron ring and a 3 TeV proton ($\sqrt{s} = 1$ TeV) with a luminosity of 0.26 $(\text{nb.s})^{-1}$ and the high-field version with 200 GeV electron beam with a 50 TeV proton beam ($\sqrt{s} = 6.32$ TeV) and a luminosity of 0.14 $(\text{nb.s})^{-1}$. We analyzed the effects of anomalous $Z\gamma Z$ couplings in radiative *ep* scattering, considering these VLHC parameters and also those of the HERA collider for comparison. For the sake of simplicity we ignore a possible $Z\gamma\gamma$ vertex.

2 $Z\gamma Z$ Trilinear Gauge Boson Vertex

The most general trilinear $Z\gamma Z$ vertex, for an on-shell photon, consistent with Lorentz and gauge invariance, is described by four anomalous couplings

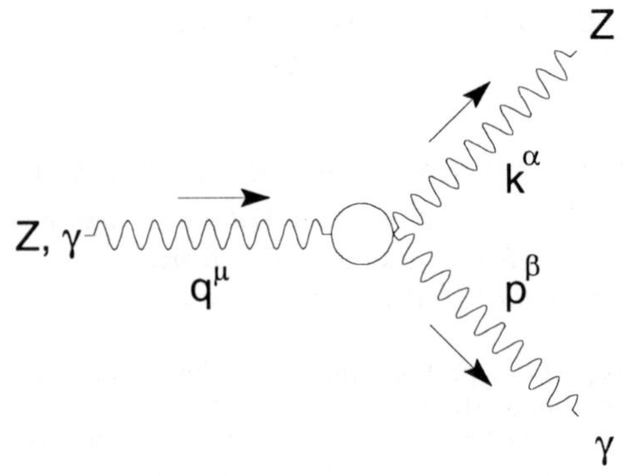

Figure 1. Neutral boson trilinear vertex.

h_i^Z $(i = 1 \ldots 4)$: [4]

$$V_{Z\gamma Z}^{\alpha\beta\mu}(k,p,q) = ie\frac{q^2-k^2}{M_Z^2}\Big\{h_1^Z(p^\mu g^{\alpha\beta} - p^\alpha g^{\mu\beta}) + \frac{h_2^Z}{M_Z^2}\big[q^\alpha(q.pg^{\mu\beta} - p^\mu q^\beta) - k^\mu(k.pg^{\alpha\beta} - p^\alpha k^\beta)\big] + h_3^Z(\varepsilon^{\mu\alpha\beta\rho}p_\rho) + \frac{h_4^Z}{M_Z^2}(\varepsilon^{\mu\beta\rho\sigma}q^\alpha q_\rho p_\sigma - \varepsilon^{\alpha\beta\rho\sigma}k^\mu k_\rho p_\sigma)\Big\}. \tag{1}$$

The couplings h_1^Z and h_2^Z are P even and violate CP, whereas h_3^Z and h_4^Z conserve CP. These four couplings h_i^Z vanish in the standard model (tree level). For the process we are considering, with an on-shell photon in the final state, the form factors may be expressed in a dipole-like form:

$$h_i^Z(q^2,k^2) = \frac{h_{i0}^Z}{\left(1 - \frac{q^2+k^2}{\Lambda^2}\right)^n}\,, \quad i = 1\ldots 4\,, \tag{2}$$

where h_{i0}^Z represents their values at low energies, and Λ is a large energy scale, around which new physics would become important. In this paper we take $n = 2$ for $h_{1,3}^Z(k^2,q^2)$ and $n = 3$ for $h_{2,4}^Z(k^2,q^2)$. With this dependence of the form factors $h_i^Z(q^2,k^2)$ on the invariants k^2 and q^2, the Bose symmetry of the two Z lines is manifest in each of the four terms of Eq. (1).

3 Monte Carlo Simulation

We studied the process $e^- + p \longrightarrow e^- + \gamma + X$, where the anomalous $Z\gamma Z$ interaction is not present in the standard model. We only considered two quark families, setting all quark masses equal to zero throughout. CTEQ5M1 parton distribution functions [5] were used to calculate all cross sections, with $\sqrt{s}$ for the parton sub-process chosen for evolution scale. The signal of interest consists of an electron, a photon and a jet produced by the scattered quark. A $p_T > 20$ GeV cut was imposed on the transverse momenta of the electron, photon and jet, along with angular cuts $|cos\,\theta| < 0.99$ on the angles of the photon and the electron with respect to the direction of the proton. We imposed isolation cuts $R = \sqrt{(\Delta\eta)^2 + (\Delta\phi)^2} > 0.5$ on the jet-photon and electron-photon separations in the pseudo-rapidity-azimuthal angle planes, which separate the photon from the electron and from the jet, and control collinear singularities.

4 Distributions and Results

The most sensitive distribution to the presence of anomalous $Z\gamma Z$ coupling turned out to be the photon transverse momentum distribution $d\sigma/p_T^\gamma$. From the $d\sigma/dp_T^\gamma$ spectra, we performed a χ^2 analysis to determine the 95% confidence level limits for the h_{i0}^Z form factors.

We considered an annual integrated luminosity of 4.42 fb^{-1} for the high-field *ep* collider and 8.20 fb^{-1} for the low-field option. The systematic error for a single measurement was taken to be $\delta = 5\%$. For a form factor scale $\Lambda = 1$ TeV, the limits and correlation coefficients on the form factor pairs are given in Table 1 [4]. Correlations are rather strong for pairs of form factors of the same CP parity. On the other hand, form factors of opposite CP parity are found to be nearly uncorrelated, suggesting that it may be possible to disentangle CP-conserving from CP-violating effects. At HERA energies the limit of the form factors are of the order of 10.

Generally speaking, experimental bounds on the form factors h_{i0}^Z obtained at the current energies and luminosities of the CERN LEP and Fermilab Tevatron are in the range $10^{-1} - 10^{-2}$ [2]. Theoretical estimates for $Z\gamma Z$ anomalous couplings using NLC [6] and LHC [7] parameters point to limits of the order of $10^{-3} - 10^{-4}$.

Acknowledgements

This work was partly supported by CNPq, FAPERJ and FINEP, Brazil.

Parameter	Negative limit	Positive limit	Correlation coefficient
h_{10}^Z	-6.8×10^{-2}	$+6.8 \times 10^{-2}$	-0.88
h_{20}^Z	-2.3×10^{-3}	$+2.3 \times 10^{-3}$	
h_{30}^Z	-7.4×10^{-2}	$+7.4 \times 10^{-2}$	-0.95
h_{40}^Z	-2.5×10^{-3}	$+2.5 \times 10^{-3}$	
h_{10}^Z	-2.2×10^{-2}	$+2.3 \times 10^{-2}$	-4.4×10^{-3}
h_{30}^Z	-2.2×10^{-2}	$+2.2 \times 10^{-2}$	
h_{10}^Z	-2.2×10^{-2}	$+2.3 \times 10^{-2}$	-2.7×10^{-3}
h_{40}^Z	-6.8×10^{-4}	$+6.8 \times 10^{-4}$	
h_{20}^Z	-6.7×10^{-4}	$+6.9 \times 10^{-4}$	-4.4×10^{-3}
h_{30}^Z	-2.2×10^{-2}	$+2.2 \times 10^{-2}$	
h_{20}^Z	-6.6×10^{-4}	$+6.9 \times 10^{-4}$	-2.8×10^{-3}
h_{40}^Z	-6.8×10^{-4}	$+6.8 \times 10^{-4}$	

Table 1. Correlated 95% C. L. limits on the form factors pairs (h_{i0}^Z,h_{j0}^Z) ($i, j = 1 \ldots 4, j > i$), for an input scale $\Lambda = 1$ TeV, and corresponding correlation coefficients.

References

1. F. M. Renard, *Phys. Lett.* **B126** (1983)59; M. Claudson, E. Fahri and R. L. Jaffe, *Phys. Rev.* **D34** (1986) 873.
2. L3 Collaboration, M. Acciarri et al., *Phys. Lett.* **B489** (2000) 55 and references threin.
3. M. Derrick et al., Proceedings of the Particle Accelerator Conference, New York, New York, USA, March 29-April2, 1999, Fermilab-Conf-99/149; J. Norem and T. Sen, ANL-HEP-PR-99-103, Fermilab-Pub-99-347.
4. Y. A. Coutinho, A. J. Ramalho. R. Walsh and S. Wulck, *Phys. Rev.* **D 64** (2001) 115008.
5. H. L. Lai et al., CTEQ Collaboration, *Eur. Phys. J.* **C12** (2000) 375.
6. R. Walsh and A. J. Ramalho, *Phys. Rev.* **D 57** (1998) 5908, and references therein.
7. G. Azuelos et al., Electroweak Physics, Proceedings of the Workshop on Standard Model Physics (and more) at the LHC, CERN, Geneva, 2000, edited by G. Altarelli and M. L. Mangano, CERN Report 2000-004, hep-ph/0003275 and references therein.

BOUNDS ON THE $Z\gamma Z$ AND $Z\gamma\gamma$ ANOMALOUS COUPLINGS IN POLARIZED *ep* COLLISIONS

Y.A. COUTINHO, A.J. RAMALHO,
R. WALSH AND S. WULCK

*Instituto de Física, Universidade Federal do Rio de Janeiro,
Rio de Janeiro, RJ, Brazil*

yara@if.ufrj.br, ramalho@if.ufrj.br, walsh@if.ufrj.br, steniow@if.ufrj.br

We discuss the feasibility of detecting effects of anomalous $Z\gamma Z$ and $Z\gamma\gamma$ couplings in radiative *ep* collisions, with longitudinally polarized electron beams at the VLHC energies.

1 Introduction

In the standard model, the strength and form of the self interaction of vector gauge bosons are determined by the underlying SU(2) x U(1) symmetry. Any deviation from the standard model expectation would point to new physics. In this paper we analyze the possible effects of nonstandard $Z\gamma Z$ and $Z\gamma\gamma$ trilinear couplings in polarized *ep* scattering at the energies proposed for the Very Large Hadron Collider (VLHC). In its high-field version, the VLHC *ep* collider [1] is expected to use a 200 GeV electron beam and a 50 TeV proton beam, to produce collisions at $\sqrt{s} = 6.32\text{TeV}$, with an estimated luminosity of 0.14 mb/s.

2 Nonstandard trilinear couplings

The most general trilinear $Z\gamma Z$ and $Z\gamma\gamma$ vertices [2] for an on-shell photon, consistent with Lorentz and gauge invariance, are described by the following expressions :

$$\begin{aligned} V_{Z\gamma Z}^{\alpha\beta\mu}(k,p,q) &= ie\frac{q^2-k^2}{M_Z^2}\Big\{h_1^Z(p^\mu g^{\alpha\beta}-p^\alpha g^{\mu\beta}) \\ &+ \frac{h_2^Z}{M_Z^2}\left[q^\alpha(q.pg^{\mu\beta}-p^\mu q^\beta)-k^\mu(k.pg^{\alpha\beta}-p^\alpha k^\beta)\right] \\ &+ h_3^Z(\varepsilon^{\mu\alpha\beta\rho}p_\rho)+\frac{h_4^Z}{M_Z^2}(\varepsilon^{\mu\beta\rho\sigma}q^\alpha q_\rho p_\sigma-\varepsilon^{\alpha\beta\rho\sigma}k^\mu k_\rho p_\sigma)\Big\}, \quad (1) \end{aligned}$$

$$V^{\alpha\beta\mu}_{Z\gamma\gamma}(k,p,q) = ie\frac{q^2}{M_Z^2}\Big\{h_5^\gamma(p^\mu g^{\alpha\beta} - p^\alpha g^{\mu\beta})$$
$$+ \frac{h_6^\gamma}{M_Z^2}\left[q^\alpha(q.pg^{\mu\beta} - p^\mu q^\beta) - k^\mu(k.pg^{\alpha\beta} - p^\alpha k^\beta)\right]$$
$$+ h_7^\gamma(\varepsilon^{\mu\alpha\beta\rho}p_\rho) + \frac{h_8^\gamma}{M_Z^2}(\varepsilon^{\mu\beta\rho\sigma}q^\alpha q_\rho p_\sigma - \varepsilon^{\alpha\beta\rho\sigma}k^\mu k_\rho p_\sigma)\Big\}, \quad (2)$$

where the form factors h_i^V are parameterized as

$$h_i^V(q^2,k^2) = \frac{h_{i0}^V}{\left(1 - \frac{q^2+k^2}{\Lambda^2}\right)^n}, \ i = 1(5)\dots 4(8) \text{ and } V = Z\ (\gamma)\ . \quad (3)$$

All the constants h_i^V, which correspond to the values of the form factors in the low-energy regime, vanish in the standard model at the tree level. The parameter Λ is some large energy scale around which the new physics would set in.

3 Limits on the anomalous couplings

In order to obtain distributions sufficiently sensitive to the presence of anomalous $Z\gamma Z$ and $Z\gamma\gamma$ couplings, we carried out a Monte Carlo simulation of radiative *ep* scattering, with longitudinally polarized e^- beams. Only two quark families were considered, and all fermion masses were neglected throughout. Cross sections were calculated with CTEQ5M1 [3] parton distribution functions , with $\sqrt{s}$ for the parton sub-process chosen for evolution scale. A $p_T > 20$ GeV cut was imposed on the transverse momenta of the final state particles, and angular cuts $|cos\theta| < 0.99$ on their momenta as well. Isolation cuts $R = \sqrt{(\Delta\eta)^2 + (\Delta\phi)^2} > 0.5$ on the jet-photon and electron-photon separations in the pseudo-rapidity-azimuthal angle planes were employed to separate the photon from the electron and from the jet, and to control collinear singularities.

The transverse momentum spectrum of the photon was found to be the most sensitive distribution to the anomalous couplings. From this transverse-momentum spectrum the confidence intervals for the form factors were determined by a χ^2 analysis. The transverse momentum interval was divided into 5 bins and the χ^2 function defined as

$$\chi^2 = \sum_{i=1}^{5}\left(\frac{N_i^{SM} - N_i^{AN}}{\Delta N_i^{SM}}\right)^2 \quad (4)$$

where N_i^{SM} and N_i^{AN} stand for the number of events in the i-th bin, as predicted by the standard model, and in the presence of anomalous couplings, respectively, and $\Delta N_i^{SM} = \sqrt{(N_i^{SM})^2 + (N_i^{SM}\delta)^2}$ stands for the total error, which combines the statistical and systematic errors in quadrature. We considered a systematic error $\delta = 5\%$ for a measurement.

4 Results

We obtained all the pair-wise correlated limits for couplings arising from the $Z\gamma Z$ and $Z\gamma\gamma$ vertices, considering separately left-handed and right-handed incoming electron beams. Fig. 1 shows the 95% C. L. contours for the form factor pair (h_3^Z, h_7^γ) for these two electron beams polarizations, as well as the unpolarized case. It is worthwhile to point out that the use of polarized beams may indeed lead to tighter limits on the form factor, as can be seen from the joint region defined by the curves in Fig.1. Table 1 shows some corrected limits for the form factor pairs, along with the correlation coefficients. We also derived the one-degree-of-freedom limits for each form factor. The results are presented in Table 2.

References

1. M. Derrick et al., FERMILAB-Conf-99/149.
2. Y. A. Coutinho, A. J. Ramalho, R. Walsh and S. Wulck, *Phys. Rev.* **D64**, (2001) 115008.
3. H. L. Lai et al., CTEQ Collaboration, *Eur. Phys. J.* **C12** (2000) 375.
4. U. Baur, L. E. Berger, *Phys. Rev.* **D47** (1993) 4889.
5. L. Lach, FERMILAB-Conf-98/264.
6. F. Cornet, R. Graciani, J. I. Illiana, ph/9609259.
7. R. Walsh and A. J. Ramalho, *Phys. Rev.* **D57** (1998) 5908.
8. R. Walsh and A. J. Ramalho, *Phys. Rev.* **D 65**, 055011.
9. Roberval Walsh, Tese de Doutorado, IF/UFRJ, 2002.

Table 1. Correlated 95% C. L. limits on some of the form factors pairs (h^V_{i0},h^V_{j0}) ($i, j = 1 \ldots 8$, $j > i$), for an input scale $\Lambda = 1$ TeV, and corresponding correlation coefficients using unpolarized electron beam.

Parameter	Negative limit	Positive limit	Correlation coefficient
h^γ_{10}	-9.34×10^{-2}	$+9.51 \times 10^{-2}$	-9.39×10^{-2}
h^γ_{20}	-2.71×10^{-3}	$+2.71 \times 10^{-3}$	
h^γ_{10}	-1.74×10^{-2}	$+2.87 \times 10^{-2}$	-2.19×10^{-2}
h^Z_{50}	-1.78×10^{-2}	$+1.53 \times 10^{-2}$	
h^Z_{50}	-7.91×10^{-2}	$+6.72 \times 10^{-2}$	-9.70×10^{-2}
h^Z_{60}	-2.11×10^{-3}	$+2.51 \times 10^{-3}$	
h^Z_{60}	-5.95×10^{-4}	$+5.78 \times 10^{-4}$	$+2.70 \times 10^{-2}$
h^Z_{70}	-1.61×10^{-2}	$+1.61 \times 10^{-2}$	

Table 2. One-degree-of-freedom bounds on the form factors. Scale $\Lambda = 1$ TeV. The predictions are for the high-field mode of VLHC *ep* collider, using unpolarized electron beam.

Parameter	Negative limit	Positive limit
h^γ_{10}	-1.34×10^{-2}	$+2.67 \times 10^{-2}$
h^γ_{20}	-5.74×10^{-4}	$+6.27 \times 10^{-4}$
h^γ_{30}	-1.96×10^{-2}	$+1.98 \times 10^{-2}$
h^γ_{40}	-5.92×10^{-4}	$+6.03 \times 10^{-4}$
h^Z_{50}	-1.45×10^{-2}	$+1.38 \times 10^{-2}$
h^Z_{60}	-4.58×10^{-4}	$+4.53 \times 10^{-4}$
h^Z_{70}	-1.40×10^{-2}	$+1.41 \times 10^{-2}$
h^Z_{80}	-4.56×10^{-4}	$+4.55 \times 10^{-4}$

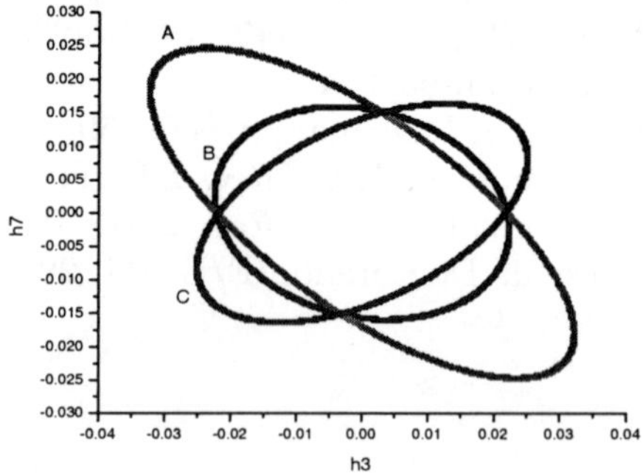

Figure 1. The 95% C. L. contours for the form factor pairs (h_3, h_7). Elipse A : Polarized beam + 1; Elipse B : unpolarized beam; Elipse C : Polarized beam - 1.

4. Decays and Medium-Energy Reactions

DALITZ-PLOT ANALYSIS OF Λ_c^+ THREE BODY DECAYS CONSIDERING THE SPIN OF THE PARTICLES

A.M.F. ENDLER AND E.C. DE OLIVEIRA
Centro Brasileiro de Pesquisas Físicas
Rua Dr. Xavier Sigaud, 150, Rio de Janeiro, 22290-180, Brazil,
endler@cbpf.br

We have used the helicity formalism [1,2] to calculate the spin dependent amplitudes for the following Λ_c^+ three body decays: $\Lambda_c^+ \to pK^-\pi^+$, $\Lambda_c^+ \to \Sigma^+\pi^-\pi^+$. We simulated events according to these amplitudes assigning values for the amplitude parameters to be later recovered by a five dimension Minuit [3] fit. We could recover the parameters used in the simulated data with reasonable agreement.

We present the results of the analysis of simulated events for (a)$\Lambda_c^+ \to pK^-\pi^+$ (50000) and for (b)$\Lambda_c^+ \to \Sigma^+\pi^-\pi^+$ (26000). We calculated a signal density for each event, fixing arbitrary values for the complex amplitudes for the decaying particles. These amplitudes are labeled by A, B, C and D with its subscripts by its spin. We have taken the amplitude D_2 of the non resonant component as reference for relative phase and magnitude. The decay amplitude can be

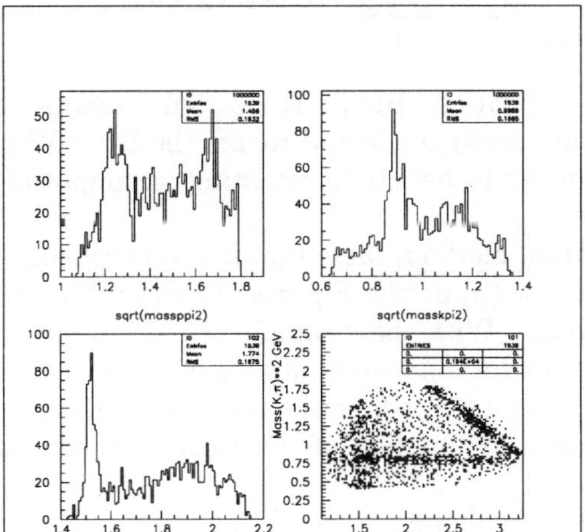

Figure 1. Dalitz plots for $pk\pi$ decay.

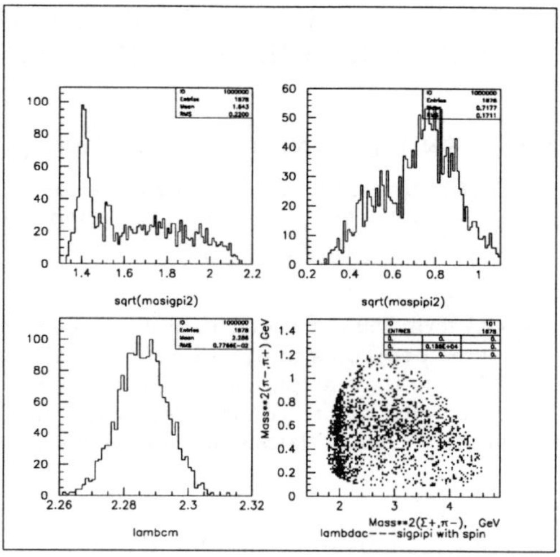

Figure 2. Dalitz plots for $\Sigma\pi\pi$ decay.

written as:

$$d\Gamma \sim \sum_{m_A} \sum_{\lambda_C \lambda_D \lambda_E} |\sum_{\lambda_B} \sum_{r} BW_r(m_B) \zeta_{B,m_A,\lambda_B,\lambda_C,\lambda_D,\lambda_B}|^2 \tag{1}$$

where, λ's are the helicities. Beside these signal amplitudes we introduced also the non resonant decay mode and we set the Breit-Wigner amplitude for the non resonant decay to be 1.0 representing an amplitude that is constant across phase space.

By the elimination method, these samples were reduced to 1939 events for (a) and 1878 events for (b) decay. Figures (1) and (2) are the Dalitz plots for these simulated events. With these sample of 1939 events for the case of $pk\pi$ decay and 1878 events for $\Sigma\pi\pi$ decay and using the CERN software package MINUIT [3], we performed a five dimensional fit of the data with the probability density function given as function of the calculated spin amplitudes as was outlined before.

As result of the likelihood fit, we obtained the values of the parameters in agreement with the values used in the simulation as we can see by the comparison Tables (1) and (2). In these Tables, the quantities labeled norm and phi are related to the complex parameters describing the amplitudes. For example $A = (a - norm)e^{i(a-phi)}$.

References

1. M.L. Perl, "High Energy Hadron Physics", John Wiley and Sons, 1974.
2. G. Otter and A.M.F. Endler, "Formalismo da helicidade e suas aplicações", CBPF-MO-002/81.
3. CERN Program Library.

$\Lambda_c^+ \to pK^-\pi^+$		
parameter	simulation	values obtained from the MINUIT fit
Λ_c^+ *mass*	2.28d0	0.22857E+01 +- 0.10846E-05
Λ_c^+ σ	0.008d0	0.79232E-02 +- 0.13377E-03
a1-norm	0.18d0	0.19573E+00 +- 0.31968E-0
a1-phi	6.0d0	0.55491E+01 +- 0.28823E+00
a2-norm	0.2d0	0.21974E+00 +- 0.39103E-01
a2-phi	-0.06d0	-0.60480E+00 +- 0.27948E+00
b1-norm	0.17d0	0.21801E+00 +- 0.43656E-01
b1-phi	4.98d0	0.48664E+01 +- 0.26557E+00
b2-norm	0.38d0	0.43802E+00 +- 0.73494E-01
b2-phi	4.88d0	0.49272E+01 +- 0.19352E+00
c1-norm	0.5d0	0.58265E+00 +- 0.97179E-01
c1-phi	-1.d0	-0.74402E+00 +- 0.33621E+00
c2-norm	0.2d0	0.23002E+00 +- 0.54995E-01
c2-phi	2.35d0	0.25643E+01 +- 0.35994E+00
c3-norm	0.20d0	0.17603E+00 +- 0.39811E-0
c3-phi	3.4d0	0.34406E+01 +- 0.30042E+00
c4 norm	0.15d0	0.12287E+00 +- 0.47462E-01
c4-phi	5.3d0	0.55475E+01 +- 0.44508E+00
d1-norm	0.46d0	0.81310E+00 +- 0.19070E+00
d1-phi	3.48d0	0.38221E+01 +- 0.32542E+00
d2-norm	1.0d0	fixed to the value 1.d0
d2-phi	0.d0	fixed to the value 0.d0
d3-norm	0.18d0	0.15804E+00 +- 0.14653E+00
d3-phi	0.75d0	-0.18981E+00 +- 0.92479E+00
d4-norm	0.94d0	0.12829E+01 +- 0.32673E+00
d4-phi	1.13d0	0.12395E+01+- 0.27490E+00

Table 1. Parameters comparison for $\Lambda_c^+ \to pK^-\pi^+$.

$\Lambda_c^+ \to \Sigma^+\pi^-\pi^+$		
parameter	simulation	values obtained from the MINUIT fit
Λ_c^+ *mass*	2.28d0	0.22857E+01 +- 0.17613E-03
Λ_c^+ σ	0.008d0	0.77585E-02 +- 0.12334E-03
a1-norm	0.17d0	0.18162E+00 +- 0.66142E-01
a1-phi	4.98d0	0.72096E+01 +- 0.73406E+00
a2-norm	0.38d0	0.14634E+00 +- 0.74025E-01
a2-phi	4.88d0	0.59820E+01 +- 0.78603E+00
b1-norm	0.17d0	0.59255E+00 +- 0.10124E+00
b1-phi	4.98d0	0.44670E+01 +- 0.64799E+00
b2-norm	0.38d0	0.50905E+00 +- 0.11603E+00
b2-phi	4.88d0	0.58636E+01 +- 0.66301E+00
c1-norm	0.5d0	0.56464E+00 +- 0.16603E+00
c1-phi	-1.d0	-0.10709E+01 +- 0.68847E+00
c2-norm	0.2d0	0.21478E+01 +- 0.18485E+00
c2-phi	2.35d0	0.31972E+01 +- 0.64952E+00
c3-norm	0.20d0	0.75107E+00 +- 0.16319E+00
c3-phi	3.4d0	0.42981E+01 +- 0.58052E+00
c4-norm	0.15d0	0.15062E+01 +- 0.20389E+00
c4-phi	5.3d0	0.61890E+01 +- 0.58589E+00
d1-norm	0.46d0	0.29395E+01 +- 0.51522E+000
d1-phi	3.48d0	0.34648E+01 +- 0.66866E+00
d2-norm	1.0d0	fixed to the value 1.d0
d2-phi	0.d0	fixed to the value 0.d0
d3-norm	0.18d0	0.53080E+01 +- 0.76870E+00
d3-phi	0.75d0	0.49317E+01 +- 0.67911E+00
d4-norm	0.94d0	0.37974E+01 +- 0.70230E+00
d4-phi	1.13d0	0.17192E+01 +- 0.58983E+00

Table 2. Parameters comparison for $\Lambda_c^+ \to \Sigma^+\pi^-\pi^+$.

KAON-NUCLEON INTERACTION IN THE QUARK-GLUON EXCHANGE FRAMEWORK

DIMITER HADJIMICHEF
Instituto de Física e Matemática, UFPel, Caixa Postal 354
96010-900, Pelotas, RS, Brazil
dimiter@ufpel.tche.br

The general meson-baryon potential obtained in the context of the Fock-Tani formalism is applied to the K^+N system. A simple spin-dependent interaction on quark level is used in order to compare with known results. The short range nature of the equivalent local potential is shown.

One of the most extensively studied interactions is the NN interaction. In the context of the Bonn potential the short range repulsion arises from an ω-meson exchange with a larger coupling constant $g^2_{NN\omega}$ than obtained from $SU(3)$ estimates[1]. In the KN-Bonn potential the K^+N repulsion also arises from an ω exchange. If one keeps the $SU(3)$ coupling a very short ranged phenomenological contribution must be added.

It has been argued [2] that a more suitable investigation of the short range part of the hadronic repulsion could be made in the KN system. The K^+N interaction is short ranged and relatively weak. The one pion exchange vertex is zero so there is no πKK vertex and 2π exchanges are weak. In the constituent quark model the $K^+ = u\,\bar{s}$ and $K^- = s\,\bar{u}$ so the simple quark exchange mechanism can be applied to the K^+N system. Many groups [3] have studied this system in the S-wave; it is necessary to extend this investigation in order to include higher partial waves, such as the P-waves.

The method we employ in order to introduce the quark-gluon degrees of freedom is known as the Fock-Tani formalism [4]. The meson-baryon potential can be obtained applying in a standard way the Fock-Tani transformed operators to the microscopic Hamiltonian [4]

$$V_{\mathrm{meson-baryon}} = \sum_{i=1}^{4} V_i(\alpha\beta;\delta\gamma)\, m^\dagger_\alpha\, b^\dagger_\beta\, m_\gamma\, b_\delta \qquad (1)$$

and

$$V_1(\alpha\beta;\delta\gamma) = -3V_{qq}(\mu\nu;\sigma\rho)\, \Phi^{*\mu\nu_2}_\alpha\, \Psi^{*\nu\mu_2\mu_3}_\beta\, \Phi^{\rho\nu_2}_\gamma\, \Psi^{\sigma\mu_2\mu_3}_\delta \qquad (2)$$

$$V_2(\alpha\beta;\delta\gamma) = -3V_{q\bar{q}}(\mu\nu;\sigma\rho)\,\Phi^{*\mu_1\nu}_{\alpha}\Psi^{*\mu\mu_2\mu_3}_{\beta}\Phi^{\sigma\rho}_{\gamma}\Psi^{\mu_1\mu_2\mu_3}_{\delta}$$
$$V_3(\alpha\beta;\delta\gamma) = -3V_{qq}(\mu\nu;\sigma\rho)\,\Phi^{*\mu\nu_2}_{\alpha}\Psi^{*\mu_1\nu\mu_3}_{\beta}\Phi^{\mu_1\nu_2}_{\gamma}\Psi^{\sigma\rho\mu_3}_{\delta}$$
$$V_4(\alpha\beta;\delta\gamma) = -6V_{q\bar{q}}(\mu\nu;\sigma\rho)\,\Phi^{*\nu_1\nu}_{\alpha}\Psi^{*\mu_1\mu\mu_3}_{\beta}\Phi^{\mu_1\rho}_{\gamma}\Psi^{\nu_1\sigma\mu_3}_{\delta}.$$

The general potential (1) can be specialized to study the KN system. The microscopic quark-quark potential and quark-antiquark can be written as [5]

$$V_{qq\text{ or }\bar{q}\bar{q}} = \sum_{a,i<j}\left(-\frac{8\pi\alpha_s}{3m_i m_j}\right)\left[\vec{S}_i\cdot\vec{S}_j\right]\left[\mathcal{F}^a_i\cdot\mathcal{F}^a_j\right]$$

The meson and baryon wave functions are

$$\Phi^{\mu\nu}_{\alpha} = \delta(\vec{p}_\alpha - \vec{k}_\mu - \vec{k}_\nu)\,\frac{\delta_{c_\mu c_\nu}}{\sqrt{3}}\,\frac{\xi^{f_\mu f_\nu}_{\alpha}}{\sqrt{2}}\,\varphi(\vec{k}_\mu,\vec{k}_\nu)$$

$$\Psi^{\mu_1\mu_2\mu_3}_{\alpha} = \delta(\vec{p}_\alpha - \vec{p}_{\mu_1} - \vec{p}_{\mu_2} - \vec{p}_{\mu_3})\,N(p_\alpha)\,\frac{\varepsilon^{c_1c_2c_3}}{\sqrt{6}}\,\frac{\chi^{m_1m_2m_3}}{\sqrt{18}}\,\phi(\vec{p}_{\mu_1})\phi(\vec{p}_{\mu_2})\phi(\vec{p}_{\mu_3}).$$

The spatial part of the meson wave function is

$$\varphi(\vec{k}_1,\vec{k}_2) = (\pi\beta^2)^{-\frac{3}{4}}\exp\left[-\frac{\left(m_1\vec{k}_1 - m_2\vec{k}_2\right)^2}{8\,\beta^2}\right],$$

with

$$m_1 = \frac{2\,m_{\bar{q}}}{m_q + m_{\bar{q}}}\;;\; m_2 = \frac{2\,m_q}{m_q + m_{\bar{q}}}\;;\; \vec{k}_1 \equiv \vec{p}_q\;;\; \vec{k}_2 \equiv \vec{p}_{\bar{q}}\,.$$

The spatial part of the baryon wave function is

$$\phi(\vec{p}) = \left(\frac{b^2}{\pi}\right)^{\frac{3}{4}}\exp\left[-\frac{1}{2}\,b^2p^2\right]\;;\; N(\vec{p}) = \left(\frac{3\pi}{b^2}\right)^{\frac{3}{4}}\exp\left[\frac{1}{6}\,b^2p^2\right].$$

In the CM coordinate system $\vec{p}_\alpha = \vec{p}$, $\vec{p}_\beta = -\vec{p}$, $\vec{p}_\gamma = \vec{p}\,'$, $\vec{p}_\delta = -\vec{p}\,'$ and using the notation of [5]:

$$m_1 = \frac{2}{1+\rho}\;;\; m_2 = \frac{2\,\rho}{1+\rho}\;;\; b = \frac{1}{\alpha}\;;\; \beta = \frac{\alpha}{\sqrt{g}}; \tag{3}$$

where $\rho = m_q/m_{\bar{q}}$, one finds, apart from an overall energy conservation δ,

$$V_1(\vec{p},\vec{p}\,') = 3\left(\frac{8\pi\alpha_s}{3m_q^2}\,\frac{1}{(2\pi)^3}\right)\eta_1\,\exp\left[-A\,(\vec{p}-\vec{p}\,')^2\right] \tag{4}$$

$$V_2(\vec{p},\vec{p}\,') = 3\left(\frac{8\pi\alpha_s}{3m_q^2}\,\frac{1}{(2\pi)^3}\right)\eta_2\,\exp\left[-B_1\,p^2 - B_2\,p'^2 + B_3\,\vec{p}\cdot\vec{p}\,'\right]$$

$$V_3(\vec{p},\vec{p}\,') = 3 \quad \left(\frac{8\pi\alpha_s}{3m_q^2}\,\frac{1}{(2\pi)^3}\right)\ \eta_3\ \exp\left[-\,C_1\,p^2 - C_2\,p\,'^2\ +\ C_3\ \vec{p}\cdot\vec{p}\,'\ \right]$$
$$V_4(\vec{p},\vec{p}\,') = 6 \quad \left(\frac{8\pi\alpha_s}{3m_q^2}\,\frac{1}{(2\pi)^3}\right)\ \eta_4\ \exp\left[-\,D_1\ (p^2\ + p\,'^2) + D_2\ \vec{p}\cdot\vec{p}\,'\right]\,,$$

with

$$\eta_1 = 1\ \ ;\ \ \eta_2 = \rho\left(\frac{6}{g+3}\right)^{3/2}$$
$$\eta_3 = \left(\frac{12g}{7g+6}\right)^{3/2}\ \ ;\ \ \eta_4 = \rho\left(\frac{12g}{(2g+3)(g+2)}\right)^{3/2}\,,$$

and

$$A = \frac{2(1+\rho)^2 + 3g}{12\alpha^2(1+\rho)^2}$$
$$B_1 = \frac{(5g+3)\rho^2 + (-2g+6)\rho + (2g+3)}{6\alpha^2(g+3)(1+\rho)^2}$$
$$B_2 = \frac{(5g+3)\rho^2 + (10g+6)\rho + (5g+3)}{6\alpha^2(g+3)(1+\rho)^2}$$
$$B_3 = \frac{(-g+1)\rho^2 + 2\rho + (g+1)}{\alpha^2(g+3)(1+\rho)^2}$$
$$C_1 = \frac{(16g+3)\rho^2 + (\ 2g+6)\rho + (21g^2+22g+3)}{12\alpha^2(7g+6)(1+\rho)^2}$$
$$C_2 = \frac{(64g+3)\rho^2 + (\ 62g+6)\rho + (21g^2+34g+3)}{12\alpha^2(7g+6)(1+\rho)^2}$$
$$C_3 = \frac{(-8g+1)\rho^2 + 2\rho +\ (7g^2+8g+1)}{2\alpha^2(7g+6)(1+\rho)^2}$$
$$D_1 = \frac{(20g^2+40g+3)\rho^2 +\ (4g^2+14g+6)\rho +\ (5g^2+10g+3)}{12\alpha^2(2g+3)(g+2)(1+\rho)^2}$$
$$D_2 = \frac{(-4g^2-8g+1)\rho^2 +\ (-4g^2-6g+2)\rho +\ (\ g^2+2g+1)}{2\alpha^2(2g+3)(g+2)(1+\rho)^2}\,.$$

In order to obtain a potential in the configuration space one first introduces the new variables $\vec{\mathcal{P}} = \frac{1}{2}(\vec{p}+\vec{p}\,')$ and $\vec{\mathcal{Q}} = (\vec{p}-\vec{p}\,')$, in the potential (4) and then uses the local approximation which consists in considering $\vec{\mathcal{P}} = 0$.

The configuration space potential is a simple Fourier transform of the momentum space potential times the spin-isospin-combinatoric factor

$$V(r) = k_{ss} \sum_{i=1}^{4} \int dQ\, e^{i\vec{Q}\cdot\vec{r}} V_i(Q) = k_{ss} \sum_{i=1}^{4} V_i(r)$$

where $k_{ss} = \pi\alpha_s/(3m_q^2\pi^2)$ and

$$V_1(r) = \left(\frac{\pi}{A}\right)^{3/2} \omega_1\, \eta_1 \exp\left(-\frac{r^2}{4A}\right)$$

$$V_2(r) = \left(\frac{4\pi}{B_1 + B_2 + B_3}\right)^{3/2} \omega_2\, \eta_2 \exp\left(-\frac{r^2}{B_1 + B_2 + B_3}\right)$$

$$V_3(r) = \left(\frac{4\pi}{C_1 + C_2 + C_3}\right)^{3/2} \omega_3\, \eta_3 \exp\left(-\frac{r^2}{C_1 + C_2 + C_3}\right)$$

$$V_4(r) = \left(\frac{4\pi}{2D_1 + D_2}\right)^{3/2} \omega_4\, \eta_4 \exp\left(-\frac{r^2}{2D_1 + D_2}\right).$$

The spin-isospin-combinatoric factors ω_i are $\left[0, \frac{1}{6}, 0, \frac{1}{6}\right]_{I=0}$ and $\left[\frac{1}{3}, \frac{1}{18}, \frac{1}{3}, \frac{1}{18}\right]_{I=1}$. The potentials are repulsive and have a range of about 0.3 fm, as one would expect for the short ranged nuclear hard core.

Acknowledgements

The author was supported by FAPERGS, Brazil.

References

1. R. Machleidt, K. Holinde, Ch. Elster, *Phys. Rep.* **149**, 1 (1987).
2. R. Büttgen, K. Holinde, D. Lohse, A. Müller-Groeling, J. Speth, P. Wyborny, *Z. Phys.* **C46** - Particles and Fields, S167 (1990).
3. J. Bender, H. G. Dosch, H. J. Pirner and H. G. Kruse, *Nucl. Phys.* **A414** (1984)359; D. Mukhopadyay and H. J. Pirner, *Nucl. Phys.* **A442** (1985)605; R. K. Campbell and D. Robson, *Phys. Rev.* **D36** (1987) 2682.
4. D. Hadjimichef, G. Krein, S. Szpigel, J. S. da Veiga, *Phys. Lett.* **B367**, (1996) 317; M. D. Girardeau, G. Krein, D. Hadjimichef, *Mod. Phys. Lett.* **A11** (1996) 1121; D. Hadjimichef, G. Krein, S. Szpigel, J. S. da Veiga, *Ann. of Phys.* **268**, number 1 (1998) 105.
5. T. Barnes, E. S. Swanson, *Phys. Rev.* **C49** (1994) 1166.

QUARK INTERCHANGE EFFECTS IN THE KN INTERACTION

D. HADJIMICHEF
Instituto de Física e Matemática, UFPel, Caixa Postal 354
96010-900, Pelotas, RS, Brazil
dimiter@ufpel.tche.br

J. HAIDENBAUER
Forschungszentrum Jülich, Institut für Kernphysik, D-52425 Jülich, Germany

G. KREIN
Instituto de Física Teórica, Universidade Estadual Paulista
Rua Pamplona, 145 - 01405-900 São Paulo, SP, Brazil

We study the short range repulsion in the KN system due to quark-gluon exchange. Phase shifts for spin-spin, color Coulomb and spin-orbit interactions are presented.

It has been argued [1] that a more suitable investigation of the short range part of the hadronic repulsion could be made in the KN system. In the constituent quark model $K^+ = u\,\bar{s}$ and $K^- = s\,\bar{u}$ while the nucleon is a *uud* or *ddu* state so, in the non-relativistic limit, simple quark exchange mechanism can be applied to the K^+N system. Many groups have studied this system in the S-wave [2]. An extension in order to include higher partial waves, such as the P-waves, is the subject of our research.

The method we employ in order to introduce the quark-gluon degrees of freedom is known as the Fock-Tani formalism [3]. The central idea in the Fock-Tani method is the change of representation concept. The operators of the composite particles are redescribed by ideal operators which obey canonical (anti)commutation relations. These ideal operators act on an enlarged Fock space which is a graded direct product of the original Fock space and an "ideal state space" [3]. The ideal operators correspond to particles with the same quantum numbers as the composite ones of the system. A change of representation is implemented by means of a unitary transformation, which transforms the single-composite states into single-ideal states. When the unitary transformation is applied to the microscopic quark-quark Hamiltonian one obtains the effective interaction. The meson-baryon potential can be obtained by applying in a standard way the Fock-Tani transformed operators to the microscopic Hamiltonian [4]:

$$V_{\text{meson-baryon}} = \sum_{i=1}^{4} V_i(\alpha\beta;\delta\gamma)\, m_\alpha^\dagger\, b_\beta^\dagger\, m_\gamma\, b_\delta \tag{1}$$

and

$$\begin{aligned}
V_1(\alpha\beta;\delta\gamma) &= -3V_{qq}(\mu\nu;\sigma\rho)\, \Phi_\alpha^{*\mu\nu_2}\, \Psi_\beta^{*\nu\mu_2\mu_3}\, \Phi_\gamma^{\rho\nu_2}\, \Psi_\delta^{\sigma\mu_2\mu_3} \\
V_2(\alpha\beta;\delta\gamma) &= -3V_{q\bar{q}}(\mu\nu;\sigma\rho)\, \Phi_\alpha^{*\mu_1\nu}\, \Psi_\beta^{*\mu\mu_2\mu_3}\, \Phi_\gamma^{\sigma\rho}\, \Psi_\delta^{\mu_1\mu_2\mu_3} \\
V_3(\alpha\beta;\delta\gamma) &= -3V_{qq}(\mu\nu;\sigma\rho)\, \Phi_\alpha^{*\mu\nu_2}\, \Psi_\beta^{*\mu_1\nu\mu_3}\, \Phi_\gamma^{\mu_1\nu_2}\, \Psi_\delta^{\sigma\rho\mu_3} \\
V_4(\alpha\beta;\delta\gamma) &= -6V_{q\bar{q}}(\mu\nu;\sigma\rho)\, \Phi_\alpha^{*\nu_1\nu}\, \Psi_\beta^{*\mu_1\mu\mu_3}\, \Phi_\gamma^{\mu_1\rho}\, \Psi_\delta^{\nu_1\sigma\mu_3}.
\end{aligned}$$

The general potential (1) can be specialized to study the KN system [5]. Details of this calculation, for the higher partial waves, will be shown elsewhere[6]. In the present calculation for the S and P waves we will use the

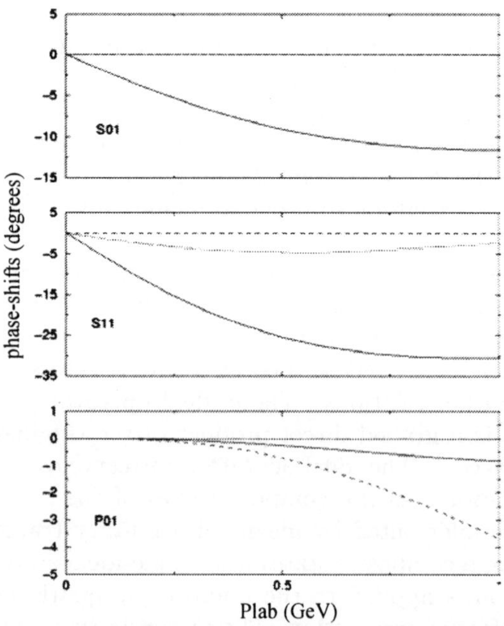

Figure 1. Spin-spin solid line, Coulomb dotted line, spin-orbit dashed line.

microscopic quark-quark potential and quark-antiquark potential written as

$$V^{\rm OGEP}_{qq \text{ or } q\bar{q}} = 4\pi\alpha_s \left[\frac{1}{q^2} - \frac{2}{3m_1m_2}\mathbf{S}_i\cdot\mathbf{S}_j + \frac{i}{q^2}\left\{ \frac{\mathbf{S}_1\cdot(\mathbf{q}\times\mathbf{p}_1)}{2m_1^2} - \frac{\mathbf{S}_1\cdot(\mathbf{q}\times\mathbf{p}_2)}{m_1m_2} - \frac{\mathbf{S}_2\cdot(\mathbf{q}\times\mathbf{p}_2)}{2m_2^2} + \frac{\mathbf{S}_2\cdot(\mathbf{q}\times\mathbf{p}_1)}{m_1m_2} \right\} \right].$$

The first term is the color Coulomb interaction, the second term is the spin-spin interaction and the last one is the spin-orbit interaction. A scattering amplitude can be obtained from the Fock-Tani formalism such that the KN OGEP (One-Gluon Exchange Potential) can give rise, in the Born approximation, to an amplitude h_{fi}^{KN} presented in ref. [7]. The phase shifts can be calculated by

$$\delta_l^{\rm KN} = -\frac{2\pi^2 P_{cm} E_K E_N}{E_K + E_N}\int_{-1}^{1} d\mu\, h_{fi}^{\rm KN} P_l(\mu)$$

where $P_l(\mu)$ is the Legendre polynomial of order l and $E_i = \sqrt{P_{cm}^2 + m_i^2}$.

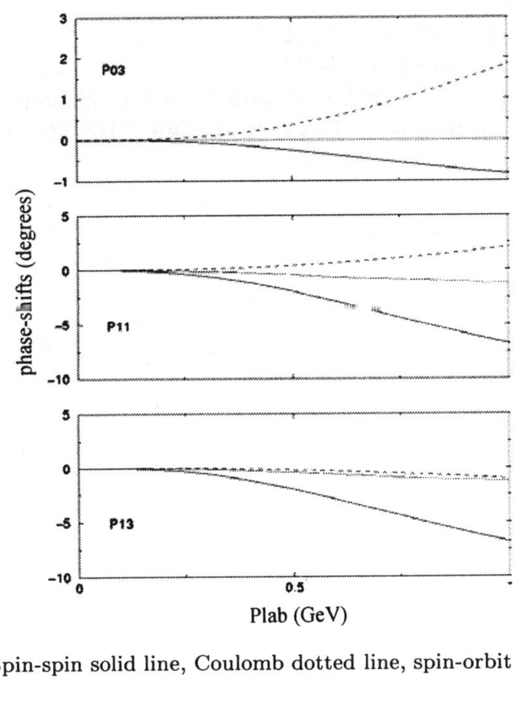

Figure 2. Spin-spin solid line, Coulomb dotted line, spin-orbit dashed line.

In Figs. 1 and 2 phase shifts are presented, as an example, for the following partial waves: S_{01}, S_{11}, P_{01} , P_{11}, P_{03}, P_{13} (where the notation $L_{I,2J}$ is used).

Acknowledgements

Financial support for this work was provided in part by the international exchange program DLR (Germany, BRA W0B 2F) - CNPq (Brazil, 910133/94-8). One of the authors (DH) is also supported by FAPERGS, Brazil.

References

1. R. Büttgen, K. Holinde, D. Lohse, A. Müller-Groeling, J. Speth, P. Wyborny, *Z. Phys.* **C46** - Particles and Fields, S167 (1990).
2. J. Bender, H. G. Dosch, H. J. Pirner and H. G. Kruse, *Nucl. Phys.* **A414**, 359 (1984); D. Mukhopadyay and H. J. Pirner, *Nucl. Phys.* **A442**, 605 (1985); R. K. Campbell and D. Robson, *Phys. Rev* **D36**, 2682 (1987).
3. D. Hadjimichef, G. Krein, S. Szpigel, J. S. da Veiga, *Phys. Lett.* **B367**, 317 (1996); D. Hadjimichef, G. Krein, S. Szpigel, J. S. da Veiga, *Ann. of Phys.* **268**, 105 (1998).
4. D. Hadjimichef, `nucl-th/9912022`.
5. D. Hadjimichef, `hep-ph/0006330`.
6. D. Hadjimichef, J. Haidenbauer, and G. Krein, in preparation.
7. T. Barnes, E. S. Swanson, *Phys. Rev.* **C49**, 1166 (1994).

HEAVY ION COLLISIONS AT INTERMEDIATE ENERGIES IN A QUARK-GLUON EXCHANGE FRAMEWORK

D.T. DA SILVA AND D. HADJIMICHEF

Instituto de Física e Matemática, Universidade Federal de Pelotas (UFPel)
CEP 96010-900 Pelotas, R.S., Brazil
neodanielsil@bol.com.br, dimiter@ufpel.tche.br

Heavy ion collisions at intermediate energies can be studied in the context of the Vlasov-Uehling-Uhlenbeck (VUU) model. One of the main features in this model is the nucleon-nucleon (NN) cross section in the collisional term. Quark interchange plays a role in the NN interaction and its effect can be observed in the cross section. We explore the possibility that quark interchange effects can appear in observables at energies lower than RHIC.

1 Introduction

In the present one of the most important subjects in hadron and particle physics is the identification of the quark-gluon plasma. It is believed that this new phase could be observed in high energy heavy ion collisions. In this regime temperature and density are sufficiently high in order to produce the deconfinement of the quarks and gluons. At lower energies heavy ion collisions can be used to probe the conditions related to the nuclear equation of state. This has been effectively studied in the context of the Vlasov-Uehling-Uhlenbeck (VUU) model [1].

On the other hand it has been shown that quark interchange effects can play a role in the nucleon-nucleon interaction at medium energies [2,3]. In this context one can speculate about the appearance of quark interchange effects in heavy ion collisions at intermediate energies.

In nuclear matter many nucleon quantities become effectives such as the radius and/or the cross section. We pretend to parameterize the NN cross section in the VUU equation by the one obtained from the quark interchange mechanism known as the Fock-Tani formalism and investigate the consequences on the transverse momentum distributions.

2 The Vlasov-Uehling-Uhlenbeck Model

The heavy ion collision is described by the classical distribution function for nuclei. This equation depends on the NN cross section and a potential U obtained from equation of state. The Vlasov-Uehling-Uhlenbeck (VUU) [1] equation is a differential equation for the classical one-body phase-space distribution function $f(\mathbf{r},\mathbf{p},t)$ corresponding to the classical limit of the Wigner function. The VUU equation takes a familiar form

$$\frac{\partial f}{\partial t} + \mathbf{v}\cdot\nabla_r f - \nabla_r U \cdot \nabla_p f = -\int \frac{d^3p_2\, d^3p_1'\, d^3p_2'}{(2\pi)^6}\, \sigma_{\rm NN}\, v_{12}\, \delta^3(p+p_2-p_1'-p_2')$$
$$\times\, [f\, f_2(1-f_1')(1-f_2') - f_1'\, f_2'(1-f)(1-f_2)]$$

where f_2' stands for $f(\mathbf{r},\mathbf{p}_2',t)$ and correspondingly for the other terms. Using the Skyrme ansatz (the potential is given by two- and three-body interactions), one finds a general form for $U = \alpha\rho + \beta\rho^\gamma$ The parameters α,β,γ are fixed[1]. The integration of the VUU equation is done by propagating an ensemble of N events in parallel, with A nucleons simulating the f function. The solution of the transport equation is not obtained analytically but in the form of Monte Carlo simulations.

3 The Fock-Tani Formalism

The central idea in the Fock-Tani method is the change of representation concept. The operators of the composite particles are redescribed by ideal operators which obey canonical anticommutation relations. These ideal operators act on an enlarged Fock space which is a graded direct product of the original Fock space and an "ideal state space". The ideal operators correspond to particles with the same quantum numbers as the composite ones of the system. A change of representation is implemented by means of a unitary transformation, which transforms the single-composite states into single-ideal states. When the unitary transformation is applied on the microscopic quark-quark Hamiltonian one obtains the effective interaction. A scattering amplitude can be obtained from this procedure and the NN OGEP (One-Gluon Exchange Potential) is described by an amplitude h_{fi} presented in Refs.[2,3]. The Fock-Tani cross-section is given by [4]

$$\sigma_{NN} = \frac{4\pi^5\, s}{s-4m^2}\int_{-(s-4m^2)}^{0} dt\, |h_{fi}|^2$$

where s and t are the Mandelstam variables and m is the nucleon mass.

4 Results and Discussion

The cross-section will depend on a parameter a which is the *rms* radius of the nucleon. In Fig. 1 this can be seen for three values of a in comparison

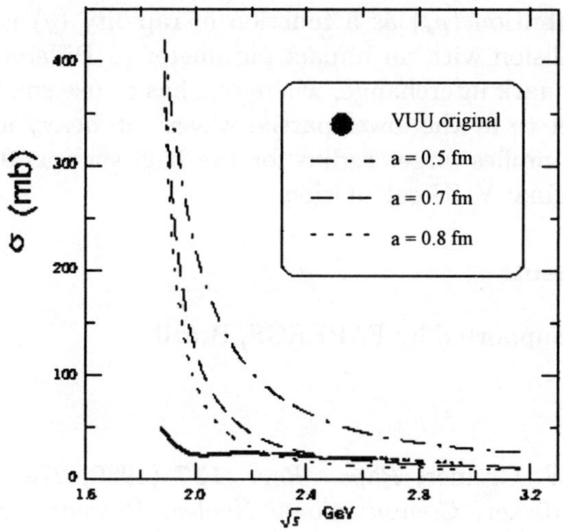

Figure 1. The pp cross-section used by the original VUU model and the modified model with the Fock-Tani cross-section.

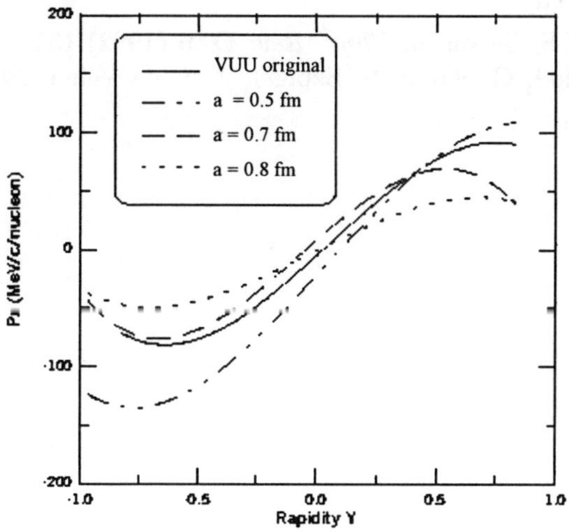

Figure 2. Nb+Nb collision: $E_{lab} = 1050$ MeV ; $b = 3$ fm.

with the original cross-section of the VUU model. The effect on the transverse momentum distribution (p_x) as a function of rapidity (y) is seen in Fig. 2 for a Nb+Nb collision with an impact parameter b. Different from free NN scattering, with quark interchange, where one has to use small a such as 0.45 to 0.6 fm in order to fit the lower partial waves[5], in heavy ion collisions the nuclear medium implies larger radius for the bag, such as 0.7 to 0.8 fm, to approach the original VUU calculation.

Acknowledgements

The authors are supported by FAPERGS, Brazil.

References

1. H. Stöcker,W. Greiner, *Phys. Rep.* **137** (1986) 278; C.Hartnack, H. Kruse, H. Stöcker, *Computational Nuclear Physics 2* K. Langanke, J. A.Maruhn, S. E. Koonin.
2. D. Hadjimichef, G. Krein, S. Szpigel, J. S. da Veiga, *Ann. Phys.* (N.Y.) **268** (1998) 105.
3. T. Barnes, S. Capstick, M.D. Kovarik and E.S. Swanson, *Phys. Rev.* **C48**(1993) 539.
4. T. Barnes, E.S. Swanson, *Phys. Rev.* **D46** (1992) 131.
5. D. Hadjimichef, G. Krein, S. Szpigel, J. S. da Veiga, *Phys. Lett.* **B**367 (1996) 317.

TEST OF NUCLEON-NUCLEON QUARK CLUSTER MODEL

J. P. B. C. DE MELO
Instituto de Física Teórica, UNESP, 01405-900, SP, Brazil
pacheco@ift.unesp.br

M. LACOMBE, B. LOISEAU AND R. VINH MAU
LPTPE, Université Piérre et Marie Curie, 4 Place Jussieu, 75252 Paris CEDEX 05, France

P. DEMETRIU
Institute of Nuclear Physics, NCSR Democritos, GR-15310 Athens, Greece

C. SEMAY
Université de Mons-Hainaut, Faculté des Sciences, 20, Place du Parc, B-7000 Mons, Belgium

The description of the short-range part of the nucleon forces in terms of quark degrees of freedom is tested by supplementing, to the short range quark cluster model, a long range mesonic force well founded theoretically.

Low energy hadron physics corresponds to the non-perturbative phase of QCD. Therefore, a description of low energy hadrons in terms of quark degrees of freedom must take into account in one way or another these non-perturbative effects. Models for these are represented by bag models, string models, soliton models or non relativistic quark models with confining potentials[1,2,3,4]. Numerous attempts to describe the nucleon-nucleon NN) interaction in terms of the quark and gluon degrees of freedom have been made within the framework of the non-relativistic quark-cluster model (QCM)[5]. The results obtained with the QCM show that this approach produces repulsive NN forces at short distances but fails to yield the intermediate-range attraction indispensable for binding nucleons in nuclei. This drawback has been amended by the introduction of meson-exchange potentials between quarks and/or between nucleons (quark clusters). This procedure improves the phase shifts and it is often concluded that the QCM provides a good description of the short-range (SR) part of the NN potential. In our opinion, however, there exist two reasons to believe that the above procedure does not provide a rigorous test of the validity of the QCM. First, the meson-exchange potentials, supplementing the interaction provided by the QCM, contain free parameters which are adjusted to reproduce the phase shifts. Second, the quality of these models are only tested by confronting their predictions with the phase shifts. In a previous paper [6], some of the present authors have studied a NN interaction, the long (LR) and medium (MR) range parts of which are given by the Paris potential[7] and the short (SR) part by a QCM potential developed by Takeuchi[8]. The predictions of this model have been compared with the *pp* observables and the fit has been found to be only very qualitative[6]. In this work, we test similar models on both *pp* and *np* observables looking at an updated set of experimental data and considering other QCM potentials for the short range part of the NN interaction (see also[9]). The interaction considered here is defined as in Ref. [6]. Its LR+MR parts are given by the theoretical part of the Paris potential, while the SR part consists of an equivalent nonlocal potential between two clusters (of three quarks) given by the the Hamiltonian kernel of the resonating-group method (RGM). The Schrödinger equation for two nucleons with mass M is given by

$$-\frac{1}{M}\vec{\nabla}^2\psi(\vec{r}\,) + V_P g(r)\psi(\vec{r}\,) + \int d\vec{r}\,'\; V_{QCM}\sqrt{f(r)f(r')}\psi(\vec{r}\,') = E\;\psi(\vec{r}\,) \quad (1)$$

where $V_P = V_P(r,E)$,$V_{QCM} = V_{QCM}(\vec{r},\vec{r}\,')$, $g(r) = 1 - f(r)$ and $E = T_{lab}/2$ is the relative energy of the two nucleons in the center-of-mass frame. V_P is the theoretical part of the Paris potential given by the one-pion-exchange contribution, the uncorrelated and correlated two-pion-exchange contributions [10]. This potential is local but energy dependent. V_{QCM} is the quark confinement plus the one-gluon-exchange contributions calculated within the RGM framework [5]. A fundamental aspect of the RGM approach in the NN interaction is that the interactions between the quark clusters is non-local. The function $f(r)$ is a cutoff function which makes a separation between the short and medium-long range interactions. The expression for $f(r)$ is[6]: $f(r) = 1/[1 + (r/R_c)^\alpha]$. In Eq. (1) the specific square-root form used for the cutoff in the integral leads to the original cutoff expression when the non-local potential reduces to a local one. The quark mass m_q, the oscillator parameter b_q, the confining strength $a_c^{(n)}$ and the strong coupling constant α_s can be adjusted to reproduce the baryon properties in the baryon sector. Here, one has been neglecting possible contributions from the meson cloud. The root mean square radius of the nucleon is equal to the range parameter of the Gaussian quark wave function, b_q. Here one should also take into account effects of the gluon and meson clouds. The most important relation to satisfy precisely is that of the nucleon stability. The parameters for the different short range potentials considered here are given in Table 1 (see[9] for details). In the work of Takeuchi *et al.* [8], where an effective meson-exchange potential is added to the QCM nucleon-nucleon potential, m_q is taken as $M/3 = 313$ MeV, the nucleon quark core size parameter b_q=0.62 fm and then α_s=1.657 is obtained from the observed $N\Delta$ mass difference. A linear confinement potential is used and the stability condition gives $a_c^{(1)} = 43.84$ MeV fm^{-1}. Yamauchi *et al.* [11] add the one-pion exchange to the quark-quark interaction; some of these changes can be found in Ref. [11] and with a choice of $m_q = M/3 = 313$ MeV and b_q=0.50 fm these authors obtained α_s=0.360 and $a_c^{(2)} = 67.27$ MeV fm^{-2}.

Table 1. QCM Parameters, with m_q in MeV,b_q in fm, $a^{(1)}$ in $MeV fm^{-1}$ and $a^{(2)}$ in $MeV fm^{-2}$.

QCM of	m_q	b_q	α_s	$a_c^{(1)}$	$a_c^{(2)}$
Takeuchi[8]	313.0	0.62	1.657	43.84	
Yamauchi[11]	313.0	0.50	0.360		67.27
Oka[12]	300.0	0.60	1.392	141.20	
Faessler[13]	355.0	0.475	0.970	61.60	

Table 2. Adjusted cutoff radius R_c for the different sets of QCM parameters.

Isospin	1	0
Model with parameters of	R_c(fm) ($^1S_0$ fit at 25 MeV)	R_c(fm) (Deuteron binding fit)
Takeuchi[8]	0.820	0.806
Yamauchi[11]	0.845	0.799
Oka[12]	0.824	0.806
Faessler[13]	0.841	0.806
Harvey[14]	0.900	0.909

In the model considered by Oka and Yazaki [12] $m_q = 300$ MeV, b_q=0.60 fm which gives α_s=1.392. The strength of the confinement potential is determined without taking into account the contribution of the OGEP to the nucleon mass stability, which then gives $a_c^{(1)} = 141.20$ MeV fm^{-1}. These authors found their results almost independent of the value of $a_c^{(1)}$. A similar conclusion is established here: very similar results will be obtained for the Takeuchi *et al.* [8] parameters and those of Oka and Yazaki [12], differing mainly in their $a_c^{(1)}$ values as can be seen in Table 1. Faessler *et al.* [13] with a choice of $m_q = 355$ MeV and b_q=0.48 fm, have α_s=0.97 from $N\ \Delta$ mass difference. The stability condition leads to $a_c^{(1)} = 61.60$ MeV fm^{-1}. The Harvey [14] parameters are taken from Ref. [15]. We have carefully checked our numerical calculation results for the short range nucleon-nucleon QCM potentials. Using the parameters of Takeuchi *et al.* [8] (see Table 1) we found necessary to modify by some numerical coefficients the Yamauchi *et al.* formulae in order to reproduce the QCM potentials of Takeuchi *et al.* [8]. Solving the Schrödinger equation (1) with QCM potentials we have checked that we obtained practically the same isospin 0 and 1 phase shifts as those calculated using the nucleon-nucleon QCM partial wave potentials provided to us by Takeuchi [16] (see Ref. [9]). In the isovector, $T = 1$ channel, R_c is adjusted to reproduce the $^1S_0$ Paris potential [7] phase shift at low energy ($T_{lab} = 25$ MeV), while in the isoscalar $T = 0$ channel, it is adjusted in order to obtain the binding energy for the deuteron. Doing so, we do not reproduce the other low-partial wave Paris Potential phases of total angular momentum $J \leq 2$ except at low energy close to 25 MeV[9]. For the $^1S_0$ phase shifts the agreement between the Takeuchi or Yamauchi models with the Paris potential is fairly good up to 150 MeV.

References

1. G. E. Brown and A. W. Thomas, Phys. Lett. **B 82**, 177 (1979).
2. A. Manohar and H. Georgi, Nucl. Phys. **B 234**, 189 (1984).
3. D. Kalafatis and R. Vinh Mau, Phys. Rev. **D 46** , 3903 (1992).
4. A. de Rújula et al., Phys. Rev. **D 12** , 147 (1975).
5. For a review, see K. Shimizu, Rep. Prog. Phys. **52**, 1-56 (1989).
6. R. Vinh Mau et al., Phys. Rev. Lett. **67**, 1392 (1991).
7. M. Lacombe et al., Phys. Rev. C **21**, 861 (1980).
8. S. Takeuchi et al., Nucl. Phys. **A 504**, 777 (1989).
9. M. Lacombe et al., Phys. Rev. **C 65**, 034004 (2002).
10. W. N. Cottingham et al., Phys. Rev. D **8**, 800 (1973).
11. Y. Yamauchi et al., Few-Body Systems **12**, 69 (1992).
12. M.Oka, K. Yazaki, Phys. lett. **90 B**, 41 (1980).
13. M.A. Faessler et al., Phys. Lett. **112B**, 201 (1982).
14. M. Harvey, Nucl. Phys. **A 352**, 326 (1981).
15. Y. Suzuki and K. T. Hecht, Phys. Rev. **C 27**, 299 (1983).
16. S. Takeuchi, private comunication.
17. R. A. Arndt, M. Pavan *et al.*, data files and instructions at http://said.phys.vt.edu (SAID).

PHOTONUCLEAR PRODUCTION OF HYPERNUCLEUS IN DELTA RESONANCE ENERGY RANGE

S.B. DUARTE, E.C. DE OLIVEIRA AND E.L. MEDEIROS

Centro Brasileiro de Pesquisas Físicas
Rua Dr. Xavier Sigaud, 150, Rio de Janeiro, Brazil, 22290-180
sbd@cbpf.br

M. GONÇALVES

Instituto de Radioproteção e Dosimetria
Av. Salvador Allende, Rio de Janeiro, Brazil, 22780-160

S. DE PINA

Instituto de Física da Universidade de São Paulo
São Paulo- Brazil, P.O.Box 66318 - CEP 05315-970

Photonuclear reactions provide an efficient mechanism of hypernucleus production in the intermediate energy range. We present here the hypernucleus formation probability, as well as the residual nuclei mass and charge distributions, as calculated in a time dependent multicollisional intranuclear cascade model. The emission of hadrons during the rapid phase of the photonuclear reaction and the characteristics of the residual nucleus formed at the end of this phase are discussed.

1 Hypernuclei formation

The main purpose of the present work is to discuss the hypernucleus formation in γ-nucleus reactions at incident energies in the range 1.0–1.4 GeV. The study is done in the context of our multicollisional intranuclear cascade model to deal with photonuclear reactions presented in Refs.[1,2,3]. First, we concentrate on estimating the chances of hypernucleus formation defining the inclusive probability of hyperons to be incorporated in the residual nucleus at the end of the rapid reaction phase as

$$\mathcal{P}_Y = \frac{\sum N_Y(Z,A)}{N_R},$$

where N_Y is the number of residual hypernuclei with charge Z and baryonic number A, occurred after N_R events generated in the Monte Carlo simulation of the photonuclear reaction. In Fig. 1, $\mathcal{P}_Y$ is displayed as a function of the incident energy for γ-carbon (solid line), γ-neon (dashed line), γ-calcium (dotted line), and γ-nickel (dot dashed line) reactions; as it is clear from the figure, this quantity increases with the incident energy, being steeper for

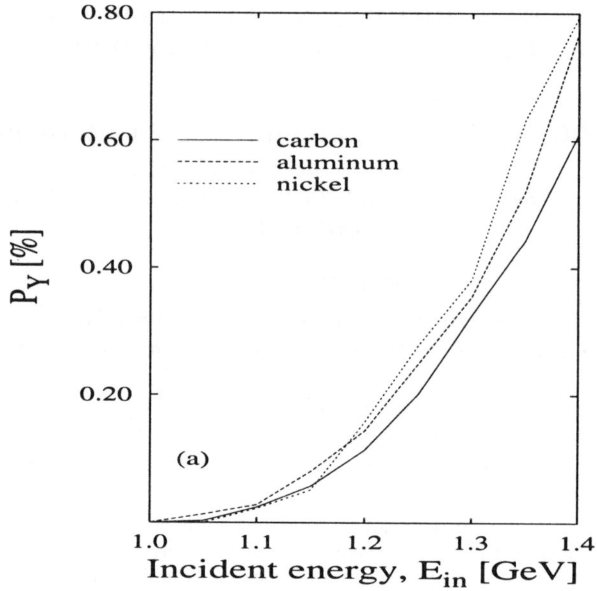

Figure 1. $\mathcal{P}_Y$, inclusive probability of hyperon having produced the residual nucleus as a function of incident energy for the reactions $^{12}C(\pi^+,K^+)$ (solid lines) $^{16}O(\pi^+,K^+)$ (dashed lines), and $^{28}Al(\pi^+,K^+)$ (dotted lines).

heavier nuclei. This result indicates that heavier nuclei are able to better confine the produced hyperon in the nuclear environment.

In an earlier work[4], we had shown that the hyperon production in the primary interaction suppresses the ejection of nucleons by the target nucleus. A stronger suppression is expected to occur when the hypernucleus production takes place. Therefore, it is important to investigate the residual nucleus mass and charge distributions at the end of the rapid phase of the reaction. The relevance of these distributions lies on the fact that the residual nucleus is the start configuration to the evaporation phase of the photonuclear reaction, and the number of baryons in the nuclear matter is determinant in deciding the clustering fissioning path chosen by the system. Also, due to the low kinetic content of the bound hyperon, we expect that this hyperon, once kept bound in the rapid phase of the reaction, will not be able to escape from the nuclear mean field during the evaporation phase. In Fig. 2 we display the distributions of residual hypernuclei in γ-nickel reactions at the incident gamma energy of 1.2 GeV. We observe in part (a) that processes with pion

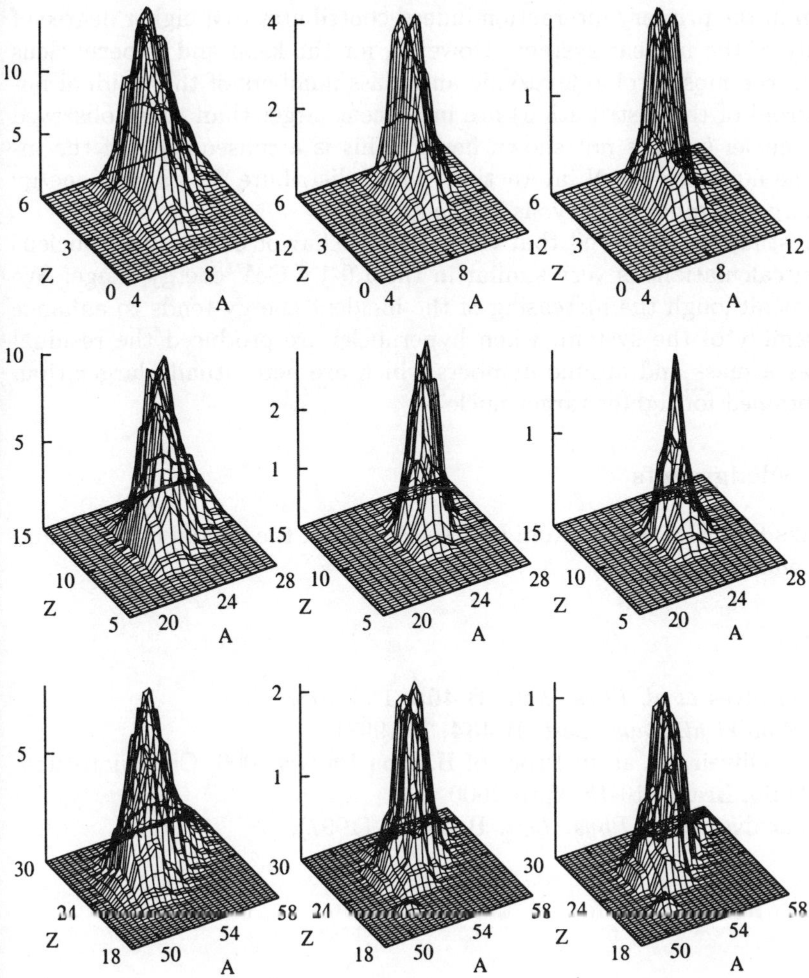

Figure 2. Residual nucleus mass and charge distributions for γ-Carbon, γ-Aluminum, and γ-Nickel reactions at 1.2 GeV incident energy (rows). In the first column the results for inclusive reactions (either a pion or a kaon is produced in the primary interaction) are displayed, in the middle we show results obtained by considering only events with kaon production, but without select the events with hypernuclei formation. The last column the distribution for the hypernucleus formed is explicitly considered. The results in columns are multiplied, respetcively, by 10^2, $2X10^4$, and $2X10^4$.

production in the primary interaction indeed contributes to a higher degree of disassembly of the nuclear system. However, for the kaon and hypernucleus production, the most probable atomic and mass numbers of the residual nucleus (centroid of the distribution) are in percent larger than those observed for lighter nuclei (results not shown here). This is a consequence of the increase in the number of Λ-N interactions which distribute the gamma energy among a larger number of baryons in the system.

In our study we observed that the general behavior of the hypernucleus production calculations is very similar in the 1.0-1.4 GeV energy range. We remark that although the increasing of the incident energy tends to enhance the disassembly of the system, when hypernuclei are produced the residual nucleus has a mass and atomic numbers which are percentually larger than the one obtained for lighter target nuclei.

2 Acknowledgments

M.G. thanks FAPERJ, CNPq, and LNCC, Brazil, for the partial financial and computational support.

References

1. M. Gonçalves *et al*, *Phys. Lett.* B **406**, 1 (1997).
2. S. de Pina *et al*, *Phys. Lett.* B **434**, 1 (1998).
3. E.C. de Oliveira et al, in Proc. of Hadron Physcs 2000, Caraguatatuba, São Paulo, Brazil, 10-15 April 2000.
4. M. Gonçalves *et al*, *Phys. Lett.* B **406**, 1 (1997).

5. Nuclear Matter, Astrophysics, Gravitation

A NUCLEAR MANY-BODY THEORY WITH TSALLIS STATISTICS

A. MESQUITA, M. RAZEIRA, C.A.Z. VASCONCELLOS
Instituto de Física, Universidade Federal do Rio Grande do Sul
91501-970 Porto Alegre, Brasil
falcon@if.ufrgs.br

H. T. COELHO
Departamento de Fsica, Universidade Federal de Pernambuco,
50670-901 Recife, Brazil
heliotcoelho@terra.com.br

M. DILLIG
Institut für Theoretische Physik III, der Universität Erlangen-Nürnberg
D91058 Erlangen, Germany
mdillig@theorie3.physik.uni-erlangen.de

F. PEREIRA
Observatório Nacional
20921-400 Rio de Janeiro, Brasil
flavio@on.br

Within the generalized statistics developed by Tsallis, we study nuclear matter at high densities in the framework of an effective many-body field theory at finite temperature. Thermodynamics properties of nuclear matter are studied considering the Sommerfeld approximation. We perform the calculations by using the nonlinear Boguta and Bodmer model, extended by the inclusion of the fundamental baryon octet and leptonic degrees of freedom. Through the integration of the Tolman-Oppenheimer-Volkoff equations we obtain, beyond the standard relations for the masses of protoneutron stars as functions of the central density, new results for these quantities as a function of temperature. The above formalism, which so far was not yet discussed in the study of many-body dense nuclear systems, predicts substantial differences from calculations based on standard statistics.

1 Introduction

The theoretical treatment of properties of compact stars is provided by relativistic phenomenological approaches based on the formalism developed by Walecka[1], quantum hadrodynamics (QHD). Walecka's formalism represents one of the most important theoretical frameworks to study the highly nonlinear behavior of QCD at the hadronic energy scales. Another important approach is the nonlinear model of Boguta and Bodmer (BB)[2] which was developed to improve the description of static properties of nuclear matter relative to the predictions of the original Walecka model. The EoS provided by model calculations and the Tolman-Oppenheimer-Volkoff (TOV) equations[3,4] characterize the structure of neutron stars and their intrinsic properties as mass, radius, crust extent and moment of inertia[5]. Conventionally, all persisting calculations derive thermodynamic properties of nuclear matter within the Boltzmann-Gibbs statistics. More recently, this framework has been extended and generalized by Tsallis[6]. The main properties associated with his generalized entropy are consistently established, particularly those corresponding to the micro-canonical and canonical ensembles; the Boltzmann-Gibbs statistics is recovered in his treatment by imposing suitable limiting conditions. In this work, we study consequences of this new approach in the study of properties of dense nuclear matter and neutron stars.

2 Nuclear Matter and Stellar Structure

Applying standard formal techniques from field theory and the mean-field approximation in the Boguta-Bodmer model[7] for nuclear matter, we obtain the EoS as a parametric equation $p = p(\varepsilon)$ which relates the energy density

$$\begin{aligned}\varepsilon = {} & \frac{1}{2}m_\sigma^2\sigma^2 + \frac{1}{2}m\omega^2\omega_0^2 + \frac{1}{2}m_\varrho^2\varrho_{03}^2 + \frac{bM}{3}\sigma^3 + \frac{c}{4}\sigma^4 \\ & + \sum_B \frac{2J_B+1}{(2\pi)^3}\int_0^\infty \sqrt{k^2 + M_B^{*2}}[n(\mu_B, T) + \bar{n}(\mu_B, T)]d^3k \\ & + \sum_\lambda \frac{2}{(2\pi)^3}\int_0^\infty \sqrt{k^2 + m_\lambda^2}[n(\mu_\lambda, T) + \bar{n}(\mu_\lambda, T)]d^3k \end{aligned} \tag{1}$$

and pressure

$$\begin{aligned}p = {} & \frac{1}{2}m_\sigma^2\sigma^2 + \frac{1}{2}m_\omega^2\omega_0^2 + \frac{1}{2}m_\varrho^2\varrho_{03}^2 + \frac{bM}{3}\sigma^3 + \frac{c}{4}\sigma^4 \\ & + \frac{1}{3}\sum_B \frac{2J_B+1}{(2\pi)^3}\int_0^\infty \frac{k^2}{\sqrt{k^2 + M_B^{*2}}}[n(\mu_B, T) + \bar{n}(\mu_B, T)]d^3k \\ & + \frac{1}{3}\sum_\lambda \frac{2}{(2\pi)^3}\int_0^\infty \frac{k^2}{\sqrt{k^2 + m_\lambda^2}}[n(\mu_\lambda, T) + \bar{n}(\mu_\lambda, T)]d^3k\,. \end{aligned} \tag{2}$$

In the expressions above, $B = n, p, \Sigma, \Xi, \Lambda$, $\lambda = e, \mu, \nu$, and J_B is the isospin baryon number. The quantities $n(\mu, T)$ and $\bar{n}(\mu, T)$ are the Fermi-Dirac distribution functions for particles and anti-particles, at finite temperature.

The protoneutron star configuration may be determined by combining the nuclear matter EoS with the TOV equations from general relativity [3,4,5], which describe the structure of a static, spherical and isotropic star with pressure $p(r)$ and energy density $\epsilon(r)$. The condition for chemical equilibrium for neutron stars is:

$$\mu_i = b_i \mu_n - q_i(\mu_l - \mu_{\nu l}) \tag{3}$$

where μ_i and μ_l stand for the baryon and lepton chemical potentials, respectively; b_i is the baryon number and the baryon and lepton electrical charges are represented by q_i.

3 Generalization of Boltzmann-Gibbs (BG) Statistics

In generalizing the BG statistics, Tsallis postulated for the entropy[6]

$$S_q = k \frac{1 - \sum_{i=1}^{W} p_i^q}{q - 1} \qquad (q \in \Re) \tag{4}$$

where k is a positive constant, $q \in \Re$ characterizes the particular statistics, W is the total number of microscopic configurations and p_i denotes the probability to find the system in state i. For the canonical ensemble, the generalized form for the entropy asymptotically yields, for $\beta(1-q) \to 0$, the relevant expectation value[8] associated with any observable $\hat{O}$

$$\begin{aligned} \langle \hat{O} \rangle_q &= Z_q^{1-q} \langle \frac{\hat{O}}{1 - (1-q)\beta \hat{H}} \rangle \\ &\sim Z_{BG}^{1-q} \langle \hat{O} \rangle_{BG} \{ 1 + (1-q)\beta \left[\frac{\langle \hat{O}\hat{H} \rangle_{BG}}{\langle \hat{O} \rangle_{BG}} + \frac{\beta}{2} \left[\langle \hat{H}^2 \rangle_{BG} - \frac{\langle \hat{O}\hat{H}^2 \rangle_{BG}}{\langle \hat{O} \rangle_{BG}} \right] \right] \} \end{aligned} \tag{5}$$

where Z_q is the partition function

$$Z_q \sim Z_{BG} 1 - \frac{1}{2}(1-q)\beta^2 \langle \hat{H}^2 \rangle_{BG} \,. \tag{6}$$

In this work our task is to obtain a qualitative idea about the degree of extensivity or non-extensivity of nuclear matter at high densities. We use as an approximation $\sqrt{\langle \hat{H}^2 \rangle_{BG}} \approx \langle \hat{H} \rangle_{BG}$. For the determination of the equation of state, we calculate the energy density and pressure of the system from the grand canonical potential

$$\varepsilon = (e^{\beta \frac{p}{\rho}})^{1-q} \varepsilon_{MFT} \left(1 + (1-q)\beta \varepsilon_{MFT}\right) \; ; \; p_q = p + \frac{1}{2}(1-q)\beta \frac{1}{\rho} p^2 \,. \tag{7}$$

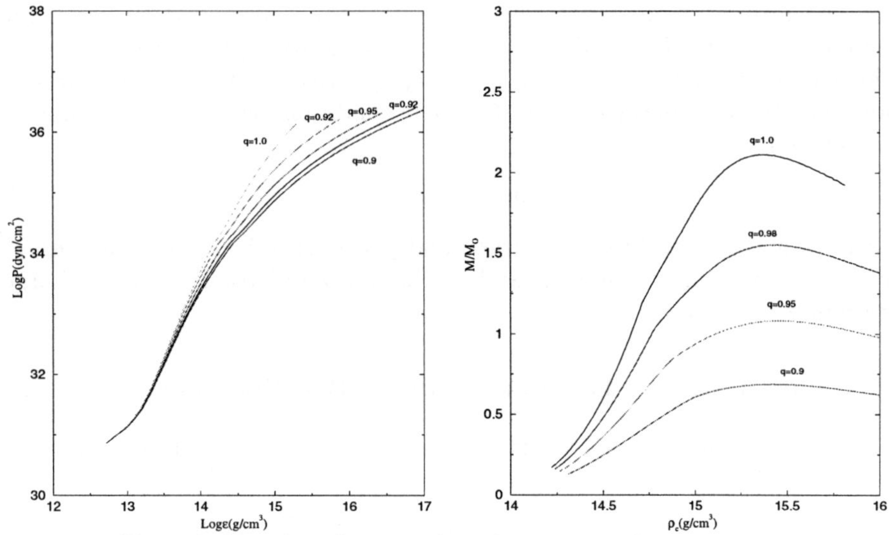

Figure 1. *Equation of state and maximum mass of neutron star.*

4 Results and Conclusions

As an example, characteristic results of our approach are presented in Fig.1. The results in the left panel indicate a significant sensibility of the equation of state in the generalized Tsallis statistics, in spite of the simple approximations adopted. The results in the right panel demonstrate a drastic variation of the maximum mass of the star for a variation of entropy index in the range $q = 0.9$ to $q = 1.0$. In calculations still going on, our primary goal is a more precise determination of $\langle \hat{H}^2 \rangle_{BG}$ as a key quantity for a better understanding of the role of Tsallis statistics in dense nuclear matter.

References

1. J.D. Walecka and B. D. Serot, *Advances in Nuclear Physics* **16** (Plenum Press, 1986).
2. J. Boguta and A. R. Bodmer, *Nucl. Phys.* **A292**, 413 (1977).
3. R.C. Tolman, *Phys. Rev.* **55**, 364 (1939).
4. J.R. Oppenheimer and G.M. Volkoff, *Phys. Rev.* **55**, 374 (1939).
5. N. K. Glendenning, *Compact Stars* (Springer-Verlang, 1997).
6. C. Tsallis, *J. Stat. Phys.* **52**, 479 (1998).
7. G.F. Marranghello, C.A.Z. Vasconcellos, M. Dillig and J.A. de Freitas Pacheco, *Int. J. Mod. Phys.* **E11**, 83 (2002).
8. C. Tsallis, F. C. Sá Barreto, Edwin E. Loh, *Phys. Rev.* **E52**, 144 (1995).

PHASE TRANSITIONS IN RELATIVISTIC MODELS: REVISITING THE NOLEN-SCHIFFER ANOMALY

D.P. MENEZES

Dep. de Física - CFM - Universidade Federal de Santa Catarina - Florianópolis - SC - CP. 476 - CEP 88.040 - 900 - Brazil

debora@server.fsc.ufsc.br

C. PROVIDÊNCIA

Centro de Física Teórica - Dep. de Física - Universidade de Coimbra - P-3004 - 516 Coimbra - Portugal

We use the non-linear Walecka model in a Thomas-Fermi approximation to investigate the effects of the $\rho - \omega$ mixing term in infinite nuclear matter and in finite nuclei. For finite nuclei the contribution of the isospin mixing term is very large as compared with the expected value to solve the Nolen-Schiffer anomaly.

1 introduction

Problems related with the liquid-gas phase transition and possible droplet formation in a vapor system of asymmetric nuclear matter at finite temperature within the framework of relativistic models [1] are of interest. The properties of the arising droplets as neutron skin thickness, surface energy, central density etc, can be evaluated. On the other hand, the Nolen-Schiffer anomaly (NSA) is known as the failure of nuclear theory in explaining the experimental mass differences between mirror nuclei. Nolen and Schiffer[2] have concluded in that if potential models are forced to fit experimental charge radii, the calculated Coulomb-energy shifts are too low by $\sim 7\%$. Many non-relativistic models have been invoked in order to explain the NSA without complete success. Some years ago, relativistic models were also used to give possible explanations for the mass difference problem, as in Refs.[3] and [4]. The $\rho - \omega$ mixing term has been held responsible for explaining about 85% of the discrepancy in the masses of mirror nuclei.

2 Non-linear Walecka model and the Thomas-Fermi approximation

In the present work we study the importance of the $\rho - \omega$ mixing in asymmetric nuclear matter and also in finite nuclei in the context of the non-linear Walecka model within the Thomas-Fermi approximation for the TM1[5] and NL3 [6] parameter sets. First of all, we discuss the effects of the $\rho-\omega$ mixing term in infinite nuclear matter. For this purpose, we perform a mean field approximation, where the fields are taken as their expectation values, in order study the influence of the $\rho-\omega$ mixing in nuclear matter properties. The nuclear bulk symmetry energy is defined [7] as $\mathcal{E}_{sym} = \frac{k_F^2}{6\epsilon_F}\rho_B + \mathcal{E}_{sym1}$, where $\epsilon_F = \sqrt{k_F^2 + M_0^2}$, $M_0 = f(M,0)$, and the last term comes from the ω and ρ meson contributions. For both parameterizations NL3 and TM1 the first order correction coming from the $\rho-\omega$ mixing term is quadratic in Λ^2. For the NL3 parameterization and to second order in Λ we get a simple expression linear in the baryonic density, $\mathcal{E}_{sym1} = \frac{g_\rho^2}{8m_\rho^2}\rho_B\left(1+\Lambda'^2\right) + \mathcal{O}(\Lambda'^4)$. For this force, the influence of the $\rho-\omega$ term on the symmetry energy is negligible, even if we consider a value of Λ ten times larger than the presently accepted experimental value. However, with the TM1, the lowest order correction in Λ has a non-linear dependence in ρ_B. It can be seen that the effect of the non-linear terms is already non-negligible for the saturation density. Next we investigate its effects on the binding energy. The binding energy in terms of the baryon density for the TM1 and NL3 parameterizations is displayed in Figure 1 for $Y_p = 0$, where the very small influence of the mixing term is seen. No influence at all is detected in symmetric nuclear matter.

We then move to finite nuclei. The coupled differential equations obtained from the Euler-Lagrange formalism are solved numerically and all relevant quantities which depend on the fields are then calculated. In order to obtain the binding energy of the nuclei, one needs to calculate the energy density, which is given in terms of the meson fields. The total energy per particle is given by $\varepsilon = \int \frac{\mathcal{E}(r)}{A} d^3r$, where A is the total number of particles in the volume under consideration, and ε_N, $\varepsilon_{\mathrm{coul.}}$ denote, respectively, the nuclear and electromagnetic contributions. The binding energy per nucleon is $B/A = \varepsilon - M$, where $A = Z + N$.

3 Results

We have studied ^{67}As, ^{79}Y, ^{81}Zr and their mirror nuclei and determined the influence of Λ on various properties of these mirror nuclei. According to our results the influence of the mixing term on the binding energy (always given in MeV/A) is not negligible. For both parameterizations, the $\rho-\omega$ mixing term gives more binding to the neutrons and less to the protons, reinforcing the effect of the Coulomb field. The difference between different nucleons binding energy $B_p/Z - B_n/N$, calculated with the inclusion of the mixing term, always increases in mirror nuclei. The effect of the mixing term on properties such as the surface energy, charge radius or neutron skin is either zero or very small. The neutron skin always decreases with the $\rho - \omega$ mixing term. We have also determined the energy difference calculated within the model, excluding the Coulomb field. The $\rho - \omega$ mixing term gives more binding to the nuclei with $N > Z$, i.e., it reinforces the Coulomb field effect. We have also determined the same values divided by the asymmetry parameter δ and the Coulomb energy difference or equivalently, the mass difference per particle $(M_{Z>} - M_{Z<})/A$ for the same mirror pairs obtained from experimental data[8]. These data are of the same order of magnitude of the data stressed before, however, do not include the Coulomb field. From our results we also conclude that the Thomas-Fermi approximation is too crude to reproduce the mass differences of mirror nuclei with parameterizations of the non-linear Walecka model obtained within the relativistic mean-field framework. However we believe the main effects of the $\rho - \omega$ term can be understood in a qualitative way. We conclude that the $\rho - \omega$ mixing is much more important in finite nuclei than in infinite nuclear matter where the influence of the mixing term for $Y_p \sim 0.5$ is very small. Hence, within our formalism, the values coming from the $\rho - \omega$ interaction are of the correct order in explaining the NSA only in infinite nuclear matter. Within a naive dimensional analysis the importance of the isospin symmetry breaking term is comparable to the non linear ω term of the TM1 parameterization. In the spirit of an effective field theory the term $\rho - \omega$, which represents a small isospin symmetry-breaking term, should be kept in the parameterization of the nuclear force within the relativistic mean-field framework.

Acknowledgments

This work was partially supported by CNPq - Brazil and CFT - Portugal under the contract POCTI/35308/FIS/2000.

References

1. C. Providência, D.P. Menezes and L. Brito, Nucl. Phys. **A** - in press - nucl-th/0108058; D.P. Menezes and C. Providência, Phys. Rev. **C 64** (2001) 044306; D.P. Menezes and C. Providência, Nucl. Phys. **A 650** (1999) 283, D.P. Menezes and C. Providência, Phys. Rev. **C 60** (1999) 024313.
2. J. A. Nolen and J. P: Schiffer, Ann. Rev. Nucl. Sci. **19** (1969) 471.
3. E.M. Henley and G. Krein, Phys. Rev. Lett. **62** (1989) 2586.
4. G. Krein, D.P. Menezes and M. Nielsen, Phys. Lett. **B 294** (1992) 7.

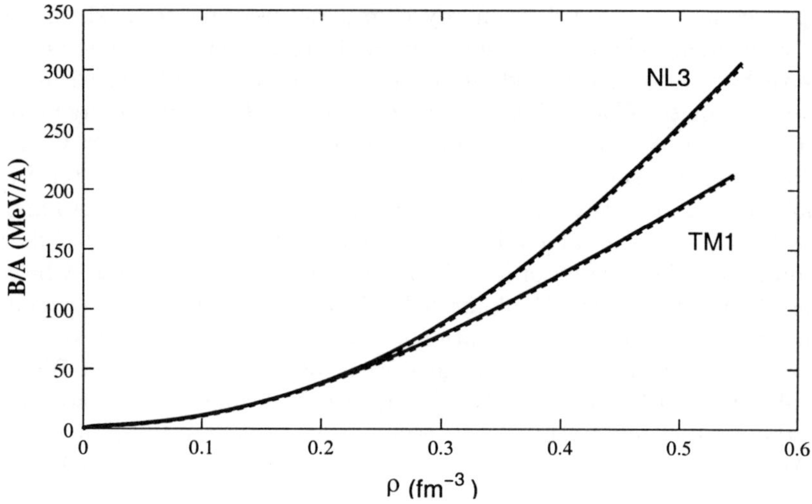

Figure 1. Binding energy with $\Lambda = 0$ (solid line) and $\Lambda = -4500$ MeV^2 (dotted line).

5. K. Sumiyoshi, H. Kuwabara, H. Toki, Nucl. Phys. **A 581** (1995) 725.
6. G. A. Lalazissis, J. König and P. Ring, Phys. Rev. **C 55** (1997) 540.
7. B.-A. Li, C. M. Ko and W. Bauer, Int. J. Mod. Phys. E **7** (1997) 147.
8. Jonghwa Chang, http://sutekh.nd.rl.ac.uk/CoN.

MODIFIED LINEAR SIGMA MODEL AT FINITE FERMIONIC DENSITY WITH DYNAMICAL SYMMETRY BREAKING

FÁBIO L. BRAGHIN
Instituto de Física, Universidade de São Paulo,
C.P. 66.318; CEP 05315-970; São Paulo - SP, Brazil
braghin@if.usp.br

The linear sigma model invariant under $O(4)$ transformations coupled to baryons and to a massive vector gauge meson is considered for the description of a finite baryonic density system. The Euler-Lagrange equations are considered such that a stability equation for a bound system is satisfied with particular prescriptions. All the bosonic components are found to have non-zero expected classical values (condensates) at finite density corresponding to dynamical symmetry breaking.

1 Introduction

Quantum Cromodynamics (QCD) has a very complex non-abelian structure and strong coupling constants at low energies being very difficult to obtain exact solutions. One is therefore lead to construct effective models which respect the main properties and symmetries of the QCD for those energy ranges. In the vacuum, the lightest strong interacting particles are known to respect, approximately at least, chiral symmetry $SU_L(2) \times SU_R(2)$ which is spontaneously broken down to $SU(2)$. Pions, whose masses are small in the hadronic scale, are viewed as the Goldstone bosons of such SSB. The vacuum acquires a non trivial structure due to the formation of a quark-anti-quark condensate $< \bar{q}q >$, the order parameter of the Chiral SSB. These features can be taken into account via sigma models which, in the linear realization with mesons, implement chiral symmetry with two fields: the (pseudo-scalars) pions and the (scalar) sigma [1]. At finite density QCD is known, and expected, to have a very complex phase diagram with the appearance of other condensates [2]. With different approaches finite density QCD and effective models have been intensively studied in the last years.

In this work we study the O(N) Linear Sigma Model (LSM) coupled to baryons which form an infinite bound stable system with a massive vector meson corresponding to a particular case of that developed in [3]. The exact field equations are truncated to allow for analytical solutions by considering particular prescriptions for the stability condition of the system.

2 Linear Sigma Model with Vector Meson at Finite Density

The Lagrangian density of the chirally symmetric Linear Sigma Model for nucleons $N(\mathbf{x})$, sigma and pions $(\sigma, \vec{\pi})$ covariantly coupled to a massive gauge vector meson V_μ is:

$$\mathcal{L} = \bar{N}(\mathbf{x}) \left(i\gamma_\mu \mathcal{D}^\mu - g_S(\sigma + i\gamma_5 \vec{\tau}.\vec{\pi}) - \mu_0\gamma_0\right) N(\mathbf{x}) + \frac{1}{2}\left(\partial_\mu\sigma\partial^\mu\sigma + \partial_\mu\vec{\pi}.\partial^\mu\vec{\pi}\right) + \\ -\frac{1}{4}F_{\mu\nu}F^{\mu\nu} + \frac{\lambda}{4}\left((\sigma)^2 + (\vec{\pi})^2 - v^2\right)^2 + \frac{1}{2}m_V^2 V_\mu V^\mu, \tag{1}$$

where the covariant gauge derivative is $\mathcal{D}^\mu = \partial^\mu - ig_V V^\mu$ and the gauge invariant tensor is $F^{\mu\nu} = \partial^\mu V^\nu - \partial^\nu V^\mu$. μ_0 is the chemical potential, g_V, g_S and λ are the coupling constants and $v = f_\pi$ is the pion decay constant in the vacuum. The coupling of the time component of the vector meson to the baryons is equivalent to the definition of a chemical potential. We consider the possible existence of classical components (condensates) for all the mesonic fields and look for solutions.

The baryonic degrees of freedom are incorporated by means of the densities: scalar, baryonic and pseudo-scalar densities (ρ_s, ρ_B and ρ_{PS}). The fermionic density can be approximately written in terms of the baryonic density as [4]:

$$\rho_f = 4\int^{k_F} \frac{d^3k}{(2\pi)^3}\sqrt{\mathbf{k}^2 + (M^*)^2}, \qquad \rho_B = 4\int^{k_F} \frac{d^3k}{(2\pi)^3}. \tag{2}$$

In these expressions k_F is the nucleon momentum at the Fermi surface.

In the homogeneous case the Euler-Lagrange equations define the minimum of the potential for these fields. The following set of equations is obtained for the sigma, pions and vector meson fields:

$$\begin{aligned} \bar{\sigma}\lambda\left(\bar{\sigma}^2 + \bar{\vec{\pi}}^2 - v^2\right) + \frac{d\rho_f}{d\bar{\sigma}} &= 0; \\ \lambda\bar{\pi}_a\left(\bar{\vec{\pi}}^2 + \bar{\sigma}^2 - v^2\right) + \frac{d\rho_f}{d\bar{\pi}_a} &= 0; \\ g_V\left(\rho_B + V_0\frac{d\rho_B}{dV_0}\right) - m_V^2 V_0 &= 0. \end{aligned} \tag{3}$$

In this way self-consistent solutions of the interacting fields can be found analytically by solving partial differential equations. We could impose still further self consistency considering mutual (implicit) dependence, as for example, by taking simultaneously $\rho_f = \rho_f(\sigma, \vec{\pi}, V_0)$ and $V_0 = V_0(\rho_B, \sigma, \vec{\pi})$ and so on. The system of coupled equations becomes hard to solve.

The equation for the classical value of the pion field (condensate) has non zero solutions whenever the derivative of the fermionic density with respect to the pion field is non zero. This occurs because

$$\frac{d\rho_f}{d\bar{\pi}_a} = \frac{d\rho_f}{d\bar{\sigma}} \frac{d\bar{\sigma}}{d\bar{\pi}_a} \neq 0. \tag{4}$$

Therefore the complete self consistency of the equations makes the pion classical field to be non zero. Besides that the effective mass is explicitly modified with the pion condensate causing a splitting between the neutron and proton masses and the possibility of oscillations between these two isospin states [3]. The physical masses for each of the bosonic quanta (ϕ_i) are found by considering the usual deviations around the minimum of each effective potential: $\phi_i \to <\phi_i> + \delta_i$. The pion mass can be set to zero establishing a relationship between $\bar{\sigma}(vac)$ and v.

The Euler-Lagrange equation for V_0 is faced as a differential equation of the baryonic density ρ_B as a function of V_0. We obtain the following solution:

$$V_0(\rho_B) = \frac{-g_V \rho_B \pm \sqrt{g_V^2 \rho_B^2 - 2C_V \rho_B m_V^2}}{m_V^2}, \tag{5}$$

where C_V is a negative constant. This constant will be the only contribution of the vector meson sector to the energy density $\mathcal{H}_V = C_V \rho_B$. In the limit of zero density we get $V_0 \to 0$.

The stability equation for nuclear matter is solved by separating the total binding energy $E/A = \mathcal{H}/\rho_B$ into three parts (fermionic, for the vector meson and for the spin zero fields) which are considered to be approximately independent in the stability equation: $\mathcal{H} = \rho_f + \mathcal{H}_V + \mathcal{H}_{LSM}$. For this expression we therefore consider

$$\begin{aligned} \frac{d\mathcal{H}_V}{d\rho_B} &= \frac{\mathcal{H}_V(\rho_0)}{\rho} \\ \frac{d\rho_f}{d\rho_B} &= \frac{\rho_f(\rho_0)}{\rho} \\ \frac{d\mathcal{H}_{LSM}}{d\rho_B} &= \frac{\mathcal{H}_{LSM}(\rho_0)}{\rho}. \end{aligned} \tag{6}$$

The solutions for these equations can also be solutions for the equations of motion depending on the constants to be fixed when solving first order differential equations. The full reliabity of this approximation is under investigation. Within this approximation the incompressibility modulus of the medium is analytically calculated [3].

With the present calculation we found non zero pion and sigma classical fields at finite baryonic (stable) density. From equations (3) an approximate value for the squared classical value of the pion can be given by

$$\bar{\vec{\pi}}^2 = \frac{\sigma^2(\sigma^2 - v^2)}{4(-\frac{\sigma^2}{2} \pm v^2)}. \tag{7}$$

As boundary conditions we use that in the vacuum $\bar{\sigma} = v$ and $\bar{\vec{\pi}}^2 = 0$, representing the chiral symmetry spontaneously broken and exact isospin symmetry. We found that the values of the classical values $\bar{\sigma}$ can increase or diminish with relation to the value of the vacuum, although the most widespread and reasonable solution is a tendency towards zero, the restoration of chiral SSB. On the other hand the pion classical field (condensate) becomes non zero breaking isospin symmetry. Extensive numerical studies will be reported elsewhere [3].

Acknowledgement

This work was supported by FAPESP, Brazil.

References

1. M. Gell Mann and M. Lévy, *Nuovo Cimento* **16**, 705 (1960).
2. R. Rapp, T. Schäfer, E. Shuryak, M. Velkovsky, *Phys. Rev. Lett.* 53 (1998); M. Alford, K. Rajagopal, F. Wilczek, *Phys. Lett.* **B 422**, 247 (1998).
3. F.L. Braghin, hep-ph/0206107, *submitted to publication* and *in preparation.*
4. B.D. Serot and J.D. Walecka, *Adv. Nucl. Phys.* **16**, 1 (1986) ; *Int. Jour. Mod. Phys.* **E6**, 515 (1997) and references therein.
5. F.L. Braghin, *Phys. Rev.* **D 64**, 125001 (2001).

FINITE TEMPERATURE EFFECTIVE FIELD THEORY FOR PROTONEUTRON STARS

G. F. MARRANGHELLO, C.A.Z. VASCONCELLOS
Instituto de Física, Universidade Federal do Rio Grande do Sul
91501-970 Porto Alegre, Brazil
gfm@if.ufrgs.br, cesarzen@if.ufrgs.br

M. DILLIG
Institut für Theoretische Physik III, der Universität Erlangen-Nürnberg
D91058 Erlangen, Germany
mdilig@theorie3.physik.uni-erlangen.de

J. A. DE FREITAS PACHECO
Observatoire de la Côte d'Azur, Nice, France
pacheco@obs-nice.fr

H. T. COELHO
Departamento de Fsica, Universidade Federal de Pernambuco,
50670-901 Recife, Brazil
heliotcoelho@terra.com.br

Thermodynamics properties of nuclear matter are studied in the framework of an effective many-body field theory at finite temperature, considering the Sommerfeld approximation. We perform the calculations by using the nonlinear Boguta and Bodmer model, extended by the inclusion of the fundamental baryon octet and leptonic degrees of freedom. Trapped neutrinos are also included in order to describe protoneutron star properties through the integration of the Tolman-Oppenheimer-Volkoff equations, from which we obtain, beyond the standard relations for the masses and radii of protoneutron stars as functions of the central density, new results of those quantities as a function of temperature. Our predictions include: the determination of the limiting mass of protoneutron stars, new structural aspects on the nuclear matter phase transition via the behaviour of the specific heat and, through the inclusion of quark degrees of freedom, the properties of a hadron-quark phase transition and hybrid protoneutron stars.

In the last few decades, studies on the evolution of compact stars became a central issue related to theoretical and experimental research on the equation of state (EoS) of dense matter. In 1967, the first pulsar was observed[1] and, based on characteristic observational features, this object was identified as a rotating neutron star. Since this direct evidence for the existence of neutron stars, nuclear models have been widely employed for the description of the internal structure, composition, dynamics and evolution of these massive compact stars. As under the pull of gravity the energy density in the core of compact stars is thought to approach or even exceed 1 GeV/fm^3, i.e. more than 6 times the density of nuclear matter, predictions on their structure depends sensitively on the equation of state for very dense matter. As the EoS describes how the energy density and pressure vary with density and temperature, it should be also able to describe different phases, from gaseous nuclear matter and liquid nuclei structure up to the deconfinement transition. The EoS provided by model calculations and the Tolman-Oppenheimer-Volkoff (TOV) equations[2,3] characterize the structure of the star, whose properties like mass, radius, crust extent, moment of inertia are then susceptible to be compared to observations[4]. Moreover, these models should also provide the description of relevant dynamical properties of compact stars as the rotational period, the emission of neutrinos and gravitational waves as well as more fundamental aspects involving the evolution, internal composition and structure of these stellar objects.

Nuclear matter models should be able to account for neutron stars at least as massive as the most massive observed pulsar, representing an important constraint on any proposed EoS for dense matter. At present, the best mass determination corresponds to the binary pulsar PSR 1913+16 ($M = 1.444 M_{\odot}$). Other estimates based on quasi-periodic oscillations (QPOs), observed in the X-ray emission of low mass binaries (LMXBs) suggest values up to $2.0 - 2.2 M_{\odot}$, but these results are quite uncertain since they are model dependent. More recently, from new radial velocity data, the mass of the neutron star associated to the X-ray source in the system Vela X-1[5] was recalculated with a larger accuracy than preceding estimates: the resulting value is $1.86 \pm 0.16 M_{\odot}$. For a given family of stars, a stiffer EoS predicts higher limiting masses and larger radii ($R > 10 - 12$ km). In particular, recent results[6] point to an EoS stiffer than those expected on the basis of a non-relativistic approaches, in better agreement with Vela X-1 results. The relativistic phenomenological approach developed by Walecka[7], quantum hadrodynamics (QHD), represents one of the most important approaches to the highly nonlinear behavior of QCD at the hadronic energy scales. QHD is a relativistic quantum field theory based on a local lagrangian density which contains the nucleon and two meson fields, the scalar-isoscalar, attractive σ and the vector-isoscalar, repulsive ω, as the relevant effective degrees of freedom. This model provides a thermodynamically consistent theoretical framework for the description of bulk static properties of strong interacting many-body nuclear systems under extreme density conditions. The nonlinear model of Boguta and Bodmer[8] was developed to improve the description of static properties of nuclear matter, compared to the original Walecka model.

With the introduction of cubic and quartic scalar meson self-interactions in the lagrangian density, the model has two additional parameters which allow and provide sufficient flexibility to reproduce, at saturation density, currently accepted values for the compression modulus of symmetric nuclear matter and the nucleon effective mass[6,9].

We investigate static properties of nuclear matter in the high density domain by using a description based on the BB model. As in its original form the model is not yet suitable for an extrapolation to higher densities, we considerably extent it by considering, at finite temperature, in a consistent way, the fundamental baryon octet, lepton degrees of freedom and trapped neutrinos, the inclusion of the ϱ-meson together with the equations for chemical equilibrium and charge neutrality. In an additional step, we surpass the limit of hadron degrees of freedom by including quarks and gluons in order to allow a possible transition to quark matter. Within this framework, we focus on a particular process in the evolution of compact stars, preceding the supernova explosion: the formation of a hot collapsed core or a *protoneutron star*[10,11], which can reach temperatures up to a few tens of MeV. These hot and dense relativistic objects are formed in a type-II supernova explosion and evolve to cold neutron stars, basically through neutrino emission. Here, the very dense and hot cores are able to trap neutrinos, imparting momentum to the outer layers and then cooling as reaching a quasi-equilibrium state. The evolution of protoneutron stars is studied, extending previous investigations to non-vanishing temperatures. Global static properties as masses and radii are then computed as a function of central densities and temperatures.

The temperature dependence of the star mass is shown in Fig.1, on the left panel, and is represented qualitatively by $M \approx M_{T=0}(1 + \pi^2\kappa^2T^2/\epsilon_F^2)$ where ϵ_F and κ are the Fermi energy and Boltzmann constant, respectively. As the novel feature we find an absolute limiting value for the protoneutron star mass, followed by a subsequent decrease of this maximum value with the increasing of temperature. A liquid-gas phase transition is predicted in various relativistic nuclear models such as Walecka's QHD-I, Boguta-Bodmer and others when symmetric nuclear matter or pure neutron matter are considered. However, for asymmetric nuclear matter with the inclusion of the ϱ meson, this phase transition disappears. In this case, the energy used to generate the liquid-gas phase transition is now used in hyperon production processes, resulting in a phase transition of asymmetric nuclear matter into hyperon matter. As a signal of this phase transition, discontinuities on the specific heat can be found as a function of temperature as shown in Fig.1, on the right panel. (For a detailed discussion of this topic see Ref.[12])

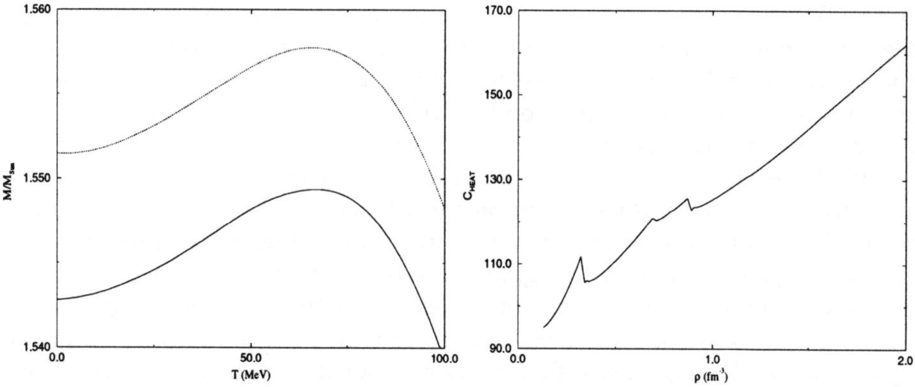

Figure 1. On the left panel: protoneutron star maximum mass as a function of temperature for lepton fraction $Y_l = 0.4$ (solid line) and $Y_L = 0$ (dotted line). On the right panel: specific heat as a function of baryon density for $T = 10\,MeV$.

References

1. A. Hewish, S. J. Bell, J. D. H. Pikington, P. F. Scott and R. A. Collins, *Nature* **217**, 709 (1968).
2. R. C. Tolman, *Phys. Rev.* **55**, 364 (1939).
3. J. R. Oppenheimer and G. M. Volkoff, *Phys. Rev.* **55**, 374 (1939).
4. N. K. Glendenning, *Compact Stars* (Springer-Verlag, 1997).
5. O. Barziv, L. Kaper, M.H. van Kerkwijk, J.H. Telting, J. van Paradijs, astro-ph/0108237.
6. A. R. Taurines, C. A. Z. Vasconcellos, M. Malheiro and M. Chiapparini, *Phys. Rev.* **C63**, 065801, 1 (2001); *Mod. Phys. Lett.* **A15**, 29, 1789 (2000).
7. J. D. Walecka and B. D. Serot, *Adv. Nucl. Phys.* **16** (Plenum Press, 1986).
8. J. Boguta and A. R. Bodmer, *Nucl. Phys.* **A292**, 413 (1977).
9. S. S. Rocha, A. R. Taurines, C. A. Z. Vasconcellos, M. Dillig and M. B. Pinto, *Mod. Phys. Lett.* **A17**, 21, 1335 (2002).
10. M. Prakash, I. Bombaci, M. Prakash, P. J. Ellis, J. M. Lattimer, J. M. and R. Korren, *Phys. Rep.* **280**, 1 (1997).
11. N. K. Glendenning, F. Weber, S. A. Moszkowski, *Phys. Rev.* **C45**, 844 (1992).
12. For a more complete version see: G. F. Marranghello, C. A. Z. Vasconcellos, M. Dillig, J. A. de Freitas Pacheco, *Int. Journ. Mod. Phys.* **E11**, 2, 83 (2002).

GRAVITATIONAL WAVE EMISSION FROM PHASE TRANSITION INSIDE NEUTRON STARS

G. F. MARRANGHELLO AND C. A. Z. VASCONCELLOS
Instituto de Física, Universidade Federal do Rio Grande do Sul
91501-970 Porto Alegre, Brazil
gfm@if.ufrgs.br, cesarzen@if.ufrgs.br

M. DILLIG
Institut für Theoretische Physik III, der Universität Erlangen-Nürnberg
D91058 Erlangen, Germany
mdilig@theorie3.physik.uni-erlangen.de

J. A. DE FREITAS PACHECO
Observatoire de la Côte d'Azur, Nice, France
pacheco@obs-nice.fr

H. T. COELHO
Departamento de Fsica, Universidade Federal de Pernambuco,
50670-901 Recife, Brazil
heliotcoelho@terra.com.br

We review the detectability of gravitational waves generated by oscillations excited during a phase transition from hadronic matter to deconfined quark-gluon matter in the core of a neutron star. Neutron star properties are computed using a Boguta-Bodmer based model and the MIT bag model. The maximum energy available to excite mechanical oscillations in the star is estimated as the energy difference between the configurations with and without a quark-gluon matter core. We find, on basis of the planned sensitivity of present laser interferometers (VIRGO or LIGO I) and those of the next generation (LIGO II), results indicating the potential detectability of neutron stars as sources of gravitational waves. From these results we conclude that the maximum distance probed by the detectors of the first generation is well beyond M31, whereas the second generation detectors will probably see phase transition implications at distances two times larger, but certainly not yet attaining the Virgo cluster.

1 Introduction

The first generation of large gravitational interferometric detectors as the French-Italian VIRGO and the American LIGO, should be fully operational within one or two years. The best signal-to-noise (S/N) ratio that can be achieved from these detectors implies the use of matched-filter techniques, that require a priori the knowledge of the signal waveform. Thus, the identification of possible sources for gravitational waves having a well defined signal is a relevant problem in the detection strategy.

Neutron stars are certainly one of the most popular potential sources of gravitational waves (GWs), since they can emit by different mechanisms for GW signals having a well known waveform. Rotating neutron stars may have a time-varying quadrupole moment and hence radiate GWs, by either having a triaxial shape or a misalignment between the symmetry and the spin axis, which produces a precessional motion[1,2,3]. Moreover, fast rotating protoneutron stars may develop different instabilities such as the so-called Chandrasekhar-Friedman-Schutz (CFS) instability[4,5], responsible for the excitation of density waves traveling around the star in the sense opposite to its rotation, or undergo a transition from axis-symmetric to triaxial shapes through the dynamical "bar-mode" instability[6]. All these mechanisms are potentially able to generate large amounts of energy in the form of GWs.

Non-radial oscillations are also a possible mechanism for neutron stars to emit GWs and this possibility was already discussed in the late sixties[7]. One of the difficulties with this mechanism concerns the energy source necessary to excite the oscillations. The elastic energy stored in the crust and released by tectonic activity was recently considered[8], but the maximum available energy is likely to be of the order of $10^{44-45}erg$. Thus, even if all this energy could be converted into non-radial modes, the maximum distance that a signal could be seen by a laser interferometer like VIRGO is only about $3.0kpc$[8]. This result depends, of course, on the adopted equation of state, which fixes the mode frequencies and damping time-scales. A considerable amount of energy would be available if the neutron star undergoes a phase transition in the core. For instance, if quark deconfinement occurs, then the star will suffer a micro-collapse, since the equation of state of quark matter is softer than that of hadronic matter and the new equilibrium configuration will be more compact, having a larger binding energy. The energy difference, reduced by the energy absorption of the phase transition, is partially used to excite mechanical modes, which have energy dissipated by different channels. The structural rearrangement suffered by a "cold" star occurs on a dynamical timescale of the order of milliseconds[9], which is much shorter than the gradual transition expected to occur in hot protoneutron stars[10]. Radial modes excited by the micro-collapse do not radiate GWs, but an important coupling with rotation exists[11] and, if the star rotates, the oscillations will be damped not only by dissipation of the mechanical energy into heat but also by the emission of GWs.

2 On the Detectability of GWs. Recent Results

In the present work, the detectability of GWs generated by oscillations excited during a phase transition in the core of a neutron star is reviewed. Neutron star properties were computed using a description of the hadronic matter based on the work of Boguta and Bodmer[12], including the fundamental baryon octet, the isovector meson ϱ and lepton degrees of freedom (see Ref.[13] for details). Hybrid models include a quark-gluon core determined via a MIT modeling and with the *same baryonic number* as the pure hadron configuration were also computed from the energy difference between both states of the maximum energy available to either excite mechanical oscillations or to be converted into heat. Then, using the planned sensitivity of present laser interferometers like VIRGO (or LIGO I) and those of the next generation (LIGO II), the maximum volume of space that can be probed by these experiments was calculated, as an indication for the potential detectability of these sources.

The maximum stable mass of one-phase objects satisfies the requirement of being higher than the mass of Vela X-1 ($M = 1.86 \pm 0.16$ $M_\odot$), determined recently[14] from a new study of its orbital motion. As one should expect that quark cores, as well as the difference between the binding energies of hybrid and one-phase objects, increase with the baryon number of the configuration, on the basis of this and from the aspect of the GW emission, it would be natural to expect that phase transitions occurring in more massive stars would be more easy to detect. However, this is not the case because, on one hand, the damping time-scale due to the emission of GWs is inversely proportional to the mass and hence massive stars have low oscillation quality factors, on the other hand, the detectors have sensitivities optimized at frequencies around $200Hz$. In comparison, non-radial oscillations (for instance $m = \ell = 2$ modes), have comparable frequencies but damping timescales one order of magnitude higher than those resulting from radial modes coupled to rotation; thus higher quality factors Q may be obtained but most of the mechanical energy will be dissipated into heat. In order to be an efficient source of GWs by the mechanism on our scenario, the star must be a fast rotator with rotation periods of the order of few milliseconds. Even in the most favorable case corresponding, in our calculations, to a baryonic mass equal to $1.32M_\odot$, even for a rotation period $4.0ms$, only 1% of the available energy will be emitted as GWs. Table 2 indicates that the maximum distance probed by detectors of first generation (VIRGO, LIGO I) is about $6.4Mpc$, well beyond M31, whereas the second generation (LIGO II) will probably see phase transition events at distances two times larger, but certainly not yet attaining the Virgo cluster. The small probed volume and the rapid rotation required for this mechanism to be efficient result in a low event rate, imposes severe limitations on the detectability of such a signal[15].

Table 1. Parameters of the stellar models: masses are in solar units and radii in km; H and HQ denote pure hadron and hybrid configurations respectively (with E given in erg). (for the details see Ref.[15])

M_{bar}	M_g (H)	M_g (HQ)	M_g (core)	$R(H)$	$R(HQ)$	$log(E)$
1.0749	1.0003	0.9625	0.1150	12.82	12.47	52.83
1.3202	1.2009	1.0953	0.3608	13.04	12.16	53.28
1.5724	1.4008	1.2468	0.6013	13.13	11.98	53.44
1.8342	1.6006	1.4164	0.8626	13.01	11.82	53.52
2.1045	1.8002	1.6141	1.1766	12.69	11.50	53.52

Table 2. Oscillation parameters: the damping timescale τ_{gw} is given for the critical period; maximum distances for VIRGO (V) and LIGO II (L) are in Mpc.

ν_0	P_{crit}	τ_{gw}	Q	D_{max}	D_{max}
(kHz)	(ms)	(ms)	–	$(VIRGO)$	$(LIGOII)$
1.62	1.64	87.0	442	4.9	10.2
1.83	1.25	27.0	155	6.4	13.5
2.06	1.13	17.0	110	6.0	12.8
2.32	1.06	11.5	84	5.1	11.1
2.72	1.00	8.4	72	3.6	5.7

References

1. A. Ferrari, R. Ruffini, *Astroph. J.* **158**, L71 (1969).
2. M. Zimmerman, E. Szedenits, *Phys. Rev.* **D 20**, 351 (1979).
3. S.L. Shapiro, S.A. Teukolsky, *in Black Holes, White Dwarfs and Neutron Stars* (Wiley, 1983).
4. S. Chandrasekhar, *Phys. Rev. Lett.* **24**, 611 (1970).
5. J.L. Friedman, B.F. Schutz, *Astroph. J.* **222**, 281 (1978).
6. D. Lai, S.L. Shapiro, *Astroph. J.* **442**, 259 (1995).
7. K.S. Thorne, A. Campolattaro, *Astroph. J.* **149**, 591 (1967).
8. J.A. de Freitas Pacheco, *A&A* **336**, 397 (1998).
9. P. Haensel, M. Prózynski, *Astroph. J.* **258**, 306 (1982).
10. T.W. Baumgarte, H.-T. Janka, W. Keil, S.L. Shapiro, S.A. Teukolsky, *Astroph. J.* **468**, 823 (1996).
11. W.-Y. Chau, *Astroph. J.* **147**, 664 (1967).
12. J. Boguta, A.R. Bodmer, *Nucl. Phys.* **A292**, 413 (1977).
13. G. F. Marranghello, C.A.Z. Vasconcellos, M. Dillig, J. A. de Freitas Pacheco, *Int. Journ. Mod. Phys.* **E11**, 2, 83 (2002).
14. O. Barziv, L. Kaper, M.H. van Kerkwijk, J.H. Telting, J. van Paradijs, astro/ph-0108237 (2001).
15. G. F. Marranghello, C.A.Z. Vasconcellos and J. A. de Freitas Pacheco, *Phys. Rev.* **D66**, 064027, 1 (2002).

GRAVITATIONAL WAVES FROM BRANEWORLD NEUTRINO OSCILLATIONS AT SUPERNOVA CORE-COLLAPSE

HERMAN J. MOSQUERA CUESTA[1,2]

[1] *Centro Brasileiro de Pesquisas Físicas, Laboratório de Cosmologia e Física Experimental de Altas Energias, Rua Dr. Xavier Sigaud 150, CEP 22290-180 Urca, Rio de Janeiro, RJ, Brazil* [2] *Centro Latinoamericano de Física (CLAF), Avenida Wenceslau Braz 71, Botafogo, Rio de Janeiro, RJ, Brazil*
hermanjc@cbpfsul.cat.cbpf.br

In braneworld-like solutions of the hierarchy problem gravitons and right-handed (sterile) neutrinos are in principle the unique non-standard model fields allowed to propagate into the bulk, thus their coupling is naturally expected. Since active-to-sterile neutrino oscillations can take place during the core bounce of a supernova collapse, then gravitational waves must be produced over the oscillation length through anisotropic neutrino flow. Because the oscillation feeds mass-energy up into (or takes it out of) the target species, the large mass-squared difference between species makes a huge amount of energy to be given off as gravity waves, which is larger than from neutrino convection and cooling, or quadrupole moments of neutron star matter. The space-time strain from these bursts would turn them the more sure supernova gravitational-wave signal detectable by interferometers, for distances out to the VIRGO cluster of galaxies.

Graviton-sterile neutrino coupling.– In *braneworld* models of string phenomenology[1,2] the neutrino sector is indicated as the most likely messenger of CPT violation to the rest of the standard model. This is because the source of dynamical or spontaneous CPT violation should lie in the bulk, while the standard model effects are expected to be visible only through couplings of SM fields, supposed to reside on branes, to suitable bulk messengers. The generic bulk messengers are the graviton and the right-handed ν. In models with not infinite extra-dimensions gravity effects can still be visible, while the sterile neutrino can still have potentially observable Dirac mass couplings[3], although the resulting neutrino oscillation physics is irrelevant to account for the solar or atmospheric neutrino problems. Thus again, because the graviton and sterile singlet neutrino may co-exist in the bulk they might couple. Consequently, oscillations of active into sterile νs must produce gravitational waves, while oscillations among active ν species cannot. In fact, in Ref.[4], and based upon most general considerations of SN core physics, I showed that this seems to be case: no GWs from active-to-active ν oscillations during a SN core bounce.

To illustrate this coupling we follow the paper by Grossman and Neubert[5] in which, for the case of a RS model; where a interesting neutrino phenomenology can develop, the action exhibiting the graviton ($g^{\mu\nu}_{vis}$) and sterile neutrino (e_{R0}) coupling reads as[5]

$$S = \int d^4x\sqrt{-g_{vis}}\{g^{\mu\nu}_{vis}\partial_\mu H_0^\dagger \partial_\nu H_0 - \lambda(|H_0|^2 - v_0^2)^2\}$$
$$+ \int d^4x\sqrt{-g_{vis}}\{\bar{L}_0\hat{\gamma}^\mu\partial_\mu L_0 + \bar{e}_{R0}\hat{\gamma}^\mu\partial_\mu e_{R0}(y_e\bar{L}_0H_0e_{R0} + h.c.)\}, \quad (1)$$

where $g^{\mu\nu}_{vis} = e^{2kr_c\pi}\eta_{\mu\nu}$ is the induced metric on the brane, with $\sqrt{g_{vis}} = det(-g^{\mu\nu}_{vis}) = e^{4kr_c\pi}$, $\hat{\gamma}^\mu = E^\mu_a(\phi = \pi)\gamma^a = e^{kr_c\pi}\gamma^\mu$. Here $H = (\phi_1, \phi_2)$ is the Higgs doublet and $L = \nu_L, e_L$ is a left-handed lepton doublet (the reader is addressed to Ref.[5] for a thorough discussion and definitions of the quantities appearing in these equations. This action is obtained after integration over the fifth dimension of the action describing the blind bulk fermion of mass m coupling minimally to gravity $S^{\rm g}_{\rm f} = \int d^4x \int d\phi\sqrt{G}\left\{E^A_a\left[\frac{i}{2}\bar{\Psi}^a(\partial_A - \bar{\partial_A})\Psi + \frac{\omega_{bcA}}{8}\bar{\Psi}^a, \sigma^{bc}\Psi\right] - m\,{\rm sgn}(\phi)\bar{\Psi}\Psi\right\}$. The coupling is mediated by the *spin connection* ω_{bcA}, in a 5-dim space of metric: $ds^2 = e^{-2kr_c|\phi|}\eta_{\mu\nu}dx^\mu dx^\nu - r_c d\phi$. To settle the remaining basics, we also recall the way the neutrino flux couples to neutron star matter, as needed for the oscillations to take place. The electroweak coupling of neutrons (N^0) and νs inside the PNS is described by the Lagrangean $L^{int}_{N^0\leftrightarrow\nu} = [\frac{G_F}{\sqrt{2}}](\bar{N}^0\gamma_\mu(1-\gamma_5)N^0)[\bar{\nu}\gamma^\mu(1-\gamma_5)\nu]$,with the ν field satisfying the time-dependent Dirac equation $\left(i\gamma^0\partial_0 + i\gamma^\alpha\partial_\alpha + \rho(t)v_\beta\gamma^\beta[\frac{(1-\gamma_5)}{2}] - m_\nu\right)\Psi = 0$. G_F is the Fermi constant, and v_β the ν 4-velocity. In dense matter (at rest) the neutron 4-vector current density $J^\mu =< \bar{N}^0\gamma_\mu(1-\gamma_5)N^0 >\equiv (\rho, 0, 0, 0)$ acquires a non-zero expectation value. Here $\rho = \frac{G_F}{\sqrt{2}}N^0_n$, with N^0_n the neutron number density.

Enlarged ν and GWs luminosity SN ν flavor conversions.– That the neutrino outflow from a PNS core bounce is a source of GWs is a well-known issue[7-11,4,6]. Numerical simulations by Müller and Janka[10] have shown that, in general, the fraction of the total binding energy emitted as GWs by pure neutrino convection is: $E^\nu_{GW} \sim [10^{-10} - 10^{-13}]M_\odot c^2$, for a ν luminosity: $L_\nu \sim 10^{53}$erg s^{-1} (see Ref.[10] for further details). Unlike GWs produced by convection of neutrinos[10], in the production of GWs by neutrino oscillations[4] from active-to-sterile there exists two main reasons for expecting a major enhancement in the GWs luminosity during the transient: a) the conversion itself, which makes it the overall luminosity of a given neutrino species to grow by a large factor: $L_\nu \leq 10\% L^{\rm total}_\nu$, see below. The enhancement stems from the mass-energy being given to, or drained from, the new species into which oscillations take place. This augment gets reflected in the species mass-squared difference (energy conservation), and their relative numbers: one species is number-depleted while the other gets its number enlarged. But, even if the energy increase is smaller, b) the abrupt conversion over the transition time, $\Delta T_{\rm OSC} \sim 10^{-4}$s, which is set up by the oscillation length, $\lambda_{\rm OSC}$, and the convective neutrino diffusion velocity[10], $\bar{V}_\nu$ ($\sim 10^9$ cms^{-1}), also magnifies transiently L_ν. Being the neutrino flavor transition the *key piece* to produce GWs by this mechanism, it is needed to estimate how many of them can actually oscillate. This quantity is measured by the transition probability: $P_{a\to s}(|\vec{x} - \vec{x}_0|)$, which we estimate next in considering the way the neutrino couples to the neutron matter so as to make it viable these $\nu_a \longleftrightarrow \nu_s$ oscillations.

ν oscillations in dense matter.– It was shown in Ref.[4] that vacuum flavor conversions among active νs cannot increase significantly the amount of escaping neutrinos from the "frozen" SN core, and thus no GWs burst is generated. However, interaction with matter may help in allowing more νs to escape if resonant conversions into sterile νs occur inside the neutrinosphere of the active νs (see Ref.[4] from where the next discussion is briefly updated). From the discussion in Ref.[4] it becomes clear that once the oscillation occurs almost all the sterile νs being created escape the star taking a huge amount of energy (and momentum) density ($\lesssim 10\%\rho^{\nu}$) away with them. The coupling energy to matter of the former active νs that convert into steriles is lost in the transition.

GWs Energetics and Detectability.– Whenever the ν oscillations do take place in the SN core, the GWs signal most likely detectable should be produced over the time interval for which the conditions for conversions to take place are maintained, i.e. $\sim (1-2)$ms, or equivalently for frequencies $f_{\rm GWs} \sim [0.5-1]$kHz. This frequency range corresponds to the optimal bandwidth for detection of ground-based detectors. For a 1ms conversions time span the neutrino luminosity reads $L_\nu \equiv \frac{\Delta E_{\nu_a \longrightarrow \nu_s}}{\Delta T_{\rm OSC}} \sim \frac{3\times 10^{52}{\rm erg}}{1\times 10^{-3}{\rm s}} = 3 \times 10^{55}\,\frac{\rm erg}{\rm s}$.

Hence, the GWs luminosity, $L_{\rm GWs}$, as a function of the ν luminosity can be obtained by relating the GWs flux: $\frac{c^3}{16\pi G}\left|\frac{dh}{dt}\right|^2 = \frac{1}{4\pi R^2} L_{\rm GWs}$, to the GWs amplitude from the neutrino outflow, h_ν, which can be computed from the neutrino luminosity after Refs.[7,10,11]: $h_{ij}^{TT} = \frac{2G}{c^4 R}\int_{\infty}^{t-R/c} dt' L_\nu(t')\alpha(t') e_i \otimes e_j$,, where $e_i \otimes e_j$ is the GW polarization tensor. This equationcan be recast in an approximate manner as: $h_\nu = \frac{2G}{c^4 R}[\Delta t \times L_\nu \times \alpha]$. The two last equations can be recast to obtain the GWs amplitude produced by the non-spherical outgoing front of the ν_s-neutrinosphere. It yields[4,7,10] $h_\nu \simeq 1.5 \times 10^{-21}[\frac{55{\rm kpc}}{R}][\frac{\Delta T}{10^{-3}{\rm s}}](\frac{L_\nu}{3\times 10^{55}\frac{\rm erg}{\rm s}})[\frac{\alpha}{4\times 10^{-1}}]{\rm Hz}^{-1/2}$, A GWs signal with this amplitude will likely be detectable by the first generation of GWs interferometers, if an event at the distance of the Large Magellanic Cloud (SN1987A) takes place in the near future. It can also detected by advanced interferometers (i.e. LIGO-II), and quantum limit operating resonant-mass detectors, over distances up to the VIRGO cluster of galaxies. Its imprint in the GWs waveform (GWs "light curve" from the PNS formed in the SN collapse) may resemble a spike of large amplitude, and timewidth of $\sim$(1-3)ms.

The GWs luminosity turns out to be: $L_{\rm GWs} = 4 \times 10^{50}\,\frac{\rm erg}{\rm s}\,[\frac{L_\nu}{3\times 10^{55}\frac{\rm erg}{\rm s}}]^2 \left(\frac{\alpha}{4\times 10^{-1}}\right)^2$, where $0.2 < \alpha < |-0.8|$ is the anisotropy parameter as defined by Burrows and Hayes[7]. It turns out that the GWs energy radiated in the process: $E_{\rm GWs} \equiv L_{\rm GWs} \times \Delta T_{\rm osc} \sim 4 \times 10^{47}$erg, is almost similar to both the one sterile neutrinos carry away: $E_\nu = 10^{57}|_{\nu_s} \times 10^4\mathrm{eV}|_{\nu_s}[\frac{10^{-33}\mathrm{gc}^2}{\mathrm{eV}}] \sim 2 \times 10^{-6}\mathrm{M}_\odot\mathrm{c}^2$, and the one estimated to be necessary to explain the proper motion of pulsars as due to pure GWs kicks: $E^{Pulsar}_{surveys}$(GWs)$\sim 4 \times 10^{-6}\mathrm{M}_\odot c^2$[13].

Discussion and Conclusions.– The quoted GWs energy also is $\sim 10^5$ larger than both the current estimates from the fluid motion of the PNS constituents[10]. With so much an energy it is possible to kick the PNS up to the observed velocities of pulsars, as shown by Mosquera Cuesta [Phys. Rev. D. 65, 061503 (2002)][6] (see also the *Gravitational wave rocket* thruster by W. B. Bonnor[14]). In summary, one can see that if neutrino flavor conversions actually take place during the SN core-bounce, the GWs signal from the process should irradiate much more GWs energy than standard mechanisms driving the neutron star dynamics at birth. This large GWs luminosity would turn these bursts the more sure detectable GWs signal from the next supernova explosion that would come to occur up to the VIRGO cluster distance, $R \sim 20$Mpc. The radiation reaction of the outbursts could be the thruster mechanism of the Galaxy runaway pulsars and black holes, as pointed out in Ref.[6].

References

1. L. Randall, R. Sundrum, Phys. Rev. Lett. 83, 4690 (1999).
2. L. Randall, R. Sundrum, Phys. Rev. Lett. 83, 3370 (1999).
3. G. Baremboim, et al., hep-ph/0108199, 30 Aug. (2001).
4. H. J. Mosquera Cuesta, Ap. J. Lett. 544, L61 (2000).
5. Y. Grossman, M. Neubert, Phys. Lett. B 474, 361 (2000).
6. H. J. Mosquera Cuesta, Phys. Rev. D. 65, 061503(R) (2002).
7. A. Burrows, J. Hayes, Phys. Rev. Lett. 76, 352 (1996).
8. A. Burrows, et al., Astrophys. J. 450, 830 (1995).
9. H.-Th. Janka, E. Müller, Astron. Astrophys. 306, 167 (1996). See also: Astron. Astrophys. 290, 290 (1994).
10. E. Müller, H.-Th. Janka, Ast. Astrophys. 317, 140 (1997).
11. R. I. Epstein, MNRAS 188, 305 (1978).
12. M. Patel, G. M. Fuller, What are sterile neutrinos good for?, report hep-ph/0003034 (v1) 5 Mar (2000).
13. S. Nazin, M. Postnov, Astron. Astrophys. 317, L79 (1997).
14. W. B. Bonnor, Class.Quant.Grav. 14:2895 (1997).

PLASMA EFFECTS IN NEUTRINO-PAIR EMISSION DUE TO COOPER PAIRING OF PROTONS IN SUPERCONDUCTING NEUTRON STARS

L. B. LEINSON

Institute of Terrestrial Magnetism, Ionosphere and Radio Wave Propagation RAS, 142190 Troitsk, Moscow Region, Russia
leinson@izmiran.rssi.ru

We analyze the effects of neutron superfluidity and/or proton superconductivity on different neutrino reactions in the neutron star core. The temperature dependence of the RPA vector weak current contribution to the neutrino emissivity in neutron stars is compared with that obtained in the free-gas approximation. We also calculate the temperature dependence of the neutrino emissivity in different reactions for β-equilibrium nuclear matter of the density $\rho = 2$ time the saturation density where the direct Urca process is forbidden. As our results indicate, under certain conditions, neutrino emission due to Cooper pairing of protons can dominate in neutrino energy losses from the neutron star. However, partial contributions of the proton and neutron pairing to the total energy losses are very sensitive to the corresponding critical temperatures.

1 Introduction

When the temperature inside a neutron star core is lower than the critical temperature T_c for nucleon pairing, the nucleon matter consists of a condensate of Cooper pairs, which has thermal excitations in the form of not paired quasi-particles. Cooper-pair formation and pair-breaking coexist in statistical equilibrium and result in additional neutrino-pair emission from the neutron star. The mechanism of neutrino emission due to the singlet-state pairing of neutrons was proposed by Flowers et al. [1] many years ago.

Specifics of the neutrino emission due to proton pairing occurs because of a smallness of the vector weak coupling of a proton. Cooper pairing of protons takes place likely in $^1S_0$-state [2]. When protons are treated non-relativistically, the total spin of the Cooper pair in the singlet-state is zero. By this reason, the axial-vector contribution of the proton weak current to the neutrino emissivity occurs only as a relativistic correction. Therefore, neutrino emission produced by the proton pairing is conventionally estimated negligible [3].

Such inference is made on the basis of calculations which ignore electromagnetic correlations among the charged particles in the QED plasma. Actually, protons are coupled to ambient electrons via the electromagnetic field. By undergoing a quantum transition to the paired state, protons polarize the medium, thus inducing the motion of electrons inside the Debye sphere

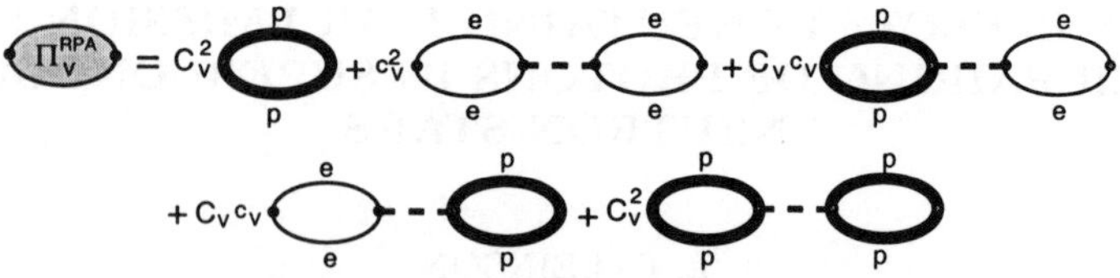

Figure 1. RPA approximation to the weak polarization of the medium. Here $C_V = 0.04$ is the constant of the proton vector weak coupling with the neutrino field; c_V is the electron coupling constant, ($c_V = 0.96$ for electron neutrinos, and $c'_V = -0.04$ for muon and tau neutrinos). The thick dashed line is the ***in-medium*** photon.

around them. The electron weak current associated to this motion generates neutrinos coherently with the weak current of protons, because the wavelength λ of radiated neutrino pairs is much larger than the electron Debye screening distance D_e (typically, $D_e^2/\lambda^2 \sim 10^{-2}$). The induced radiation from ambient electrons many times exceeds the neutrino radiation from the initial proton pair [4], [5].

2 Neutrino Emissivity

To demonstrate this we study the neutrino emissivity, caused by the proton pairing, by the use of the fluctuation-dissipation theorem. We consider the total energy which is emitted into neutrino pairs per unit volume and time. For one neutrino flavor, the emissivity is given by the following formula:

$$Q^{\mathrm{RPA}} = \frac{G_F^2}{2} \int \frac{\omega}{\exp\left(\frac{\omega}{T}\right) - 1} 2 \,\mathrm{Im}\, \tilde{\Pi}_R^{\mu\nu}(K)\, \mathrm{Tr}\left(j_\mu j_\nu^*\right) \frac{d^3 k_1}{2\omega_1 (2\pi)^3} \frac{d^3 k_2}{2\omega_2 (2\pi)^3},$$

where the integration goes over the phase volume of neutrino and antineutrino of the total energy $\omega = \omega_1 + \omega_2$ and the total momentum $\mathbf{k} = \mathbf{k}_1 + \mathbf{k}_2$. In this formula G_F is the Fermi coupling constant, $\tilde{\Pi}_R^{\mu\nu}$ is the retarded polarization tensor of the medium which has ends at the weak vertex, and the neutrino weak current is of the standard form

$$j_\mu = \bar{\nu} \gamma_\mu (1 + \gamma_5)\, \nu. \tag{1}$$

To incorporate the collective plasma effects, one should take into account exchange of photons between charged particles in the plasma. This can be done in the Random Phase Approximation (RPA). In this case the weak polarization tensor of the medium is given by the diagrams shown in Fig. 1.

Here the first term describes the contribution, which comes directly from the pairing protons. The other terms are caused by the plasma polarization.

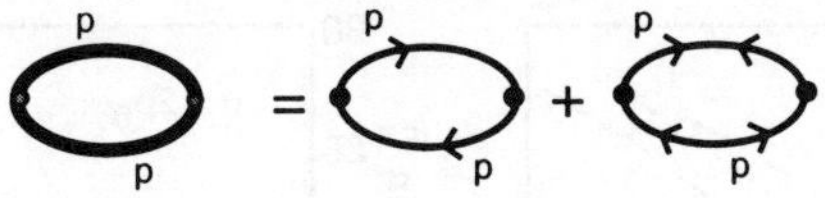

Figure 2. Free-gas approximation to the weak polarization of protons.

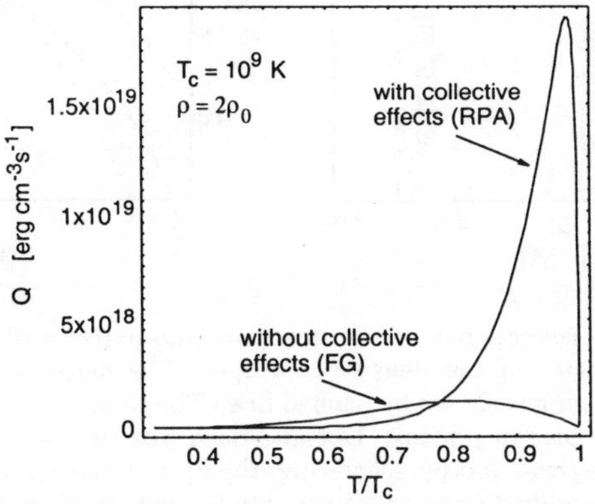

Figure 3. Temperature dependence of the RPA vector weak current contribution to the neutrino emissivity in comparison with that obtained in the free-gas approximation (FG). The emissivities are shown versus the dimensionless temperature $\tau = T/T_c$ for beta-equilibrium nuclear matter.

3 Results

To demonstrate efficiency of the collective effects, the Q^{RPA} is plotted in Fig. 3 versus the dimensionless temperature $\tau = T/T_c$ together with that obtained in the free-gas approximation (FG) without the collective effects. One can see that the collective effects substantially enhance the $\nu\bar{\nu}$ emissivity caused by the vector weak current of protons.

The effect of neutron superfluidity and/or proton superconductivity on different neutrino reactions in the neutron star core is very complicated. Different neutrino production mechanisms can dominate at different cooling stages depending on the temperature, and the matter density, and vary along with the chosen parameters T_{cp}, T_{cn}, which are the critical temperatures for the proton and neutron pairing. Fig. 4 shows efficiency of different reactions at the baryon density $\rho = 2\rho_0$, where the direct Urca process is forbidden. Neutrino emissivities due to the singlet-state proton pairing, and the triplet-state neutron pairing are plotted in logarithmic scale against the temperature together with the total emissivity of two branches of the modified Urca processes, and the total bremsstrahlung emissivity caused by nn-, np-, and

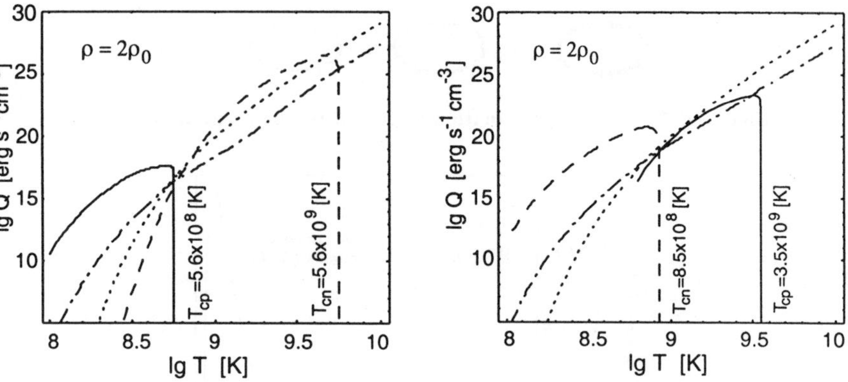

Figure 4. Temperature dependence of the neutrino emissivity in different reactions for β-equilibrium nuclear matter of the density $\rho = 2\rho_0$. The neutrino emissivity due to the triplet-state neutron pairing is shown by dashed line. The solid line is the neutrino emissivity due to the singlet-state proton pairing. Dot-and-dash line shows the total bremsstrahlung emissivity caused by nn-, np-, and pp-scattering; the dotted line exhibits the total emissivity of two branches of the modified Urca processes. On the left panel we took $T_{cp} = 5.6 \times 10^8$ K, and $T_{cn} = 5.6 \times 10^9$ K. On the right panel we assume $T_{cp} = 3.5 \times 10^9$ K, and $T_{cn} = 8.5 \times 10^8$ K.

pp-scattering, which are suppressed due to the neutron superfluidity and/or proton superconductivity. The two panels of Fig. 4 differ only by different choice of parameters T_{cp} and T_{cn}. On the left panel we took $T_{cp} = 5.6 \times 10^8$ K, and $T_{cn} = 5.6 \times 10^9$ K. On the right panel we assume $T_{cp} = 3.5 \times 10^9$ K, and $T_{cn} = 8.5 \times 10^8$ K. As one can see, under certain conditions, neutrino emission due to Cooper pairing of protons can dominate in neutrino energy losses from the neutron star. However, partial contributions of the proton and neutron pairing to the total energy losses are very sensitive to the corresponding critical temperatures.

References

1. E. Flowers, M. Ruderman, P. Sutherland, *ApJ* **205** (1976) 541.
2. T. Takatsuka, R. Tamagaki, *Progr. Theor. Phys. Suppl.* **112** (1993) 27.
3. A. D. Kaminker, P. Haensel, and D. G. Yakovlev, *A&A* **345** (1999) L14.
4. L. B. Leinson, *Phys. Lett.* **B473** (2000) 318.
5. L. B. Leinson, *Nucl. Phys.* **A687** (2001) 489.

PAIRING CORRELATIONS AND THE NUCLEAR SURFACE DIFFUSENESS IN THE RELATIVISTIC HARTREE APPROACH

M.L. CESCATO

Departamento de Física-CCEN-UFPB, Caixa Postal 5008, 58051-970 João Pessoa-PB, Brazil

mcescato@fisica.ufpb.br

The relativistic Hartree-Bogolyubov equations are derived in a suitable basis for the description of semi-infinite nuclear matter system. The pairing force is considered in the zero-range approach.

1 Introduction

Pairing correlations depend in a sensitive way on several effects, changing if one is leaving the valley of β-stability: a) they depend on the level density, which is given by the shell structure and b) they depend on the density of the nuclear medium. Therefore they play a crucial role in the nuclear surface region and are expected to change with the increasing surface diffuseness. In calculations of finite nuclei both effects, which depend on each other, cannot be disentangled in a simple fashion.

In order to understand the influence of surface diffuseness on the pairing properties one can investigate these properties in the simpler semi-infinite nuclear matter system. In the present contribution we develop the relativistic Hartree-Bogolyubov equations in a convenient basis for analyzing this system. How to solve these equations in a self-consistent way is also described.

2 Relativistic Hartree-Bogolyubov Equations

We start from a model [1] in which the nucleons are treated as Dirac spinors ψ interacting by the exchange of several mesons. Here we consider a model Lagrangian which includes scalar isoscalar σ-mesons and vector isoscalar ω-mesons. See Ref. 2 for the explicit form of the Lagrangian.

The relativistic mean field theory is constructed, using the no-sea approximation and removing the fluctuations of the meson fields, through substitution by their expectation values.[2] Nevertheless this theory cannot account for pairing correlations, since the resulting Dirac Hamiltonian contains only single-particle field operators with the structure $\psi^\dagger\psi$, and pairing correlations can only be described in a generalized single-particle theory by field operators

$\psi^\dagger\psi^\dagger$ and $\psi\psi$, which do not conserve the particle number. A self-consistent treatment of mean-field and pairing correlations can be done through the relativistic extension of the Hartree-Fock-Bogolyubov theory.[3] In this framework, starting from the model Lagrangian and performing the quantization also of the meson fields, the Hartree-Fock-Bogolyubov equations are derived by the use of Green's function techniques

$$\begin{pmatrix} h-\lambda & \Delta \\ -\Delta^\dagger & -h^*+\lambda \end{pmatrix} \begin{pmatrix} U \\ V \end{pmatrix}_\alpha = E_\alpha \begin{pmatrix} U \\ V \end{pmatrix}_\alpha, \tag{1}$$

where λ is the chemical potential and E_α are the quasi-particle energies. In Eq. (1), h and Δ are the single-nucleon Dirac Hamiltonian and the relativistic pairing field respectively. Considering time reversal symmetry in the Dirac Hamiltonian and neglecting the Fock terms, this equation represents the Relativistic Hartree-Bogolyubov (RHB) equation. The relativistic density matrix ρ and the relativistic pairing tensor κ are given by $\rho = V^*V^T$ and $\kappa = V^*U^T$.

A generic matrix element of the relativistic pairing field can then be written as

$$\Delta_{a_1p_1a_2p_2} = \frac{1}{2}\sum_{a_3p_3a_4p_4} \langle a_1p_1a_2p_2|V^{pp}|a_3p_3a_4p_4\rangle_a \, \kappa_{a_3p_3a_4p_4}, \tag{2}$$

where $\langle V^{pp}\rangle_a$ is the relativistic anti-symmetrized two-body matrix element, $p_i = \{+,-\}$ are the relativistic indices, that label large and small components and the indices a_i stand for all quantum numbers necessary to specify the basis vectors. The indices i specifies the particle.

We suppose a phenomenological relativistic pairing originated from the exchange of the mesons σ and ω and use a zero-range approximation, obtaining in this way[4]

$$V_\delta^{pp}(\vec{r}_1,\vec{r}_2) = c_0\left\{-C_s^2\gamma_{13}^0\gamma_{24}^0 + C_v^2\mathbf{I}_{13}\mathbf{I}_{24} - C_v^2\vec{\alpha}_{13}\cdot\vec{\alpha}_{24}\right\}\delta(\vec{r}_1-\vec{r}_2), \tag{3}$$

where $\mathbf{I}$ is the identity matrix, $\gamma^\mu = (\gamma^0,\vec{\gamma})$ are the usual gamma matrices, $\gamma^0 = \beta$, $\vec{\gamma} = \beta\vec{\alpha}$ and the coupling constants $C_s \equiv g_\sigma/m_\sigma$ and $C_v \equiv g_\omega/m_\omega$. The constant c_0, appearing in Eq. (3), is usually adjusted to reproduce the pairing energy of a given nucleus obtained with other HFB calculations.

3 Application to Semi-infinite Nuclear Matter

3.1 The Problem Without Pairing

Semi-infinite nuclear matter is a system in which translation invariance is assumed in directions transverse to the normal to a surface. The only spatial dependence in all physical quantities is on the z coordinate on the direction of this normal. In this system it is convenient to work in a basis defined by the eigenvectors of the operator $\hat{K}$, given by [5]

$$\hat{K} = \begin{pmatrix} \sigma_y p_x - \sigma_x p_y & 0 \\ 0 & -(\sigma_y p_x - \sigma_x p_y) \end{pmatrix}, \tag{4}$$

where $\vec{\sigma} : \{\sigma_x, \sigma_y, \sigma_z\}$ are the Pauli matrices and $\vec{p} : \{p_x, p_y, p_z\}$ is the linear momentum. With the eigenvectors of $\hat{K}$ we can construct the transformation T, which diagonalizes this operator, containing in its columns the eigenvectors of $\pm(\sigma_y p_x - \sigma_x p_y)$. With this transformation the Dirac Hamiltonian can be transformed to its representation in the basis where $\hat{K}$ is diagonal. The eigenvalue equation for this matrix furnishes two sets of equations, one for each value of $\nu = \pm 1$,

$$(\frac{d}{dz} - \nu p_\perp) G_n^\nu(z) = \quad (\varepsilon_n + M^* - V) F_n^\nu(z), \tag{5}$$

$$(\frac{d}{dz} + \nu p_\perp) F_n^\nu(z) = -(\varepsilon_n - M^* - V) G_n^\nu(z), \tag{6}$$

from whose solutions we construct the eigenvectors corresponding to the eigenvalues ε_n. The label ν specifies the eigenvector χ^ν of the operator $\sigma_y p_x - \sigma_x p_y$, corresponding to the eigenvalue $\nu p_\perp$, with $p_\perp \equiv \sqrt{p_x^2 + p_y^2}$. The G_n^ν and F_n^ν functions are respectively the big and the small components of the eigenvector $|\psi_{n,\nu}(z)\rangle$. The complete solution for the Dirac Hamiltonian is $|\psi_{k_\perp,n}^\nu(x,y,z)\rangle = |\psi_n^\nu(z)\rangle \, \chi^\nu \, e^{\vec{k}_\perp \cdot \vec{r}_\perp}$.

3.2 Inclusion of Pairing

We now should consider the full problem including pairing interactions. By writing the eigenvalue equation, Eq. (1), in the new basis, using the transformed Dirac Hamiltonian, and assuming for the vectors $|\psi_\alpha^\nu\rangle$ and for the Δ matrix structures consistent with the symmetries, we get two sets of equations for the wave-functions $G_U{}_\alpha^\nu, F_U{}_\alpha^\nu, G_V{}_\alpha^\nu$ and $F_V{}_\alpha^\nu$, one for each value of $\nu = \pm 1$

$$(\frac{d}{dz} + \nu p_\perp) F_U{}_\alpha^\nu - \Delta_{++}^\nu G_V{}_\alpha^\nu - \nu \Delta_{+-}^\nu F_V{}_\alpha^\nu = (-E_\alpha + M^* + V) G_U{}_\alpha^\nu, \tag{7}$$

$$(\frac{d}{dz} - \nu p_\perp) G_U{}_\alpha^\nu - \nu \Delta_{-+}^\nu G_V{}_\alpha^\nu + \Delta_{--}^\nu F_V{}_\alpha^\nu = (+E_\alpha + M^* - V) F_U{}_\alpha^\nu, \tag{8}$$

$$(\frac{d}{dz} + \nu p_\perp) F_V{}_\alpha^\nu + \Delta_{++}^\nu G_U{}_\alpha^\nu - \nu \Delta_{-+}^\nu F_U{}_\alpha^\nu = (+E_\alpha + M^* + V) G_V{}_\alpha^\nu, \tag{9}$$

$$(\frac{d}{dz} - \nu p_\perp) G_{V\alpha}^{\ \nu} - \nu \Delta_{+-}^{\nu} G_{U\alpha}^{\ \nu} - \Delta_{--}^{\nu} F_{U\alpha}^{\ \nu} = (-E_\alpha + M^* - V) F_{V\alpha}^{\ \nu}. \quad (10)$$

These equations should be solved self-consistently, using the Δ operator defined by the considered pairing force. The pairing relativistic field, Eq. (2), for the zero-range pairing, Eq. (3), we consider here, should be written in the new basis. Using the T transformation we obtain for the relativistic pairing field

$$\Delta_{p_1 p_2} = \Delta^{s}_{p_1 p_2} + \Delta^{\omega^0}_{p_1 p_2} + \Delta^{\vec{\omega}}_{p_1 p_2}, \quad (11)$$

with

$$\Delta^{s}_{p_1 p_2}(z) = \frac{1}{2} c_0 \, C_s^2 \begin{cases} -\bar{\kappa}_{p_1 p_2}(z), & p_1 = p_2, \\ +\bar{\kappa}_{p_1 p_2}(z), & p_1 \neq p_2, \end{cases} \quad (12)$$

$$\Delta^{\omega^0}_{p_1 p_2}(z) = \frac{1}{2} c_0 \, C_v^2 \ \bar{\kappa}_{p_1 p_2}(z), \quad (13)$$

$$\Delta^{\vec{\omega}}_{p_1 p_2}(z) = -\frac{1}{2} c_0 \, C_v^2 \begin{pmatrix} \bar{\kappa}^{+1,+1}_{-p_1 -p_2}(z) & -3\,\bar{\kappa}^{+1,-1}_{-p_1 -p_2}(z) \\ -3\bar{\kappa}^{-1,+1}_{-p_1 -p_2}(z) & \bar{\kappa}^{-1,-1}_{-p_1 -p_2}(z) \end{pmatrix}, \quad (14)$$

where we have introduced the notation $\bar{\kappa}_{p_1 p_2}$ to denote antisymmetrization of $\kappa_{p_1 p_2}$ relative to the particle indices, that should be done in the case of no-sea approximation we use in this work.

4 Conclusion

We have derived the Relativistic Hartree-Bogolyubov equations for the description of the semi-infinite nuclear matter system. A convenient basis for analyzing this system was introduced for the first time in the presence of the pairing force. In the next step of this investigation, these equations will be numerically solved. Work in this direction is now in progress.

References

1. B.D. Serot and J.D. Walecka, Adv. Nucl. Phys. **16**, 1 (1986).
2. P. Ring, Progr. Part. Nucl. Phys. **37**, 193 (1996).
3. H. Kucharek and P. Ring, Z. Phys. **A339**, 23 (1991).
4. M. Serra, Physics Doctor Thesis, Technische Universität München, 2001.
5. D. Hofer and W. Stocker, Nucl. Phys. **A492**, 637 (1989).

INITIAL-CONDITIONS PROBLEM FOR A NAMBU-JONA-LASINIO SYSTEM

P.L. NATTI

Departamento de Matemática, Universidade Estadual de Londrina,
C. P. 6001, 86051-970, Londrina, Paraná, Brazil
plnatti@uel.br

E.R. TAKANO NATTI

Centro Politécnico, Universidade Norte do Paraná
Rua Tietê, 1208, 86025-230, Londrina, Paraná, Brazil
erica.takano@uol.com.br

We study an off-equilibrium, spatially uniform many-quark system described by Nambu-Jona-Lasinio model. On the basis of the general dynamics of the field, we derive dynamical equations for the set of one-body variables in Gaussian mean-field approximation by a general time-dependent projection technique.

1 Introduction

The focus of this work is to describe a mean-field approximation, for fermionic fields, in terms of a time-dependent projection approach developed earlier for the non-relativistic nuclear many-body dynamics by Nemes and de Toledo Piza [1]. This approach allows for the formulation of a mean-field expansion for the dynamics of the two-point correlation function from which one recovers the results of the Gaussian mean-field approximations in lowest order. This approach was recently applied for the solution of non-relativistic systems [2] and, in the context of QFT, for bosonic $\lambda\phi^4$ system in (1+1) dimensions [3], for fermionic chiral Gross Neveu system [4], and for relativistic scalar plasma system in 3+1 dimensions [5].

In this work, we consider the case of an off-equilibrium, spatially uniform, relativistic many-fermion system described by Nambu-Jona-Lasinio model, with Hamiltonian density [6] given by

$$\mathcal{H}_{\text{NJL}} = -i\bar{\psi}(x)\gamma.\nabla\psi(x) - g\left\{\left[\bar{\psi}(x)\psi(x)\right]^2 - \left[\bar{\psi}(x)\gamma_5\psi(x)\right]^2\right\} \quad . \qquad (1)$$

This is an interesting non-trivial, non-renormalizable, massless fermion theory in (3 + 1) dimensions with quartic interaction for which many results are available in the literature, so that it offers suitable testing ground for the proposed approach.

The idea of our approach is to focus on the time evolution of a set of observables which are bilinear (Gaussian observables) in the creation and annihilation parts of the fermionic fields $\psi(x)$,

$$\psi(\mathbf{x},t) = \sum_{\mathbf{k},s}\left(\frac{M}{k_0}\right)^{1/2}\frac{1}{\sqrt{V}}\left[u_1(\mathbf{k},s)b^{(1)}_{\mathbf{k},s}(t)e^{i\mathbf{k}.\mathbf{x}} + u_2(\mathbf{k},s)b^{(2)\,\dagger}_{\mathbf{k},s}(t)e^{-i\mathbf{k}.\mathbf{x}}\right] \,,$$

$$\bar{\psi}(\mathbf{x},t) = \sum_{\mathbf{k},s}\left(\frac{M}{k_0}\right)^{1/2}\frac{1}{\sqrt{V}}\left[\bar{u}_1(\mathbf{k},s)b^{(1)\,\dagger}_{\mathbf{k},s}(t)e^{-i\mathbf{k}.\mathbf{x}} + \bar{u}_2(\mathbf{k},s)b^{(2)}_{\mathbf{k},s}(t)e^{i\mathbf{k}.\mathbf{x}}\right] \,, \tag{2}$$

where $b^{(1)\,\dagger}_{\mathbf{k},s}(t)$ and $b^{(1)}_{\mathbf{k},s}(t)$ [$b^{(2)\,\dagger}_{\mathbf{k},s}(t)$ and $b^{(2)}_{\mathbf{k},s}(t)$] are fermion creation and annihilation operators associated with positive [negative]-energy solutions $u_1(\mathbf{k},s)$ [$u_2(\mathbf{k},s)$].

In general, the state of the system is given in terms of a many-body density operator $\mathcal{F}$ in the Heisenberg picture, a time independent, non-negative, Hermitian operator with unit trace. The extended one-body density [7]

$$\mathcal{R}_{\mathbf{k},s} = \begin{bmatrix} \langle b^{(\lambda')\,\dagger}_{\mathbf{k},s}(t)b^{(\lambda)}_{\mathbf{k},s}(t)\rangle & \langle b^{(\lambda')}_{-\mathbf{k},s}(t)b^{(\lambda)}_{\mathbf{k},s}(t)\rangle \\ \langle b^{(\lambda')\,\dagger}_{-\mathbf{k},s}(t)b^{(\lambda)\,\dagger}_{\mathbf{k},s}(t)\rangle & \langle b^{(\lambda')}_{\mathbf{k},s}(t)b^{(\lambda)\,\dagger}_{\mathbf{k},s}(t)\rangle \end{bmatrix} \,, \tag{3}$$

summarizes all information on the Gaussian observables and provides an adequate starting point for our kinetic treatment.

The first step is standard and consists in reducing the extended one-body density to diagonal form. This can be achieved by subjecting the creation and annihilation operators to a canonical transformation of the Bogolyubov type with coefficients defined by the eigenvalue problem

$$\mathcal{X}^{\dagger}_{\mathbf{k},s}(t)\mathcal{R}_{\mathbf{k},s}(t)\mathcal{X}_{\mathbf{k},s}(t) = \mathcal{Q}_{\mathbf{k},s}(t) \,, \tag{4}$$

where $\mathcal{Q}_{\mathbf{k},s}(t)$ is a diagonal matrix written as

$$\mathcal{Q}_{\mathbf{k},s} = \begin{bmatrix} \nu_{\mathbf{k},s} & 0 \\ 0 & I_2 - \nu_{\mathbf{k},s} \end{bmatrix} = \begin{bmatrix} \langle \beta^{(\lambda')\,\dagger}_{\mathbf{k},s}(t)\beta^{(\lambda)}_{\mathbf{k},s}(t)\rangle & \langle \beta^{(\lambda')}_{-\mathbf{k},s}(t)\beta^{(\lambda)}_{\mathbf{k},s}(t)\rangle \\ \langle \beta^{(\lambda')\,\dagger}_{-\mathbf{k},s}(t)\beta^{(\lambda)\,\dagger}_{\mathbf{k},s}(t)\rangle & \langle \beta^{(\lambda')}_{\mathbf{k},s}(t)\beta^{(\lambda)\,\dagger}_{\mathbf{k},s}(t)\rangle \end{bmatrix} \tag{5}$$

and $\nu_{\mathbf{k},s}$ is the Bogolyubov quasiparticle occupation matrix given by

$$\nu_{\mathbf{k},s} = \langle \beta^{(\lambda')\,\dagger}_{\mathbf{k},s}(t)\beta^{(\lambda)}_{\mathbf{k},s}(t)\rangle = \begin{bmatrix} \nu^{(1)}_{\mathbf{k},s}(t) & 0 \\ 0 & \nu^{(2)}_{\mathbf{k},s}(t) \end{bmatrix} \,. \tag{6}$$

The unitary matrix $\mathcal{X}_{\mathbf{k},s}(t)$ which diagonalizes $\mathcal{R}_{\mathbf{k},s}(t)$, relates the fermion operators $b^{(\lambda)}_{\mathbf{k},s}(t)$ to the new quasi-fermion operators $\beta^{(\lambda)}_{\mathbf{k},s}(t)$ for $\lambda = 1, 2$ as

$$\begin{bmatrix} b^{(1)}_{\mathbf{k},s} \\ b^{(2)}_{\mathbf{k},s} \\ b^{(1)}_{-\mathbf{k},s}{}^{\dagger} \\ b^{(2)}_{-\mathbf{k},s}{}^{\dagger} \end{bmatrix} = \begin{bmatrix} \cos\varphi_{\mathbf{k}} & 0 & 0 & -\sin\varphi_{\mathbf{k}} e^{-i\gamma_{\mathbf{k}}} \\ 0 & \cos\varphi_{\mathbf{k}} & \sin\varphi_{\mathbf{k}} e^{-i\gamma_{\mathbf{k}}} & 0 \\ 0 & -\sin\varphi_{\mathbf{k}} e^{i\gamma_{\mathbf{k}}} & \cos\varphi_{\mathbf{k}} & 0 \\ \sin\varphi_{\mathbf{k}} e^{i\gamma_{\mathbf{k}}} & 0 & 0 & \cos\varphi_{\mathbf{k}} \end{bmatrix} \begin{bmatrix} \beta^{(1)}_{\mathbf{k},s} \\ \beta^{(2)}_{\mathbf{k},s} \\ \beta^{(1)}_{-\mathbf{k},s}{}^{\dagger} \\ \beta^{(2)}_{-\mathbf{k},s}{}^{\dagger} \end{bmatrix} . \tag{7}$$

Observe that the general Bogolyubov transformation breaks both chiral and charge symmetries, but we restrict the development to a special Bogolyubov transformation (to be called Nambu transformation) which breaks the chiral symmetry of our system only. Observe that the elements of (7) are spin-independent. We next describe a projection technique which permits to obtain closed approximations to the equations of motion of our system.

2 Projection Technique and Approximation Scheme

In this section we introduce the time-dependent projection technique [1]. It allows for the formulation of a mean-field expansion for the dynamics of the two-point correlation function from which one recovers the results of the Gaussian mean-field approximations in lowest order. We begin by decomposing the full density $\mathcal{F}$ as

$$\mathcal{F} = \mathcal{F}_0(t) + \mathcal{F}'(t) \ , \tag{8}$$

where $\mathcal{F}_0(t)$ is a Gaussian ansatz which achieves a Hartree-Fock factorization of traces involving more than two field operators. The Gaussian density $\mathcal{F}_0(t)$ is a time-dependent object, which acquires a particularly simple form when expressed in terms of the Bogolyubov quasi-fermion operators [8]

$$\mathcal{F}_0(t) = \prod_{\mathbf{k},s,\lambda} \left[\nu^{(\lambda)}_{\mathbf{k},s} \beta^{(\lambda)}_{\mathbf{k},s}{}^{\dagger} \beta^{(\lambda)}_{\mathbf{k},s} + (1 - \nu^{(\lambda)}_{\mathbf{k},s}) \beta^{(\lambda)}_{\mathbf{k},s} \beta^{(\lambda)}_{\mathbf{k},s}{}^{\dagger} \right] \ . \tag{9}$$

The "remainder" density $\mathcal{F}'(t)$, defined by Eq.(8), is a traceless, pure correlation density. If carried to higher orders it allows for the inclusion and evaluation of higher dynamical correlation corrections to the simplest mean-field approximation [2]. In this work we will consider only the lowest (mean-field) approximation, corresponding to $\mathcal{F}'(t) = 0$. Collisional correlations will be treated elsewhere.

3 Kinetics of the Nambu-Jona-Lasinio System

We next consider the initial conditions problem for Nambu-Jona-Lasinio system. Taking the time derivative of Eq.(4), substituting Eqs. (5-7) and calculating the traces, we obtain the equations which describe the time evolution of our system, namely

$$\dot{\nu}^{(1)}_{\mathbf{k},s} = 0 \quad \text{and} \quad , \dot{\nu}^{(2)}_{\mathbf{k},s} = 0 \tag{10}$$

$$[i\dot{\varphi}_{\mathbf{k}} + \dot{\gamma}_{\mathbf{k}} \sin\varphi_{\mathbf{k}} \cos\varphi_{\mathbf{k}}]\, e^{-i\gamma_{\mathbf{k}}} = \frac{Tr\left([\beta^{(1)}_{-\mathbf{k},s}\beta^{(2)}_{\mathbf{k},s}, H_{\mathrm{CGNM}}]\mathcal{F}_0\right)}{(1-\nu^{(1)}_{\mathbf{k},s} - \nu^{(2)}_{\mathbf{k},s})} . \tag{11}$$

Equation (10) shows that the occupation numbers of the paired natural orbitals are constant, i.e., we recover the general isoentropic character of the mean-field approximation. The complex equation of motion (11) describes the time evolution of the Nambu parameters. Writing the NJL Hamiltonian in the Nambu basis using Eqs.(7), and substituting this Hamiltonian in Eq.(11), we obtain the explicit form of this equation.

References

1. A.F.R. de Toledo Piza, in *Time-Dependent Hartree-Fock and Beyond*, ed. K. Goeke and P.-G. Reinhardt, *Lectures Notes in Physics* **171** (Springer-Verlag, Berlin, 1982); M.C. Nemes and A.F.R. de Toledo Piza, *Phys. Rev.* **C27**, 862 (1983); M.C. Nemes and A.F.R. de Toledo Piza, *Physica* **A137**, 367 (1986).
2. L.C. Yong and A.F.R. de Toledo Piza, *Mod. Phys. Lett.* **A5**, 1605 (1990); E.R. Takano Natti and A.F.R. de Toledo Piza, *Physica* **A236**, 321 (1997).
3. L.C. Yong and A.F.R. de Toledo Piza, *Phys. Rev.* **D46**, 742 (1992).
4. P.L. Natti and A.F.R. de Toledo Piza, *Phys. Rev.* **D54**, 7867 (1996); P.L. Natti and A.F.R. de Toledo Piza, *Phys. Rev.* **D55**, 3403 (1997).
5. E.R. Takano Natti, C-Y. Lin, A.F.R. de Toledo Piza and P.L. Natti, *Phys. Rev.* **D60**, 125013 (1999).
6. Y. Nambu and G. Jona-Lasinio, *Phys. Rev.* **122**, 345 (1961).
7. A. Kerman and T. Troudet, *Ann. Phys. (N.Y.)* **154**, 456 (1984).
8. J. des Cloiseaux, in *The Many-Body Physics*, edited by C. de Witt and R. Balian (Gordon and Breach, New York, 1968).

6. Structure Functions and Sum Rules

SEEKING HIGH DENSITY EFFECTS IN LEPTON-A PROCESSES

A.L. AYALA, V.P. GONÇALVES, D.R. GRATIERI, M.L.L. DA SILVA AND K.S. TORRES

Instituto de Física e Matemática, Universidade Federal de Pelotas
Caixa Postal 354, CEP 96010-090 Pelotas, RS - Brazil
ayala@ufpel.tche.br

In this contribution we calculate the nuclear structure functions F_2^c and F_L^A and the F_2^A slope in the kinematic region which could be explored in the future eA colliders, considering the high density (saturation) effects in the nuclear gluon distribution. We verify that the high parton density implies large modifications in the behavior of these observables, allowing to explicit the saturation effects in a future experimental analysis.

1 The High Density Effects

The colliders RHIC and LHC will be able to probe new regimes of dense quark matter and the possible Quark Gluon Plasma (QGP) formation. The initial conditions of the QGP will be determined by the nuclear gluon distribution $xG^A(x,Q^2)$ at small x and large A. In this region, the main uncertainty comes from the QCD description of parton evolution and high density (saturation) effects.

The parton saturation effects in $xG^A\left(x,Q^2\right)$ can be reasonably described by the Glauber-Mueller approach in the kinematic region that will be probed in the future eA colliders. The gluon distribution xG^A can be described, in the nuclear target rest frame, considering a virtual gluon probe that emits a gluon pair $G^* \to GG$, which multiple scatters inside the nucleus. In this so called partonic Glauber-Mueller approach, the gluon distribution is given by[1]

$$xG^A\left(x,Q^2\right)[GM] = \frac{Q^2}{4\pi^2\alpha_s}\int_0^1 dz \int \frac{d^2r_t}{\pi}\left|\Psi_t^{g^*}\right|^2 \sigma^{gg+A}\left(z,r_t^2\right) , \qquad (1)$$

where the GG cross section with the nucleus is written

$$\sigma^{gg+A}\left(z,\vec{r_t}\right) = \int \frac{d^2\vec{b}_t}{\pi} 2\left[1 - e^{-\frac{1}{2}\sigma^{gg+N}(z,\vec{r}_t)S(\vec{b}_t)}\right] . \qquad (2)$$

In the above expression, r_t is the transverse separation os the gluon pair, z is the gluon momentum fraction and $S(\vec{b}_t)$ is the nuclear profile, which is a function of the impact parameter $\vec{b_t}$. Expanding the exponential, the first term describes the low density regime and corresponds to the DGLAP evolution in DLA limit. The full eikonal expression describes the multiple scattering interference, which reduces the cross section and takes into account the high density regime in parton evolution.

2 Improved Nuclear Gluon Distribution

We investigated the role of the high density effects in the charm component of the nucleus $F_2^{c\,A}(x,Q^2)$, in the longitudinal structure function $F_L^A(x,Q^2)$ and in the derivative $dF_2^A/d\ln Q^2$. These deep inelastic eA observables can be measured in future experiments. In our approach we assume that the medium effects are determined by a combination of non-perturbative and perturbative effects[2]. The non-perturbative medium effects are described by the so called EKS analysis [3] (Eskola, Kolhinen and Salgado) which describes the current nuclear data (F_2^A and DY) taking into account the DGLAP evolution with suitable initial condition. The parton saturation effects at small x are estimated by the Glauber-Mueller approach, derived in perturbative QCD. In this approach the improved nuclear gluons distribution is given by

$$xG_A(x,Q^2) \;=\; xG_A(x,Q^2)[GM] \;-\; xG_A(x,Q^2)[DLA] \;+\; xG_A(x,Q^2)[EKS]\,.$$

Taking into account this gluon distribution, we predict the behavior of the eA observables. The charm component of the nuclear structure function $F_2^{c\,A}$ is described by the photon gluon fusion mechanism and is closely related with the gluon content of the nucleus:

$$\frac{1}{x}F_2^{c\,A}(x,Q^2,m_c^2) = 2e_c^2\frac{\alpha_s(\mu'^2)}{2\pi}\int_{ax}^{1}\frac{dy}{y}C_{g,2}^c\left(\frac{x}{y},\frac{m_c^2}{Q^2}\right)g^A(y,\mu'^2)\,, \tag{3}$$

where $\alpha_s = 1+4m_c^2$, $\mu' = 4m$ and $C_{g,2}^c$ describes the γ^*-gluon fusion with charm pair production.

Fig. (1) shows the relative contribution of the charm production to the eA cross section, $F_2^{c\,A}/F_2^A$ as a function of the parton momentum fraction x and the virtual photon momentum Q^2. We see that the parton saturation effects will be important for $x \leq 10^{-4}$ and for 5 GeV^2 $< Q^2 <$ 10 GeV^2. The longitudinal distribution F_L^A is obtained from the F_2^A and from the gluon distribution xG^A by

$$F_L^A = \frac{\alpha_s(Q^2)x^2}{2\pi}\int_x^1\frac{dy}{y^3}\{\frac{8}{3}F_2(y,Q^2) + 4\sum_q e_q^2(1-\frac{x}{y})yG(y,Q^2)\} \tag{4}$$

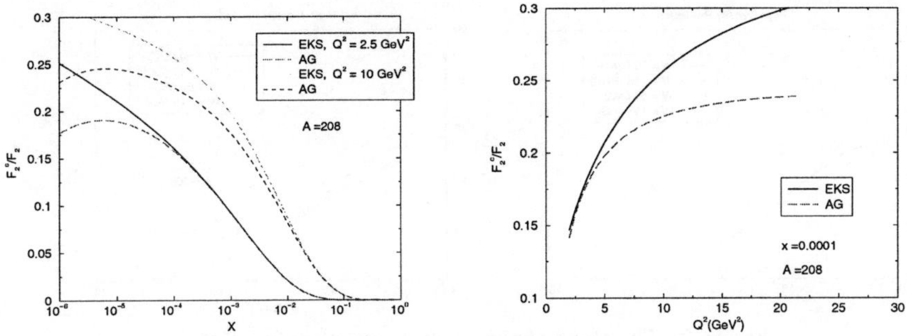

Figure 1. The charm component of the nuclear structure function.

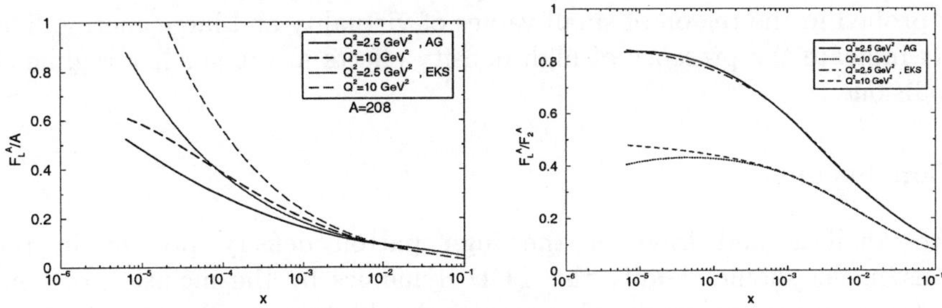

Figure 2. The nuclear longitudinal structure function F_L^A.

Fig. (2) shows the normalized F_L^A and the relative contribution of F_L to F_2 in a nucleus. The parton saturation effects are sizeable for $x < 10^{-3}$ and for $2.5\ \mathrm{GeV}^2 < Q^2 < 10\ \mathrm{GeV}^2$ which includes almost all kinematic region explored in the future colliders. The contribution of the parton saturation effects to the F_L/F_2 ratio are sizable just for $x < 10^{-4}$ and $Q^2 \approx 10\,GeV^2$. This behavior is related to the y integration, which includes contribution from large x region, where the effects associated to the high parton density can be disregarded.

In the DGLAP approach, the derivative $\frac{dF_2^A}{d\ln Q^2}$ is also xG^A dependent and is given by

$$\frac{dF_2^A}{d\ln Q^2} = \frac{5\alpha_s(Q^2)}{9\pi}\int_0^{1-x} dz[(1-z)^2 + z^2]\frac{x}{1-z}G(\frac{x}{1-z}, Q^2). \tag{5}$$

The behavior of the nuclear structure function slope is analyzed in Fig. (3)

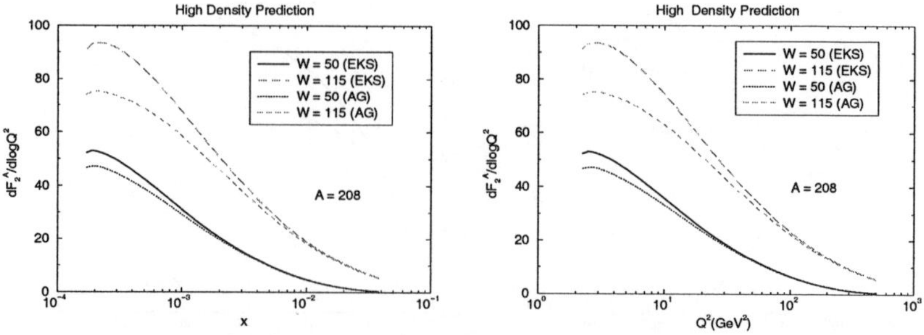

Figure 3. The slope of the nuclear structure function.

for different center of mass energies W, considering the constraint $x = \frac{Q^2}{W^2}$, which is considered in the HERA data (the Caldweel plot). Smaller values of x are probed in the region of small values of virtuality and large energy. The results indicate the presence of high density effects in the small x regime of eA collisions.

3 Conclusions

At small x and large A the high parton density present in the hadrons/nuclei should modify the QCD dynamics by the inclusion of non-linear terms in the evolution. These terms should diminish the growth of the nuclear gluon distribution, leading to the saturation of the parton distributions. The presence of parton saturation effects shall modify the behavior of xG_A and the observables directly dependent. Our analysis of F_L^A, $F_2^{c\,A}$ and F_2^A slope indicate that these effects are large in the kinematic region which will be probed in the future eA colliders.

We conclude by emphasizing the importance of determining gluon shadowing in eA processes to fix theoretically the initial conditions in AA collisions, since the early evolution of these collisions is governed by the dominant role of gluons as well as the signatures of the QGP.

References

1. A.L. Ayala, M.B. Gay Ducati and E.M. Levin, *Nucl. Phys.* B **439**, 305 (1997).
2. A. L. Ayala Filho, V. P. Gonçalves. *Eur. Phys. J.* **C20** (2001) 343.
3. K. J. Eskola, V. J. Kolhinen, C. A. Salgado. *Eur. Phys. J.* **C9** (1999) 61; K.J. Eskola, V. J. Kolhinen, P. V. Ruuskanen. *Nucl. Phys.* **B535** (1998) 351.

ρDD FORM FACTOR

A. LOZÉA, M.E. BRACCO AND M. CHIAPPARINI,
Instituto de Física, Universidade do Estado do Rio de Janeiro,
R. São Francisco Xavier 524, Maracanã, 20559-900 Rio de Janeiro-RJ, Brazil.
albertoloze@bol.com.br

M. NIELSEN AND F.S. NAVARRA.
Instituto de Física, Universidade de São Paulo. C.P. 66318,
05389-970 São Paulo-SP, Brazil.

We calculated the form factor and the coupling constant for the vertex ρDD in the cases where the mesons ρ and D are off-shell, using the techniques of the QCD Sum Rules. We found that the moment dependences are very different in both cases, but they give some coupling constant in the vertex ρDD.

1 Introduction

In the present work we will further investigate form factors involving heavy mesons. In particular, we want to check whether the coupling constant is the same regardless of which particle is off-shell. For these purposes we consider the vertex ρDD and compute form factors for the cases where the D is off-shell, the ρ is off-shell and then compare the results and extract the coupling constant. This coupling constant is important to determine some cross-section that can estimate the J/Ψ suppression.

2 Calculation

Following the QCDSR formalism described in previous works [1] we write the three-point correlation function associated with a ρDD vertex, which is given by

$$\Pi_{\mu}^{(D)}\left(p,p'\right) = \int d^4x\, d^4y \langle 0|T\left\{j_+(x)j_0(y)j_{\mu}^{\dagger}(0)\right\}|0\rangle_{\rm QCD}\, e^{ip'.x}\, e^{-i(p'-p).y}, \qquad (1)$$

for a D off-shell meson, and by

$$\Pi_{\mu}^{(\rho)}\left(p,p'\right) = \int d^4x\, d^4y \langle 0|T\left\{j_+(x)j_{\mu}^{\dagger}(y)j_0(0)\right\}|0\rangle_{\rm QCD}\, e^{ip'.x}\, e^{-i(p'-p).y}, \qquad (2)$$

for the ρ off-shell meson, where $j_+ = i\bar{d}\gamma_5 c$, $j_{\mu}^{\dagger} = \bar{u}\gamma_{\mu}d$ and $j_0 = i\bar{c}\gamma_5 u$ are the interpolating fields for D^+, D and ρ^+, respectively, with u, d and c being the up, down and charm quark fields. The general expression for the vertex function in Eqs. (1) and (2) has two independent structures. Writing Π_{μ} in terms of the invariant amplitudes associated with these two structures, we have:

$$\Pi_\mu\left(p,p',q^2\right) = \Pi_p\left(p^2,p'^2,q^2\right)p_\mu + \Pi_{p'}\left(p^2,p'^2,q^2\right)p'_\mu. \tag{3}$$

We can write a double dispersion relation in the QCD side for each one of the invariant amplitudes $\Pi_i(i=p,p')$, over the virtualities p^2 and p'^2 holding $Q^2=-q^2$ fixed:

$$\Pi_i(p^2,p'^2,Q^2) = -\frac{1}{4\pi^2}\int_{u_{min}}^{\infty} du \int_{s_{min}}^{\infty} ds\, \frac{\rho_i(s,u,Q^2)}{(s-p^2)(u-p'^2)}, \tag{4}$$

where $\rho_i(s,u,Q^2)$ equals the double discontinuity of the amplitude $\Pi_i(p^2,p'^2,Q^2)$ on the cuts $s_{min}=0$ and $u_{min}=m_c^2[1+s/(Q+m_c^2)]$ in the case of the D off-shell and $s_{min}=m_c^2[1+Q^2/(u-m_c^2)]$ and $u_{min}=m_c^2[1+Q^2/(s-m_c^2)]$ in case of ρ off-shell, the dispersion relation being now written in terms of the ρ and the D meson momenta.

The phenomenological side of the vertex function is obtained by considering the contribution of the ρ and one D meson, or the two D meson states to the matrix element in Eqs. (1) and (2), respectively:

$$\begin{aligned}\Pi_\mu^{phen}\left(p,p',Q^2\right) = &-\frac{m_D^4 f_D^2 m_\rho^2}{m_c^2 g_\rho}\frac{g_{\rho DD}^{(D)}\left(Q^2\right)}{(p'^2-m_D^2)(p^2-m_\rho^2)(Q^2+m_D^2)}\\ &\times\left(-2p'_\mu + \frac{m_D^2+m_\rho^2+Q^2}{m_\rho^2}p_\mu\right) + \text{higher resonances,}\end{aligned} \tag{5}$$

$$\begin{aligned}\Pi_\mu^{phen}\left(p,p',Q^2\right) = &-\frac{m_D^4 f_D^2 m_\rho^2}{m_c^2 g_\rho}\frac{g_{\rho DD}^{(\rho)}\left(Q^2\right)}{(p'^2-m_D^2)(p^2-m_D^2)(Q^2+m_\rho^2)}\\ &\times\left(p'_\mu + p_\mu\right) + \text{higher resonances,}\end{aligned} \tag{6}$$

where m_D, m_ρ, f_D and g_ρ are the masses and decay constants of the mesons D and ρ, respectively, and $g_{\rho DD}^{(M)}(Q^2)$ is the form factor at the ρDD vertex when the meson M is off-shell. The contribution of higher resonances in Eqs. (5) and (6) will be taken into account as usual in the standard form of continuum contribution from the thresholds s_0 and u_0 [2]. Finally we perform a double Borel transformation [2] in both variables $P^2=-p^2\to M^2$ and $P'^2=-p'^2\to M'^2$, for each invariant amplitude. Using the Borel region $3\leq M^2\leq 5\,\text{GeV}^2$ (ρ off-shell) and $0.3\leq M^2\leq 0.9\,\text{GeV}^2$ (D off-shell) we found a good stability for $g_{\rho DD}^{(M)}(Q^2)$ at $Q^2=1\,\text{GeV}^2$. In Fig. 1 we show for both cases, the stability.

Fixing $M'^2=4.7\,\text{GeV}^2$ (with $M^2=0.8\,\text{GeV}^2$ in the D off-shell case) we show, in Fig. 2, the momentum dependence of the form factor (circles). We can note that the Q^2 dependence of the form factor can be well reproduced by the monopole parameterization (dashed line)

$$g_{\rho DD}^{(D)}(Q^2) = \frac{37.41}{Q^2+12.12}, \tag{7}$$

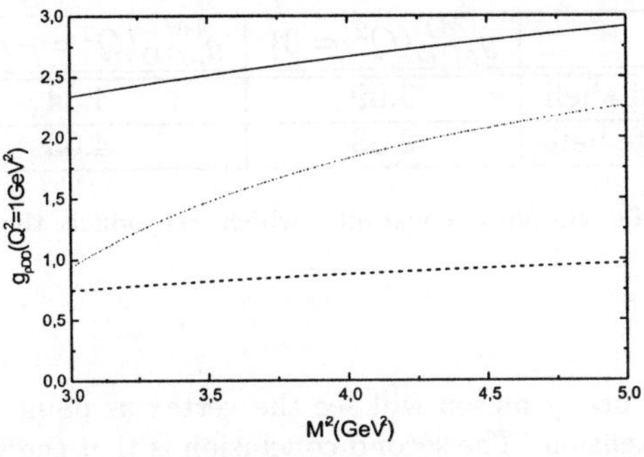

Figure 1. M'^2 dependence of the $g^{(M)}_{\rho DD}(Q^2)$. The dashed line gives the QCDSR result for ρ off-shell case and the dot-dashed and solid lines give the results for D off-shell case in the p_μ and p'_μ structures respectively.

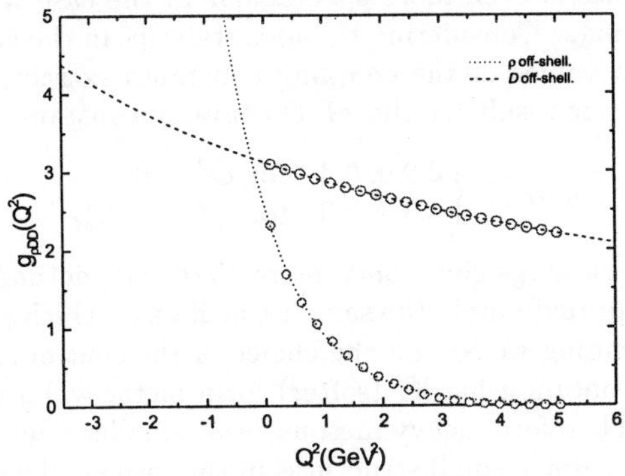

Figure 2. Momentum dependence of the ρDD form factor. The lines give the parameterizations of the QCDSR results.

for D off-shell case and can be well reproduced by the exponential parameterization (dot line)

$$g^{(\rho)}_{\rho DD}(Q^2) = 2.53\, e^{-\frac{Q^2}{0.98}} \tag{8}$$

for ρ off-shell case.

In order to extract the coupling constant $g_{\rho DD}$ from the form factor, we need first to define it in both cases. In Ref. [3] the coupling constant is defined as the value of the form factor at $Q^2 = 0$. However, in Ref. [4] the coupling is defined as the value of the form factor at $Q^2 = -m_M^2$, where m_M is the mass of the off-shell meson. The coupling constants resulting of the parameterizations are given in Table 1. There are two very important conclusions that we can draw from the results in Table 1. First of all is the fact that the form factor is harder if the off-shell meson is heavy, implying that the size of the vertex depends on the exchanged

	$g^{(M)}_{\rho DD}(Q^2 = 0)$	$g^{(M)}_{\rho DD}(Q^2 = -m^2_M)$
D off-shell	3.09	4.34
ρ off-shell	2.53	4.63

Table 1. Values of the coupling constants which reproduce the QCDSR results for $g^{(M)}_{\rho DD}(Q^2)$.

meson. Therefore, a heavy meson will see the vertex as point-like, whereas a light meson will see its extension. The second conclusion is that the value of the coupling constant extracted from the QCDSR results depends, of course on its definition. Having in mind that the inherent precision off the method is of order of 20%, we can say that the results nearly coincide for the two cases considered. This agreement between the two values is even more spectacular in the case where the coupling is extracted at $Q^2 = -m^2_M$. Considering the uncertainties in the continuum threshold, and the difference in values of the coupling constants extracted when the D or ρ mesons are off-shell, our result for the ρDD coupling constant is:

$$g_{\rho DD} = \begin{cases} 2.9 \pm 0.4 & \text{for } Q^2 = 0 \\ 4.3 \pm 0.3 & \text{for } Q^2 = -m^2_M \end{cases} . \tag{9}$$

Our results for the couplings show once more that this method is robust, yielding numbers which are approximately the same regardless of which particle we choose to be off-shell and depending weakly on the choice of the continuum threshold. As for the form factors, we obtain a harder (softer) form factor when the off-shell particle is heavier (lighter). Therefore, heavy mesons "see" smaller sizes in the light mesons, but the latter cannot resolve small structures in the former. These results are better shown in [5], what agrees with our previous works [6]. Our program will continue and studies of other vertices are in progress.

References

1. M.A. Shifman, A.I. Vainshtein, V.I. Zakharov, *Nuc. Phys.* B **147** (1979) 385-447 and M.A. Shifman, A.I. Vainshtein, V.I. Zakharov, *Nuc. Phys.* B **147** (1979) 519-534
2. B. L. Ioffe, A. V. Smilga, *Nucl. Phys.* B**216** (1983), 373-407.
3. Z. Lin, C.M. Ko, *Phys. Rev.* **C62**, 034903 (2000).
4. Z. Lin, C.M. Ko, B. Znang, *Phys. Rev.* **C61**, 024904 (2000).
5. M. E. Bracco, M. Chiapparini, A. Lozéa, F. S. Navarra and M. Nielsen, *Phys. Let.* **B 521**, 1-6 (2001).
6. M. E. Bracco, M. Chiapparini, C. L. Schat, F. S. Navarra and M. Nielsen, *Phys. Let.* **B 489**, 319-328 (2000) and F. S. Navarra, M. Nielsen and M. E. Bracco, *Phys.Rev.* **D65** 037502 (2002).

CHARMONIUM - PION CROSS SECTION FROM QCD SUM RULES

F.S. NAVARRA, M. NIELSEN,

Instituto de Física, Universidade de São Paulo, C.P. 66318, 05315-970 São Paulo, SP, Brazil

navarra@if.usp.br

R.S. MARQUES DE CARVALHO AND G. KREIN

Instituto de Física Teórica, Universidade Estadual Paulista, Rua Pamplona 145, 01405-900 São Paulo, SP, Brazil

The $J/\psi\ \pi \to \bar{D}\ D^*$, $D\ \bar{D}^*$, $\bar{D}^*\ D^*$ and $\bar{D}\ D$ cross sections as a function of $\sqrt{s}$ are evaluated in a QCD sum rule calculation. We study the Borel sum rule for the four point function involving pseudoscalar and vector meson currents, up to dimension four in the operator product expansion. We find that our results are smaller than the $J/\psi\ \pi \to$ charmed mesons cross sections obtained with models based on meson exchange, but are close to those obtained with quark exchange models.

Charmonium-hadron cross sections are of crucial importance in the context of quark-gluon plasma physics [1]. Small J/ψ-hadron dissociation cross sections may favor an interpretation of the recent Pb + Pb data in terms of the production of a new phase of matter. Part of these interactions happens in the early stages of the nucleus - nucleus collisions and therefore at high energies ($\sqrt{s} \simeq 10 - 20$ GeV) and one may try to apply perturbative QCD. However, even in this regime, non-perturbative effects may be important [2].

In this work we use the QCD sum rules (QCDSR) technique [3,4] to study the $J/\psi - \pi$ dissociation. In view of our relatively poor understanding of J/ψ reactions in nuclear matter and considering the large discrepancies between different model estimates, we believe that our work adds to a better understanding of this important topic.

We consider all four channels $J/\psi\ \pi \to$: $\bar{D}\ D^*$, $D\ \bar{D}^*$, $\bar{D}\ D$ and $\bar{D}^*\ D^*$. Let us start with the the four-point function for the process $J/\psi\ \pi \to \bar{D}\ D^*$:

$$\Pi_{\mu\nu} = i \int d^4x\ d^4y\ d^4z\ e^{-ip_1.x}\ e^{ip_3.y}\ e^{ip_4.z} \times \langle 0|T\{j_\pi(x) j_\nu^{D^*}(y) j_\mu^{\psi}(0) j_D(z)\}|0\rangle\ , \tag{1}$$

with the currents given by $j_\pi = \overline{d} i\gamma_5 u$, $j_\nu^{D^*} = \overline{u}\gamma_\nu c$, $j_\mu^\psi = \overline{c}\gamma_\mu c$ and $j_D = \overline{c} i\gamma_5 d$ [4], where c, u and d are the charm, up and down quark fields respectively, and

p_1, p_2, p_3 and p_4 are the four-momenta of the mesons π, J/ψ, D^* and D respectively, with $p_1 + p_2 = p_3 + p_4$.

The phenomenological side of the correlation function, $\Pi_{\mu\nu}$, is obtained by the consideration of J/ψ, π, D and D^* state contribution to the matrix element in Eq. (1):

$$\Pi_{\mu\nu}^{phen} = -\frac{m_\pi^2 F_\pi}{m_u + m_d} \frac{m_D^2 f_D}{m_c} \frac{m_{D^*} f_{D^*} \, m_\psi f_\psi \, \mathcal{M}^{\alpha\beta}}{(p_1^2 - m_\pi^2)(p_4^2 - m_D^2)} \times \frac{g_{\mu\alpha} - p_{2\mu}p_{2\alpha}/m_\psi^2}{p_2^2 - m_\psi^2} \frac{g_{\nu\beta} - p_{3\nu}p_{3\beta}/m_{D^*}^2}{p_3^2 - m_{D^*}^2} + \text{h. r.}\,, \tag{2}$$

where h.r. means higher resonances and the hadronic amplitude for the process $J/\psi\ \pi \to \bar{D}\ D^*$ is given by

$$\mathcal{M} = \mathcal{M}_{\mu\nu}(p_1, p_2, p_3, p_4)\ \epsilon_2^\mu\ \epsilon_3^{*\nu}\,. \tag{3}$$

We note that one has $1/p_1^2$ pole in Eq. (2) in the limit of a vanishing pion mass. Following [4,5,6,7], we can write a sum rule at $p_1^2 = 0$ and single out the leading terms in the operator product expansion (OPE) of Eq. (1) that match the $1/p_1^2$ term. The perturbative diagram does not contribute with $1/p_1^2$ and, up to dimension four, only the diagrams proportional to the quark condensate contribute. After collecting the $1/p_1^2$ terms on the theoretical side and taking the limit $p_{1\mu} \to 0$ in the residue of the pion pole, one obtains for the contribution of these two graphs

$$\Pi_{\mu\nu}^{<\bar{q}q>} = -\frac{2m_c\langle\bar{q}q\rangle}{p_1^2} \frac{p_{1\nu}(p_{1\mu} + p_{2\mu} - 2p_{3\mu}) - p_{1\mu}p_{2\nu}}{(p_3^2 - m_c^2)(p_4^2 - m_c^2)}\,. \tag{4}$$

Contracting the hadronic amplitude with the numerators of J/ψ and D^* propagators in Eq. (2) and comparing with Eq. (4), the structure defining $\mathcal{M}_{\mu\nu}$ in Eq. (3) is easily identified. Therefore, defining

$$\mathcal{M}_{\mu\nu} = \Lambda_{DD^*}\ (p_{1\mu}p_{1\nu} - p_{1\mu}p_{2\nu} - 2p_{1\nu}p_{3\mu})\,, \tag{5}$$

we can write a sum rule for Λ_{DD^*} in any of the three structures appearing in Eq. (5). To improve the matching between the phenomenological and theoretical sides we follow the usual procedure and make a single Borel transform, with all the external momenta (except p_1^2) taken to be equal: $-p_2^2 = -p_3^2 = -p_4^2 = P^2 \to M^2$. Next, we calculate Λ_{DD^*} (for details see Ref.[8]). We can proceed in a similar way to consider the process $J/\psi\ \pi \to \bar{D}\ D$ ($J/\psi\ \pi \to \bar{D}^*\ D^*$). In this case we have to change the current $j_\nu^{D^*}$ (j_D) in Eq. (1) to $\bar{u}i\gamma_5 c$ ($\bar{c}\gamma_\alpha d$).

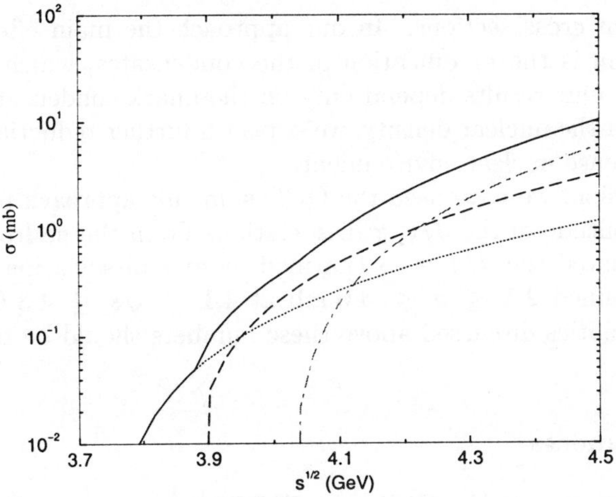

Figure 1. Total cross sections of the processes $J/\psi\ \pi \to \bar{D}\ D^* + D\ \bar{D}^*$ (dashed line), $\bar{D}\ D$ (dotted line) and $\bar{D}^*\ D^*$ (dot-dashed line). The solid line gives the total $J/\psi\ \pi$ dissociation cross section.

Having the QCD sum rule results for the amplitude of the three processes $J/\psi\ \pi \to \bar{D}\ D^*$, $\bar{D}\ D$, $\bar{D}^*\ D^*$, we can evaluate the differential cross section. We show, in Fig. 1, the cross section for the $J/\psi\ \pi$ dissociation. It is important to keep in mind that, since our sum rule was derived in the limit $p_1 \to 0$, we can not extend our results to large values of $\sqrt{s}$. Also, since the perturbative contribution is absent in our calculation, we were not able to properly disentangle the continuum contribution and our cross section may include contributions from higher states. Whereas they are certainly not important in the case of the pion, they may give some contribution to the heavy currents. Therefore our $J/\psi\ \pi$ cross section may implicitly include (at least partially) the process $\psi'\ \pi$. For this reason, our numbers might be regarded as upper bounds.

At higher energies, due to our low energy approximation, our approach gradually loses validity. In the fiducial region, close to threshold, $4.1 \leq \sqrt{s} \leq 4.3$ GeV, we find $2.5 \leq \sigma \leq 4.0$ mb and these values are much smaller than those obtained with the effective Lagrangians without form factors in the hadronic vertices, but agree in order of magnitude with the quark model calculations of [9].

Finally, we should mention that we have been studying the dissociation processes of the J/ψ in vacuum and the quantities relevant for QGP physics are in medium cross sections. In our approach the main effect introduced by the medium is the modification of the condensates, which is thought to be very mild. Our results depend only on the quark condensate and since it decreases with the nuclear density, we expect a further reduction in our cross section in a dense nuclear environment.

In conclusion, we have used the QCD sum rule approach to evaluate the hadronic amplitude of the $J/\psi\ \pi$ dissociation. From the hadronic amplitude we have evaluated the $J/\psi\ \pi \rightarrow$ charmed mesons dissociation cross section, and have obtained $2.5 \leq \sigma \leq 4.0$ mb at $4.1 \leq \sqrt{s} \leq 4.3$ GeV. In view of the uncertainties discussed above these numbers should be taken as upper limits.

Acknowledgments

This work was supported by CNPq and FAPESP (contract numbers 98/06590-2, 99/12987-5, 00/04422-7 and 00/12619-5).

References

1. For recent reviews see B. Müller, *Nucl. Phys.* **A661** (1999) 272 ; T. Barnes, E.S. Swanson and C.Y. Wong, nucl-th/0006012.
2. H.G. Dosch, F.S. Navarra, M. Nielsen and M. Rueter, *Phys. Lett.* **B466** (1999) 363.
3. M.A. Shifman, A.I. Vainshtein and V.I. Zakharov, *Nucl. Phys.* **B120** (1977) 316.
4. L.J. Reinders, H. Rubinstein and S. Yazaki, *Phys. Rep.* **127** (1985) 1.
5. S. Choe, M.K. Cheoun and S.H. Lee, *Phys. Rev.* **C53** (1996) 1363.
6. M.E. Bracco, F.S. Navarra and M. Nielsen, *Phys. Lett.* **B454** (1999) 346.
7. F.S. Navarra, M. Nielsen, M.E. Bracco, M. Chiapparini and C.L. Schat, *Phys. Lett.* **B489** (2000) 319.
8. F.S. Navarra, M. Nielsen, R.S. Marques de Carvalho and G. Krein, *Phys. Lett.* **B529** (2002) 87.
9. Cheuk-Yin Wong, E. S. Swanson and T. Barnes, *Phys. Rev.* **C62** (2000) 045201, and references therein.

A QCD SUM RULE APPROACH TO THE κ RESONANCE

H.G. DOSCH, E.M. FERREIRA, F.S. NAVARRA AND M. NIELSEN

Instituto de Física, Universidade de São Paulo, C.P. 66318, 05389-970 São Paulo, SP, Brazil

navarra@if.usp.br

We investigate semileptonic decays of D-mesons into scalar hadronic states. We consider two extreme cases: a) the meson decays directly into an uncorrelated scalar state of two mesons and b) the decay proceeds via resonance formation. QCD sum rules including instanton contributions are used to calculate total and differential decay rates under the two assumptions.

In a recent analysis, done by I. Bediaga and his E-791 collaborators (see in these proceedings) of D- and D_s-meson nonleptonic decays distinct signals for strong enhancements in S-wave $\pi\pi$ and $K\pi$ channels have been observed, reviving the interest for these states. In the following we shall refer to the signal in the $K\pi$ chanel as κ. The enhancements can be well described by a Breit-Wigner-type resonance form, with the correct threshold behaviour $\rho_{X,BW}(s) = \Gamma_X(s) m_X / [(s - m_X^2)^2 + m_X^2 \Gamma_X(s)^2]$ where the subscript X stands for κ. For the S-wave $K\pi$ channel, $m_a = m_K$, $m_b = m_\pi$, the following values were found: $m_\kappa = 0.797 \pm 0.019 \pm 0.042$ GeV; $\Gamma_{0\kappa} = 0.410 \pm 0.043 \pm 0.085$ GeV.

We consider two limiting cases: i) the observed two-meson final state couples to the corresponding quark-antiquark state only through the κ resonance, and: ii) the quark-antiquark state couples to uncorrelated final mesons in an S-wave. We refer to our paper for details and clarifications [1]. We investigate the semileptonic decays $D \to X \ell \bar{\nu}_\ell$ where X might be the κ or an uncorre-

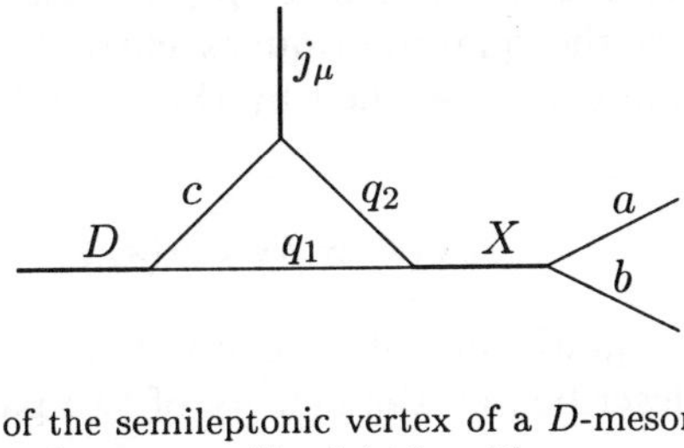

Figure 1. Schematic picture of the semileptonic vertex of a D-meson decay. c, q_1, q_2 are quark lines, j_μ denotes the weak current. X might be either a resonance or represent an uncorrelated two-meson state of mesons a anf b.

lated $K\pi$ pair in the S-state, as depicted in Fig. 1. The semileptonic decay of a D-meson with momentum p_D into a scalar state with total momentum p_X and invariant mass $\sqrt{s} = \sqrt{p_X^2}$ is described by the two form factors of the matrix element of the weak current

$$\langle X|j_\mu|D\rangle = (p_D + p_X)_\mu f_{+X}(t) + (p_D - p_X)_\mu f_{-X}(t)\,, \tag{1}$$

where $t = (p_D - p_X)^2$. The differential semileptonic decay rate is given by

$$\frac{d^2\Gamma(s,t)}{ds\;dt} = \frac{G_F^2|V_{cq_2}|^2}{192\pi^3 m_D^3}\lambda^{3/2}(m_D^2, s, t) f_{+X}^2(t)\frac{\rho_X(s)}{\pi}\,, \tag{2}$$

where G_F is the Fermi coupling constant and V_{cq_2} the CKM transition element from the charmed quark to the quark q_2. The D meson in the initial state is interpolated by the pseudoscalar current $j_D(x) = \bar{c}(x)i\gamma_5 q_1(x)$ and the final hadronic state X is interpolated by the scalar current $j_X(x) = \bar{q}_1(x)q_2(x)$. where q_1 and q_2 represent light or strange quark fields. The semileptonic decay rate is obtained from the time ordered product of the two interpolating fields and the weak current $j_\mu^W = \bar{q}_2\gamma_\mu(1-\gamma_5)c$

$$T_{\mu X}(p_D^2, p_X^2, t) = i^2 \int d^4x d^4y\, \langle 0|\mathrm{T}[j_D(x) j_\mu^W(0) j_X(y)]|0\rangle e^{i(p_D\cdot x - p_X\cdot y)}\,. \tag{3}$$

On the phenomenological side we concentrate on the relevant form factor f_+ and introduce the factor T_+ which multiplies the vector $(p_D + p_X)_\mu$

$$T_{+X}^{phen}(p_D^2, p_X^2, t) = \frac{1}{\pi^2}\int ds_D\, ds_X\, A_D A_X \frac{\rho_D(s_D)}{s_D - p_D^2}\frac{\rho_X(s_X)}{s_X - p_X^2} f_{+X}(t, s_D, s_X)\,. \tag{4}$$

For the D-meson the density $\rho_D(s_D)$ is given by $\rho_D(s_D) = \pi\delta(s_D - m_D^2)$ and we obtain $A_D = f_D m_D^2/m_c$. f_D is the D-meson decay constant in the conventional notation. For the density ρ_X we use the two extreme *ansätze* mentioned above: i) the quark-antiquark current j_X in the $J = 0^+$ state couples to the κ-resonance described by the Breit-Wigner distribution given above

$$\rho_X(s) = \rho_{X,BW}(s)\,. \tag{5}$$

or: ii) the quark-antiquark current j_X couples to an uncorrelated meson pair, the density being described by the density of two-particle phase space

$$\rho_X(s) = \frac{\pi}{16\pi^2}\frac{\lambda(s, m_a^2, m_b^2)^{1/2}}{s}\,, \tag{6}$$

where m_a, m_b are the masses of the mesons in the final state. The three-point function can be evaluated by perturbative QCD with nonperturbative power corrections:

$$T_{+X}^{theor}(p_D^2, p_X^2, t) = \frac{1}{\pi^2}\int_{s_{D,th}} ds_D \int_{s_{X,th}} ds_X \frac{\sigma_{+X}(s_D, s_X, t)}{(s_D - p_D^2)(s_X - p_X^2)} + \frac{C_{ij}\,\langle O_{ij}\rangle}{(p_D^2)^i (p_X^2)^j}\,. \tag{7}$$

The perturbative contribution is contained in the double spectral function σ_{+X}. The Wilson coefficients C_{ij} multiplying the power corrections ($\langle O_{ij}\rangle$) can be evaluated in perturbative QCD. By equating the Borel transforms of the phenomenological expression in (4) and that of the "theoretical expression", Eq. (7), we obtain the desired sum rule. The final sum rule for the form factor is obtained by inserting the QCDSR expressions for f_D and A_X:

$$f_{+X}(t, m_D^2, \bar{s}_X) =$$
$$e^{m_D^2/M_D^2}\left(\frac{m_D^2}{m_c} f_D(M_D'^2) A_X(M_X'^2)\frac{1}{\pi}\int_{(m_a+m_b)^2}^{s_{0X}} ds_X\,\rho_X(s_X) e^{-s_X/M_X^2}\right)^{-1}$$
$$\times\left(\int_{s_{D,th}}^{s_{0D}} ds_D \int_{s_{X,th}}^{s_{0X}} ds_X\,\sigma_{+X}(s_D, s_X, t) e^{-s_D^2/M_D^2}\, e^{-s_X/M_X^2}\right)\,. \tag{8}$$

The strange and charm masses are $m_s = 0.16$ GeV, $m_c = 1.3$ GeV, the up and down quark masses are put to zero. We take for the non-strange quark condensate $\langle\bar{q}q\rangle = -(0.24)^3$ GeV^3 , for the strange quark condensate $\langle\bar{s}s\rangle = 0.8\,\langle\bar{q}q\rangle$, and for the mixed quark-gluon condensate $\langle\bar{q}\sigma Gq\rangle\langle\bar{q}q\rangle = m_0^2\langle\bar{q}q\rangle$ with $m_0^2 = 0.8$ GeV^2.

We start with the decay $D \to \kappa\ell\bar{\nu}_\ell$ and first consider the case where the scalar quark current couples directly to the κ-signal through a Breit-Wigner distribution (5). In the range $0 \le t \le 0.5$ GeV^2 $f_{+\kappa}$ can be very well approximated by a monopole expression $f_{+\kappa}(t) = f_{+\kappa}(0)/[1 - \frac{t}{M_P^2}]$ and extrapolated to the full physical region. For 1.4 GeV$^2 \le s_{0\kappa} \le 1.8$ GeV^2 we find for the form factor at $t = 0$: $0.48 \le f_{+\kappa}(0) \le 0.55$; and for the pole mass: 1.9 GeV $\le M_P \le 2.2$ GeV. For the total semileptonic decay width we obtain

$$\Gamma(D \to \kappa\ell\nu) = (5.5 \pm 1.0) \times 10^{-15}\ \text{GeV} \tag{9}$$

where we have used $V_{cs} = 0.97$. The same calculation done in an – unjustified – zero width approximation would yield a total semileptonic decay width which is about 20 % larger. The spectral distribution $d\Gamma/d\sqrt{s}$, Eq. (2), where $\sqrt{s}$ is the invariant mass of the πK state, is given in Fig. 2, solid line. Next we investigate the same decay under the assumption that the scalar current does not couple to a resonance, but to an uncorrelated πK pair in an S-state.

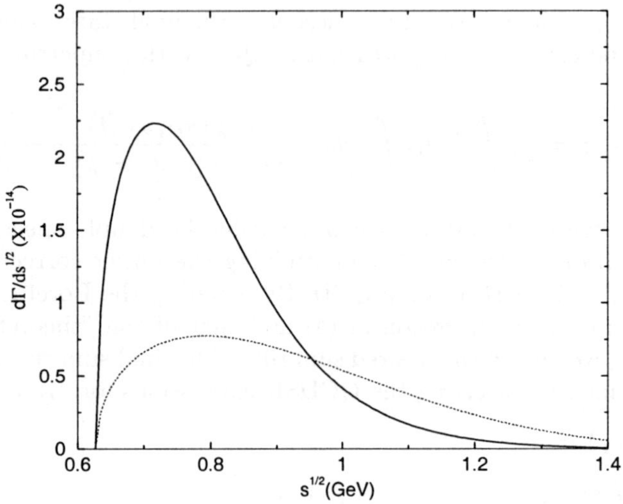

Figure 2. Invariant mass distribution $d\Gamma(\sqrt{s})/d\sqrt{s}$ of the hadronic final state in the decay $D \to (K\pi)_S \ell\nu$. Solid: resonance formation; dots: uncorrelated $K\pi$.

The masses of the pole fit to the t-dependence are practically the same as for the resonance case; for the form factor at $t = 0$ we obtain $5.0\,\mathrm{GeV}^{-1} f_{+\kappa}(0) \leq 7.1\,\mathrm{GeV}^{-1}$ and the total width comes out for the case of an uncorrelated πK pair in an S state as

$$\Gamma(D \to (K\pi)_S \ell\bar{\nu}_\ell) = (3.7 \pm 1.1) \times 10^{-15}\ \mathrm{GeV}. \qquad (10)$$

The spectral distribution for this case is also shown in Fig. 2, with a dotted line. Although there is no resonance formation the resulting distribution shows a maximum at approximately the mass of the κ, which is an effect of the decrease of the total phase space near the kinematical limits.

References

1. H.G. Dosch, E.M. Ferreira, F.S. Navarra and M. Nielsen, hep-ph/0203225.

QCD SUM RULES AND BROWN-RHO SCALING LAW

L.A. BARREIRO
Departamento de Física e Química - UNIFEI, Itajubá, MG, Brazil
Departamento de Física - UNESP, Rio Claro, SP, Brazil
barreiro@rc.unesp.br

The scalar and vector self-energies obtained through QCD sum rules are introduced in Quantum Hadrodynamics (QHD) equations. The results indicate that the ratios of coupling constants to respective meson mass have a very small dependence on density. Then, an extended Brown-Rho scaling law is conjectured.

1 Introduction

It is well established that in-medium nucleon (N) properties, need not to be the same as their free space values[1]. The origin of this changes must be reflected in modifications of the N-N interaction, which is predominantly one-boson exchange. The parameter in meson exchange models are meson coupling constants and their masses. So, in recent years the density dependence of meson masses and meson-baryon coupling constants have received increasing interest[2,3,4]. Medium modification of hadron properties can be seen as a manifestation of quark substructure and the correct theory underlying physics of quarks (and gluons) is quantum chromodynamics (QCD). However, it is very difficult to use QCD to study low- and medium-energy nuclear phenomena. This is because straightforward perturbation theory in the effective QCD coupling constant fails. Nevertheless, in most problems, the relevant phenomena are confined to a specific length scale, and thus it is not necessary to explicitly include dynamics at significantly shorter length scales. This is the case for Quantum Hadrodynamics (QHD)[5,6]. The objective of this work is introduce some content of QCD in QHD and the natural way to make that connection take into account the nucleon self-energy[3] and the well-known method of QCD sum rules[7].

2 QHD Model

To investigate in-medium meson-nucleon coupling constants and meson masses, it is first necessary to solve the nuclear ground state within Relativistic

Hartree Approximation. So, self-energies are given by[5]

$$\Sigma_s = M^* - M = -\frac{g_s^2}{m_s^2}(\rho_s + \Delta_{VF}) \quad \text{and} \quad \Sigma_v = \frac{g_v^2}{m_v^2}\rho_B\,, \tag{1}$$

where baryon and scalar densities are given by

$$\rho_B = \frac{\gamma}{6\pi^2}k_F^3, \tag{2}$$

$$\rho_s = \rho_s(M^*) = \frac{\gamma M^*}{4\pi^2}\left[k_F\sqrt{k_F^2 + M^{*2}} - M^{*2}\ln\left(\frac{k_F + \sqrt{k_F^2 + M^{*2}}}{M^*}\right)\right] \tag{3}$$

and the vacuum fluctuation term produced by counterterms present in $\delta\mathcal{L}$ is

$$\Delta_{VF} = \Delta_{VF}(M^*) = \frac{1}{\pi^2}\left[M^{*3}\ln\left(\frac{M^*}{M}\right) - M^2\Sigma_s - \frac{5}{2}M\Sigma_s^2 - \frac{11}{6}\Sigma_s^3\right]. \tag{4}$$

Here, M^* is the effective nucleon mass in the medium, M is the free nucleon mass with value $M \approx 939$ MeV, k_F is the Fermi momentum, γ is the spin-isospin degeneracy, g_s and g_v are, respectively, σ- and ω-N coupling constants and m_s and m_v are, respectively, σ- and ω-meson masses. The connection between QHD and QCD can be realized through the introduction of density dependence on coupling constants and meson masses. So, using asterisks to indicate in-medium quantities, we can write

$$\frac{g_s^{*2}}{m_s^{*2}} = f_s(\rho_B)\frac{g_s^2}{m_s^2} \quad \text{and} \quad \frac{g_v^{*2}}{m_v^{*2}} = f_v(\rho_B)\frac{g_v^2}{m_v^2} \tag{5}$$

where the dependence on density has been explicitly represented through $f_s(\rho_B)$ and $f_v(\rho_B)$. To determine these functions, the self-energies (1) are rewritten as

$$\frac{g_s^{*2}}{m_s^{*2}} = -\frac{M^*_{QCD} - M}{\rho_s(M^*_{QCD}) + \Delta_{VF}(M^*_{QCD})} \quad \text{and} \quad \frac{g_v^{*2}}{m_v^{*2}} = \frac{\Sigma_v^{(QCD)}}{\rho_B} \tag{6}$$

where M^*_{QCD} and $\Sigma_v^{(QCD)}$ can be obtained with the QCD sum rules method[8].

3 QCD Sum Rules

QCD sum rules results for an in-medium nucleon are given in Ref.[8]. For the purpose of this work we will consider just the leading term and we will not take into account continuum contributions as well as perturbative corrections. Then, we obtain

$$\frac{M^*_{QCD}}{M} = \frac{\langle\bar{q}q\rangle_{\rho_B}}{\langle\bar{q}q\rangle_{vac}} \quad \text{and} \quad \frac{\Sigma_v^{(QCD)}}{M} = -\frac{4}{3}\frac{c_2}{c_1}\frac{\langle q^\dagger q\rangle_{\rho_B}}{\langle\bar{q}q\rangle_{vac}}\,. \tag{7}$$

where we defined the following terms:

$$c_1 \equiv 7 - 2t - 5t^2 \quad \text{and} \quad c_2 \equiv 5 + 2t + 5t^2. \tag{8}$$

The in-medium quark condensate obtained with a model independent (MI) calculation is[9]:

$$\frac{\langle \bar{q}q \rangle_{\rho_B}}{\langle \bar{q}q \rangle_{vac}} = 1 - \frac{\sigma_N \rho_B}{m_\pi^2 f_\pi^2} + \cdots \quad \text{with} \quad \langle \bar{q}q \rangle_{vac} = \frac{m_\pi^2 f_\pi^2}{2m_q} \tag{9}$$

and where $\langle q^\dagger q \rangle_{\rho_B} = \frac{3}{2}\rho_B$ represents the quark density. Using these results in (6) , we obtain for the scalar and vector sectors,

$$\frac{g_s^{*2}}{m_s^{*2}} = \frac{\sigma_N M}{m_\pi^2 f_\pi^2} \frac{\rho_B}{\rho_s(M_{QCD}^*) + \Delta_{VF}(M_{QCD}^*)} \quad \text{and} \quad \frac{g_v^{*2}}{m_v^{*2}} = -4\frac{c_2}{c_1}\frac{m_q M}{m_\pi^2 f_\pi^2} \tag{10}$$

respectively. The first thing we can observe is: *the vector ratio is independent of density.* As a consequence, the functions defined in (5) can be written as

$$f_s(\rho_B) = \frac{\rho_B}{\rho_s(M_{QCD}^*) + \Delta_{VF}(M_{QCD}^*)} \quad \text{and} \quad f_v(\rho_B) = 1 \tag{11}$$

In Figure 1, $f_s(\rho_B)$ is plotted as a function of density. As a comparison, other models for the quark condensate have been used. The dashed line represents the result obtained with Nambu-Jona-Lasinio (NJL) model and the dotted line uses Gell-Mann-Lévy (GML) model[9]. In all calculation performed here we adopted $\sigma_N = 30$ MeV. The results indicate a small variation of f_s up to the equilibrium density of nuclear matter. The results (12) permit one to extend the Brown-Rho (BR) scaling law[2], by including coupling constants[4]:

$$\frac{M_N^*}{M_N} = \frac{m_v^*}{m_v} = \frac{m_s^*}{m_s} = \frac{g_v^*}{g_v} = f_s^{-2}\frac{g_s^*}{g_s} = \frac{1}{1 + \alpha\rho_B/\rho_0}, \tag{12}$$

with $\alpha = 0.28$. Due to f_s^{-2}, g_s^*/g_s obey a different scale for which $\alpha \approx 0.23$.

4 Conclusion

Our principal result is: *the ratios of coupling constants to respective meson masses are almost constant.* Since meson masses obey a scaling law, then coupling constants should follow almost that same law. Therefore, we can extend the scaling law to include coupling constant. However, the results indicate that for high densities the scalar coupling constant obey a different scale indicating an increase in the attraction. The preliminary results of nuclear matter calculation with these scales indicate a decrease in the compressibility, as is requested by the experimental result.

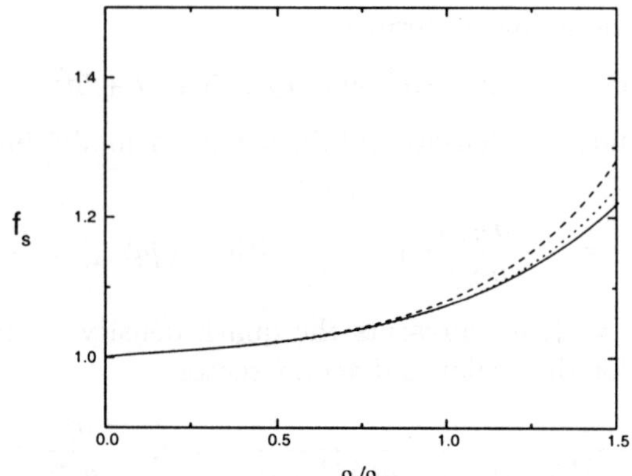

Figure 1. f_s, Eq. (12), vs nucleon density with MI calculation (solid line), NJL model (dashed line) and GML model (dotted line) [9].

Acknowledgements

I would like to thank FAPESP (Fundação de Amparo à Pesquisa do Estado de São Paulo, Brazil) for the financial support.

References

1. T. Alm, G. Ropke and M. Schmidt, *Phys. Rev.* **C 50** (1994) 931.
2. G.E. Brown and M. Rho, *Phys. Rev. Lett.* **66** (1991) 2720.
3. R. Brockmann and H. Toki, *Phys. Rev. Lett.* **68** (1992) 3408.
4. C. Song, G.E. Brown, Dong-Ph Mih and M. Rho, *Phys. Rev.* **C 56** (1997) 2244
5. B. D. Serot and J. D. Walecka,*Adv. Nuc. Phys.* **16** (1986) 1.
6. B. D. Serot and J. D. Walecka, *Int. J. Mod. Phys.* **E 6** (1997) 515.
7. M. A. Shifman, A. I. Vainshtein, and V. I. Zakharov, *Nucl. Phys.* **B 147** (1979) 385.
8. T. D. Cohen, R. J. Furnstahl, D. K. Griegel, and Xuemin Jin, *Prog. Part. Nucl. Phys.* **35** (1995) 221.
9. T. D. Cohen, R. J. Furnstahl and D. K. Griegel, *Phys. Rev.* **C45** (1992) 1881.

THE POLARIZED STRUCTURE FUNCTION WITH AN EFFECTIVE POTENTIAL

L.A. TREVISAN
Universidade Estadual de Ponta Grossa,
84010-330, Ponta Grossa, Paraná, Brazil
latrevis@uepg.br

We use a quark model, with an effective potential, to obtain the polarized structure function of the quarks in the nucleon. In the present approach, we consider the harmonic oscillator potential for the quarks.

The simplest way to implement the two cornerstones of the Quantum Chromodynamics, the confinement and asymptotic freedom, is through an effective potential for the quarks. The most popular and simple effective potential that includes the above properties, the MIT bag model[1], was used by Jaffe[2] to obtain the nucleon structure function.

Among the many other relativistic potentials for the quarks, that have been considered to obtain different properties of the hadrons, we can mention the pioneering works of Refs.[3,4] (linear and/or harmonic oscillator, with equal scalar plus vector parts)[5,6]. For a more general discussion about effective potentials in QCD, see also[7].

One common characteristic of the references cited above is that, when modeling the hadrons, with two or three valence quarks, they consider the same kind of potential for the quarks. This is an important concern of the present work. When we consider, for the up and down quarks, potentials with the same strength, the wave-functions will be exactly the same; so, in such models, if we calculate the structure functions of the neutron (F^{2n}) and proton (F^{2p}) in the momentum space, we will obtain a constant ratio between them. For instance, if we wish to fit the ratios F^{2n}/F^{2p}, in such simple quark models, the strengths of the potentials must consider different energies for the ground states of the up and down quarks, as pointed out in Ref.[8].

Another point to be considered is the spin as an additional degree of freedom in QCD. Our purpose here is to verify how the present quark model can fit the experimental data for the polarized structure function. We start by considering the effective potential for the quarks.

There are two advantages in using a relativistic confining potential for the quarks with a scalar plus vector (S+V) structure: first, the Dirac equation becomes similar to the non-relativistic Schrödinger equation; second, one avoids the Klein paradox[9] (the possibility of tunnelling even with an infinity potential barrier). The choice of a harmonic oscillator potential is for convenience in obtaining analytical solutions.

The S+V harmonic oscillator potential, for a quark q at a distance r from the center of the nucleon, can be written as $V(r) = (1+\beta)\frac{1}{2}K_q r^2$, where K_q is the strength of the potential and β the Dirac matrix.

The corresponding Dirac equation is written as [5]:

$$\left[i\partial_\nu \gamma^\nu + \frac{1}{2}(1+\beta)K_q r^2\right]\psi(\vec{r},t) = 0 \tag{1}$$

where we are considering the current mass of the up (u) and down (d) quarks equal to zero.

After replacing the Dirac matrices and considering the spherical symmetry, we define:

$$\psi(\vec{r}) = \frac{1}{r}\begin{pmatrix} g(r)\Omega_{jj_3l} \\ if(r)\Omega_{jj_3l'} \end{pmatrix} = \frac{1}{r}\begin{pmatrix} g(r) \\ -i\vec{\sigma}.\vec{n}f(r) \end{pmatrix}\Omega_{jj_3l}, \tag{2}$$

where Ω_{jj_3l} are the spherical wave functions.

After solving the system for $g(r)$ and $f(r)$, we obtain the following equations:

$$g''(r) - \left\{(E_q+m_q)(K_q r^2 - E_q + m_q) + \frac{l(l+1)}{r^2}\right\} g(r) = 0, \tag{3}$$

$$f(r) = \frac{1}{E_q+m_q}\left(g'(r) + \frac{k}{r}g(r)\right) \tag{4}$$

where $k = (j+\frac{1}{2})(-1)^{j-l+1/2}$ is the Dirac quantum number.

The energy levels are given , for the case $m_q = 0$, by:

$$E_{nl} = \left[2\sqrt{K_q}\left(2n + l - \frac{1}{2}\right)\right]^{2/3}.$$

We choose the K_q in such a way that $E_u = 0.35 \times 939MeV$ and $E_d = 0.30 \times 939MeV$ because this helps to fit the structure function ratio, as already mentioned. We used only the ground state.

The solutions for the $g(r)$ are :

$$g(r) = N_{nl}\frac{r}{r_o}e^{-r^2/r_o^2}L_{n-1}^{l+3/2}\left(\frac{r^2}{r_o^2}\right).$$

In the momentum space, the wave-functions became, for the ground state:

$$G(p) = 2pe^{-p^2c}$$

and the lower component:

$$F(p) = \frac{2}{E}p^2e^{-p^2c} + \frac{4}{Ec}e^{-p^2c}$$

where $c = \frac{1}{2\sqrt{E_qK_q}}$.

To obtain the structure functions $u(x)$ or $d(x)$ we have to integrate the probability density in the null plane variable $p_\perp$, considering that the other null plan variable is given by $p_z = M_n(x - E_q/M_n)$.

In this work we use only the ground state, therefore there is no need to indicate the quantum numbers n and l when we mention the energy, only the kind of quark. The polarized structure functions are obtained in the same way, observing that the lower and upper components of the quark wave-functions must be integrated separately.

Another important point is the model we are considering for the quarks in the nucleons: a) when the two u quarks have positive helicity, we sum them first, subtracting the helicities of the quark d ; b) when the u quarks have opposite helicity, the spin results from the quark d. We take the mean value, in both of the hypotheses. Making the integration of the polarized structure function, for the proton, we obtain

$$\int_0^1 g_1^p(x) = 0.148,$$

that is in quite good agreement with previous theoretical and experimental results reported in Ref.[10].

References

1. A.Chodos et al, *Phys. Rev.* **D9**, 3471 (1974).
2. R.L.Jaffe, *Phys.Rev.* **D11**, 1953 (1975).
3. L.A. Trevisan et al, *Eur. Phys. J.* **C11** , 351 (1999).
4. P.L. Ferreira et al, *Nuovo Cim.* **A55**, 215 (1980).

5. R.Tegen, *Ann. of Phys,* **197**, 439 (1990).
6. B.E.Palladino and P.L.Ferreira, *Revista Brasileira de Física* **18** , 478 (1980).
7. E.Laermann, Talk given at Workshop on QCD- 20 years later- Aachen -Germany Jun 9-13 Aachen QCD Workshop 1992, 455-480 (1992); see also *Nucl. Phys. Proc. Suppl.* **20**, 394-398 (1991).
8. L.A. Trevisan and L.Tomio, *Nucl. Phys.* **A689**, 485c-488c (2001).
9. O.Klein, *Z.Phys.* **53** , 157 (1929); P.M.Fishbane, S.G.Gasiorowicz, D.C. Johannsen and P.Kaus, *Phys. Rev.* **D27**, 2433 (1983).
10. E.Reya, Talk given at Workshop on High Energy Phenomenology II, Calcutta, 1991, World Scientific, Singapore.

THE NUCLEON STRUCTURE FUNCTION AND THE QUARK EFFECTIVE POTENTIAL

L.A. TREVISAN

Departamento de Matemática e Estatística, Universidade Estadual de Ponta Grossa, CEP 84010-330 Ponta Grossa, PR, Brazil

latrevis@uepg.br

L. SALVI AND LAURO TOMIO

Instituto de Física Teórica, Universidade Estadual Paulista, Rua Pamplona 145, 01405-900 São Paulo, SP, Brazil

tomio@ift.unesp.br

A statistical quark model is considered in order to reproduce the nucleon structure function, by using a temperature parameter to adjust the Gottfried sum rule. In the present statistical model, the quark levels in the nucleon are generated by a Dirac equation with harmonic scalar plus vector potential. We note that a good fit for the ratio between the structure functions of the neutron and proton, F_2^n/F_2^p, can be obtained if different strengths are used for the effective confining potentials of the up and down quarks.

1 Introduction

Recently, a statistical thermal model was considered for the structure function of the nucleon, in which the quarks are confined by an effective linear potential [1]. The model was successful in describing the strangeness content of the nucleon and also gives a reasonable result for the differences of the structure functions of the proton and neutron. The quark levels are generated by using a Dirac equation, where the scalar and vector parts of the potential have the same analytical form; with zero mass and the same potential strength for the quarks up (u) and down (d); with non-zero mass and a distinct strength for the strange (s) quark. Once obtained the quark energy levels, with the Dirac equation, a Fermi-Dirac distribution was considered with different chemical potentials for the u and d quarks, in order to reach the corresponding normalized wave-functions. Obviously, such model gives exactly the same wave-functions and energy levels for the u and d quarks; the difference that occurs between the neutron and proton structure functions is due to the statistical weight. Another parameter, a critical temperature, is adjusted to reproduce the violation of the Gottfried sum rule [2]. In spite of the success of the model in describing observables as the strangeness and the difference between the structure functions F_2^n and F_2^p, it fails to give a good description of the ratio between the structure functions, F_2^n/F_2^p, that results to be almost constant when plotting it as a function of the momentum fraction carried by the quarks, x. This result is far from the one given by experimental data.

In order to improve the model and reach a better agreement with data, a natural way is to consider different strengths for the effective confining potentials of the zero-mass u and d quarks. This further consideration can easily be included in a new parameterization of the model in order to not affect the previous good results that were obtained in [1]. Presently, instead of the linear potential used in [1], we are considering a central harmonic potential for the quark interactions, following Ref. [3].

Our preliminary approach, considering the ground-state levels of the quarks, already shows a good improvement in the agreement with experimental data when the ratio between the strengths of the potentials for the up and down quarks is approximately given by 1.58 (See Fig. 1). This result is being considered in order to improve the statistical quark model presented in Ref.[1], where the Pauli principle was incorporated using the Fermi-Dirac statistics. In our new parameterization of the model, we are also considering the recent experimental results for the asymmetry in the sea-quarks of the nucleon.

2 Quark Dynamics

We choose the relativistic harmonic oscillator potential for the quark dynamics, with scalar and vector parts having the same analytical form, but flavor dependent strengths K_q [4,3]:

$$V(r) = (1+\beta)\,\frac{1}{2}K_q r^2. \tag{1}$$

The above scalar plus vector choice of the potential has the following advantages: (i) the corresponding coupled Dirac equation can be reduced to a single Schrödinger equation; (ii) the Klein's paradox [5], that is, tunneling effect that happens even if there is an infinite potential, is avoided. This effect can occur if the vector component of the potential is dominant.

The structure functions are obtained for each flavor using the wave functions in the momentum space, considering relativistic harmonic potential with different values of the parameter K_q. Once these functions are obtained for the up and down quarks, one can easily build the corresponding structure functions for the proton and neutron, as well as the ratio F_n^2/F_p^2.

Using the null-plane variables p^+ and the fraction of the total momentum x, $(p^+ = xM_n)$, where M_n is the nucleon mass, the longitudinal component of the momentum is given by $p_z = p^+ - E$, or, $p_z = M_n(x - E/M_n)$. [Note that $p = \sqrt{p_z^2 + p_\perp^2}$, such that we have to integrate in the $p_\perp$ variable to obtain a function that depends only on the Bjorken scale x.] The components of the wave-function in the above null-plane variables, $\widetilde{F}(p^+, p_\perp)$ and $\widetilde{G}(p^+, p_\perp)$, are obtained after the corresponding Fourier transforms of the upper and lower Dirac components of the wave-function.With these wave-functions, the quark structure functions are given by

$$q(x) = 2\pi \int_0^\infty \left[\left| F(p^+, p_\perp) \right|^2 + \left| G(p^+, p_\perp) \right|^2 \right] dp_\perp . \tag{2}$$

The integrals in $p_\perp$ can be solved, resulting in an (almost) analytical formula, that depends on the x variable. We define $f_q = E_q/M_n$ as the more probable energy fraction for each quark. Next, we fitted $f_u = 0.35$, (E_u =328.65 MeV) and $f_d = 0.30$, (E_d =281.7 MeV). The corresponding values obtained for K_q are: K_u =101.2865 MeV fm^{-2}, K_d =63.792 MeV fm^{-2}. Note that, if we consider the same f_q for each quark, as the energy levels and wave functions are the same and the only difference between the up and down quarks is in the normalization of the corresponding wave-function, the ratio F_n^2/F_p^2 will be practically constant, as in Ref. [1]. The total momentum fraction, carried out by each particle flavor u and d, results in $p_u = 0.7012$ and $p_d = 0.3006$, consistent with the total sum very close to one.

3 Results

We have studied an effective potential model, with the quarks confined by a central force, following the model presented in [1]. In the present work, considering analytical convenience, we have used the relativistic harmonic oscillator, with the same analytical expression for the vector and scalar parts of the potential. In order to obtain different wave functions for the u and d quarks we suppose different strengths in the confining potential. In this way, the corresponding results for the strengths that we have obtained are K_u =101.2865 MeV fm^{-2} and K_d =63.792 MeV fm^{-2}. With these values for the strengths, the corresponding ground state energies are E_u =328.65 MeV for the quark u, and E_d =281.7 MeV for the quark d. With these energy levels, each u quark carries out approximately 0.35 of the total momentum; and the d quark carries out 0.30. With such values we have obtained a reasonable good fit for the ratio of the structure functions for the neutron and proton, F_n^2/F_p^2. Actually, the choice of K_u and K_d was consider in order to fit the ratio, according to the experimental data given in reference [2]. In this communication, we have noted that the statistical quark model presented in [1] can be improved by considering flavor dependent strengths for the light quarks u and d presented in the nucleon.

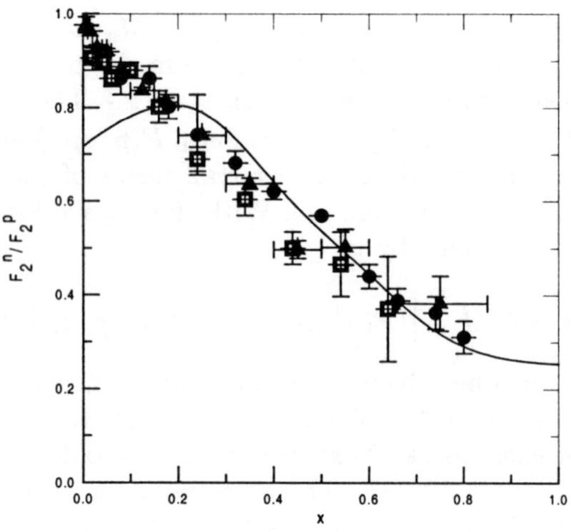

Figure 1. Ratio for the structure functions of the neutron and proton. The model is given in solid line. Experimental data are taken from EMC collaboration[6] (circles), SLAC[7] (squares), and NMC[8] (triangles).

Acknowledgments

Work partially supported by Fundação de Amparo à Pesquisa do Estado de São Paulo and CNPq of Brazil.

References

1. L.A. Trevisan, T. Frederico and L. Tomio, *Eur. Phys. J.* **C11** (1999) 351.
2. A.D. Martin, R.G. Roberts, W.J. Stirling and R.S. Thorne, *Eur. Phys. J.* **C4** (1998) 463.
3. R.Tegen, *Annals of Physics* **197** (1990) 439.
4. P.L. Ferreira, J.A. Helayel, and N. Zagury, *Nuovo Cim.* **A55** (1980) 215.
5. O.Klein, *Z. Phys.* **53** (1929) 157.
6. J.J. Aubert, et al., *Nucl.Phys.* **B293** (1987) 740.
7. S. Stein, et al., *Phys. Rev.* **12** (1975) 1884; A. Bodek, et al., *Phys. Rev. Lett.* **30** (1973) 1087.
8. NMC Collaboration: P. Amaudruz, et al., *Phys. Rev. Lett.* **66** (1991) 2712; M. Arneodo, et al., *Phys. Rev.* **D50** (1994) R1.

QCD SUM RULES FOR THE $G_{KK^*\pi}(Q^2)$ FORM FACTOR

L.B. HOLANDA, R.S. MARQUES DE CARVALHO

Instituto de Física Teórica, Universidade Estadual Paulista, Rua Pamplona 145, 01405-900 São Paulo, SP, Brazil
holanda@ift.unesp.br, raquel@ift.unesp.br

M. NIELSEN

Instituto de Física, Universidade de São Paulo, C.P. 66318, 05315-970 São Paulo, SP, Brazil
mnielsen@if.usp.br

The QCD Sum Rules have been used to evaluate the form factor in the vertex $KK^*\pi$. The method of QCD Sum Rules is based on the duality principle in which it is assumed that the hadrons can simultaneously be described in two levels: quarks and hadrons. This work showed that the axial current, used to describe the meson K is not appropriated to study the form factor.

We start from the three point correlation function

$$A_{\mu\nu}(p,p',q) = \int d^4y \; d^4x \; e^{ip'.x} e^{iq.y} \langle 0|T\{j_\nu(x) j_5(y) j_\mu^\dagger(0)\}|0\rangle \; , \tag{1}$$

with $p' = p - q$.

In the phenomenological side we can insert intermediate states of the K, K^* and π mesons and get:

$$A_{\mu\nu}(p,p',q) = -\frac{i \, m_{K^*} \, m_\pi^2 \, f_\pi \, f_K \, f_{K^*} \, p'_\nu \, q^\alpha}{(m_u + m_d)(p'^2 - m_K^2)(p^2 - m_{K^*}^2)} \times \frac{-g_{\mu\alpha} + p_\mu \, p_\alpha / m_{K^*}^2}{(q^2 - m_\pi^2)} \, g_{KK^*\pi}(q^2) + \text{excited states} \; . \tag{2}$$

We can rewrite eq.(2) as

$$A_{\mu\nu}^{fen}(p,p',q) = -\frac{i \, m_{K^*} \, m_\pi^2 \, f_\pi \, f_K \, f_{K^*}}{(m_u + m_d)(p'^2 - m_K^2)(p^2 - m_{K^*}^2)} \frac{g_{KK^*\pi}(q^2)}{(q^2 - m_\pi^2)} \times \left[- p'_\nu \, p'_\mu + \frac{-q^2 + m_K^2 + m_{K^*}^2}{2m_{K^*}^2} \, p'_\nu \, p_\mu \right] + \text{excited states} \; . \tag{3}$$

The two lower order processes that contribute for the QCD side are the perturbative term and the s quark condensate term. The diagrams which are proportional to u (and d) quark condensate will be suppressed by the double Borel transform.

In the QCD side, starting from Eq.(1), using the currents

$$j_\nu(x) = \overline{s}(x)\gamma_\nu\gamma_5 u(x) \ , \ j_5(y) = i\ \overline{u}(y)\gamma_5 d(y) \ , \ j_\mu^\dagger(0) = \overline{d}(0)\gamma_\mu s(0)$$

we get

$$A_{\mu\nu}(p,p',q){=}i\int d^4y d^4x e^{ip'.x} e^{iq.y}\langle 0|T\{\overline{s}(x)\gamma_\nu\gamma_5 u(x)\overline{u}(y)\gamma_5 d(y)\overline{d}(0)\gamma_\mu s(0)\}|0\rangle \ . \tag{4}$$

Defining $S^q_{ab}(x-y) = \langle 0|T\{q_a(x)\overline{q}_b(y)\}|0\rangle$, we obtain

$$A_{\mu\nu}(p,p',q) = -i\int d^4y\ d^4x\ e^{ip'.x}e^{iq.y}\mathrm{Tr}[\gamma_\nu\gamma_5 S^u_{ab}(x-y)\gamma_5 S^d_{bc}(y)\gamma_\mu S^s_{ca}(-y)] \tag{5}$$

and with some manipulations:

$$A_{\mu\nu}(p,p',q) = -\frac{i}{(2\pi)^4}\int d^4k\mathrm{Tr}[\gamma_\nu\gamma_5 S^u_{ab}(p'-k)\gamma_5 S^d_{bc}(p-k)\gamma_\mu S^s_{ca}(-k)] \ . \tag{6}$$

The perturbative fermion propagator in the momentum space can be written, for small masses :

$$S^q_{ab}(p) = i\frac{\not p + m_q}{p^2} \ , \tag{7}$$

and, evaluating the trace,

$$A_{\mu\nu}(p,p',q) = -\frac{12m_s}{(2\pi)^4}\int\frac{d^4k}{(p'-k)^2(p-k)^2k^2}\{[(p+p').k - k^2 - p.p']g_{\mu\nu} + q_\mu k_\nu - q_\nu k_\mu - p'_\nu p_\mu + p_\nu p'_\mu\}. \tag{8}$$

We can rewrite this correlation function as a double dispersion relation for each invariant structure. In $p_\mu p'_\nu$ structure we get:

$$A_{p\ p'}(p,p',q) = -\frac{1}{4\pi^2}\int ds\ du \frac{DD[A_{p\ p'}]}{(p^2-s)(p'^2-u)} \ , \tag{9}$$

where

$$DD\Big[A_{p\ p'}(p,p',q)\Big] = -\frac{3im_s}{\lambda^{3/2}(s,u,t)}t(s-t+u) \tag{10}$$

with

$$\lambda(s,u,t) = s^2 + u^2 + t^2 - 2su - 2st - 2ut \ . \tag{11}$$

We also need to calculate the s quark condensate contribution where we use $S^s_{ab}(-y) = -\frac{1}{12}\langle\overline{s}s\rangle\delta_{ab}$. So,

$$A_{\mu\nu} = -\frac{i\ \langle\overline{s}s\rangle}{p^2\ p'^2}[p_\mu p'_\nu + p.p' g_{\mu\nu} - p'_\mu p_\nu] \ . \tag{12}$$

Performing the double Borel transform in Eq.(9) and for the $p_\mu p'_\nu$ structure in Eq.(12) and comparing with the double Borel transform of the term which is proportional to the same structure in the phenomenological side, Eq.(3), we have

$$g_{KK^*\pi}(Q^2,M^2,M'^2)=-\frac{m_{K^*}^2(m_u+m_d)(Q^2+m_\pi^2)}{2\pi^2 m_\pi^2 f_\pi f_K(M'^2)f_{K^*}(M^2)}\frac{e^{m_{K^*}^2/M^2}e^{m_K^2/M'^2}}{m_{K^*}(Q^2+m_K^2+m_{K^*}^2)}$$
$$\times\left\{3m_s\left[\int_0^{s_0}ds\int_0^{u_0}due^{-s/M^2}\,e^{-u/M'^2}\frac{t(s-t+u)}{\lambda^{3/2}(s,u,t)}\right]-4\pi^2\langle\bar{s}s\rangle\right\}. \qquad (13)$$

In this equation we are using the sum rules for the decay constants of K and K^* mesons. In the case of π meson, we are using the experimental value, since that π meson is off-shell. The sum rule for f_{K^*} is

$$f_{K^*}^2(M^2)=\left(\frac{M^2}{4\pi^2}E_0+\frac{4m_s\langle\bar{q}q\rangle}{3M^2}-\frac{m_s\langle\bar{s}s\rangle}{3M^2}+\frac{\langle g_s^2\,G^2\rangle}{48\pi^2M^2}\right)e^{m_{K^*}^2/M^2}\,. \qquad (14)$$

The sum rule for f_K using the pseudo-vector current is

$$f_K^2(M^2)=\left[\frac{M^2}{4\pi^2}E_0-\frac{M^4\;E_1}{4\pi^2 m_{k_1}^2}-\frac{m_s\langle\bar{q}q\rangle}{m_{k_1}^2}+\frac{m_s\langle\bar{s}s\rangle}{M^2}+\frac{\langle g_s^2G^2\rangle}{2^4 3\pi^2}\left(\frac{1}{M^2}+\frac{1}{m_{k_1}^2}\right)\right]e^{m_K^2/M^2}. \qquad (15)$$

Using 1.6 $GeV^2 \leq M^2 \leq$ 4.0 GeV^2 , the decay values involved[1], and fixing Q^2 in 2.0 GeV^2 we show our results in Fig.1. From this Figure we see that the term which is proportional to the quark condensate is the predominant one. This is because when using the pseudo-vector current for K meson the perturbative term is proportional to the s quark mass. The perturbative term is much more stable as function of M^2, while the final result is not so stable.

We expected that $g_{KK^*\pi}(Q^2)$ form factor to be approximately constant as a function of the Borel mass. In this way, we could fix any value of the Borel mass to study the variation of $g_{KK^*\pi}(Q^2)$ with Q^2. However, we see that the $g_{KK^*\pi}(Q^2)$ form factor is not constant, Fig.1. These two facts: no stability of $g_{KK^*\pi}(Q^2)$ and non-perturbative contribution bigger than the perturbative one, tell that the expansion (OPE) in this case is not under control and we should have to consider necessarily more terms in the expansion. The axial current describes very well the two-point function while the pseudoscalar one does not. In the three-point function the axial current is not appropriate since the perturbative term is smaller than the next term of the series. When we use QCD sum rules it is expected that the perturbative term is predominant and the non-perturbative ones are corrections of it.

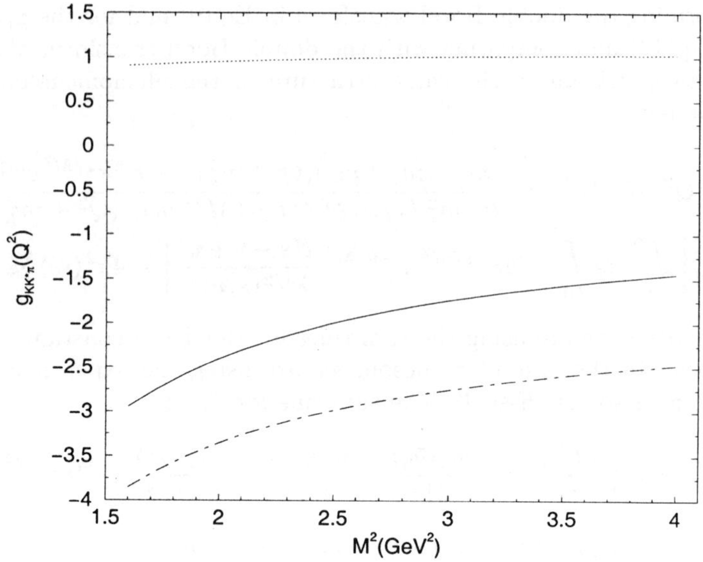

Figure 1. Solid line: $g_{KK^*\pi}(q^2)$. Dotted line: perturbative term. Dotted-dashed line: s quark condensate.

Perhaps the use of the pseudoscalar current with the inclusion of instantons contributions is the best way to solve the problem, since it is known that the two point function with the instanton contribution gives a good description of f_K [2,3].

Acknowledgments

This work was supported by CNPq and FAPESP, Brazil.

References

1. Particle Data Group, *Eur. Phys. Jour.* **C15**, 1 (2000).
2. E. V. Shuryak, *Nucl. Phys.* **B214**, 237 (1983).
3. E. V. Shuryak, *Nucl.Phys.Proc.Suppl.* **34**, 107 (1994).

$D^*D\pi$ AND $B^*B\pi$ VERTICES FROM QCD SUM RULES

M. NIELSEN, F.S. NAVARRA
Instituto de Física, USP, C.P. 66318, 05389-970 São Paulo, SP, Brazil
mnielsen@if.usp.br

M.E. BRACCO
Instituto de Física, UERJ, Rua São Francisco Xavier 524, Maracanã, 20559-900, RJ, Brazil
bracco@uerj.br

The $D^*D\pi$ form factor is evaluated in a QCD sum rule calculation for both D and π off-shell mesons. We study the Borel sum rule for the three point function of one pseudoscalar, one axial and one vector meson currents. We find that the momentum dependence of the form factors is very different if the D or the π meson is off-shell, but they lead to the same coupling constant in the $D^*D\pi$ vertex.

Of all experimental observables that are sensitive to the presence of the quark-gluon plasma (QGP), charmonium suppression [1] is among the most promising ones. However, in order to use the J/ψ suppression as a signature for the formation of the QGP in heavy ion collisions it is important to have a better knowledge on the interactions between charmonium states and co-moving hadrons. Various approaches have been used in evaluating the charmonium absorption cross section by hadrons, and some of them use meson exchange models based on hadronic effective lagrangians. However, the couplings and form factors needed in these effective lagrangians are not phenomenologically known.

Here we report the use of the QCD sum rule (QCDSR) method [2], based on the three-point function, to evaluate the $D^*D\pi$ and $B^*B\pi$ hadronic form factors and coupling constants. Following the QCDSR formalism, the three-point function associated with an $HH'M$ vertex, where H and H' are the external hadrons and M is the exchanged meson, is

$$\Gamma_M(p,p') = \int d^4x\, d^4y\, \langle 0|T\{j_{H'}(x)j_M(y)j_H^\dagger(0)\}|0\rangle\, e^{ip'.x}\, e^{-i(p'-p).y}\ , \qquad (1)$$

where j_i is the interpolating fields of the hadron i, written in terms of the quark content and with the same quantum numbers of these hadrons.

The phenomenological side of the vertex function, $\Gamma_M(p,p')$, is obtained by the consideration of H and H' state contribution to the matrix element in

Eq. (1):

$$\Gamma_M^{phen}(p,p') = \lambda_H \lambda_{H'} \lambda_M \, g_{HH'M}(q^2) \, \Delta_{H'}(p) \Delta_H(p') \Delta_M(q) \, , \qquad (2)$$

with Δ_i being the propagator of the hadron i, and $q = p - p'$. The matrix element $\langle 0|j_H|H\rangle = \lambda_H$ defines the couplings of the currents with the respective hadronic states, and $g_{HH'M}(q^2)$ is the form factor we want to determine.

The QCD side, or theoretical side, of the vertex function is evaluated by performing Wilson's operator product expansion (OPE) of the operator in Eq. (1). Writing Γ_M in terms of the invariant amplitudes, we can write a double dispersion relation for each one of the invariant amplitudes, over the virtualities p^2 and p'^2 holding q^2 fixed:

$$\Gamma(p^2, p'^2, q^2) = -\frac{1}{4\pi^2} \int_{m_H^2}^{s_0} ds \int_{m_{H'}^2}^{u_0} du \frac{\rho(s,u,q^2)}{(s-p^2)(u-p'^2)} \, , \qquad (3)$$

where $\rho(s,u,q^2)$ equals the double discontinuity of the amplitude $\Gamma(p^2, p'^2, q^2)$.

As usual, our goal is to make a match between the two representations of the correlation function. To improve the matching a Borel transform [2] is performed in both sides of the sum rule in both variables $-p^2 = P^2 \to M^2$ and $-p'^2 = P'^2 \to M'^2$. The Borel transform suppresses higher dimension operators in the OPE side by a factorial factor, and changes a power-law suppression to an exponential suppression of the higher mass states in the phenomenological side. The double Borel transform also suppresses single pole contribution, associated with the pole-continuum transitions [3].

We first discuss the $D^*D\pi$ form factor with an off-shell pion. Fixing $M^2 = 3.5$ GeV^2 we show, in Fig. 1, the momentum dependence of the form factor (squares) in the interval $2 \leq Q^2 \leq 5\,\text{GeV}$. Since the present approach cannot be used at $Q^2 = 0$, to extract the $g_{D^*D\pi}$ coupling from the form factor we need to extrapolate the curve to $Q^2 = 0$ (in the approximation $M_\pi^2 = 0$).

In Ref. [4] the Q^2 dependence of the form factor, represented by the squares, was parameterized by a gaussian form (dashed line)

$$g_{D^*D\pi}^{(\pi)}(Q^2) = 5.7\, e^{-Q^4/9.17} \, . \qquad (4)$$

However, as can be seen by the dot-dashed line, the Q^2 dependence of the QCDSR results for $g_{D^*D\pi}(Q^2)$ for an off-shell pion can also be well reproduced by the exponential parameterization

$$g_{D^*D\pi}^{(\pi)}(Q^2) = 15.5\, e^{-Q^2/1.48} \, . \qquad (5)$$

Off course, the two parameterizations in Eqs. (4) and (5) lead to very different values for the $D^*D\pi$ coupling constant, defined as the value of the form factor

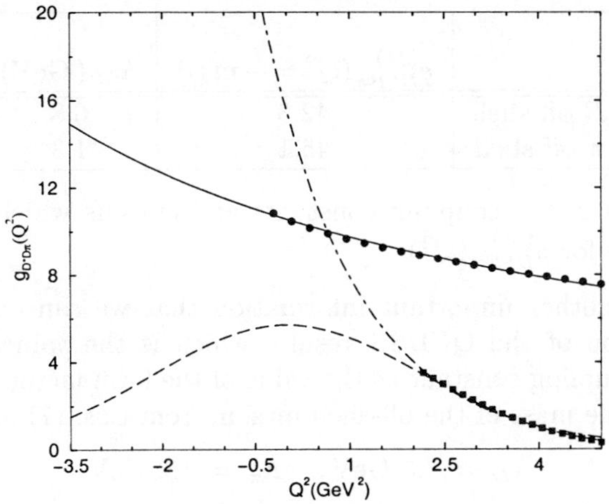

Figure 1. Momentum dependence of the $D^*D\pi$ form factor. The solid, dashed and dot-dashed lines give the parameterization of the QCDSR results through Eq. (7) for the circles, and Eqs. (4) and (5) for the squares.

at the pole of the off-shell meson ($Q^2 = -M_\pi^2 \sim 0$ in the case of the off-shell pion):

$$g_{D^*D\pi} = \begin{cases} 5.7 & \text{with the gaussian parameterization} \\ 15.5 & \text{with the exponential parameterization} \end{cases} \tag{6}$$

To solve this problem we analyze the $D^*D\pi$ form factor for an off-shell D meson. In Fig. 1 we also show, through the circles, the momentum dependence of the $D^*D\pi$ form factor for an off-shell D meson, in the interval $-0.5 \leq Q^2 \leq 5\,\text{GeV}$, where we expect the sum rules to be valid (since in this case the cut in the t channel starts at $t \sim m_c^2$ and thus the Euclidian region stretches up to that threshold). From this figure we can see that the Q^2 dependence of the form factor represented by the circles can be well reproduced by the monopole parameterization (solid line)

$$g_{D^*D\pi}^{(D)}(Q^2) = \frac{126.1}{Q^2 + 11.95} \,. \tag{7}$$

From the parameterization in Eq. (7) we can also extract the $D^*D\pi$ coupling constant, which now is defined as the value of the form factor at the D pole ($Q^2 = -M_D^2$). We get $g_{D^*D\pi} = 14.9$, in an excellent agreement with the exponential parameterization of $g_{D^*D\pi}^{(\pi)}(Q^2)$, but it is still smaller than the experimental value [5]: $g_{D^*D\pi} = 17.9 \pm 0.3 \pm 1.9$.

	$g^{(M)}_{B^*B\pi}(Q^2 = -m^2_M)$	Λ_M (GeV)
B off-shell	42.3	6.8
π off-shell	45.1	1.3

Table1. Values of the coupling constants and cut-offs which reproduce the QCDSR results for $g^{(M)}_{B^*B\pi}(Q^2)$.

There is another important information that we can extract form the parameterization of the QCDSR results which is the value of the cut-off. Defining the coupling constant as the value of the form factor at $Q^2 = -M^2_M$, where M_M is the mass of the off-shell meson, from Eqs. (7) and (5) we get [6]

$$\Lambda_D = 3.5 \text{ GeV}, \quad \Lambda_\pi = 1.2 \text{ GeV}. \tag{8}$$

Therefore, the form factor is harder if the off-shell meson is heavy, implying that the size of the vertex depends on the exchanged meson. This means that a heavy meson will see the vertex as point-like, whereas a light meson will see its extension.

The same calculation can be done for the $B^*B\pi$ form factor and one has only to change the D^*, D and quark c masses by B^*, B and quark b masses, that we take as: $M_{B^*} = 5.33$ GeV, $M_B = 5.28$ GeV and $m_b = 4.7$ GeV. We get the couplings and cut-offs shown in Table I.

To summarize, we have used the method of QCD sum rules to compute form factors and coupling constants in $D^*D\pi$ and $B^*B\pi$ vertices. Our results for the couplings show once more that this method is robust, yielding numbers which are approximately the same regardless of which particle we choose to be off-shell and depending weakly on the choice of the continuum threshold[6]. As for the form factors, we obtain a harder (softer) form factor when the off-shell particle is heavier (lighter). Therefore we can say that heavy mesons "see" smaller sizes in the light mesons, but the latter cannot resolve small structures in the former.

References

1. T. Matsui and H. Satz, *Phys. Lett.* **B178**, 416 (1986).
2. M.A. Shifman, A.I. Vainshtein and V.I. Zakharov, *Nucl. Phys.* **B120**, 316 (1977).
3. B.L. Ioffe and A.V. Smilga, *Nucl. Phys.* **B232**, 109 (1984).
4. F.S. Navarra et al., *Phys. Lett.* **B489**, 319 (2000).
5. S. Ahmed et al., (CLEO Collaboration), hep-ex/0108013.
6. F.S. Navarra, M. Nielsen, M.E. Bracco, *Phys. Rev.* **D65** (2002) 037502.

QUARK CONDENSATES AND HIGH DENSITY QUARK MATTER

P.K. PANDA, G. KREIN AND R.S. MARQUES DE CARVALHO
Instituto de Física Teórica, Universidade Estadual Paulista, São Paulo, Brazil
panda@ift.unesp.br

We consider a Coulomb gauge quark model which includes an explicit construct for a nontrivial vacuum structure in QCD. The dynamics is described by a Hamiltonain that contains a linearly rising confining potential and longitudinal and transverse Coulomb-type interactions. The Coulomb potential gives rise to ultraviolate divergences which are non-perturbatively renormalized by adding appropriate counter terms to the Hamiltonian. The equation of state for u and d quark matter at zero temperature is derived in the Hartree-Fock approximation.

The study of high density and high temperature hadronic matter is of interest for the understanding of many aspects of the physics of superdense stars and relativistic heavy-ion collisions. Properties of high density matter have been studied using perturbative QCD and semi-phenomenological models, like bag models. Validity of a perturbative analysis is unfortunately limited to asymptotically high density and the perturbative series becomes ill-behaved at densities a few times larger than the saturation density of nuclear matter. Here we sketch a calculation performed in the context of a field theoretical Coulomb-gauge quark model using a non-perturbative method with an explicit construct of a chiral symmetry breaking vacuum involving quark-antiquark condensates. Our interest in such a study is the derivation of a quark-matter equation of state at zero temperature, which might be of relevance for studies of superdense stars.

We start with the Coulomb gauge QCD Hamiltonian[1,2]

$$\mathcal{H}(\vec{x}) = \psi_\alpha(\vec{x})^\dagger(-i\vec{\alpha}\cdot\vec{\nabla} + \beta m_q)_{\alpha\beta}\psi_\beta(\vec{x}) + \frac{1}{2}\int d\vec{y}\;\psi(\vec{x})^\dagger T^a\Gamma\psi(\vec{x})\;V_\Gamma(\vec{x}-\vec{y})\psi(\vec{y})^\dagger T^a\Gamma\psi(\vec{y}), \tag{1}$$

where $T^a = 1/2\lambda^a$, $a = 1,\cdots,8$ are the generators of the color SU(3) group, Γ is one, or a combination of Dirac matrices and V_Γ is a potential. Here we use the following Dirac structure for the interaction

$$(\psi^\dagger\psi)(\psi^\dagger\psi) - \xi(\psi^\dagger\vec{\alpha}\psi)\cdot(\psi^\dagger\vec{\alpha}\psi), \tag{2}$$

where ξ is a free parameter. For $\xi = 1$, one has a Dirac structure in the form of a covariant current-current interaction, $(\bar{\psi}\gamma^\mu\psi)\,(\bar{\psi}\gamma_\mu\psi)$. The field operators

$\psi(\vec{x})$ may be expanded as

$$\psi(\vec{x}) = \frac{1}{(2\pi)^{3/2}} \int d\vec{k} \left[U_r(\vec{k}) c_r(\vec{k}) + V_s(-\vec{k}) \tilde{c}_s(-\vec{k}) \right] e^{i\vec{k}\cdot\vec{x}}, \tag{3}$$

where U and V are spinors. A trial vaccum state is constructed as

$$| \Omega \rangle = U \mid 0 \rangle \equiv \exp(B^\dagger - B) \mid 0 \rangle, \tag{4}$$

with

$$B^\dagger = \int \varphi(\vec{k}) c_r(\vec{k})^\dagger \chi_r^\dagger (\vec{\sigma}.\hat{k}) \tilde{\chi}_s \tilde{c}_s(-\vec{k}) d\vec{k}, \tag{5}$$

where $\hat{k} = \vec{k}/|k|$, χ is a Pauli spinor, $\tilde{\chi} = -i\sigma^2 \chi$ and $\varphi(\vec{k})$ is a trial function associated with the quark anti-quark condensate. We note that while c and $\tilde{c}^\dagger$ annihilate the perturbative vaccum, $c \mid 0 \rangle = 0 = \tilde{c}^\dagger \mid 0 \rangle$, the operators

$$b(\vec{k}) = U c(\vec{k}) U^{-1} \tag{6}$$

annihilate the non-perturbative $|\Omega\rangle$. The energy density of the non-perturbative vacuum with respect to the perturbative vacuum is then given as

$$\Delta\mathcal{E} = \langle \Omega \mid \mathcal{H} \mid \Omega \rangle(\varphi) - \langle \Omega \mid \mathcal{H} \mid \Omega \rangle(\varphi = 0) \equiv \mathcal{T} + \mathcal{V}_{pot} + \mathcal{V}_{self}, \tag{7}$$

where

$$\mathcal{T} = \frac{12N_f}{(2\pi)^3} \int d\vec{k}\; |\vec{k}| \sin^2 \varphi(k) - \frac{6N_f}{(2\pi)^3} \int d\vec{k}\; m_q \;\sin 2\varphi(k), \tag{8}$$

$$\mathcal{V}_{pot} = -\frac{2N_f}{(2\pi)^6} \int d\vec{k}_1\; d\vec{k}_2 \tilde{V}(\mid \vec{k}_1 - \vec{k}_2 \mid) \Big[(1 + 3\xi) \sin 2\varphi(k_1) \sin 2\varphi(k_2)$$

$$+ 4\hat{k}_1.\hat{k}_2\; (1 + \xi) \sin^2 \varphi(k_1) \sin^2 \varphi(k_2) \Big], \tag{9}$$

$$\mathcal{V}_{self} = \frac{8N_f}{(2\pi)^6} \int d\vec{k}_1\; d\vec{k}_2\; \tilde{V}(\mid \vec{k}_1 - \vec{k}_2 \mid)\; (\hat{k}_1 \cdot \hat{k}_2)(1 + \xi) \sin^2 \varphi(k_1). \tag{10}$$

In the above, $\mathcal{V}_{self}$ is divergent for a Coulomb-like potential. The pairing function $\varphi(\vec{k})$ is determined by the gap equation, which is obtained by minimizing the energy functional of Eq. (7) with respect to $\varphi(\vec{k})$. The gap equation so obtained is given by

$$|\vec{k}| \sin 2\varphi(k) - m_q \cos 2\varphi(k) = \frac{2}{3} \cdot \frac{1}{(2\pi)^3} \int d\vec{k}'\; \tilde{V}(|\vec{k} - \vec{k}'|)$$

$$\times \Big[(1 + 3\xi) \cos 2\varphi(k) \sin 2\varphi(k') - \hat{k} \cdot \hat{k}'(1 + \xi) \cos 2\varphi(k') \sin 2\varphi(k) \Big]. \tag{11}$$

We choose to work with a potential of the type Coulomb plus linear confinement,

$$\tilde{V}(\vec{k}) = -\frac{4\pi\alpha(\vec{k}^2)}{k^2} - \frac{8\pi\sigma}{(k^2)^2}, \tag{12}$$

where $\alpha(\vec{k}^2)$ is a momentum dependent coupling and σ is the string tension. For such a potential, the term proportional to $\hat{k} \cdot \hat{k}'$ in the gap equation Eq. (11) is ultraviolet divergent. The divergence can be canceled by adding to the Hamiltonian a counterterm of the form[3,4]

$$\mathcal{H}_{ct}(\vec{x}) = (Z-1)\bar{\psi}(\vec{x})\vec{\gamma}.(i\vec{\nabla})\psi(\vec{x}). \tag{13}$$

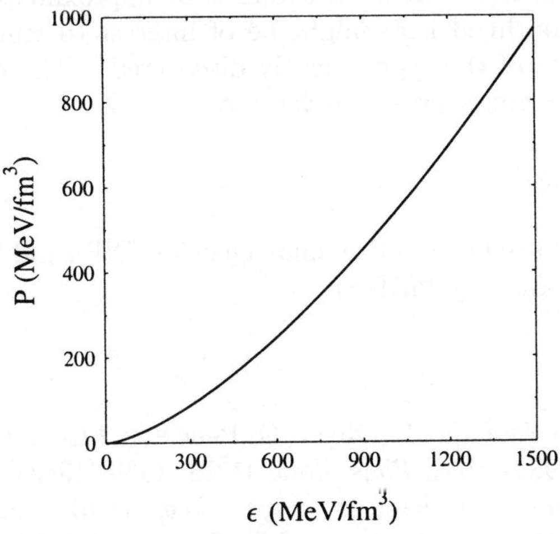

Figure 1. Pressure P versus energy density ϵ.

We now proceed with the numerical calculations. First, we consider $\alpha(\vec{k}^2)$ in the form

$$\alpha(\vec{k}^2) = \frac{12\pi}{(11N_c - 2N_f)\ln(\tau + \vec{p}^2/\Lambda^2)}, \tag{14}$$

which is taken from the one-loop renormalization group result - the N_f-dependence of $\alpha(\vec{k})$ comes from quark loops. We shall work in the zero current quark mass limit, $m_q = 0$. In this limit, the quark-antiquark condensate is given by

$$\langle \bar{q}q \rangle = -\frac{3}{\pi^2} \int dp\, p^2 \;\; \sin \varphi(p). \tag{15}$$

Taking $\sqrt{\sigma} = 180$ MeV, $\xi = 0.02$, and $\tau = 2.7$ the quark-antiquark condensate turns out to be

$$\langle \bar{q}q \rangle = (226MeV)^3. \tag{16}$$

This numerical value is largely insensitive to values of $\tau > 1$.

Once the parameters are fixed, and the counterterms at zero density are determined, the energy density of quark matter at zero temperature is calculated in the Hartree-Fock approximation[5]. Figure 1 presents the pressure (P) versus the energy density (ϵ), after subtracting the non-perturbative vacuum energy - or 'bag constant', which turns out to be approximately $(290\,\mathrm{MeV})^4$.

The approach outlined here might be of interest to studies of compact astrophysical objects of the type recently discovered[6]. The extension of the approach to finite temperatures is underway.

Acknowledgements

Work partially supported by the Brazilian agencies CNPq and FAPESP, Brazil (grants 99/08544-0 and 00/12619-5).

References

1. A. Amer, A. Le Yaouanc, L. Oliver, O. Pene and J.C. Raynal, *Phys. Rev. Lett.* **50**, 87 (1983); ibid, *Phys. Rev.* D **28**, 1530 (1983).
2. P.J.A Bicudo and J.E. Ribeiro, *Phys. Rev.* D **42**, 1611 (1625), 1635 (1990); P.J.A Bicudo, G. Krein, J.E. Riberio and J.E. Villate, *Phys. Rev.* D **45**, 1673 (1992), ibid, *Phys. Rev.* C **64**, 025202 (2001).
3. S.L. Adler and A.C. Davis, *Nucl. Phys.* B **244**, 469 (1984); R. Alkofer and P. A. Amundsen, *Nucl. Phys.* B **306**, 305 (1988).
4. A. Mishra, H. Mishra, S.P. Misra, *Z. Phys.* C **57**, 241 (1993); A. Mishra, H. Mishra, V. Sheel, S.P. Misra, P.K. Panda *Int. J. Mod. Phys.* E **5**, 93 (1996).
5. A. Kocić, *Phys. Rev.* D **33**, 1785 (1986).
6. J.J. Drake et al., *Astrophys. J.* **572**, 996 (2002).

THE $J/\psi DD$ VERTEX IN QCD SUM RULES

R.D. MATHEUS, F.S. NAVARRA, M. NIELSEN
AND R. RODRIGUES DA SILVA
*Instituto de Física, Universidade de São Paulo, C.P. 66318,
05389-970 São Paulo, SP, Brazil*
navarra@if.usp.br

The $J/\psi DD$ form factor is evaluated in a QCD sum rule calculation for both D and J/ψ off-shell mesons. We study the double Borel sum rule for the three point function of two pseudoscalar and one vector meson current. We find that the momentum dependence of the form factors is different if the D or the J/ψ meson is off-shell, but they lead to the same coupling constant in the $J/\psi DD$ vertex.

In the present work we will investigate form factors involving heavy mesons. We also want to better estimate the uncertainties in the procedure of determining coupling constants with our techniques. For these purposes we consider the vertex $J/\psi DD$ and compute form factors and coupling constants for the cases where the D is off-shell, the J/ψ is off-shell and then compare the results.

Following the QCDSR formalism described in our previous works [1,2,3] we write the three-point function associated with a $J/\psi DD$ vertex, which is given by

$$\Gamma_\mu^{(D)}(p,p') = \int d^4x d^4y \langle 0|T\{j_D(x) j_D^\dagger(y) j_\mu^\dagger(0)\}|0\rangle e^{ip'.x} e^{-i(p'-p).y} , \quad (1)$$

for a D off-shell meson, and by

$$\Gamma_\mu^{(J/\psi)}(p,p') = \int d^4x d^4y \langle 0|T\{j_D(x) j_\mu^\dagger(y) j_D^\dagger(0)\}|0\rangle e^{ip'.x} e^{-i(p'-p).y} , \quad (2)$$

for a J/ψ off-shell meson, where $j_D = i\bar{d}\gamma_5 c$, $j_D^\dagger = i\bar{c}\gamma_5 d$ and $j_\mu^\dagger = \bar{c}\gamma_\mu c$ are the interpolating fields for the outgoing D^+, incoming D^+ and J/ψ respectively with d and c being the down, and charm quark fields respectively.

The general expression for the vertex function, or correlator, in Eqs. (1) and (2) has two independent structures. We can write Γ_μ in terms of the invariant amplitudes associated with these two structures:

$$\Gamma_\mu(p,p') = \Gamma_1(p^2, p'^2, q^2) p_\mu + \Gamma_2(p^2, p'^2, q^2) p'_\mu \, . \quad (3)$$

As usual we shall write the correlators (1) and (2) in terms of hadron and quark-gluon degrees of freedom. These two representations, also called phenomenological and theoretical side, are then identified one with the other yielding a sum rule. On the theoretical side we make an operator product expansion (OPE), keeping the first terms. We can write a double dispersion relation for each one of the invariant amplitudes $\Gamma_i^{pert}(i = 1,2)$, over the virtualities p^2 and p'^2 holding $Q^2 = -q^2$ (with $q = p' - p$) fixed:

$$\Gamma_i^{pert}(p^2,p'^2,Q^2) = -\frac{1}{4\pi^2}\int_{s_{min}}^{\infty} ds \int_{m_c^2}^{\infty} du \frac{\rho_i^{pert}(s,u,Q^2)}{(s-p^2)(u-p'^2)}, \tag{4}$$

where $\rho_i^{pert}(s,u,Q^2)$ equals the double discontinuity of the amplitude $\Gamma_i(p^2,p'^2,Q^2)$.

Since we are dealing with heavy quarks, we expect the perturbative contribution to be dominant on the OPE side. However, for the J/ψ off-shell it turns out that the contribution of the quark condensate is also important and must be included:

$$\Gamma_\mu^{<\bar{q}q>(J/\psi)}(p,p') = \frac{m_c\langle\bar{q}q\rangle}{(p^2-m_c^2)(p'^2-m_c^2)}(p_\mu + p'_\mu) = \Gamma^{<\bar{q}q>}(p_\mu + p'_\mu). \tag{5}$$

The phenomenological side of the vertex function is obtained by considering the contribution of the J/ψ and one D meson, or the two D mesons states to the matrix element in Eqs. (1) and (2) respectively. In doing so, we introduce the decay constants f_D and $f_{J/\psi}$, which are defined by the matrix elements $\langle 0|j_D|D\rangle = \frac{m_D^2 f_D}{m_c}$ and $\langle J/\psi|j_\mu^\dagger|0\rangle = m_{J/\psi}f_{J/\psi}\epsilon_\mu^*$ where ϵ_μ is the polarization of the vector meson. We also make use of the matrix element $\langle J/\psi(q)D(p')|D(p)\rangle = -g_{J/\psi DD}(Q^2)(p_\mu + p'_\mu)\epsilon^\mu$. Altogether these steps lead to:

$$\Gamma_\mu^{phen(D)}(p,p') = -\frac{m_D^2 f_D}{m_c} m_{J/\psi} f_\psi g_D(Q^2) \frac{1}{p^2-m_{J/\psi}^2}\frac{1}{p'^2-m_D^2} \times$$

$$\left(-2p'_\mu + \frac{m_D^2+m_{J/\psi}^2+Q^2}{m_{J/\psi}^2}p_\mu\right), \tag{6}$$

$$\Gamma_\mu^{phen(J/\psi)}(p,p') = -\frac{m_D^4 f_D^2}{m_c^2} g_{J/\psi}(Q^2)\frac{1}{p^2-m_D^2}\frac{1}{p'^2-m_D^2}\left(p'_\mu + p_\mu\right), \tag{7}$$

where

$$g_D(Q^2) = \frac{m_D^2 f_D}{m_c}\frac{g_{DDJ/\psi}^{(D)}(Q^2)}{Q^2+m_D^2}, \tag{8}$$

and

$$g_{J/\psi}(Q^2) = m_{J/\psi} f_\psi \frac{g^{(J/\psi)}_{DDJ/\psi}(Q^2)}{Q^2 + m^2_{J/\psi}} \,, \tag{9}$$

where m_D, $m_{J/\psi}$ are the masses of the mesons D and J/ψ respectively, and $g^{(M)}_{DDJ/\psi}(Q^2)$ is the form factor at the DDJ/ψ vertex when the meson M is off-shell. We will concentrate in the p'_μ structure, which we found to be the more stable one. The sum rules in this structure read:

$$\Gamma_2^{phen(D)}(p^2, p'^2, q^2) = \Gamma_2^{pert(D)}(p^2, p'^2, q^2) \tag{10}$$

for a D off-shell and

$$\Gamma_2^{phen(J/\psi)}(p^2, p'^2, q^2) = \Gamma_2^{pert(J/\psi)}(p^2, p'^2, q^2) + \Gamma^{<\bar{q}q>}(p^2, p'^2, q^2) \tag{11}$$

for a J/ψ off-shell.

Inserting in these equations the corresponding expressions for the perturbative, quark condensate and phenomenological terms and performing a double Borel transformation in both variables $P^2 = -p^2 \rightarrow M^2$ and $P'^2 = -p'^2 \rightarrow M'^2$ we obtain the final expressions for the sum rules.

The parameter values used in all calculations are $m_q = (m_u + m_d)/2 = 7MeV$, $m_c = 1.3GeV$, $m_D = 1.87GeV$, $m_{J/\psi} = 3.1GeV$, $\langle\bar{q}q\rangle = -(0.23)^3 GeV^3$. Fixing $M^2 = 11GeV^2$ we calculate the momentum dependence of the form factor $g^{(D)}_{DDJ/\psi}(Q^2)$. Our numerical calculations can be well reproduced by the gaussian parameterization $g^{(D)}_{DDJ/\psi}(Q^2) = 16.4e^{-[\frac{(Q^2+16.2)^2}{228}]}$. We stress here that it was not possible to fit our results with a monopole form!

We define the coupling constant as the value of the form factor at $Q^2 = -m_M^2$, where m_M is the mass of the off-shell meson. In the case of an off-shell D, this leads to

$$g^{(D)}_{DDJ/\psi} = 8.05 \,. \tag{12}$$

Along the same lines we choose $M^2 = 8\text{GeV}^2$ and calculate the momentum dependence of the form factor $g^{(J/\psi)}_{DDJ/\psi}(Q^2)$. The numerical results can be fitted by the monopole parameterization $g^{(J/\psi)}_{DDJ/\psi}(Q^2) = \frac{1069.76}{Q^2+143.18}$ which, at the J/ψ pole leads to:

$$g^{(J/\psi)}_{DDJ/\psi} = 7.98 \,. \tag{13}$$

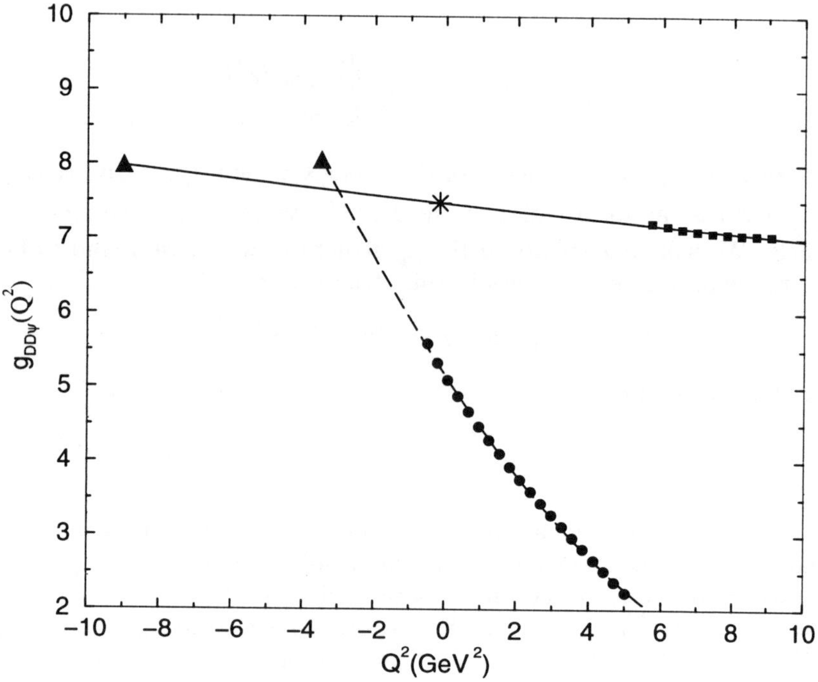

Figure 1. Momentum dependence of the DDJ/ψ form factor. Circles and squares represent our numerical calculations for the D and J/ψ off-shell respectively. The dashed and solid lines give the parameterization of the QCDSR results through a gaussian for the circles and monopole for the squares. The triangles give the form factors at the poles of the particles (which we identify with the coupling constant). The star shows the form factor at $Q^2 = 0$.

All our numerical results and parameterizations are shown in Fig.1. The circles and squares correspond to the numerical results for the D and J/ψ off-shell respectively. These points are fitted by the dashed (gaussian) and solid lines (monopole) respectively and extrapolated to the meson poles which are represented by the triangles, which give the numbers quoted in (12) and (13).

References

1. F.S. Navarra, M. Nielsen, M.E. Bracco, M. Chiapparini and C.L. Schat, *Phys. Lett.* **B489** (2000) 319.
2. F.S. Navarra, M. Nielsen and M.E. Bracco, *Phys. Rev.* **D65** (2002) 037502.
3. M.E. Bracco, M. Chiapparini, A. Lozea, F.S. Navarra and M. Nielsen, *Phys. Lett.* **B521** (2001) 1.

7. Field Theories, Formal Developments

BOUND STATES OF A RELATIVISTIC SCALAR PLASMA SYSTEM

E.R. TAKANO NATTI
Centro Politécnico, Universidade Norte do Paraná
Rua Tietê, 1208, Cep 86025-230, Londrina, Paraná, Brazil
erica.takano@uol.com.br

P.L. NATTI
Departamento de Matemática, Universidade Estadual de Londrina,
C. P. 6001, Cep 86051-970, Londrina, Paraná, Brazil
plnatti@uel.br

C.-Y. LIN
Department of Physics, National Dong Hwa University,
Hua-lien, Taiwan, Republic of China
lcyong@mail.ndhu.edu.tw

We investigate the physics of elementary excitations for the so called relativistic scalar plasma system. Following the standard many-body procedure we have obtained the RPA equations for this model by linearizing the TDHFB equations of motion around equilibrium and shown that these oscillation modes give one-meson and two-fermion state of the theory. The resulting equations have a closed solution, from which one can examinate the conditions for the existence of bound states.

1 Introduction

In a previous work [1] we have presented a framework to investigate the initial-value problem in the context of interacting fermion-scalar field theories. The method allows one to obtain a set of self consistent equations for the expectation values of linear and bilinear forms of field operators. These dynamical equations acquire the structure of kinetic type, where the lowest order of the approximation corresponds to the usual gaussian mean-field approximation.

In this work we follow a recent work by Kerman and Lin [2,3] in order to investigate the near equilibrium dynamics around the stationary solution. We shall show that one-meson and two-(quasi) fermion physics can be studied from the linear approximation of the mean-field equations. In particular, one can solve these equations in a closed form and find scattering amplitude as well as the conditions for the two-fermion bound state.

For the scalar plasma model, the dynamics is governed by the hamiltonian density

$$\mathcal{H} = -\bar{\psi}(i\vec{\gamma}.\vec{\partial} - m)\psi - g\bar{\psi}\phi\psi + \frac{1}{8\pi}\left[\frac{(4\pi)^2}{1+Z}\Pi^2 + (1+Z)|\partial\phi|^2 + \mu^2\phi^2\right] + \mathcal{H}_c \ , \tag{1}$$

where ψ is a spin-$\frac{1}{2}$ field while ϕ is a scalar field. The parameters m and μ are, respectively, the mass of fermion and scalar particles and g is the coupling constant. The last term of this expression is the counterterms necessary to remove the infinities occurring later [1,4],

$$4\pi\mathcal{H}_C = \frac{A}{1!}\phi + \frac{\delta\mu^2}{2!}\phi^2 + \frac{C}{3!}\phi^3 + \frac{D}{4!}\phi^4 \ , \tag{2}$$

where the coefficients A, $\delta\mu^2$, C, D, and Z are infinity constants to be definided later.

In order to study the dynamics of the system, we focus on the selected set of observables related to the expectation values of linear and bilinear forms of field operators. The time evolution of these quantities obeys the Heisenberg equation. We describe the pairing correlations by using a unitary Bogoliubov transformation. In doing so, we obtain [1]

$$\dot{\varphi}_{\mathrm{k}} = g\langle\phi\rangle\frac{|\mathbf{k}|}{k_0}\sin\gamma_{\mathrm{k}} \tag{3}$$

$$\dot{\gamma}_{\mathrm{k}} = 2k_0 - 2gm\langle\phi\rangle\frac{1}{k_0}\left(1 - \frac{|\mathbf{k}|}{m}\cot 2\varphi_{\mathrm{k}}\cos\gamma_{\mathrm{k}}\right) \tag{4}$$

$$\langle\dot{\phi}\rangle = \frac{4\pi}{(1+Z)}\langle\Pi\rangle \tag{5}$$

$$\langle\dot{\Pi}\rangle = -\left(\frac{A}{4\pi} + \frac{C}{2}G(\mu)\right) - \left(\frac{\mu^2}{4\pi} + \frac{\delta\mu^2}{4\pi} + \frac{D}{2}G(\mu)\right)\langle\phi\rangle \tag{6}$$

$$-\frac{C}{8\pi}\langle\phi\rangle^2 - \frac{D}{24\pi}\langle\phi\rangle^3 - \frac{2g}{(2\pi)^3}\left[I_1(m) + I_2(m)\right]$$

where

$$G(\mu) = \frac{1}{(2\pi)^3}\int d^3\mathrm{p}\frac{1}{2(\mathbf{p}^2+\mu^2)^{1/2}} \tag{7}$$

$$I_1(m) = \int d^3\mathrm{k}\frac{m}{(\mathbf{k}^2+m^2)^{1/2}}\cos 2\varphi_{\mathrm{k}} \tag{8}$$

$$I_2(m) = \int d^3\mathrm{k}\frac{|\mathbf{k}|}{(\mathbf{k}^2+m^2)^{1/2}}\sin 2\varphi_{\mathrm{k}}\cos\gamma_{\mathrm{k}} \ . \tag{9}$$

The equations above describe the real-time evolution for the relativistic scalar plasma system in the gaussian mean-field approximation.

2 Near equilibrium dynamics

First we consider the static problem. Recalling (3)-(6), the stationary conditions require $\dot{\gamma}_{\mathbf{k}} = \dot{\varphi}_{\mathbf{k}} = \langle\dot{\phi}\rangle = \langle\dot{\Pi}\rangle = 0$. Investigation of the near equilibrium motion proceeds by considering flutuactions around the stationary point, expanding (3)-(6) up to first order in these fluctuations. In doing so, we obtain second order equations. Thus, the small oscillation dynamics is described by coupled linear oscillator equations as usual in RPA treatment. In particular, these modes decouple when $g = 0$ yielding two simple oscillator equations.

3 RPA Equations as a scattering problem

We shall show in this that these coupled linear oscillator equations can be solved analytically giving the wavefunctions and the spectrum of excitations. Since the oscillation amplitudes play the roles of wavefunctions of quantum particles, it is then more convenient to treat this system as a coupled channel scattering problem with appropriate boundary conditions [3].

In this case, we define a new variable $\Psi_{\mathbf{k}} = \sqrt{\bar{k}_0}\delta\varphi_{\mathbf{k}}$ and can solve the coupled equations as follows:

$$\left(\omega^2 - 4\bar{k}_0^2\right)\Psi_{\mathbf{k}} = \frac{-32\pi g^2}{\omega^2 - \mu^2 + Z\omega^2 - \Sigma + i\eta}\frac{|\mathbf{k}|}{\sqrt{\bar{k}_0}}\int_{\mathbf{k}'}\frac{|\mathbf{k}|}{\sqrt{\bar{k}_0}}\Psi_{\mathbf{k}'} \;, \tag{10}$$

Notice that the potential is separable in the sense that [5]

$$\langle\mathbf{k}|V|\mathbf{k}'\rangle = v(\mathbf{k})v(\mathbf{k}') = \frac{|\mathbf{k}|}{\sqrt{\bar{k}_0}}\frac{|\mathbf{k}'|}{\sqrt{\bar{k}_0'}} \;. \tag{11}$$

so that it is convenient to rewrite (10) as an integral equation,

$$\Psi(\mathbf{k},\mathbf{q};\omega) = \alpha\delta(\mathbf{q}-\mathbf{k}) \tag{12}$$

$$+\frac{1}{[\omega^2 - 4\bar{k}_0^2 + i\epsilon]}\frac{-32\pi g^2}{[\omega^2 - \mu^2 + Z\omega^2 - \Sigma + i\eta]}v(\mathbf{k})\int_{\mathbf{k}'}v(\mathbf{k}')\Psi(\mathbf{k}',\mathbf{q};\omega) \;,$$

where $\mathbf{q}$ is the relative momentum for two incident quasi-fermions and α is an overall phase factor. We have used the boundary condition of the outgoing wave condition $(+i\epsilon)$ as solution of Eq.(11).

Now we integrate this expression with respect to $\mathbf{k}$. Substituting this result back into (12), and using the renormalization procedure [1], we obtain

$$\Psi(\mathbf{k},\mathbf{q};\omega) = \alpha\delta(\mathbf{q}-\mathbf{k}) + \frac{1}{\omega^2 - 4\bar{k}_0^2 + i\epsilon}\frac{\alpha v(\mathbf{q})}{\Delta^+(\omega)} \tag{13}$$

where

$$\Delta^+(\omega) = -\frac{\pi\mu^2}{g^2 m^2} + F(\omega) - \theta(\omega - 4\bar{m}^2)\frac{i}{8\pi}\left[\omega^2 - 4\bar{m}^2\right] \tag{14}$$

with the finite term $F(\omega)$, when $0 \leq \omega^2 \leq 4\bar{m}^2$, given by

$$F(\omega) = \left(\omega^2 - 6\bar{m}^2\right) \log\left(\frac{\bar{m}}{2m}\right) + \frac{2(4\bar{m}^2 - \omega^2)^{3/2}}{\omega} \tan^{-1} \sqrt{\frac{\omega^2}{4\bar{m}^2 - \omega^2}} \, . \quad (15)$$

We have, thus, found an analytical solution for the elastic channel of two-fermion scattering problem. The special form of interacting potential in this case allows one also to get easily a closed expression for the scattering matrix

$$T(\mathbf{k}, \mathbf{k}'; \omega) = v(\mathbf{k}) \frac{1}{\Delta^+(\omega)} v(\mathbf{k}') \, . \quad (16)$$

We investigate the conditions for the existence of bound states in a system of scalar plasma, using the framework developed in the previous sections [6]. Discussion of the problem consists now in solving the equation

$$\Delta^+(\omega) = 0. \quad (17)$$

Depending on the value of ω the system has different dynamical behavior. Solution of interest here is in the interval of $0 < \omega^2 < 4\bar{m}^2$. In this case, the system may present a stable bound state if one finds ω_B such that $\Delta^+(\omega_B) = 0$. A natural way to interprete this result is that the meson mass determines the range of the Yukawa potential for a coupling. When μ of the system is large, it is more dificult to the fermions to interact and consequently decreases the probability of forming a bound state, and vice-versa.

In summary, this work has investigated the physics of elementary excitations for the so called relativistic scalar plasma model. To reach this goal we have derived RPA equations for this sistem by linearizing the TDHFB equations obtained in a previous work. In this context, the amplitudes of excitations are identified with quantum particles and the resulting equations allow one to study scattering processes nonperturbatively. In particular, our results indicate existence of bound state for certain region of phase diagram.

References

1. E.R. Takano Natti, C.-Y. Lin, A.F.R. de Toledo Piza and P.L. Natti, *Phys. Rev. D* **60**, 125013 (1999).
2. A. Kerman and C.-Y. Lin, *Ann. Phys.(N.Y.)* **241**, 185 (1995).
3. A. Kerman and C.-Y. Lin, *Ann. Phys.(N.Y.)* **269**, 55 (1998).
4. J.D. Alonso and R. Hakim, *Phys. Rev. D* **38**, 1780 (1988).
5. R. Newton, *Scattering Theory of Waves and Particles*, (Springer-Verlag,Berlin, 1982).
6. P.L. Natti and A.F.R. de Toledo Piza, *Phys. Rev. D* **55**, 3403 (1997).

SPONTANEOUS $SO(4,2)$ SYMMETRY BREAKING AND ITS IMPLICATIONS ON MASS TERMS

S. MITTMANN DOS SANTOS, B. E. J. BODMANN

Universidade do Vale do Rio dos Sinos - UNISINOS, Av. Unisinos 950, 93022-000 São Leopoldo, RS, Brasil

mittmann, bardo@exatas.unisinos.br

The present work is a step toward an understanding of the relation between specific mass spectra and the structure of the underlying Quantum Field Dynamics. Upon interpreting the mass shell equation as a six dimensional zero mass equation the infinitesimal mass term may be defined by a normalization times an asymmetry from the additional dimensions. We show that a generator may be defined by the commutator of the Poincaré and the conformal translation, which transforms a massless vacuum state into a massive one and thus implements spontaneous $SO(4,2)$ symmetry breaking.

1 Introduction

Specific symmetries (chiral symmetry or dilatation) are considered as a central issue when discussing the question of particle mass spectra in the realm of Quantum Hadrodynamics [1,2] or Quantum Chromodynamics [3,4]. A recent article [5] showed, that starting from the hypothesis of spontaneous dilatation symmetry breaking, one may obtain self reciprocal mass ratios from massless Quantum Electrodynamics in curved energy-momentum space. Thus there arise the following questions: What is the underlying mechanism that implies in energy-momentum curvature, which might turn massless quantum theories massive ones? As a promising candidate we consider spontaneous breaking of $SO(4,2)$ symmetry. Is the dynamical structure of the Quantum Field Theory sufficient to imply in its specific spectrum, once an arbitrary scale normalization is adopted? At the present stage of the work we believe that internal symmetries may render the role of Quantum Field Theory to impose a "discrete filter" on the mass continuum.

2 Transformations

In literature [6] one finds discussions of space-time transformations embedded in a higher dimensional (> 4) flat hyperspace. In order to indicate the relevant transformations, which preserve causality, the infinitesimal mass shell equation $dm^2 = g_{\mu\nu}dp^\mu dp^\nu$ serves as a convenient starting point. Any transformation in momentum is then generically given by $p^\mu \to p^\mu + \epsilon q^\mu(p)$. Upon inserting the momentum change into the mass shell equation yields the infinitesimal change in the metric tensor $\delta(dm^2) = \epsilon\left(\frac{\partial q_\mu}{\partial p_\nu} + \frac{\partial q_\nu}{\partial p_\mu}\right) dp_\mu dp_\nu = \epsilon G^{\mu\nu} dp_\mu dp_\nu$. Taking into account the causality constraint one determines the modified metric $G^{\mu\nu}$ in terms of $\sigma g^{\mu\nu} = G^{\mu\nu}$ with $\sigma = \frac{1}{4} g_{\mu\nu} G^{\mu\nu}$, which in turn defines the constraints for the most general momentum dependence of the infinitesimal transformation by $q^\mu(p)$. The specific form of the transformation may be determined using a power expansion $q^\mu = {}_0a^\mu + {}^1_1a^\mu{}_\nu p^\nu + {}^2_1a^\nu{}_\nu p^\mu + {}^1_2a^\mu{}_{\nu\lambda} p^\nu p^\lambda + {}^2_2a^\nu{}_{\nu\lambda} p^\lambda p^\mu + {}^2_2a^\lambda{}_{\nu\lambda} p^\nu p^\mu + O(p^3)$. The symmetry conditions read then for terms that go with a specific power in $(p^\mu)^n$: (1) No restriction on $(p^\mu)^0$ represents the Poincaré translation in energy-momentum space. (2) For $(p^\mu)^1$ the scalar coefficient ${}^2_1a$ is an arbitrary factor, reflecting the dilatation transformation. In addition one gets $0 = {}^1_1a^{\mu\nu} + {}^1_1a^{\nu\mu} - \frac{1}{2} g^{\mu\nu} {}^1_1a^\lambda{}_\lambda$, which may be identified with the Lorentz transformation. (3) From the terms that go with $(p^\mu)^2$, one obtains, $0 = 2\,{}^1_2a^{\mu\nu}{}_\lambda p^\lambda + 2\,{}^1_2a^{\nu\mu}{}_\lambda p^\lambda - {}^1_2a^\kappa{}_{\lambda\kappa} g^{\mu\nu} p^\lambda + 2\left({}^2_2a^\nu p^\mu + {}^2_2a_\lambda g^{\mu\nu} p^\lambda + {}^2_2a^\mu p^\nu\right)$. Upon contracting the expression by $g_{\mu\nu}$ eliminates all terms except for the one in parenthesis and thus implies in ${}^2_2a_\lambda \equiv 0$. Additionally one may rewrite the coefficient in terms of an arbitrary vector c^μ and the metric, ${}^1_2a^{\mu\nu\lambda} = g^{\mu\nu} c^\lambda + g^{\mu\lambda} c^\nu - g^{\nu\lambda} c^\mu$, which has got the characteristics of a conformal translation. (4) Terms with higher powers in p^μ vanish identically, thus the most general admissible form of the infinitesimal transformation is composed by the Poincaré, the Lorentz Dilatation and the conformal transformation, $q^\mu = b^\mu + \omega^\mu_\nu p^\nu + \lambda p^\mu + (2c_\lambda p^\lambda p^\mu - c^\mu p^\lambda p_\lambda)$. From the finite form of the conformal translation one recognizes, that these transformations may turn singular in a sub-manifold, where the denominator vanishes. Therefore the transformation has to be restricted in c^μ such as to define diffeomorphisms in the physically relevant region of energy-momentum space.

3 A topological mass

Since we have to consolidate the conformal group with a broken symmetry and a normalization, we extend the dimension of energy-momentum space by two and represent the group by linear transformations of the pseudo-orthogonal group $SO(4,2)$ with $S_6 = \{p | g_{\alpha\beta} p^\alpha p^\beta = 0\}$. The second fundamental form of Gauß contains all curvature properties and permits to interpret the normal vector on an oriented four dimensional energy-momentum hyper-surface as a reciprocal normalization N^{-1}, which is fixed but may be arbitrarily chosen, and an asymmetry A, which is then a function of energy-momentum. One possibility is to define the normalization and asymmetry by $N^{-1} = p^4 + p^5$ and $A = p^4 - p^5$, and the mass equation is then $m^2 = N^{-1}A = p^\mu p_\mu$. For convenience we define unit-less momenta $\pi^\mu = Np^\mu$ with the scale invariant mass shell equation $\mu^2 = N^{-2}m^2 = \pi^\mu \pi_\mu = N\,A$.

4 $SO(4,2)$ symmetry breaking

In order to find the generator which produces the desired asymmetry we analyze the subgroups of $SO(4,2)$: (1) Maintaining the normalization and the asymmetry constant, the transformation reduces to Lorentz $\pi^{\mu\prime} = \omega^\mu_\nu \pi^\nu$. (2) The subgroup which leaves the normalization invariant defines the translation in energy-momentum space with $\pi^{\mu\prime} = \pi^\mu + Nb^\mu$ and $A' = A + 2\pi^\mu b_\mu + Nb^\mu b_\mu$. (3) The subgroup with four parameters represents conformal translations and leaves the asymmetry invariant with $\pi^{\mu\prime} = \pi^\mu - Ac^\mu$ and $N' = N - 2\pi^\mu c_\mu + Ac^\mu c_\mu$. (4) The one parameter subgroup defines the dilatation with $\pi^{\mu\prime} = \pi^\mu$, $A' = \lambda A$ and $N' = \lambda^{-1} N$. The Poincaré as well as the conformal translation are candidates to break $SO(4,2)$ symmetry, since in a specific system with $A = 0$, upon transformation, the asymmetry may turn non-zero.

One may now construct an operator, which transforms a vacuum state with $\langle\psi|\pi_\mu\pi^\mu|\psi\rangle = 0$, i.e. $A = 0$ into a non-vanishing one. To this end one may "transport" the momentum π^μ first by a Poincaré displacement P followed by a conformal translation C and then return by the inverse sequence. Thus the momentum change after "transport" is $[C^\mu_\nu, P^\nu_\lambda]\pi^\lambda = Nb^\mu - Ac^\mu$, where $[\cdot,\cdot]$ is the usual commutator, which plays the role of a generator and transforms a specific vacuum state into another equivalent state. The change in the vacuum expectation value is then

$$\langle\psi|\pi^\nu[C^\lambda_\nu, P^\kappa_\lambda]g_{\kappa\mu}\pi^\mu|\psi\rangle = \langle\psi|Nb_\mu\pi^\mu - Ac_\mu\pi^\mu|\psi\rangle \tag{1}$$

Even in the limit of a vanishing asymmetry $A \to 0$, there remains the scale invariant term $Nb_\mu\pi^\mu$, which may be non-zero, however p_4 and p_5 shall be finite. The fact that we have found a generator, which transforms a zero state into a non-zero state may be understood as an implementation of a spontaneous $SO(4,2)$ symmetry breaking.

5 Conclusions

From ref. [5] and from the present discussion we believe to have done another step which in future might show how symmetry, its spontaneous breaking and the structure of quantum field dynamics imply in a specific discrete mass spectrum. In order to prepare the playground for Quantum Field theory with its usually linear operators we have chosen the group representation by linear transformations of the pseudo-orthogonal group $SO(4,2)$. We found indeed a generator defined by the commutation of the Poincaré with the conformal translation which allows one to change the zero mass vacuum state into a massive one. Thus breaking of diffeomorph $SO(4,2)$ symmetry breaks as a consequence dilatation as well as chiral symmetry.

Since, it is an accepted practice in renormalization prescriptions of Quantum Field Theories to introduce a dimensional scale into the theory, consequently such a theory is no longer scale invariant. The present prescription showed that there exists at least a mechanism to understand the Lorentz invariant momentum cutoff on a natural basis, i.e. from normalization and asymmetry, and thence it is sufficient to define only the Quantum Field theory. Further the already proposed combination of the always necessary mass-scale normalization with the renormalization procedure for Quantum Fields of massless particles [5] is feasible, in a self-consistent fashion. Hence there will be no need to change the usual dynamical description of the space-time manifold carrying bosonic degrees of freedom, but it is the curved energy-momentum hyper-surface which puts a symmetry condition on the allowed solutions of the fermion field equations.

We are completely aware of the fact, that our discussion at the present status is restricted to the question how spontaneous symmetry breaking in a curved energy-momentum space may explain mass terms in a *ad inicio* massless theory. Our reasoning is far from leading to a theory, which is applicable to the complete set of elementary particles and their interactions. As a future work we will elaborate the equivalent symmetry considerations for internal symmetries $SU(2,2)$ as a continuation of the present work, in order to investigate the role of the specific interaction structure (abelian, non-abelian gauge structure) on the mass solutions.

Acknowledgments

The authors would like to thank FAPERGS, CNPq and CAPES, Brazil.

References

1. J. M. Eisenberg, G. Kaelbermann *Int. J. Mod. Phys.* **E 5** 423, (1996).
2. B. E. J. Bodmann, Th. A. J. Maris, C. A. Z. Vasconcellos, M. Dillig *Il Nuovo Cim.* **111 A**, 223 (1998).
3. J. Bijnens, G. Colangelo, G. Ecker *JHEP* **20**, 9902 (1999).
4. H. Leutwyler *Nucl. Phys. Proc. Suppl.* **94**, 108 (2001).
5. B. E. J. Bodmann, Th. A. J. Maris *Phys. Lett.* **B 495**, 98 (2000).
6. D. Buchholz, R. Verch *Rev. Math. Phys.* **7**, 1195 (1995).

OUTLINES

OUTLINES OF LECTURES ON HIGH-PRECISION NON-PERTURBATIVE QCD

G. P. LEPAGE

HIGH-PRECISION NON-PERTURBATIVE QCD

G. P. LEPAGE
Cornell University
Ithaca, New York State, USA
gpl@mail.lns.cornell.edu

Ten years ago realistic lattice QCD simulations seemed hopelessly unattainable. Following major innovations in the 1990's, and particularly in the last two years, such simulations are now possible for the first time. A landmark in the history of non-perturbative quantum field theory, this development is also of great importance for QCD phenomenology, particularly for heavy-quark mesons. The talks here outline the basic concepts of lattice QCD, describes the revolutionary changes that occurred in the 1990's, and look into future applications within QCD and beyond.

Why High-Precision and Nonperturbative?

Essential for Standard Model

E.g., CKM weak interaction parameters
ρ and η from:

B-$\bar{B}$ mixing
$B \longrightarrow \pi l \nu$
K-$\bar{K}$ mixing
...

→ Nonpert've QCD Part × Weak Int'n Part

CKM today . . . . . . and with 2–3% theory errors.

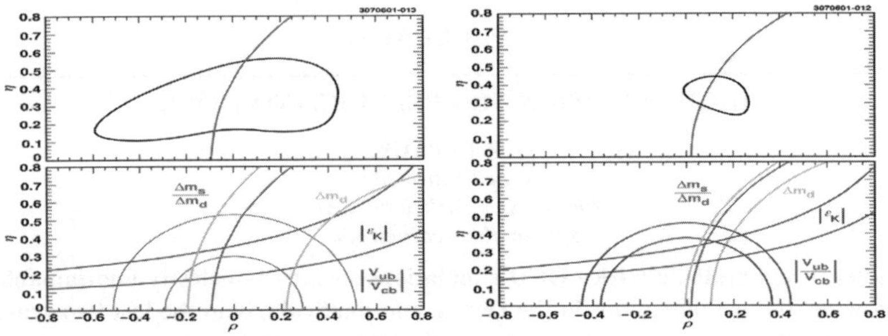

95% confidence levels; CLEO-c (2001).

Essential Beyond the S.M.?

Strongly coupled field theories are an outstanding challenge to all theoretical physics.

- Field theory is generic; weak coupling is not.

2 of 3 known interactions are strongly coupled: QCD, gravity.

Asymp. freedom $\Rightarrow$ strong coupling at low E
. . . + log. evolution $\Rightarrow$ large mass hierarchies.

- E.g., in QCD:

$$\alpha_s(M_{\text{planck}}) = 0.02 \quad \text{and} \quad \alpha_s(m_{\text{hadron}}) \approx 1$$

$$\Rightarrow \quad m_{\text{hadron}}/M_{\text{planck}} \approx 10^{-19}.$$

Thus strong coupling is *natural* in particle physics.

Strong coupling is possible (likely?) at the LHC and/or beyond.

- Generic at low energies in non-abelian gauge theories...
- ... unless gauge symmetry spontaneously broken ⇒ dynamical symmetry breaking ⇒ strong coupling.
- Critical near-term need for reliable, generic techniques for strong coupling.

What is Lattice QCD?

Lattice Approximation

Continuous Space & Time ⟶

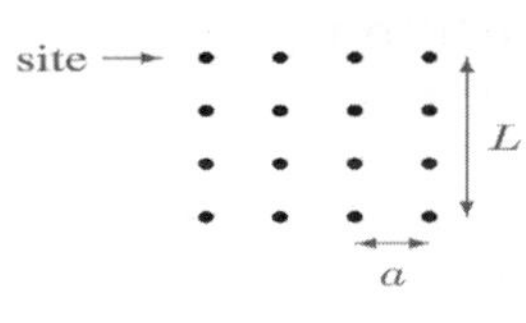

⇒ Fields $\psi(x)$, $A_\mu(x)$ specified only at grid sites; interpolate for other points.

$\Rightarrow$ **QCD $\rightarrow$ multidimensional integration.**

$$\int \mathcal{D}A_\mu \ldots \mathrm{e}^{-\int L \mathrm{d}t} \longrightarrow \int \prod_{x_j \in \text{grid}} \mathrm{d}A_\mu(x_j) \ldots \mathrm{e}^{-a \sum L_j}.$$

$\Rightarrow$ **Millions of integration variables.**

$\Rightarrow$ **Numerical Monte Carlo integration.**

N.B. Cost $\propto (1/a)^\omega$ where $\omega \geq 6$ implies must keep a as large as possible!

Fall & Rise of LQCD

- Invented in 1974; "explains" confinement.
- Stalls for almost 20 years.
  - Ken Wilson declares it dead! (1989)
- Renaissance in 1990's.
  - Perturbation theory fixed.
  - Effective field theories for c, b's.
  - Improved discretizations $\Rightarrow$ larger a's.
  - Unquenching! (2000)
- First high-prescision nonperturbative results.
  - $\alpha_s(M_Z)$, $M_b \ldots$ to few %.
  - Ken Wilson retracts. (1995)

QCD Revolution

Traditional wisdom	$\Rightarrow$	need $a \leq 0.05$ fm.
New simulation results	$\Rightarrow$	$a = 0.1$–0.4 fm works.

Simulation cost $\propto (1/a)^6$

$\Rightarrow$ new simulations cost 10^2–10^6 times less!

Quantum Field Theory on a Lattice

Approximate Derivatives

Numerical Analysis $\Rightarrow$

$$\frac{\partial\psi(x_j)}{\partial x} = \Delta_x\psi(x_j) + \mathcal{O}(a^2)$$

$$\Delta_x\psi(x_j) = \frac{\psi(x_j+a) - \psi(x_j-a)}{2a}$$

$\Rightarrow$ uses only ψ's at grid sites.

Large $a \Rightarrow$ need *improved discretizations*.

E.g.

$$\frac{\partial\psi}{\partial x} = \Delta\psi - \frac{a^2}{6}\,\Delta^3\psi + \mathcal{O}(a^4)$$

$\frac{a^2}{6}\Delta^3\psi$: 10–15% for $a = 0.4\,\mathrm{fm}$

$\mathcal{O}(a^4)$: 1–2% for $a = 0.4\,\mathrm{fm}$

$\Rightarrow a = 0.4\,\mathrm{fm}$ okay?

Ultraviolet Cutoff

$\lambda_{\text{min}} = 2a$ is smallest wavelength.

E.g.) $\psi = +1 \quad -1 \quad +1 \quad -1 \quad +1$

$\bullet \quad \bullet \quad \bullet \quad \bullet \quad \bullet$

$\Rightarrow$ all quark and gluon states with $p > \pi/a$ are excluded by the lattice since $p = 2\pi/\lambda$.

But $\forall\, p$s important in quantum field theory! (Consider ultraviolet divergences.)

Renormalization Theory $\Rightarrow$ mimic effects of $p > \pi/a$ excluded states by adding extra a-dependent *local* terms to the field equations, Lagrangian, currents, etc.

$$\Rightarrow \qquad \partial\psi \rightarrow \Delta\psi + c(a)\, a^2\, \Delta^3\psi + \cdots$$

where

$$c(a) = -\frac{1}{6} + \text{Contribution for } p > \pi/a \text{ physics}$$

$-\frac{1}{6}$: Numerical Analysis

Contribution for $p > \pi/a$ physics: Theory & context specific $\Rightarrow$ not universal!

Bad News: Need a^2 corrections when a large, but *Numerical Recipes* won't tell you values of $c(a)\ldots$

Good News: $p > \pi/a$ **QCD is perturbative if** a ***small*** **enough (asymptotic freedom).**

$\Rightarrow$ **compute** $c(a)\ldots$ **using perturbation theory.**

Perturbation theory fills in gaps in lattice;
$\Rightarrow$ **continuum results without** $a \to 0$**!**

E.g.,

$$\begin{aligned}\mathcal{L}^{(a)} &= Z(a)\,\overline{\psi}(\Delta\cdot\gamma - m(a))\psi \\ &\quad + c(a)\,a^2\,\overline{\psi}\Delta^3\cdot\gamma\psi \\ &\quad + \cdots\end{aligned}$$

where

$$c(a) = -\frac{1}{6} + c_1\,\alpha_s(\pi/a) + \cdots$$

$-\frac{1}{6}$: Numerical Analysis

$c_1\,\alpha_s(\pi/a)$: Mimics effects of $p > \pi/a$ states excluded by grid.

Asymptotic freedom in QCD $\Rightarrow$

- short-distance physics simple (perturbative);
- long-distance physics difficult (nonperturbative).

Lattice separates "short" from "long":

- $p > \pi/a$ QCD $\rightarrow$ corrections $\delta\mathcal{L}$ computed in perturbation theory (determines a);
- $p < \pi/a$ QCD $\rightarrow$ nonperturbative, numerical Monte Carlo integration.

Computing Lattice Operators

Compute on-shell, low-momentum ($p_i \ll \pi/a$) scattering amplitudes for:

Continuum: $T =$

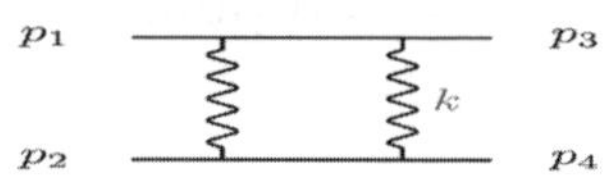

Lattice: $T^{(a)} =$

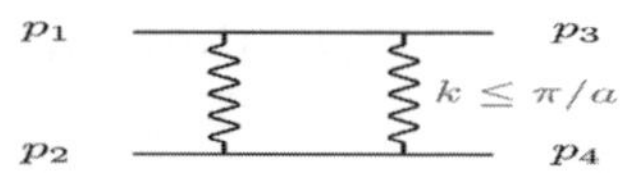

$T - T^{(a)}$ = what is omitted by lattice QCD.

$T - T^{(a)}$ dominated by $k > \pi/a \gg p_i$

$\Rightarrow$ Taylor expand in powers of $p_i/k \approx p_i a \ll 1$:

$$\begin{aligned} T - T^{(a)} &= c(a)\, a^2\, \overline{u}(p_2)\gamma_\mu u(p_1)\, \overline{u}(p_4)\gamma^\mu u(p_3) \\ &+ c_A(a)\, a^2\, \overline{u}\gamma_\mu\gamma_5 u\, \overline{u}\gamma^\mu\gamma_5 u \\ &+ d(a)\, a^4\, (p_1 - p_2)^2\, \overline{u}\gamma_\mu u\, \overline{u}\gamma^\mu u + \cdots. \end{aligned}$$

Dimensionless Taylor coefficient; depends upon $\alpha_s(\pi/a)$, $m\,a$.

Expansion is in powers of $p_i a \ll 1$

Correct lattice $\mathcal{L}$: add new, local interactions that generate the missing pieces in $T - T^{(a)}$:

$$\begin{aligned} \delta\mathcal{L}^{(a)} &= \tfrac{1}{2}\,c(a)\,a^2\,\overline{\psi}\gamma_\mu\psi\,\overline{\psi}\gamma^\mu\psi \\ &+ \tfrac{1}{2}\,c_A(a)\,a^2\,\overline{\psi}\gamma_\mu\gamma_5\psi\,\overline{\psi}\gamma^\mu\gamma_5\psi \\ &+ d(a)\,a^4\,\overline{\psi}D^2\gamma_\mu\psi\,\overline{\psi}\gamma^\mu\psi \\ &+ \cdots. \end{aligned}$$

Intermediate particles in $T - T^{(a)}$ have $k > \pi/a \gg p_i$

- $\Rightarrow$ highly virtual ($k^2 > 1/a^2$);
- $\Rightarrow$ propagate only short distances ($\approx a$);
- $\Rightarrow$ corrections must be local;
- $\Rightarrow$ only finite number to a given order in $p_i a$.

Correction terms needed for $\mathcal{L}$, currents, and all other operators. (C.f., effective field theory.)

Perturbation Theory

Improved discretizations and larger as — old ideas.

But perturbation theory is essential.

- $\Rightarrow$ a must be small enough so that $p \approx \pi/a$ QCD is perturbative.
- $\Rightarrow$ Before 1992: $a < 0.05$ fm.
- $\Rightarrow$ After 1992: $a < 0.4$ fm works.

Test by comparing short-distance quantities from:

- perturbation theory;
- numerical Monte Carlo integration ($\Rightarrow$ exact result).

E.g., Wilson loops:

$$W(\mathcal{C}) \equiv \langle 0|\tfrac{1}{3}\,\mathrm{Re\ Tr\ P}\,\mathrm{e}^{-ig\oint_{\mathcal{C}} A\cdot \mathrm{d}x}|0\rangle,$$

$\mathcal{C}$ = small, closed path.

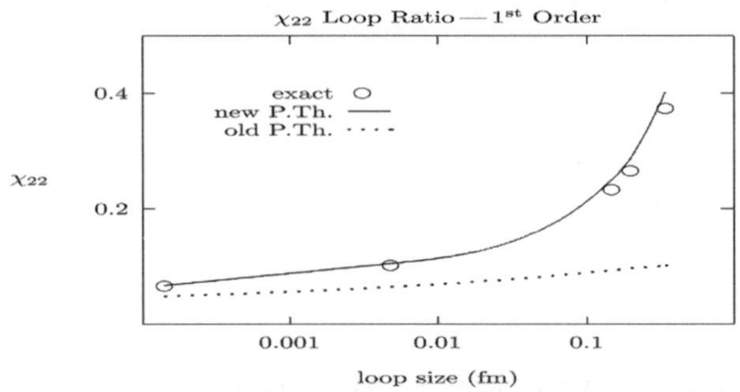

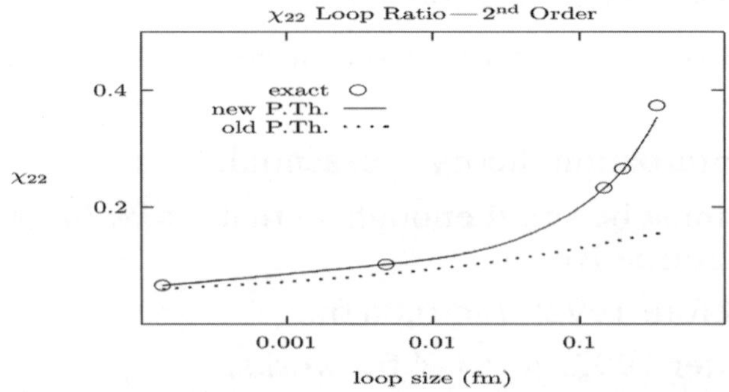

Running coupling constant:

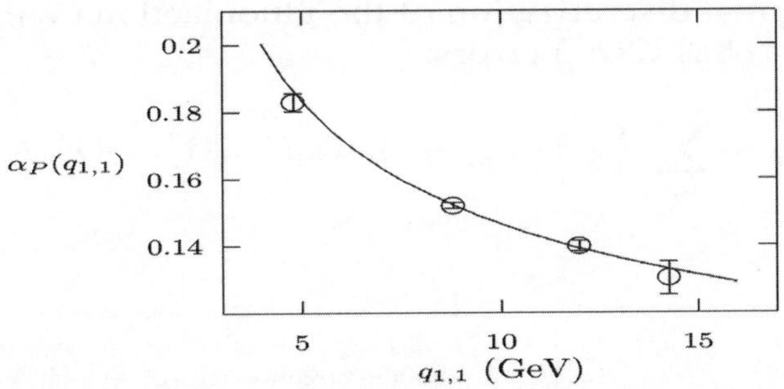

LQCD Tour

Recap

Expect errors $<$ few % from simulations with $a < 0.4$ fm.

Improved discretizations essential for speed and precision.

Questions:

- Do the improvements work?
- Are the simulations faster?

Gluons

Original discretization of the gluon action (Wilson, 1974) has $\mathcal{O}(a^2)$ errors:

$$\mathcal{L}_{\text{Wil}} \approx \sum_{\mu,\nu} \left\{ \tfrac{1}{2} \operatorname{Tr} F_{\mu\nu}^2 + \tfrac{a^2}{24} \operatorname{Tr} F_{\mu\nu}(\mathrm{D}_\mu^2 + \mathrm{D}_\nu^2) F_{\mu\nu} \cdots \right\}.$$

$\mathcal{O}(a^2)$ error violates rotation/Poincaré invariance (due to lattice); removed by adding correction terms.

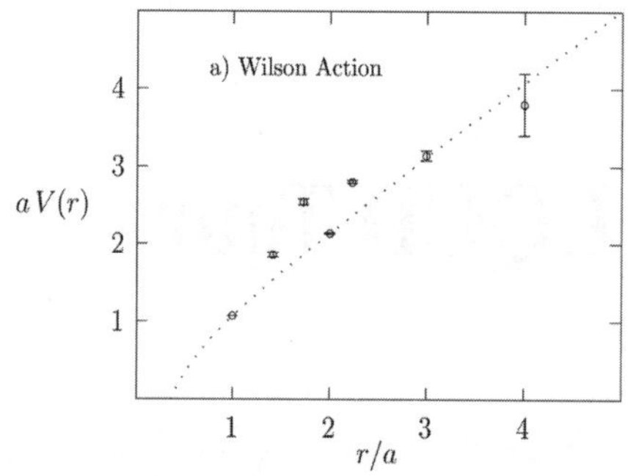

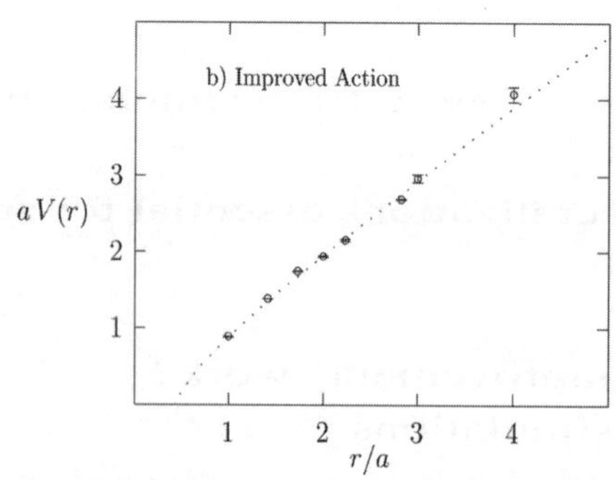

Quarks

The standard discretization of the quark action has $\mathcal{O}(a^2)$ errors:

$$\mathcal{L}_{\text{lat}} \approx \overline{\psi}(D\cdot\gamma + m)\psi + \frac{a^2}{6}\sum_{\mu}\overline{\psi}D_\mu^3\gamma^\mu\psi + \cdots$$

$\mathcal{O}(a^2)$ error violates rotation/Poincaré invariance; removed by adding correction term.

Test by computing

$$c^2(\mathbf{p}) \equiv \frac{E^2(\mathbf{p}) - m^2}{\mathbf{p}^2};$$

Lorentz invariance implies:

$$c^2(\mathbf{p}) = 1 \qquad \forall\,\mathbf{p}.$$

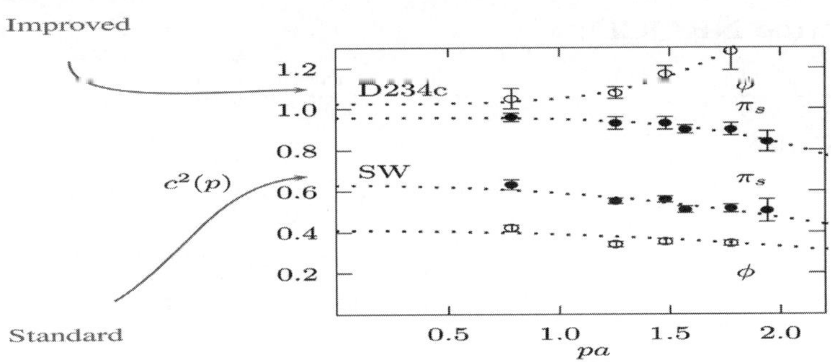

Alford et al (1997).

Heavy Quarks

Lattice errors $\propto (a\,E)^n$, $(a\,p)^n$

$\Rightarrow$ need $a \ll 1/M$ where M = hadron mass.

$$\begin{aligned} B, \Upsilon \ldots \quad &\Rightarrow \quad \text{Need } a \to a/10 \\ &\Rightarrow \quad \text{Cost} \to 10^6 \text{ cost!} \\ &\Rightarrow \quad \text{Impossible?} \end{aligned}$$

No! b quark is nonrelativistic:

$$\frac{v^2}{2} \approx \frac{\Delta M}{M} \approx \frac{0.5\,\text{GeV}}{10\,\text{GeV}}$$

$\Rightarrow$ $v^2 \approx 0.1$;

$\Rightarrow$ don't use Dirac; use effective field theory.

$$\begin{aligned} \text{Schrödinger} \quad &+ \quad \mathcal{O}(a, a^2) \text{ corrections} \\ &+ \quad \mathcal{O}(v^2, v^4) \text{ corrections} \\ &+ \quad \cdots. \end{aligned}$$

Lattice NRQCD:

Schrödinger:

$$H_0 \sim -\frac{\mathbf{D}^2}{2M_0} + ig\,A_0,$$

Corrections:

$$\begin{aligned} \delta H \quad \sim \quad & -c_1 \frac{(\mathbf{D}^2)^2}{8M_0^3}\left(1 + \frac{aM_0}{2n}\right) + c_2 \frac{a^2 \sum_i \mathrm{D}_i^4}{24M_0} \\ & -c_3 \frac{g}{2M_0}\,\sigma \cdot \mathbf{B} + c_4 \frac{ig}{8M_0^2}\,(\mathbf{D}\cdot\mathbf{E} - \mathbf{E}\cdot\mathbf{D}) \\ & -c_5 \frac{g}{8M_0^2}\,\sigma \cdot (\mathbf{D}\times\mathbf{E} - \mathbf{E}\times\mathbf{D}). \end{aligned}$$

where perturbation theory $\Rightarrow c_i = 1 + c_{i1}\alpha_s(\pi/a) + \cdots$;

$\Rightarrow$ only two parameters: α_s and M_0.

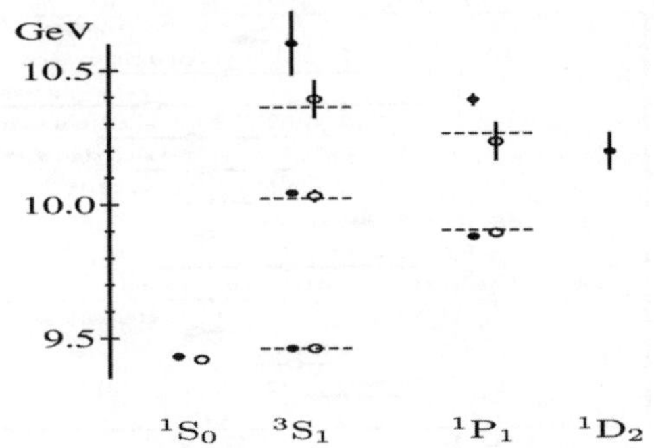

Davies et al. (1997).

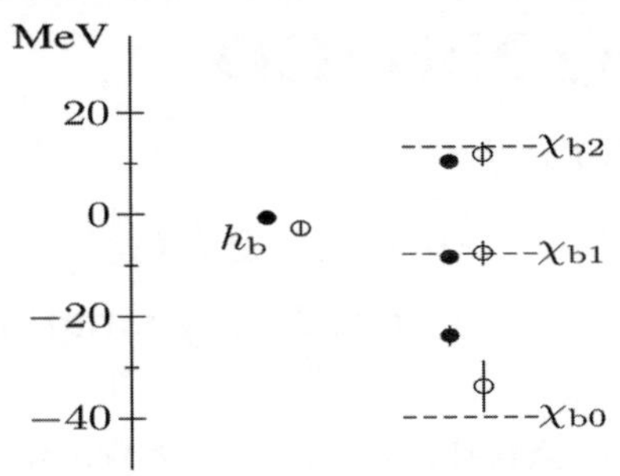

Davies et al. (1997).

Tune two parameters to reproduce data

$\Rightarrow$ few % accurate results for M_b and $\alpha_{\overline{\rm MS}}(M_Z)$.

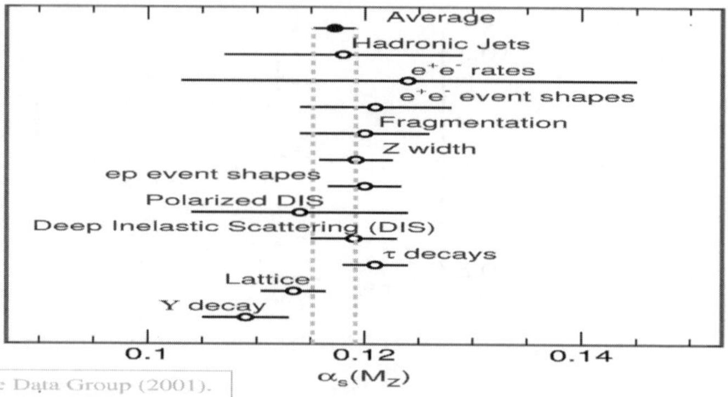

Particle Data Group (2001).

Moving NRQCD

A Problem:

$B \to \pi\, l\, \nu \quad \Rightarrow p_\pi \to 2.5\,\mathrm{GeV},$

$\Rightarrow$ need very small a so that $a p_\pi \ll 1$.

Solution: Analyze in π rest frame

$\Rightarrow$ B carries large momentum.

Moving B's

$\Rightarrow$ Speed of b quark $\approx$ speed of B meson.

$\Rightarrow$ $p_b \equiv M_b\, u + k$ where u^μ is 4-velocity of B and $k \ll p_b$. ($k \approx p_b/10$ for B's.)

$\Rightarrow$ Light-quark $p_q \approx k$.

$\Rightarrow$ Treat $M_b\, u$ exactly; discretize only k, p_q.

$\Rightarrow$ 10 times larger a for B's.

mNRQCD — Moving NRQCD

$$\psi_b(x) = \frac{e^{-iMu\cdot x}}{\sqrt{2(1+u_0)}} \begin{pmatrix} 1+u_0 & \boldsymbol{\sigma}\cdot\mathbf{u} \\ \boldsymbol{\sigma}\cdot\mathbf{u} & 1+u_0 \end{pmatrix} \begin{pmatrix} \chi(x) \\ \frac{-i\boldsymbol{\sigma}\cdot\tilde{\mathbf{D}}}{2M}\chi(x) \end{pmatrix}$$

Remove momentum $M\,u$ from χ.

Boost to moving frame.

Compute χ on lattice using low-energy equation:

$$(iD_t + i\mathbf{v}\cdot\mathbf{D} - H)\chi = 0$$

"Unquenching"

Unquenched LQCD

"Quenched" QCD ≡ QCD without quark vacuum polarization.

- ⇒ 15–20% errors in most calculations;
- ⇒ *the* major limitation of LQCD *until 2000*.

Naive/staggered quarks + improved discretization

- ⇒ 10–100 times faster
  & smallest finite-a errors
  & best behavior in chiral limit!
- ⇒ high-precision (1%) LQCD possible *now!*

Naive/Staggered Quarks

Simplest discretization of light quarks,

$$\mathcal{L} = \overline{\psi}(x)(\Delta \cdot \gamma + m)\psi(x)$$

$\Rightarrow$ an exact "doubling" symmetry:

$$\begin{aligned}\psi(x) \quad \longrightarrow \quad \tilde{\psi}(x) &\equiv i\gamma_5\gamma_\rho\,(-1)^{x_\rho/a}\,\psi(x)\\ &= i\gamma_5\gamma_\rho \exp(i\,x_\rho\pi/a)\,\psi(x).\end{aligned}$$

$\Rightarrow$ Any low-energy mode ψ $\equiv$ Another mode, $\tilde{\psi}$, with $p_\rho \approx \pi/a$.

→ max. p on lattice

General case:

$$\psi(x) \longrightarrow \mathcal{B}_\zeta(x)\,\psi(x)$$

where

$$\mathcal{B}_\zeta(x) \equiv \prod_\rho (i\gamma_5\gamma_\rho)^{\zeta_\rho} \exp(i\,x \cdot \zeta\,\pi/a)$$

→ $\zeta = (1,0,0,0), (0,1,0,0)\ldots(1,1,0,0)\ldots$, 15 in all.

$\Rightarrow$ 1 field $\psi(x)$ creates 16 *different* but *exactly equivalent* flavors of quark ($p \approx \zeta\pi/a$)!

16 flavors is bad!

Two traditional options:

1. (Wilson, SW ...) Break doubling symmtry by adding $-a\overline{\psi}(D \cdot \gamma)^2\psi/2$ to $\mathcal{L}$; destroys all chiral symmetry.
2. (Kogut-Susskind ...) Live with the 16 flavors by inserting factors of $1/16$ in strategic places.

→ Follow option 2!

E.g., 16 equivalent B mesons constructed from a b quark and 16 flavors of u antiquark: e.g.,

Ignore all but first (equivalent) by limiting total momentum:

$$P_B \equiv p_b + p_u < \frac{\pi}{2a}.$$

$\Rightarrow$ P_B distinguishes between flavors.

Light hadrons harder.

Bad News: Flavor-changing strong interactions —

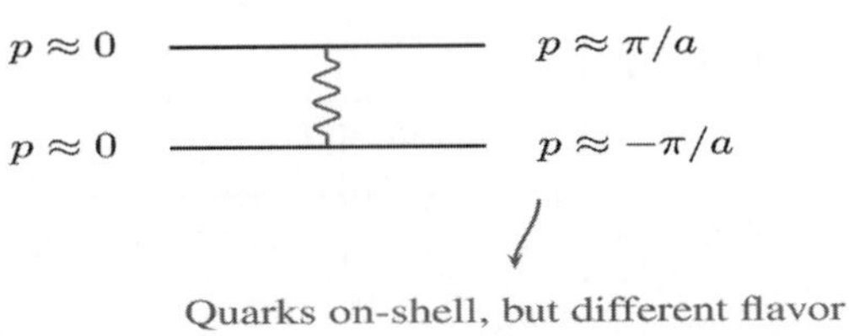

Good News: The gluon carries the largest lattice momentum, π/a;

- $\Rightarrow$ highly virtual and perturbative;
- $\Rightarrow$ can remove by (local) perturbative modifications to $\mathcal{L}$ to any order in $\alpha_s(\pi/a)$.

The new idea!
$\Rightarrow$ Four quark operators + $a^2\Delta^3$ correction as before.
$\Rightarrow$ Most accurate discretization.

See Lepage (1998); MILC (1999).

An amazing fact:

$$\Omega(x) \equiv \prod_{\mu} (\gamma_\mu)^{x_\mu/a}$$

$$\Rightarrow \qquad S_F(x,y;A_\mu) = g(x,y;A_\mu)\,\Omega(x)\,\Omega^\dagger(y)$$

Propagator.

Dirac scalar for any gluon field A_μ.

Dirac structure is completely independent of A_μ!

$\Rightarrow$ 10–100 times faster than all other alternatives!

Domain-Wall Fermions

An approach for chiral fermions?

- Embed 4-D space-time in 5-D: (x^μ, s).
- Use Wilson lattice Dirac equation in 5-D,

$$\left(D\cdot\gamma + D_s\gamma_5 - \frac{a}{2}(D^2 + D_s^2) + m(s)\right)\,\Psi = 0,$$

but with

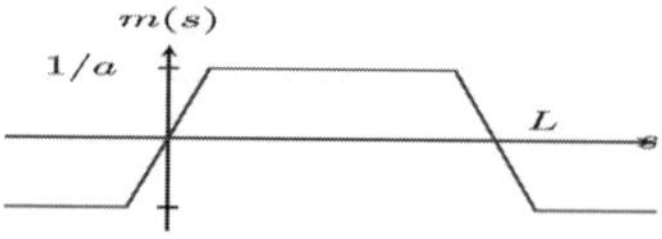

See Kaplan (1992).

- Separable $\Rightarrow$ chiral $\Psi_\pm$, such that

$$D\cdot\gamma\,\Psi_\pm = 0,$$

localized where $m(s) = 0$ — domain walls:

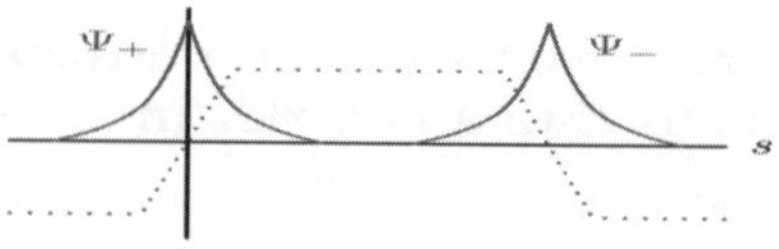

- "Keep" only $s = 0$ or L solution $\Rightarrow$ chiral theory $\Rightarrow$ wide range of new applications?
- Key issue: Do the two solutions communicate?

Why are Quarks so Hard?

Anomalies!? Chiral symm. + massless quarks $\Rightarrow$

- Quark helicity ($\pm$) is Lorentz invariant.
- $\partial^\mu j_{5\mu} = 0$ (classically)
  $\Rightarrow$ $n_+ - n_-$ conserved.
  $\Rightarrow$ No interaction in $\mathcal{L}$ changes $-$ to $+$.
- Quantum effects (anomaly)
  $\Rightarrow$ $\partial \cdot j_5 \propto \mathbf{E} \cdot \mathbf{B}$
  $\Rightarrow$ $n_+ - n_-$ can change!

E.g., vacuum Fermi levels shift as $\mathbf{E} \cdot \mathbf{B}$ varies:

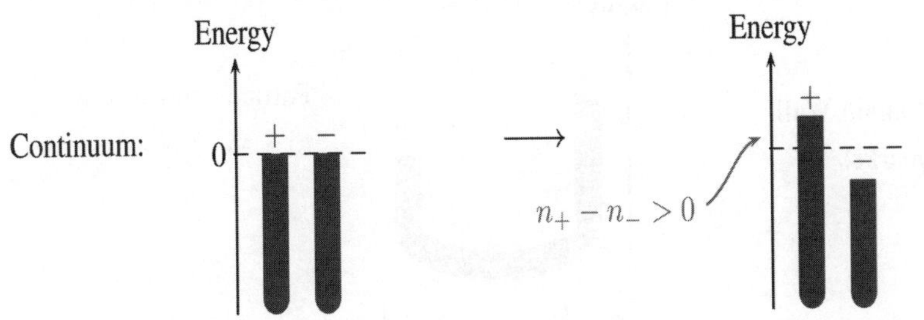

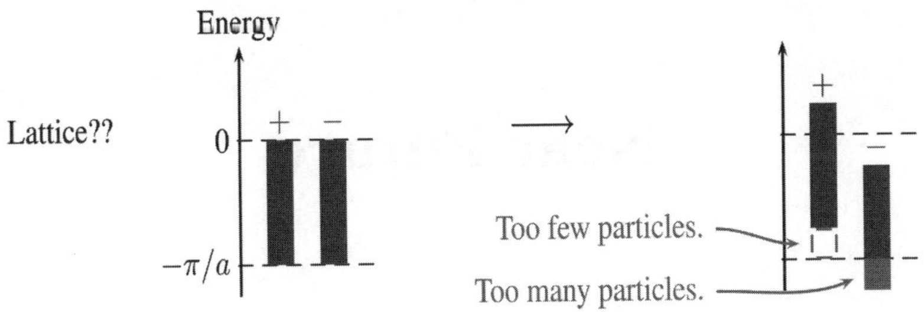

Naive/Staggered Quarks:

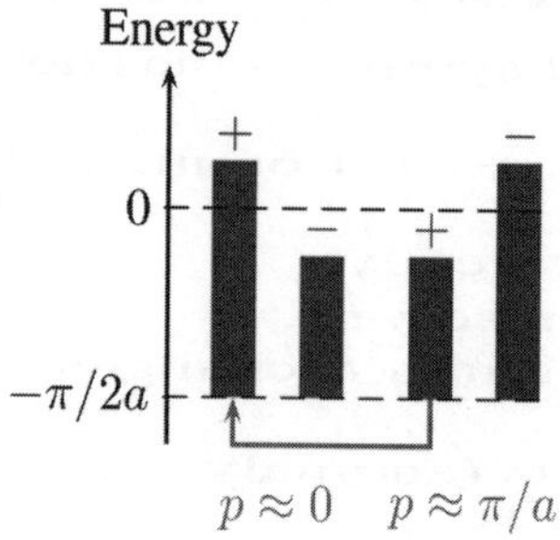

$n_+ - n_- = 0$, still;
$\Rightarrow$ anomalies cancel between flavors.

Domain Wall Quarks:

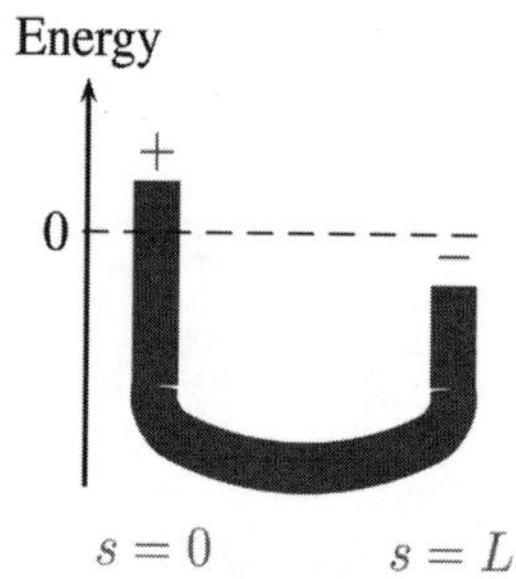

Particles move from − to + via 5th dimension.

Near Future

Algorithmic Advances in 1990's

1992 LQCD perturbation theory.

1992 Effective field theory (e.g., NRQCD, Fermilab).

1995 Improved discretizations $\Rightarrow$ larger a.

1999 Improved staggered/naive quarks:

- $\mathrm{Det}(D \cdot \gamma + m) > 0$ as in continuum (small m okay).
- Flavor-changing perturbative $\Rightarrow$ remove systematically.
- Naive quarks $\Rightarrow$ simple analyses, operators.
- No $\mathcal{O}(a)$ errors *and* much faster unquenching!

$\Longrightarrow$ Lattice QCD is in revolution:

- 100–∞% errors in 1990.
- 10–20% errors today for wide range of masses, form factors ...
- 1–3% possible now and in next few years.

High-Precision Possible Now

Few % accuracy for "gold-plated" calculations:
(Cornell Workshop, 2001)

- Masses, decay constants, semileptonic form factors, and mixing amplitudes for D, D_s, D^*, D_s^*, B, B_s, B^*, B_s^*, and baryons.
- Masses, leptonic widths, electromagnetic form factors, and mixing amplitudes for any meson in ψ/Υ families below D/B threshold.
- Masses, decay constants, electroweak form factors, charge radii, magnetic moments and mixing for low-lying light-quark hadrons.
- High-precision $\Rightarrow$ masses and amplitudes with at most one hadron in the initial and/or final state, for stable or nearly stable hadrons ($\Gamma <$ 10–20 MeV).

Uses current techniques.

- Progress driven by improved methods.
- Future pace will be much faster than pace of hardware evolution.

HPQCD Plan

Compute dozens (?) of gold-plated quantities to few percent over next few years.

- Unquenched $n_f = 6$ with improved staggered quarks.
- NRQCD and Fermilab actions for c, b quarks through v^6.
- Actions, operators corrected through order a^2.
- One and two-loop perturbation theory (automated).
- PC cluster is optimal (simulations, pert'n theory ...).

Focus on B and D physics $\Rightarrow$ maximum impact on experimental community (BaBar, Belle, CLEO ...).

- Gold-plated quantities for every off-diagonal CKM element.
- Extensive cross-checks for error calibration.

So far ...

- Automated one-loop perturbation theory.
  - $\Lambda_V/\Lambda_{\mathrm{latt}}$ for improved gluons
  - f_π and f_K, B_K, and m_s.
  - Flavor-changing in naive/staggered quarks.
  - Anisotropy and "speed of light" on anisotropic lattices
- Tree-level spin-independent NRQCD through v^6 and spin-dependent through v^8; one-loop soon.
- Operator design with naive quarks.
- High-β techniques for nonperturbative perturbation theory.
- Constrained/Bayesian curve fitting.

HPQCD Impact

CKM today ...   ... and with 2–3% theory errors.

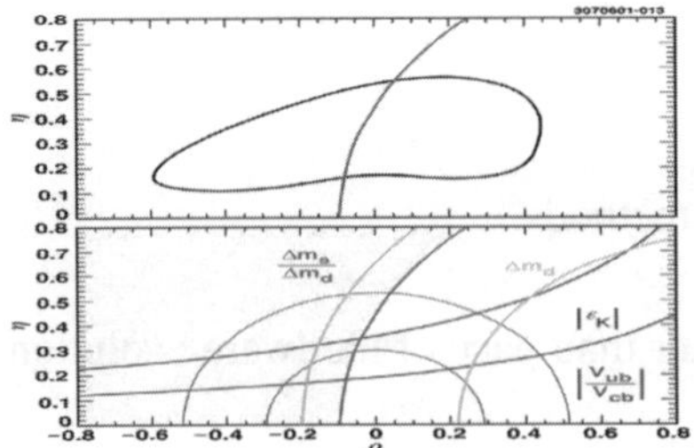

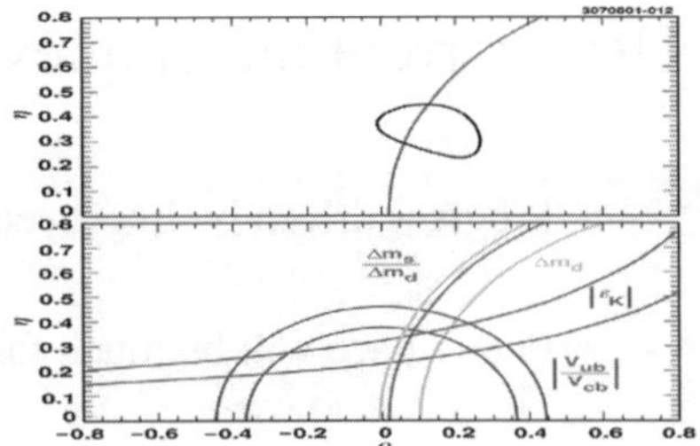

95% confidence levels; CLEO-c (2001).

HPQCD case study: f_D errors

Perturbation theory:

- Connects lattice to continuum (fills in gaps): for f_D use

$$J_{\text{cont}} = Z\,J_{\text{latt}} + a^2\,\Delta J$$

where

$$Z = 1 + c_1\alpha_s(\mu) + c_2\alpha_s^2 + \cdots$$

and $\mu \approx 2/a$ is typical (for α_V and $n_f = 3$).

- Current work uses 1st-order results; relative error is $\mathcal{O}(\alpha_s^2) \approx 7\%$ for $a = 0.1$ fm. This is the dominant error.
- Next generation will use 2nd-order results, giving relative errors of $\mathcal{O}(\alpha_s^3) \approx 1.6\%$ at $a = 0.1$ fm.

HPQCD case study: f_D errors

Finite-lattice spacing errors:

- E.g., on lattice

$$p^2 \to (\sin(pa)/a)^2 = p^2\left(1 - (pa)^2/3 + \cdots\right).$$

- Remove a^2, a^4 ... errors (improved discretizations).
- $(pa)^3/3 \approx 0.7\%$ for $p \approx 300\,\text{MeV}$ and $a = 0.1\,\text{fm}$.
- $\mathcal{O}(a^2)$ improvement crucial for high-momentum form factors, and for suppressing flavor-changing interactions.

HPQCD case study: f_D errors

$1/M$ errors:

- Effective field theory (e.g., NRQCD) essential for heavy quarks $\Rightarrow 1/M$ expansion.
- Current work accurate through $\mathcal{O}(1/M)$ errors:
  - $\mathcal{O}(1/M^2) \approx 2\%$ or less for f_D;
  - $\mathcal{O}(\alpha_s/M) \approx 3.6\%$ for f_D;
  - 3–10 times smaller for B mesons.
- Future work accurate through $\mathcal{O}(\alpha_s/M, 1/M^2)$; relative error is $\mathcal{O}(\alpha_s^2/M) \approx 0.9\%$.

HPQCD case study: f_D errors

Unquenching:

- Most past work is quenched: $m_{u,d,s} \to \infty$ for sea quarks (i.e., no vacuum polarization) $\Rightarrow$ errors of 10–20%.
- Current simulations use realistic m_s and $m_{u,d} = m_s/2, m_s/4 \ldots$.
  - Chiral perturbation theory $\Rightarrow$ error estimates/corrections.
  - Relative errors $= \mathcal{O}(15\% \times (m_{u,d}/m_s)^2) \approx 1\%$ for $m_{u,d} = m_s/4$.
- Simulations with $a = 0.1$ fm, $n_f = 6$, $m_{u,d} = m_s/4$ require $\approx$ 3 months on 200-node PC cluster for 1% statistical errors.
  - Use improved staggered quarks.
  - No more quenched analyses!

Challenge for lattice QCD

Demonstrate reliability at the level of 1–3% errors, given past history of 10–20% errors.

- Requires comparison with wide variety of highly accurate experimental data.
  - High precision $\Rightarrow$ differentiate QCD from models.
  - Wide variety $\Rightarrow$ independent tests of all components.
- Must test:
  - Heavy-quark actions (NRQCD, Fermilab, etc.).
  - Light-quark actions (improved stagg. quarks).
  - Gluon action.
  - High-order perturbation theory.
  - Techniques for computing spectra, form factors ....

A problem for lattice QCD

There is very little high-precision QCD data from experiment.

Solution: new CLEO-c experiment!

CLEO-c: High-Precision QCD

2002 Υ family:

- Masses, spin splittings, widths, form factors, mixing ... to few %.
- Richest testing ground for heavy-quark actions ⇒ *independent* calibration/test of b-quark action used in B simulations.
- Overconstrain b-quark action.

2004/5 D, D_s mesons:

- Leptonic, semileptonic widths and form factors to few %.
- Calibrate LQCD on analogues of crucial B processes.
- V_{cd} and V_{cs}, new physics (?).

2006 ψ/J family, glueball spectrum.

⇒ LQCD in race to *predict* CLEO-c results!

Conclusion

Superb opportunity for LQCD to have an impact on particle physics.

- LQCD essential to high-precision B/D physics at BaBar, Belle, CLEO-c ...
- *Predicting* CLEO-c, BaBar/Belle results ⇒ much-needed credibility for LQCD.
- Landmark in history of quantum field theory: quantitative verification of nonperturbative technology (c.f., 1950's).
- Ready for beyond the Standard Model!

OUTLINES OF SEMINAR ON HEAVY-ION COLLISIONS AT LHC WITH THE ATLAS DETECTOR

H. TAKAI

HEAVY-ION COLLISIONS AT LHC WITH THE ATLAS DETECTOR

H. TAKAI
Brookhaven National Laboratory
Upton, New York, USA
takai@sun2.bnl.gov

The ATLAS detector has been designed for determining mainly the source of electroweak symmetry breaking in proton-proton collisions. The physics pursued by the collaboration is vast and includes: Higgs boson search, searches for SUSY, and other scenarios beyond the Standard Model. Many of its components could serve to measure the properties of nucleus-nucleus and proton-nucleus collisions.

Lead Lead collisions at LHC will take place at over a *1,000 TeV* total center-of-mass energy, which is well over an order of magnitude higher in center of mass energy than achievable at the Relativistic Heavy Ion Collider (RHIC). It is expected that energy densities of up to *30 GeV/fm*^{3} can be reached allowing for the exploration of forms of matter not previously seen in the laboratory. The planning for a nuclear physics program for the ATLAS detector has just started. The program will focus mainly in exploring the hot QCD matter via the hard probes. The detector is extremely well suited for jet physics. At LHC energy there is an excellent opportunity to study jet properties because of the higher collision energy compared to RHIC. In particular quenching effects could be studied in detail. We intend to study proton nucleus collisions as well.

The large coverage of the ATLAS detector is ideal for this type of physics. The proton-nucleus physics at the LHC will focus on the perturbative production of gluons and the modifications of the gluon distribution in the nucleus at low x. It will also provide a baseline for the understanding of nucleus-nucleus physics. A third physics topic is the Ultra Peripheral Collision. The highly Lorentz contracted electric field of heavy nuclei is a source of high energy photon-photon and photon nucleon collisions. Heavy quark photo-production may allow for the determination of the saturation scale within the context of color glass condensate model. In the presentation we review and discuss the status of our program. We hope to generate enough interest to attract both experimental and theoretical physicists to participate in the program.

Heavy Ion Collisions at LHC with the ATLAS Detector

Helio Takai
takai@bnl.gov
Brookhaven National Laboratory
(J. Nagle, B. Cole and S. White)

Hadron 2002
Bento Gonçalves, April 2002

BROOKHAVEN
NATIONAL LABORATORY

Before we start

✓First of all my sincere thanks to the organizers !

✓After I accepted the invitation I got worried !

What can I talk about? I decided to talk about heavy ion physics with the ATLAS detector that it is still far in the future. However it has been our experience in ATLAS that <u>help from our theoretical colleagues in planning the experiments has been a big plus</u>. So this is the spirit of this presentation.

Before we start

Another reason why I got worried is because I have been away from **nuclear physics** for a long time - almost 8 years!

So I will behave myself, and make no jokes about Our Lady of Perpetual Motion, the Church of Christ Geologist or the Jojoba Witnesses. However one can't pass the opportunity to mention that everybody missed one good heavy ion experiment:

Plan for this presentation

i. Flirting with heavy ions
ii. The ATLAS detector
iii. Heavy Ion Physics at LHC energies
iv. A nuclear physics program for ATLAS
v. What is going to happen next
vi. Summary and conclusions

ATLAS and Heavy Ions

✓ATLAS is an experiment designed to study the origins of the electroweak symmetry breaking in proton-proton collisions, search for SUSY, and exotica - e.g., extra dimensions.

✓ATLAS is designed to acquire data at full LHC design luminosity of $L=10^{34}$ $cm^{-2}s^{-1}$, or 40 MHz collision rate.

✓The detector is designed to cover a range of $|\eta|<5.0$ in rapidity with inner tracking, calorimetric coverage and muon spectrometer.

✓ATLAS is also a large collaboration with over 2,000 physicists from 157 institutions worldwide, including Brazil.

ATLAS and Heavy Ions

✓Recent RHIC data suggests that jets may be quenched.

✓CERN theorists have been sponsoring a one year long workshop on hard probes in heavy ion physics.

✓This has sparked renewed interest by a group of physicists within ATLAS to revisit the heavy ion physics program with the detector.

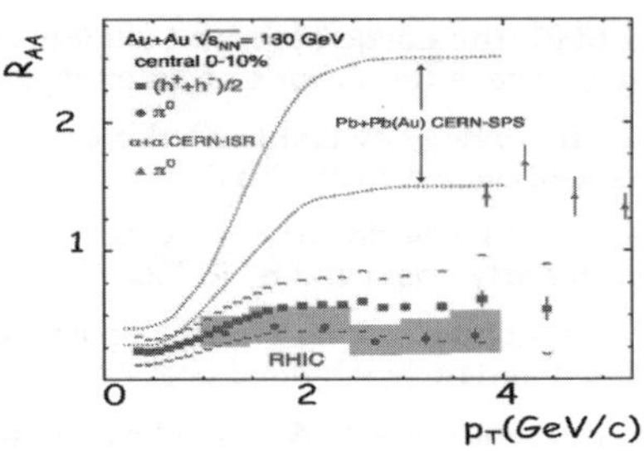

ATLAS and Heavy Ions

✓Large calorimetric coverage detector are well suited for jet physics

✓Jets at LHC heavy ion energies are similar to Tevatron jets.

✓For about one year, since the workshop in BNL, we have been developing a nuclear physics program for the ATLAS detector.

✓However there are many hurdles to overcome ... people, funding, etc...

The people

The institutions involved in ATLAS heavy ions working group are:

- ✓Brookhaven National Laboratory
- ✓CERN
- ✓Columbia University
- ✓Prague
- ✓Rio de Janeiro
- ✓University of Geneva
- ✓Helsinki

LHC and heavy ions

✓LHC, the Large Hadron Collider, is now under construction and will accelerate heavy ions at an energy of 2.75 TeV/nucleon

✓The energy available at the center of mass of a Lead-Lead collision is well over a 1,000 TeV.

✓The energy density is expected to be about 30 times what is currently observed at RHIC

✓New and exciting physics could be observed by probing the hot QCD matter with hard probes.

✓But what is ATLAS exactly? Where is the experiment?

CERN

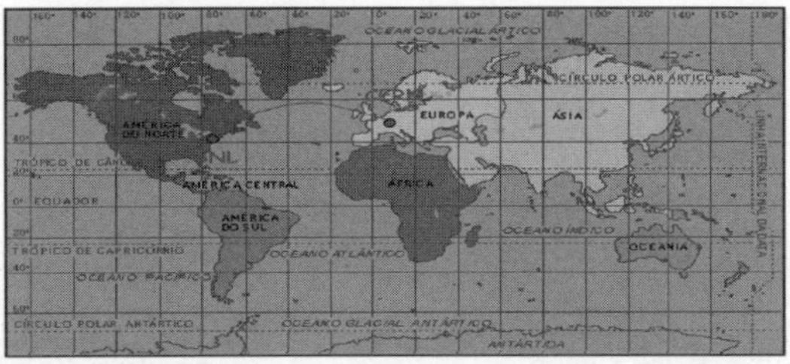

Where is CERN? It is in Switzerland, not Swaziland...
and about 6,000 km from New York.

CERN

The ATLAS detector

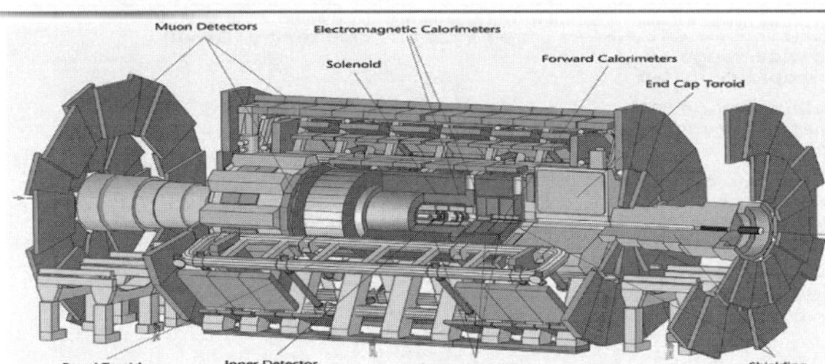

Inner Detectors

The ATLAS inner detector is composed of three systems: Pixel, SCT and TRT

The Semiconductor Tracker are also silicon devices.

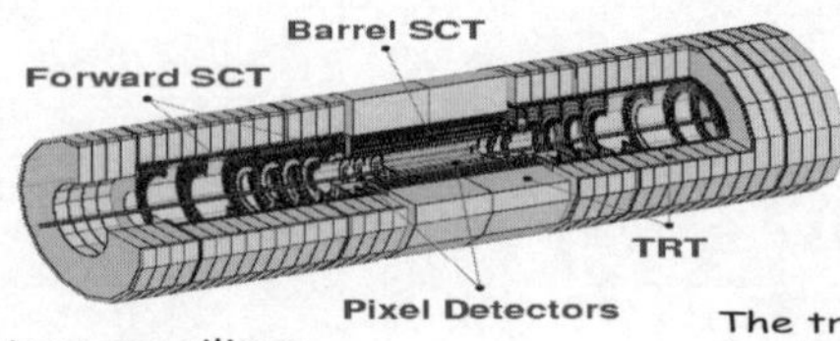

The pixel detectors are silicon devices and are located near to the collision point.

The transition radiation tracker are gas detectors and used to identify electrons.

Inner Detectors

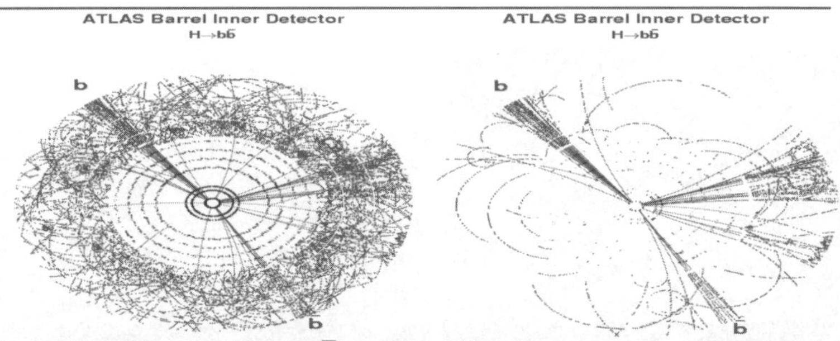

The same event, H $b\bar{b}$ in full and low luminosity running conditions.

Calorimeters

Calorimeters in ATLAS cover a wide range of pseudo-rapidity, | |<5.

The electromagnetic calorimeter is realized in liquid argon technology

The hadronic calorimeter is implemented as a iron-scintillator device, except in the forward direction.

The very forward region is covered by axial drift liquid argon calorimeter.

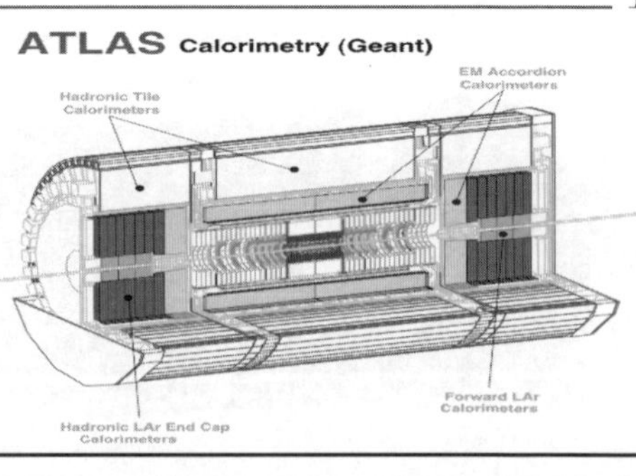

Electromagnetic Calorimeter

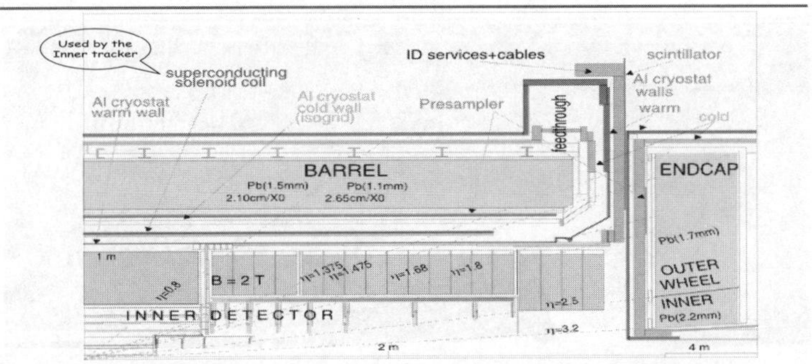

Electromagnetic Calorimeter Segmentation

The Electromagnetic Calorimeter is segmented longitudinally and transversely.

Will be used for ^{0} identification.

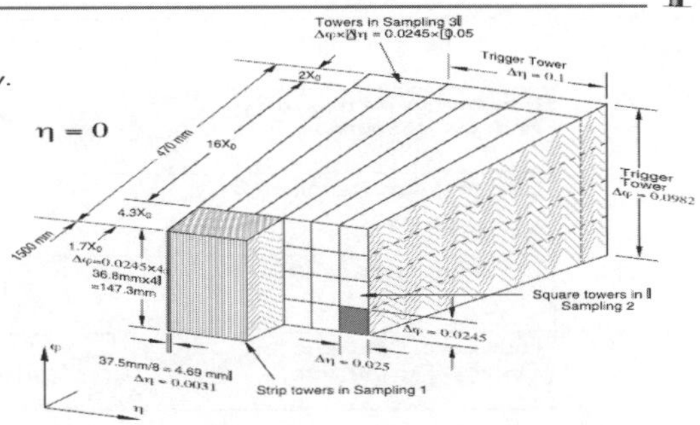

Segmentation helps the identification of in the background of ^{0}s.

Hadronic Tile Calorimeter

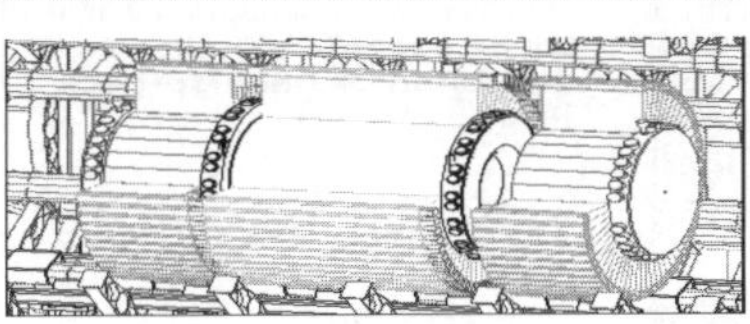

The hadronic tile calorimeter 'hugs' the liquid argon electromagnetic calorimeter. It is built in iron-scintillator technology and readout by WLS optical fibers.

Hadronic Tile Calorimeter

Calorimeter Characteristics

Electromagnetic Cal. Energy Resolution	$\frac{\sigma_E}{E} = \frac{10\%}{\sqrt{E}} \oplus 0.7\%$
EM Angular Resolution	$\sigma_\theta = \frac{60mrad}{\sqrt{E}}$
EM Timing Resolution	$\sigma_\tau = \frac{4.15}{E} ns \bullet GeV$
Hadronic Calorimeter Energy Resolution	$\frac{\sigma_E}{E}(\pi) = \frac{47\%}{\sqrt{E}} \oplus 2\%$

Jet Energy Resolution

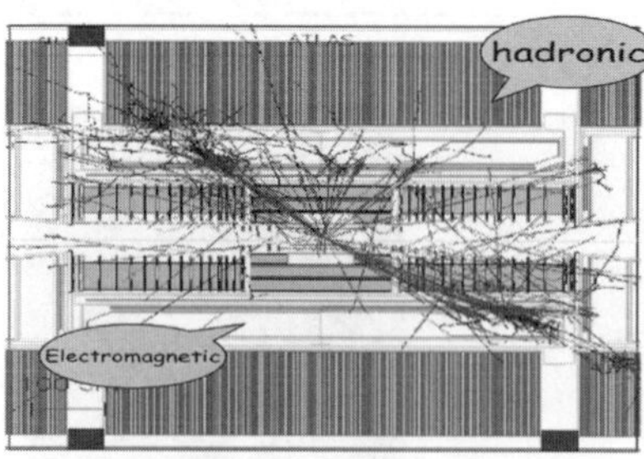

Di-jet event in ATLAS

Large acceptance allows for detection of back-to-back jets.

Jet Energy Resolution :

$$\frac{\sigma_E}{E} = \frac{50\%}{\sqrt{E}} \oplus 2\%$$

Ha! But this is for pp collisions!!!

Muon Spectrometer

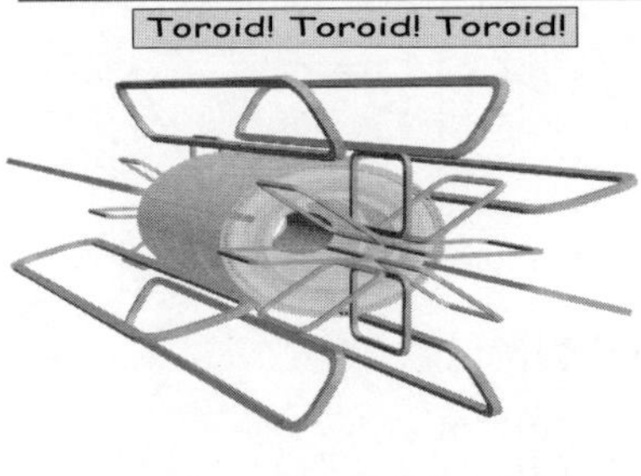

Air Core Toroidal Magnet System

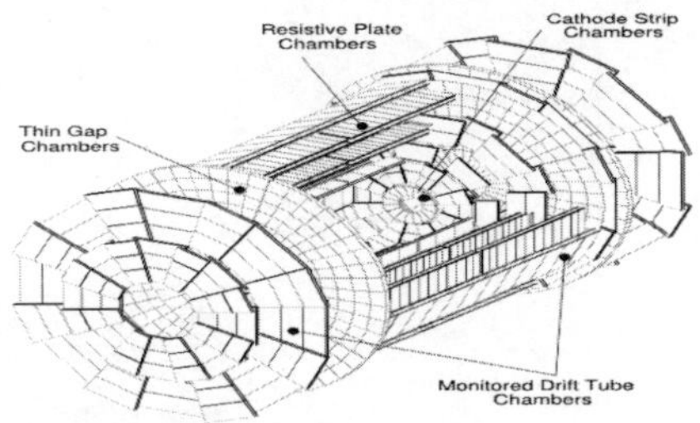

Muon detectors

Muon Spectrometer

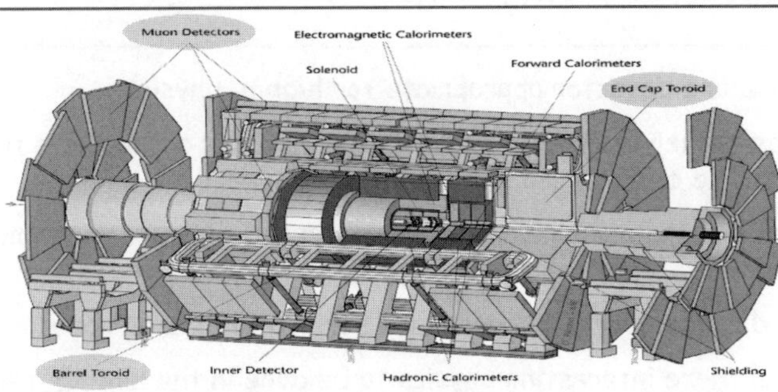

Muon spectrometer

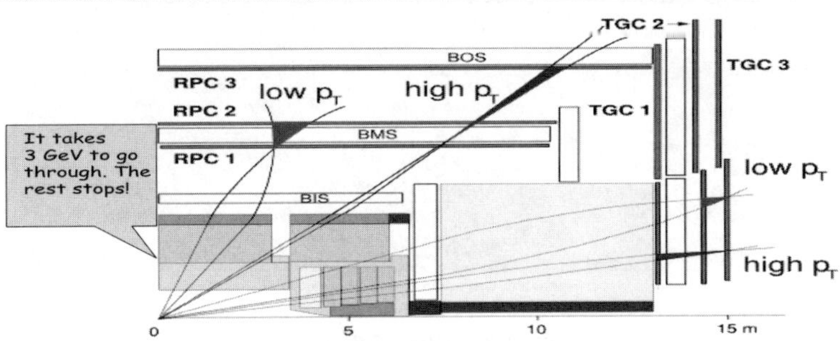

Muon Spectrometer Performance

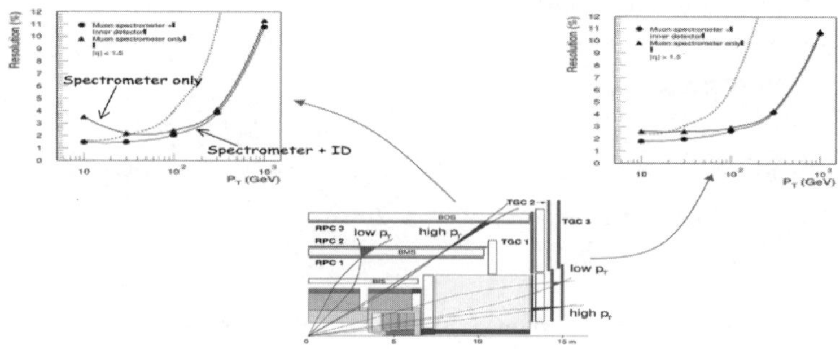

The case for Heavy Ions

✓ATLAS is a detector appropriate for high p_T physics.

✓It has a finely segmented calorimeter which is appropriate for jet physics. The energy resolution for jets is superb.

✓It has a stand alone muon spectrometer with good momentum resolution.

✓It is designed for high data rates for proton-proton collisions

✓But, is there interesting physics to be done in the high Q^2 regime?

Heavy Ion collisions

...the picture says it all !!!!

Gluon Densities

HERA experiments have observed a dramatic increase in the gluon density at low x.

This increase must end at some point when the gluon density saturates.

Large Hadron Collider Pb-Pb collisions probe the gluon structure below x~10^{-3} - 10^{-5}.

Note that xg(x) is enhanced by $A^{1/3}$ ~ 6 in Pb over the proton.

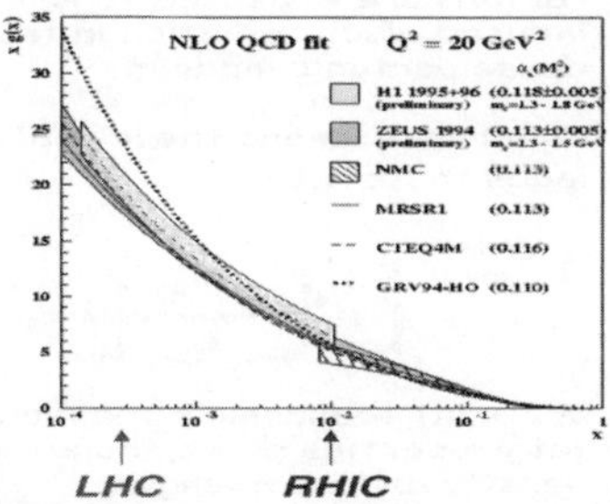

LHC and RHIC

The saturation scale is much larger at the LHC than at RHIC.

Thus, the initial partonic state may be dominated by the saturation region (described as a color glass condensate).

Also, the cross section for high p_T processes is much larger, thus yielding better pQCD calibrated probes of the possible gluon plasma.

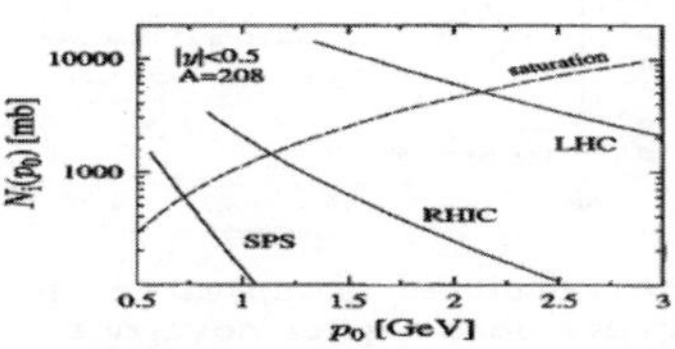

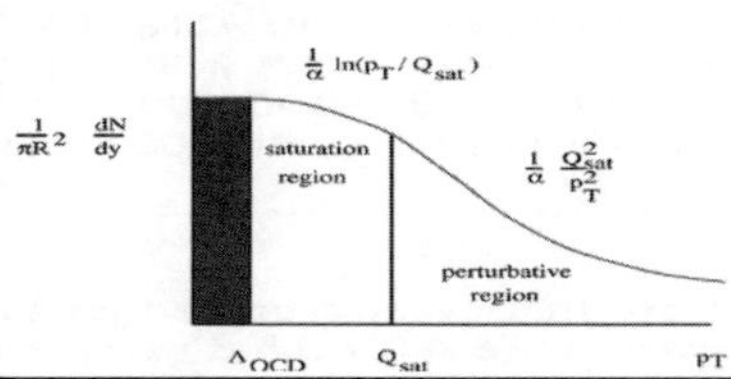

Freeing the Gluons (the QLF)

In a future Electron-Ion Collider (EIC) one can probe the low-x gluon structure one gluon at a time.

At the LHC, tens of thousands of gluons, quark and antiquarks are made physical in the laboratory in every collision !

Very complementary physics.

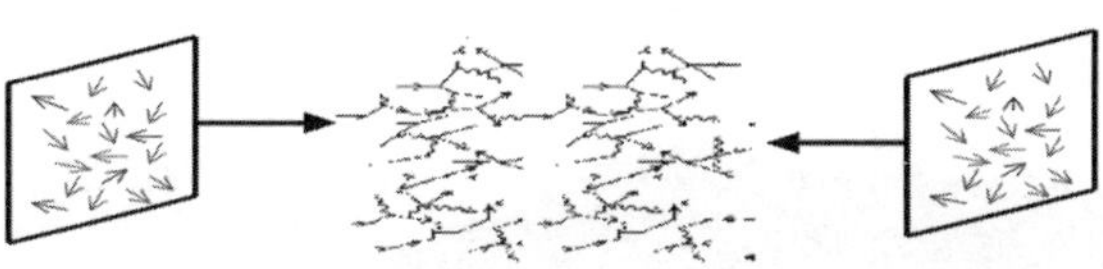

Then we can study the nature of this very hot bath of partons (QGP) !
The plasma should be hotter and live longer than at RHIC.

Jet Probes of the Plasma

Partons are expected to lose energy via induced gluon radiation in traversing a dense partonic medium.

Coherence among these radiated gluons leads to $\Delta E \propto L^2$

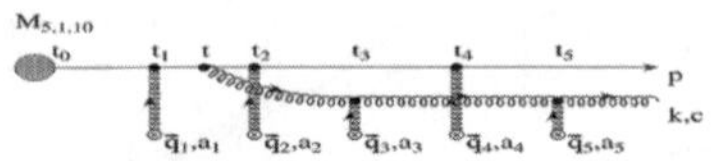

We want to measure the modification of jet properties as we change the gluon density and path length.

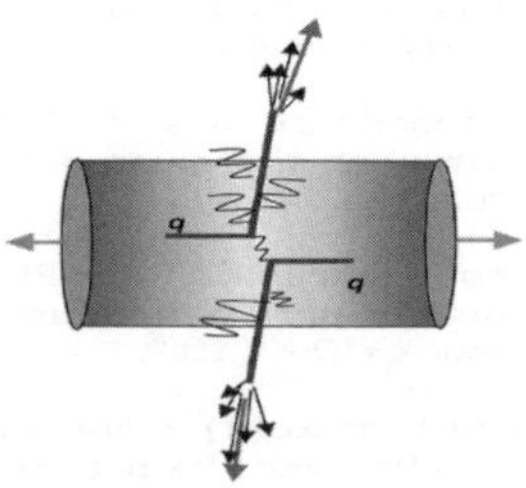

Baier, Dokshitzer, Mueller, Schiff, hep-ph/9907267
Gyulassy, Levai, Vitev, hep-pl/9907461
Wang, nucl-th/9812021
and many more.....

ATLAS Jet Rates

In one month of Pb-Pb running with three experiments at LHC, ATLAS will measure an enormous number of jets.

ATLAS accepted jets for central Pb-Pb

Jet p_T > 50 GeV	**30 million !**
Jet p_T > 100 GeV	**1.5 million**
Jet p_T > 150 GeV	**190,000**
Jet p_T > 200 GeV	**44,000**

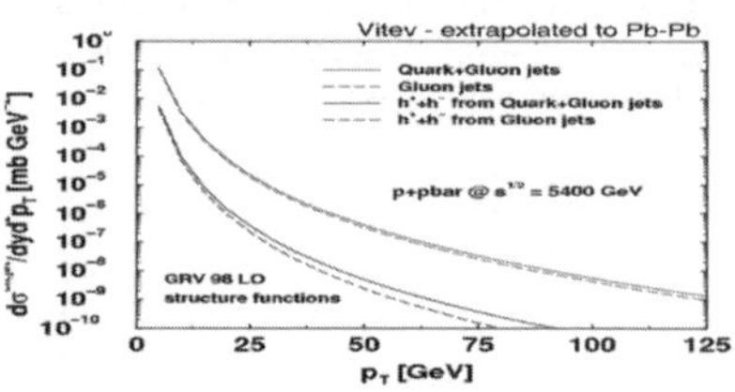

Note that every accepted jet event is really an accepted jet-jet event since ATLAS has nearly complete phase space coverage !

ATLAS Jet Measurements

ATLAS can measure jets with E > 70 GeV with reasonable resolution and efficiency in the highest multiplicity central Pb-Pb events. More detailed studies are currently underway.

Substantial background reduction can be achieved by simultaneously finding back-to-back jets.

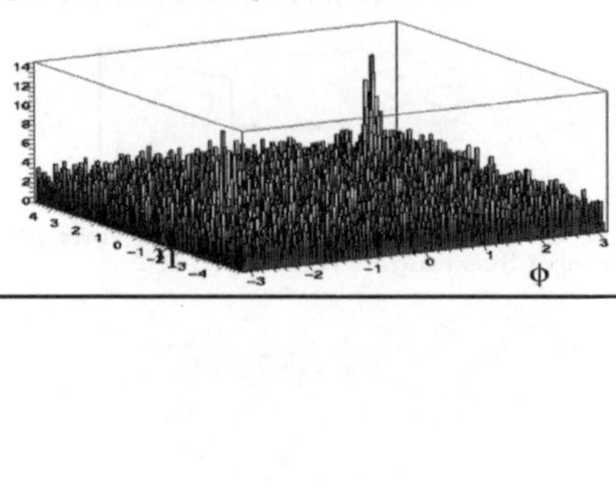

200 GeV jet overlay on central Pb-Pb event with ATLAS segmentation

Jet Profile Analysis

The induced gluon radiation may be measurable due to the broader angular energy distribution than from the jet.

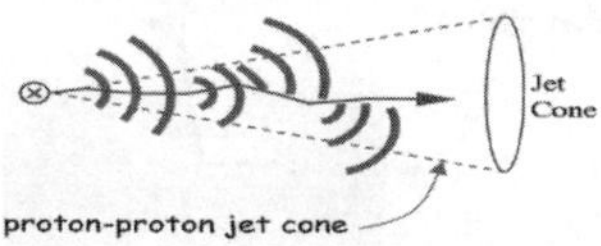

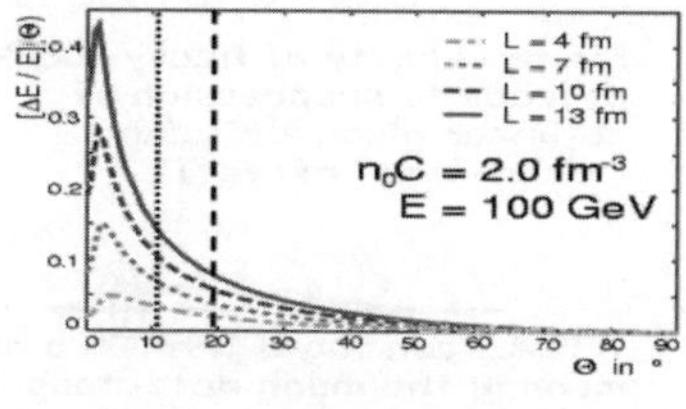

Possible observation of reduced "jet" cross section from this effect.

U.A. Wiedemann, hep-ph/0008241.
BDMS, hep-ph/0105062.

Fragmentation Functions

ATLAS can measure identified π^0 and η mesons via photons.
Excellent energy and timing resolution will help to limit background.

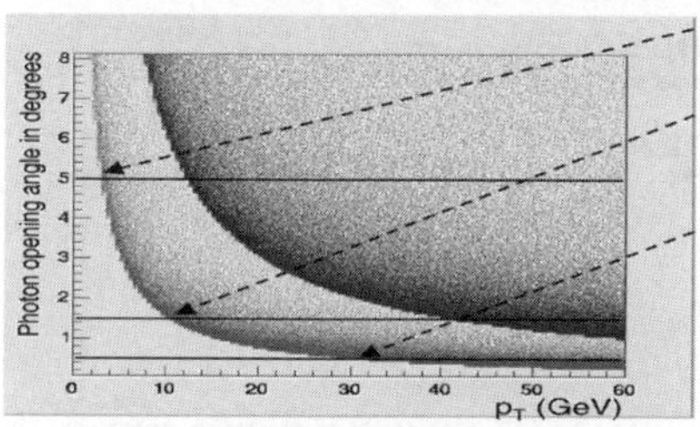

2nd sample 3x5 cluster invariant mass calculation for identification.

2nd sample cluster two γ shower shape identification.

$\Delta\eta \times \Delta\phi = 0.025 \times 0.025$

1st sample cluster two γ separation and total energy measure in 2nd sample.

$\Delta\eta \times \Delta\phi = 0.003 \times 0.1$

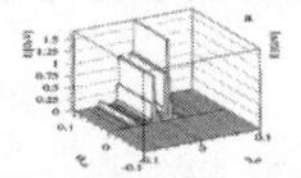

Background and resolution studies are underway.

γ-Jet Physics

ATLAS γ-jet rate is very large, thus allowing for detailed studies.

In one month, over 1000 events with γ energy = 60 GeV in a 1 GeV bin !

Above a certain p_T~30 GeV, ATLAS can no longer cleanly separate single γ from two γ resulting from a π^0 decay.

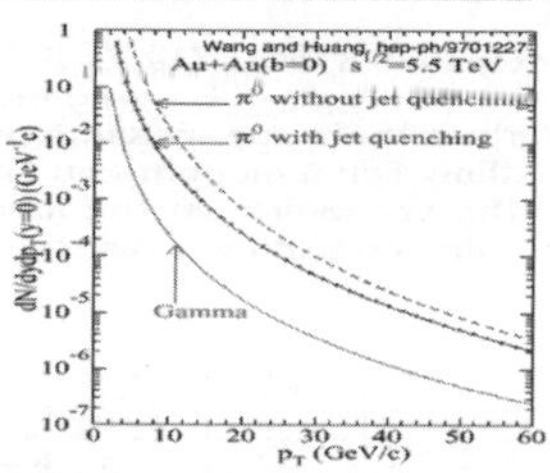

We are investigating whether with isolation cuts on a single high energy shower and an opposite side jet, which process dominates
(1) γ-jet events
(2) jet-jet events with one jet with a high z fragmentation

Beauty Jets

Radiative quark energy loss is qualitatively different for heavy and light quarks.

Finite velocity of heavy quarks at finite p_T leads to suppression of co-linear gluon emission ("dead-cone" effect).

ATLAS can tag B jets via a high p_T muon in the muon detectors.

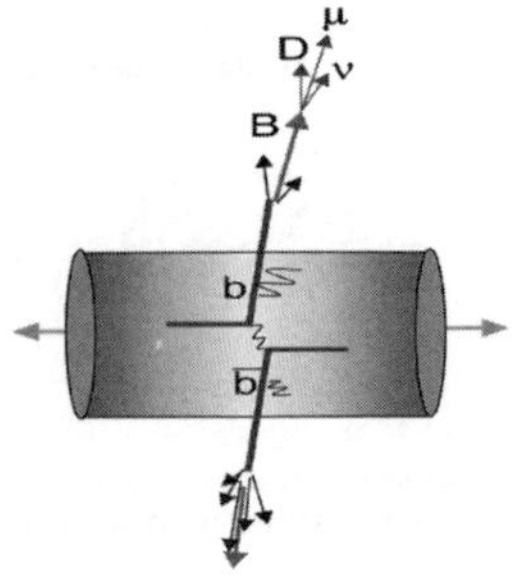

Y.L.Dokshitzer and D.E. Kharzeev, hep-ph/0106202

γ^*-Jet Physics

One can also study virtual photon-jet events, where $\gamma^* \rightarrow \mu+\ \mu-$.

Rate is down two orders of magnitude from γ-jet. Good muon coverage makes this possible.

In one month in central Pb-Pb, ATLAS would accept ~ 10,000 events with $p_T > 40$ GeV.

Z^0-jet reconstruction is possible, but less than 500 total Z^0 events per month.

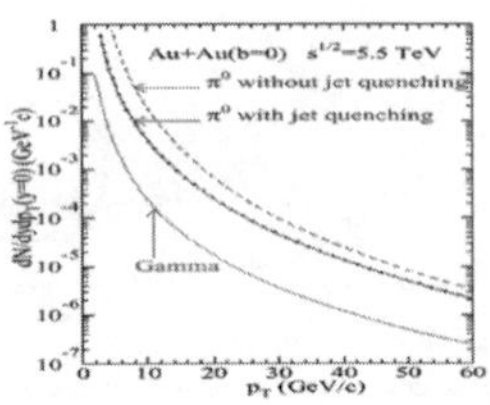

Probes of Deconfinement

Upsilon states (1s,2s,3s) span a large range in binding energy and thus their suppression pattern may allow for a mapping on the onset in the screening on the long range color confining potential.

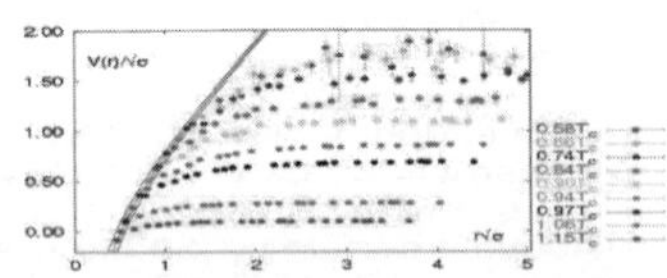

state	J/ψ	χ_c	ψ'	$\Upsilon(1s)$	χ_b	$\Upsilon(2s)$	χ_b'	$\Upsilon(3s)$
Mass [GeV]	3.096	3.415	3.686	9.46	9.859	10.023	10.232	10.355
B.E. [GeV]	0.64	0.2	0.05	1.1	0.67	0.54	0.31	0.2
T_d/T_c	---	0.74	0.15	---	---	0.93	0.83	0.74

ATLAS is currently investigating the mass resolution in the muon system with alternate track matching algorithms.

Correlated Global Measures

ATLAS will measure many global observables and have high statistics for correlating them with high p_T probes.

1) Transverse energy
2) Charged particle multiplicity
3) Zero degree energy
4) Reaction plane

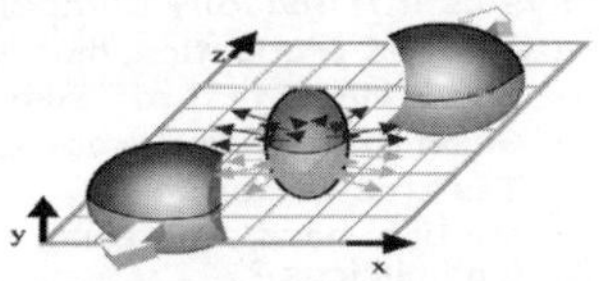

Jet observables as a function of reaction plane
Azimuthal distribution of high p_T π^0 and η
Coverage over a broad range of pseudorapidity

Unique Opportunity at LHC

A parton propagating through the hot QCD media will radiate gluons loosing energy. The overall energy is, however, conserved therefore the measured jet properties will be modified.

The modification of the jet fragmentation function is the most sensitive physical quantity. Other expected modifications are the cone radius, etc.

Whatever measurement is done, it is worthwhile balancing the jet energy by the opposite γ, γ^* or Z^0. Therefore large coverage is beneficial.

Heavy Quarks

Dokshitzer and Kharzeev have suggested that heavy quarks would not radiate as much as light quarks when propagating through the media.

Therefore b-jets should not be quenched as much as light quark jets.

b-jets can be identified by the dislocated vertex or by tagging jets with the associated muon.

t-quarks although interesting do not live long enough to 'feel' the media.

p-A Physics in ATLAS

- Study of p-A collisions is essential @ LHC
  - To provide baseline for heavy ion measurements.
  - Physics intrinsically compelling
    - Mini-jet production, multiple semi-hard scattering.
    - Shadowing – test of "Eikonal" QCD.
    - Gluon saturation – probe QCD @ high gluon density.
    - Test factorization.
    - Multiple hard scattering – Measure parton correlations in nucleon (and nucleus ?)
- ATLAS is ideal detector for p-A studies
  - η coverage, calorimeter performance, b tagging, lepton identification, inner tracking.

Test Perturbative QCD & Factorization

- ATLAS (& CMS) will provide the most comprehensive detectors ever built for study of hard physics in p-A.
- Simultaneously study di/γ/Z-jet, Z, W$\pm$, b-bbar, DY, ...
  - Check that same nuclear PDF's can reproduce all results.
  - Ask questions like:
    - Does jet fragmentation change in "bath" of dense gluons from nucleus ?
    - Are hard radiative processes the same in p-A (A-A) collisions ?
- Study double parton scattering events
  - CDF γ+3 jet data (twice the expected rate) & interpretation:
    - Stronger than expected parton correlations in nucleon.
    - Correlation between valence quark separation & gluon+sea distribution ??
- Study @ LHC using W^+W^+/W^-W^-, double b-bbar, γ+3 jet, W+2 jet
  - Different processes involve different valence-sea combinations.
  - In both p-p and p-A (p-A relaxes correlations in one nucleon).
  - Estimated cross-section for double b-bbar, 10 μBarn in p-p (10^{-3}/p-Pb ev)

Summary of pA physics

- ATLAS detector provides an ideal tool for the study of perturbative & high gluon density QCD in p-A collisions.
- Measurements of mini-jet production and gluon shadowing are critical for heavy ion collisions @ LHC.
- Measurements will complement RHIC & EIC
  - Can match RHIC Q_s (2 GeV ?) in cleaner events.
    - Use azimuthal & rapidity coverage to test saturation model predictions.
  - Comparing to EIC data – can test $\frac{d}{d\log Q^2} \frac{d^2N}{dk_T^2 dx}$
    - Equal only at leading twist but saturation → all twist terms important.
    - McLerran: Not valid for $Q^2 << \alpha_s^2 Q_s^2$.
- Make detailed study of pQCD in nuclear environment.
- Study parton correlations in proton/nucleus w/ double parton scattering events.

RHIC and LHC, γ+hadron colliders

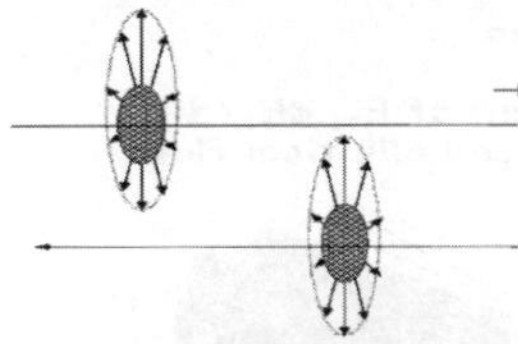

γ_{beam}^{Rhic}=100

S=2$m_A\omega$'

=>Nucleus at rest,effective Lorentz $\gamma_{eff}=2*\gamma_{beam}^2-1$

Heavy Ions

$$\frac{1}{2}\frac{dN(\omega)}{d\omega}=\frac{2\alpha Z^2}{\pi}\ln(\frac{0.68\,\hbar c\gamma_{eff}}{R_{nucleus}\bullet\omega})$$

e-Hadron collider

$$\frac{dN(\omega)}{d\omega}=\frac{2\alpha}{\pi}\ln(\frac{m_e\bullet\gamma_{eff}}{\omega})$$

Ultra Peripheral Collisions

Heavy Ion Physics= Opportunities with a tool that we are just learning to exploit (c.f. e^+e^- physics)

LHC energy scale

Equivalent γ flux Up to 100's of TeV

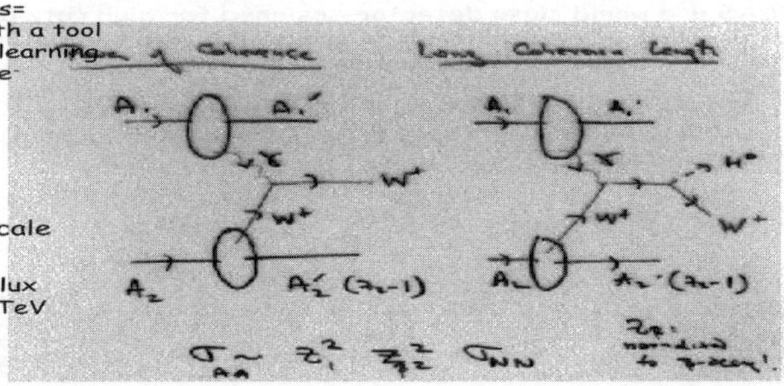

Ultra Peripheral Collisions

The photon spectrum:

The photon spectrum depends on the rigorous treatment of the form factors.

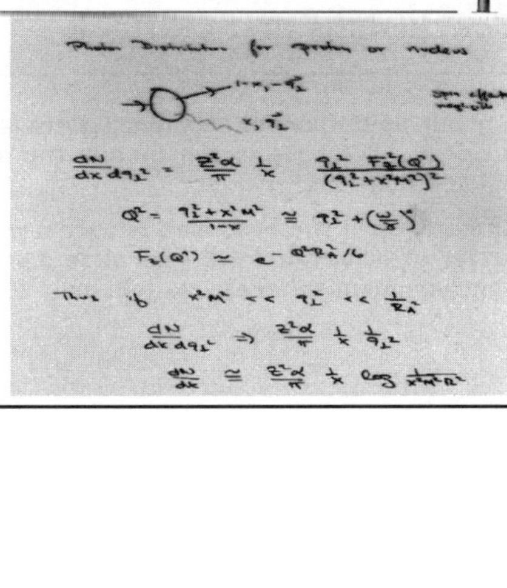

Summary of the physics program

The ATLAS Nuclear Physics program includes

Global variable measurements: measurement of E_T, $dE_T/d\eta$, charged particle multiplicity N, $dN/d\eta$, and elliptical flow.

Jet Physics : γ+jet, γ^*+jet, Z^0+jet

Heavy Quarks : beauty jets

Quarkonia : Υ suppression

p+A physics

Ultra Peripheral Collisions

"I think you should be more explicit here in step two."

Conclusions and Outlook

ATLAS is a world class detector designed for high rate proton-proton collisions and will provide great opportunity to study high Q^2 physics in nucleus nucleus collisions.

The ATLAS heavy ion working group is being formed and simulation is under way. Input from Theorist and Experimentalists are very welcome

Conclusions and Outlook

Strong input from phenomenologists is the ticket for a successful program. We have experience in the proton-proton program and has made a huge difference.

In early april the proposed program was presented to DOE as a letter of intent and will formalize a proposal to LHCC late fall (in the northern hemisphere) - so join in!!!

OUTLINES OF SEMINAR ON FIRST RESULTS FROM RHIC

M. ROSATI

FIRST RESULTS FROM RHIC

M. Rosati
Iowa State University, Ames, Iowa, USA
mrosati@iastate.edu

Theoretical studies of the thermodynamics of strongly interacting matter at high temperatures and/or baryon densities unambiguously predict the existence of the quark-gluon plasma, a state of matter in which hadrons have dissolved into their constituent quarks and gluons. The primary goal of the four experiments at the Relativistic Heavy Ion Collider (RHIC) is the identification and study of this new phase of nuclear matter. In my talk, I will present an overview of the results obtained in the first year of data taking at RHIC.

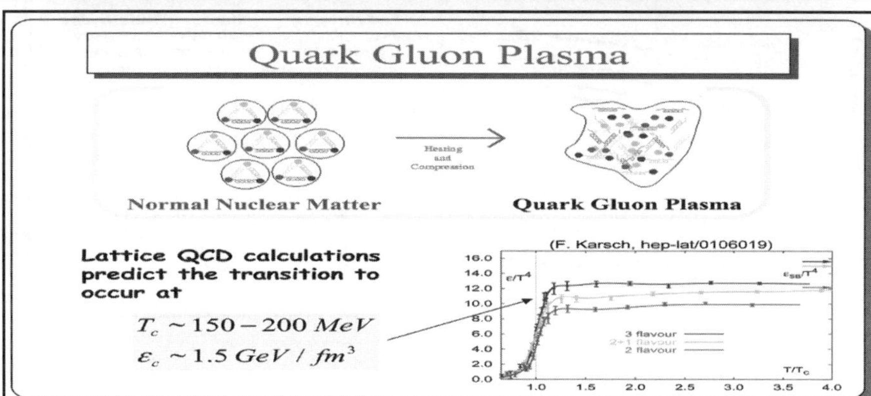

Some QGP Diagnostics

	Why	What
Global Observables	Is initial state dense enough?	Particle Multiplicities Baryon Density Energy Density
Collective Behavior	Is QGP a thermalized state?	Elliptic Flow
Hard Probes	Formed early, probe medium	Energy loss of jets

The RHIC Complex

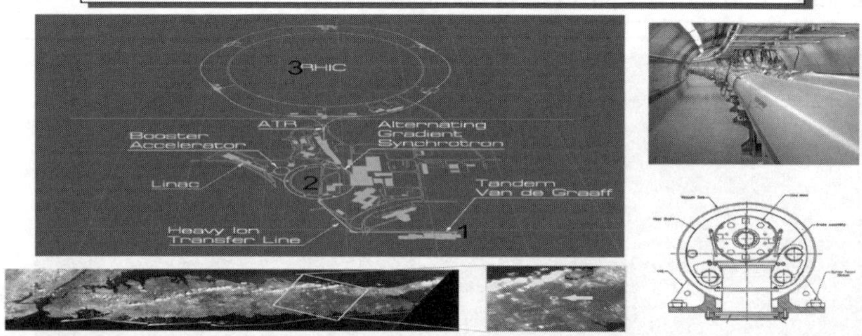

RHIC Capabilities & Run I

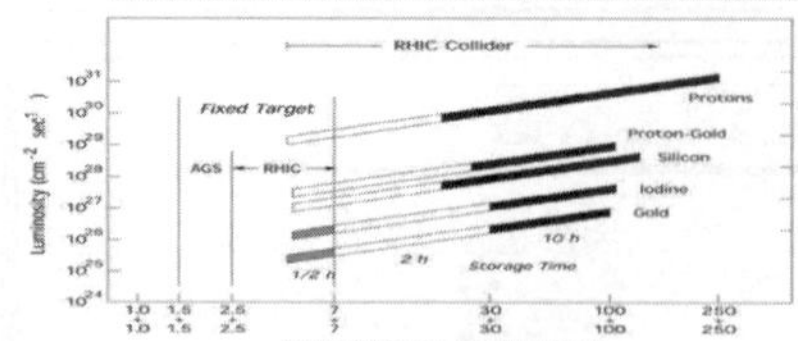

Performance	Run1	RHIC Design
$\sqrt{s_{nn}}$	130 GeV	200 GeV
L [cm^{-2} s^{-1}]	~ 2 x 10^{25}	2 x 10^{26}
Interaction rates	~ 100 Hz	1400 Hz

Run1 $\int L dt = 6.5 \mu b^{-1}$

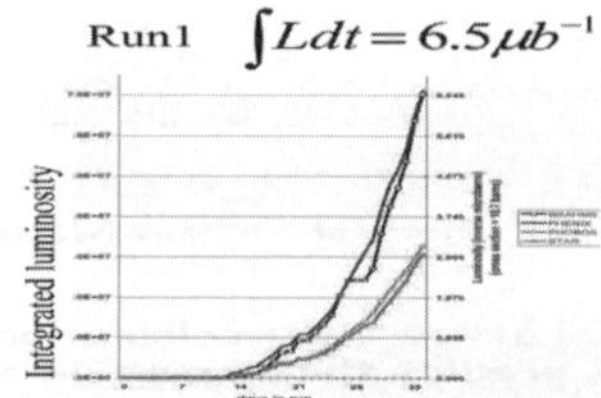

- **Nucleus-nucleus (AA) collisions up to $\sqrt{s_{NN}} = 200$ GeV**
- **Polarized proton-proton (pp) collisions up to $\sqrt{s_{NN}} = 450$ GeV**

RHIC is an International Operation

- **There are more than 1000 scientists participating in the 4 RHIC experiments from all over the world.**

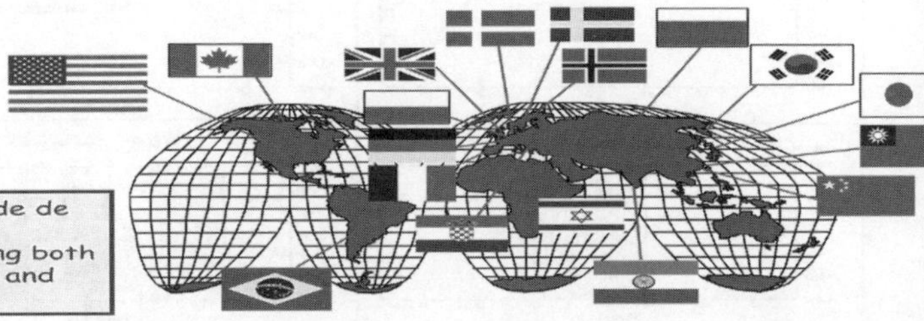

Universidade de Sao Paulo participating both in PHENIX and STAR

How is RHIC Different?

- **It's a collider**
  - Detector systematics independent of ECM
- **It's dedicated**
  - Heavy ions will run 20-30 weeks/year
- **It's high energy**
  - Access to perturbative phenomena
    - Jets
    - Non-linear dE/dx
- **Its detectors are comprehensive**
  - ~All final state species measured with a suite of detectors that nonetheless have significant overlap for comparisons

RHIC Experiments

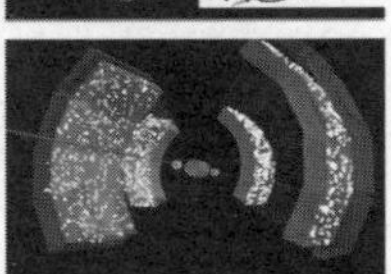
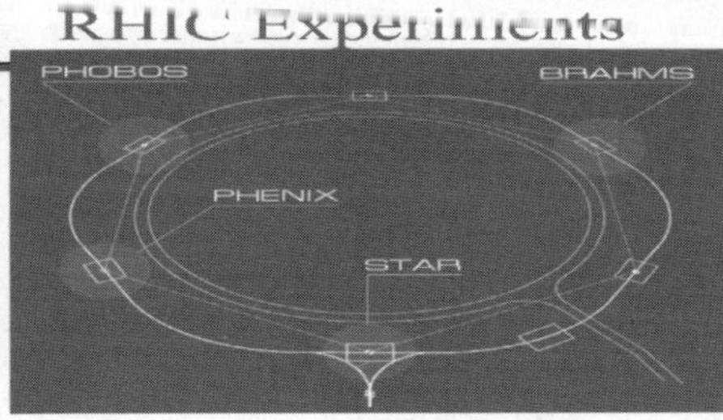

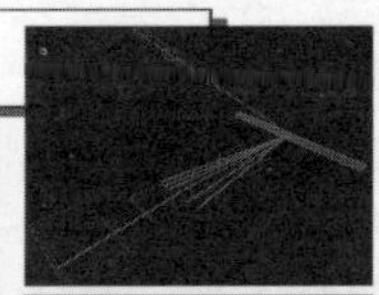
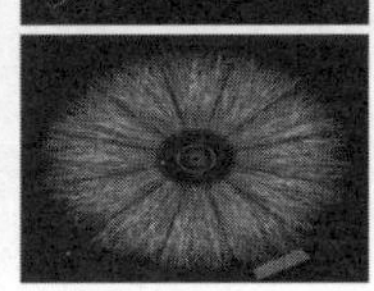

Constraints on design
- **High multiplicity events**
- **High rate needed**
- **Low-cost required**

Choices made
- **Two large, flexible (& expensive!) experiments**
- **Two small, optimized (& inexpensive!) experiments**

(PID) Acceptances

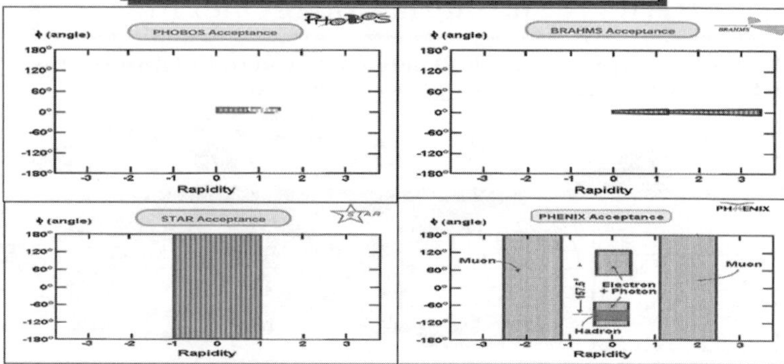

Results (1)

How many particles are produced?

- **What is the (charged) particle multiplicity emitted in the direction perpendicular to the collision axis?**
- **Equivalently, what is the value of dn/dy at y=0?**

PHOBOS

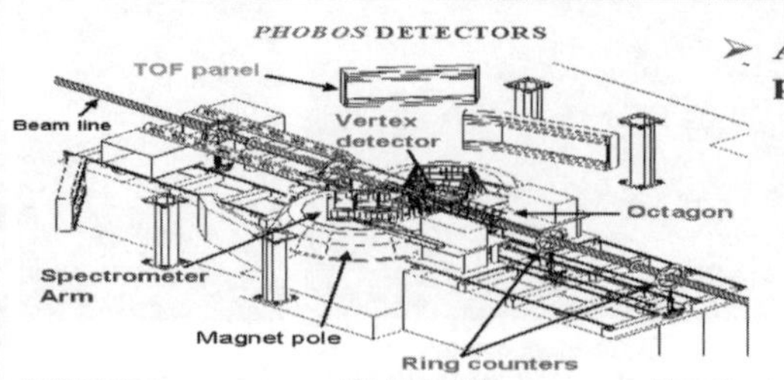

- **An experiment with a philosophy:**
  - Global phenomena
    - ⇨ large spatial sizes
    - ⇨ small momenta
  - Minimize the number of technologies:
    - ⇨ All Si-strip tracking
    - ⇨ PMT-based TOF
  - Unbiased global look at very large number of collisions ($\sim 10^9$)

Geometry of Nuclear Collisions

- **Nuclei are extended**
  - $R_{Au} \sim 6.4$ fm $R_p \sim 0.8$ fm
- **Impact parameter (b) determines**
  - N_{coll} – binary collisions
  - N_{part} – 1 or more collisions
  - $N_{spect} = 2A - N_{part}$
- **Proton-Proton**
  - $N_{coll} = 1$, $N_{part} = 2$
- **Proton-Nucleus:**
  - $N_{coll} = N_{part} - 1$
- **Nucleus-Nucleus**
  - $N_{coll} \propto N_{part}^{4/3}$

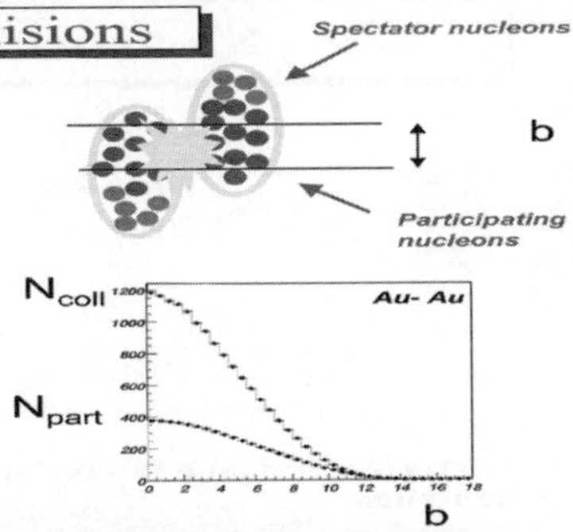

Useful quantities to compare Au+Au to p+p collisions!

Charged Particle Multiplicity

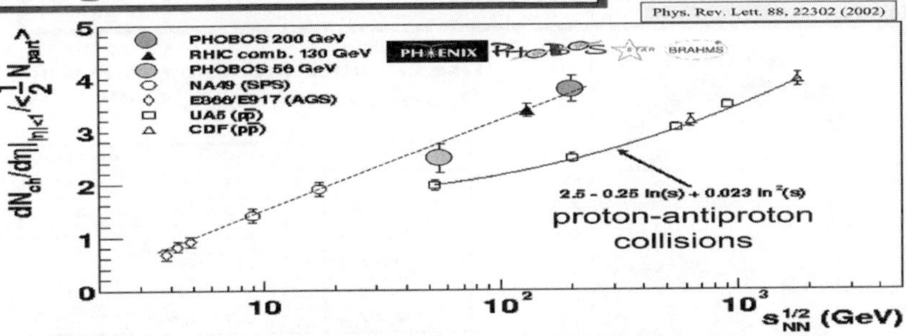

- **Au+Au collisions produce more particles per participant pair than proton-proton collisions at same $\sqrt{s}$**

Global Event Characterization

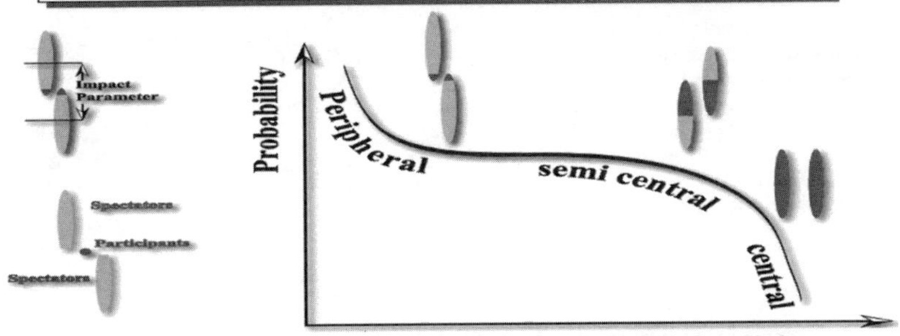

Measuring Centrality

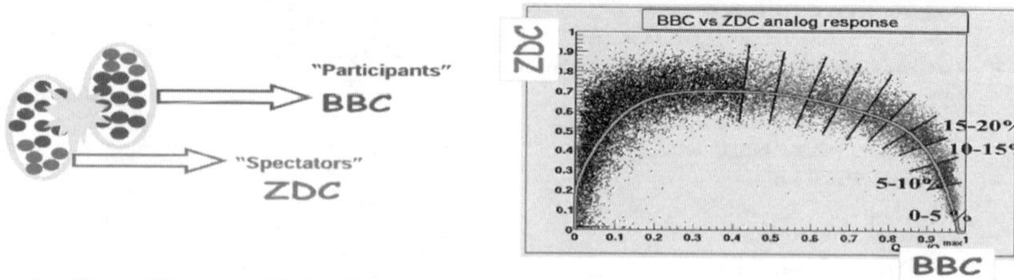

- **Zero Degree Calorimeters (ZDC) are sensitive to spectator neutrons**
- **Using a combination of the ZDC's and BBC's we can define Centrality Classes**

From Centrality to N_{binary} and $N_{participants}$

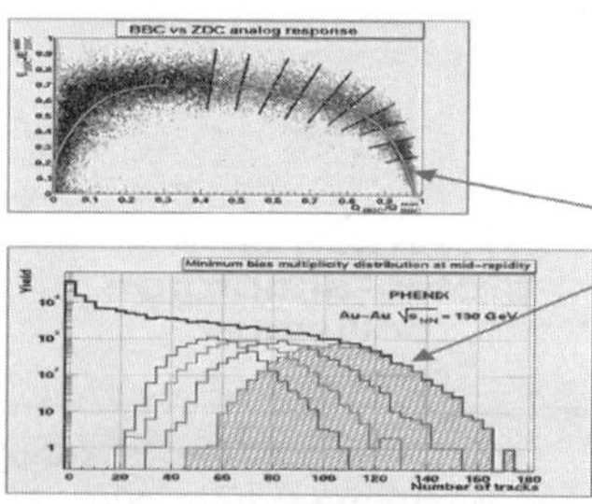

Use Glauber modeling to extract $N_{participants}$ and N_{binary} Collisions

Centrality	Collisions	Participants
0-5%	945 ± 15%	347 ± 15%
5-15%	673 ± 15%	271 ± 15%
15-30%	383 ± 15%	178 ± 15%
30-60%	123 ± 15%	76 ± 15%
60-80%	19 ± 60%	19 ± 60%
80-92%	3.7 ± 60%	5 ± 60%

Particle Density for Central Collision

- **Does not (quite) distinguish between**
  - "Saturation" models, dominated by gg →g
  - "Cascade" models, dominated by gg →gg, gg →ggq

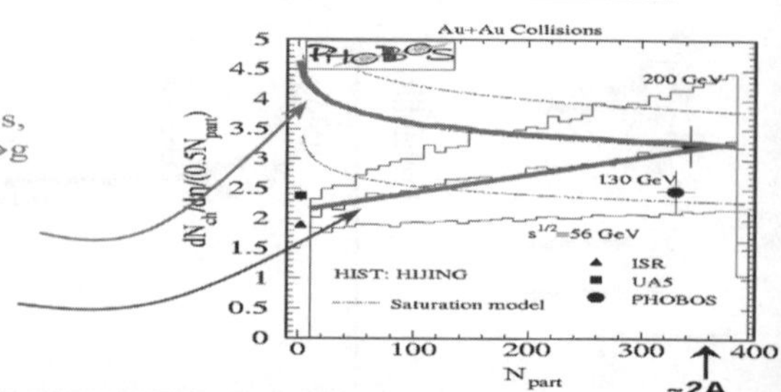

(X.N. Wang and M. Gyulassy, nucl-th/0008014)

Particle Density vs. Centrality

➢ **Trend**

- ↬ Qualitatively consistent with HIJING (Wang & Gyulassy Phys. Lett. B 86,3496 ,2001)
- ↬ Incompatible with final-state gluon saturation model (Eskola et al., Phys. Lett.B 439, 37,2001)
- ↬ Good agreement with model based on *initial-state* saturation (Kharzeev and Nardi, nucl-th/0012025)

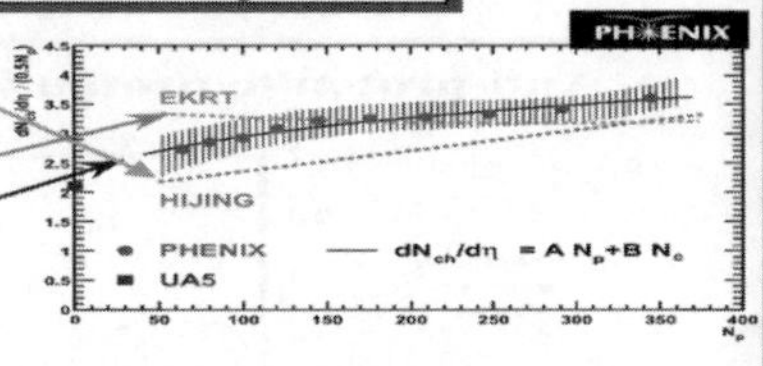

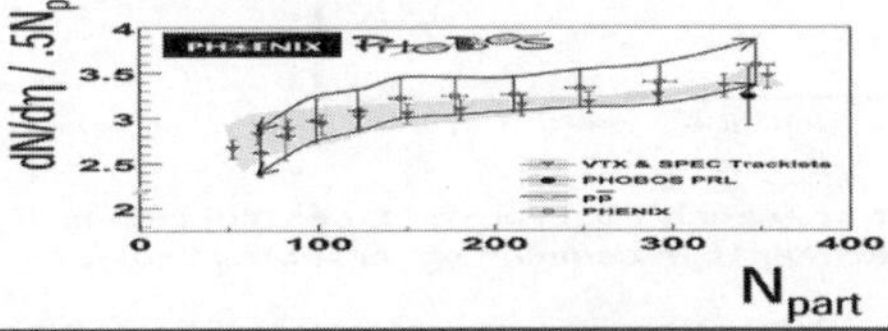

- **Excellent agreement between PHENIX and PHOBOS analyses of this systematic variation with nuclear overlap.**

Results (2)

Have we made contact with the early universe?

➢ **What is the ratio of anti-protons to protons?**

➢ **How does this ratio vary with rapidity?**

BRAHMS

An experiment with an emphasis:

- ↬ Quality PID spectra over a broad range of rapidity and p_T
- ↬ Special emphasis:
  - ⇨ Where do the baryons go?
  - ⇨ How is directed energy transferred to the reaction products?
- ↬ Two magnetic dipole spectrometers in "classic" fixed-target configuration

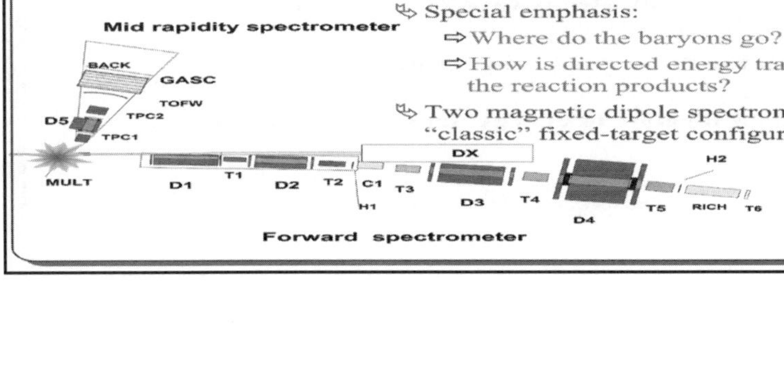

Antiproton to Proton Ratio

➢ **Anti-proton to proton yields as a function of rapidity:**

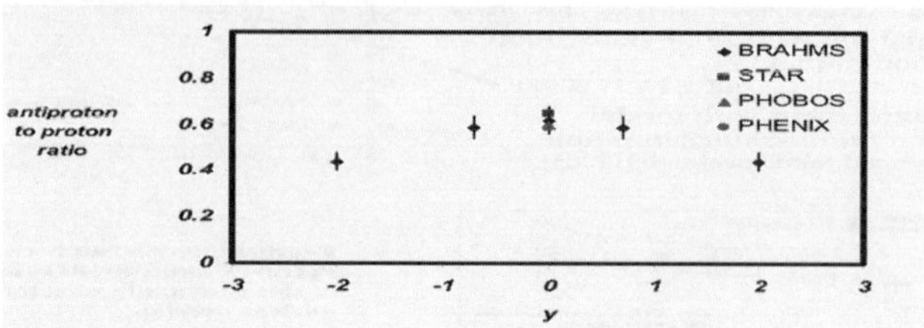

✓ *Clear evidence for development of (nearly) baryon-free central region*
✓ *Excellent agreement of all experiments in common observables provides confidence.*

Approaching the Early Universe

➢ **Early Universe:**
  ⇨ Anti-proton/proton = 0.999999999

- We've created "pure matter" approaching this value
- For the first time in heavy ion collisions, more baryons are pair-produced than brought in from initial state

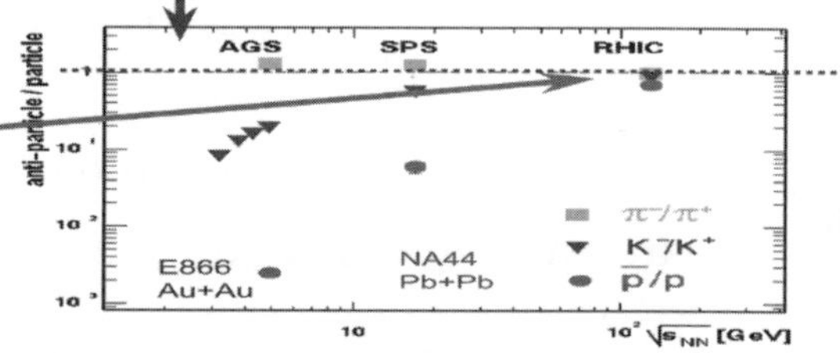

Results (3)

Have we made sufficiently high energy densities for the phase transition?

➢ **What is the "transverse energy density"?**

➢ **How does it relate to energy density in collision?**

PHENIX Design

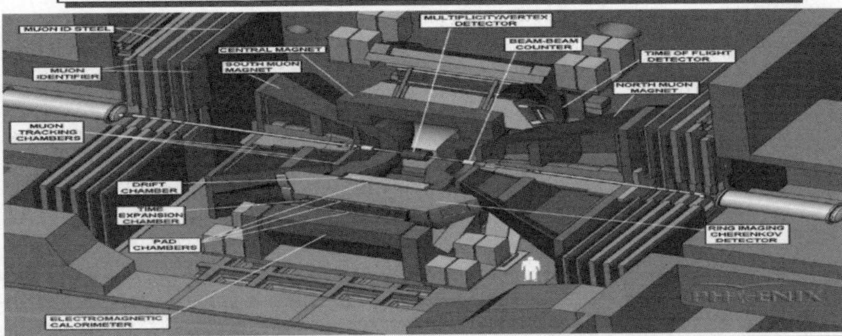

Virtual Tour of PHENIX

How to estimate initial energy density?

- **Assuming**
  - ~ 1-d hydrodynamic expansion
  - Invariance in y along "central rapidity plateau"

  [from J.D. Bjorken, Phys. Rev. D27, 140 (1983)]

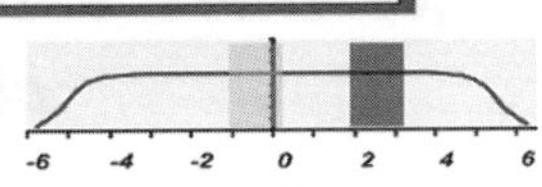

πR^2

$dz = c\tau_0 dy$

$$\varepsilon_{Bj} = \frac{E}{V} \sim \frac{dE_T}{\pi R^2 \cdot dz}$$

$$\varepsilon_{Bj} = \frac{1}{\pi \mathbf{R}^2} \frac{1}{c\tau_0} \left(\frac{d\mathbf{E}_T}{dy} \right)$$

Measured in PHENIX

6.88 fm (hard sphere radius)

Time to thermalize the system

Energy Density at RHIC

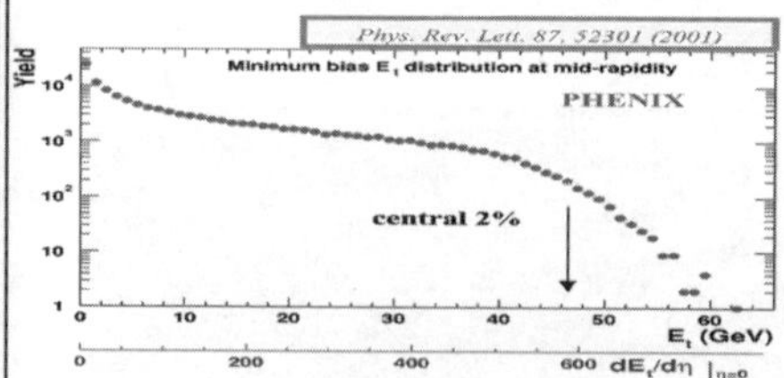

$$\frac{dE_T}{dy} = \left(578^{+26}_{-39}\,GeV\right)\times(1.19 \pm 0.01)$$

$$\varepsilon_{Bj} = \frac{1}{\pi R^2}\frac{1}{c\tau_0}\frac{dE_T}{dy}$$

formation time: 0.2 - 1 fm

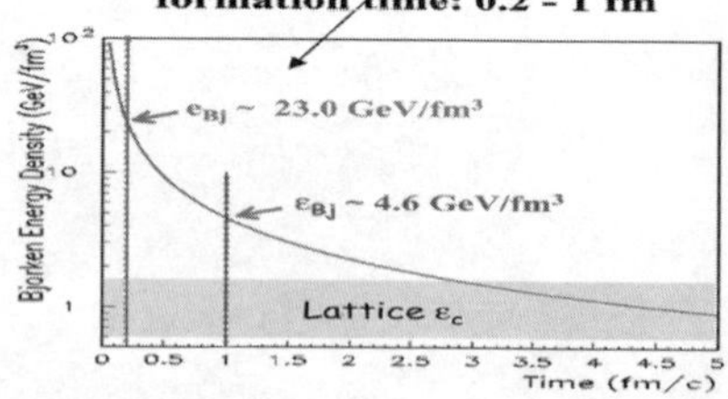

Energy Density

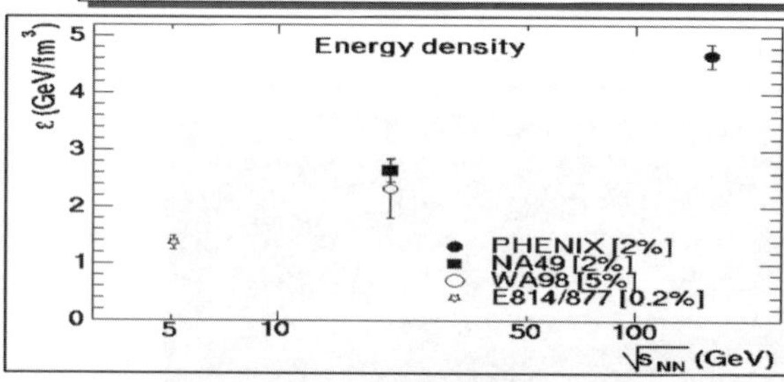

Energy density

$$\varepsilon = \frac{1}{\pi R^2 \tau}\frac{dE_T}{dy}$$

$$R = 1.18A^{1/3}$$
$$\tau_{SPS} = 1\ fm/c$$
$$\tau_{RHIC} = 0.2 - 1\ fm/c$$

From SPS to RHIC: at least 70% increase in ε

Results (4)

Is there evidence that the dense matter behaves collectively?

Does the initial spatial anisotropy in non-central collisions get translated into momentum space distributions?

Are the results consistent with "elliptic flow"?

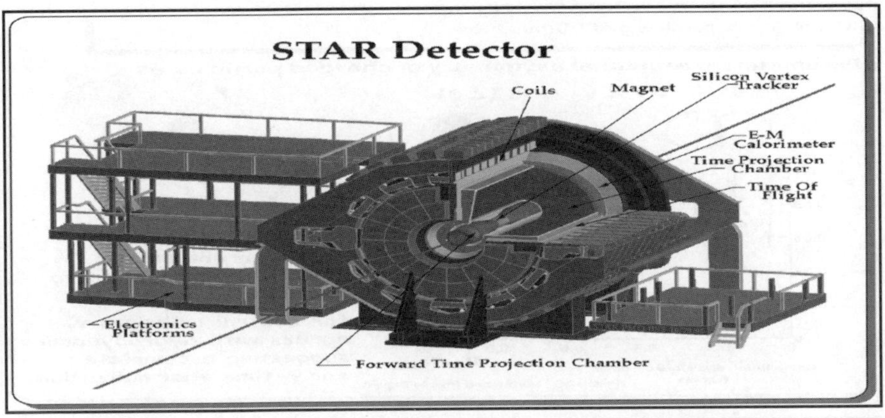
STAR Detector
Coils
Magnet
Silicon Vertex Tracker
E-M Calorimeter
Time Projection Chamber
Time Of Flight
Electronics Platforms
Forward Time Projection Chamber

STAR

STAR Event
Data Taken June 25, 2000.
Pictures from Level 3 online display.

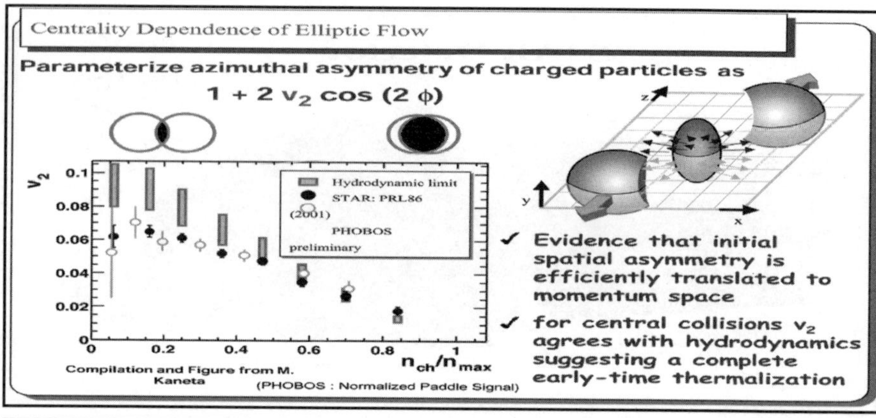
Centrality Dependence of Elliptic Flow
Parameterize azimuthal asymmetry of charged particles as
1 + 2 v2 cos (2 ϕ)
v2
Hydrodynamic limit
STAR: PRL86 (2001)
PHOBOS preliminary
nch/nmax
Compilation and Figure from M. Kaneta
(PHOBOS : Normalized Paddle Signal)
✓ Evidence that initial spatial asymmetry is efficiently translated to momentum space
✓ for central collisions v2 agrees with hydrodynamics suggesting a complete early-time thermalization

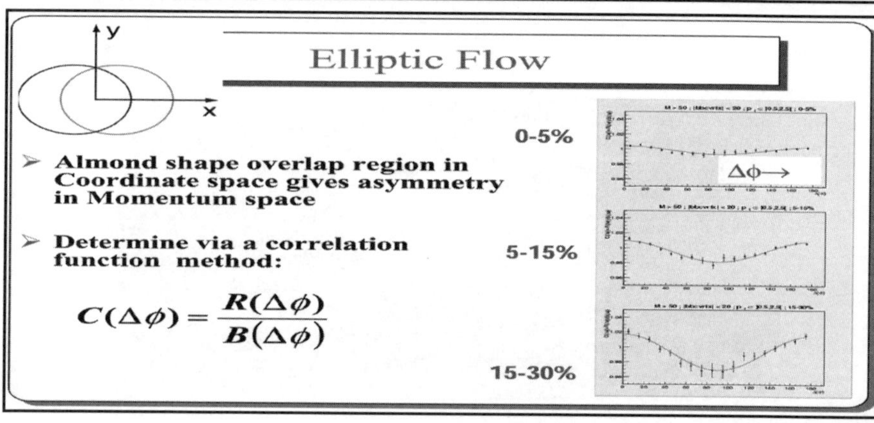
Elliptic Flow
y
x
Almond shape overlap region in Coordinate space gives asymmetry in Momentum space
Determine via a correlation function method:
C(Δϕ) = R(Δϕ) / B(Δϕ)
0-5%
5-15%
15-30%
Δϕ→

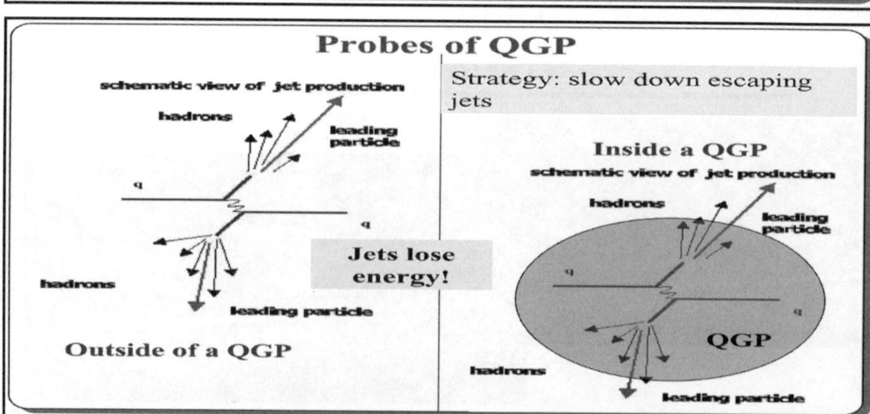
Probes of QGP
schematic view of jet production
hadrons
leading particle
q
q
hadrons
leading particle
Outside of a QGP
Jets lose energy!
Strategy: slow down escaping jets
Inside a QGP
schematic view of jet production
hadrons
leading particle
q
q
QGP
hadrons
leading particle

Rare processes at RHIC

Both PHENIX and STAR have measured charged particle spectra out to "small" cross sections

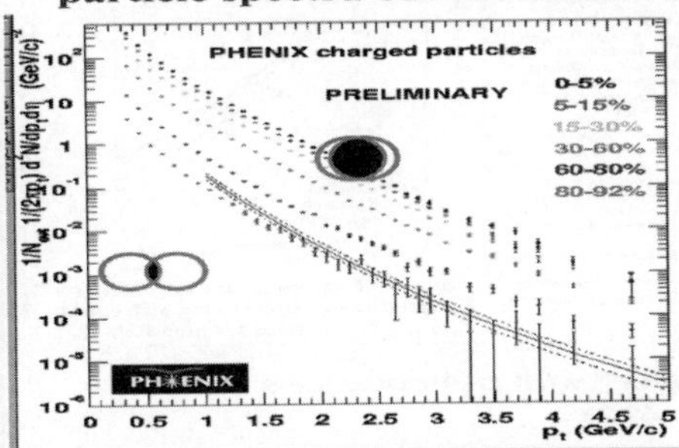

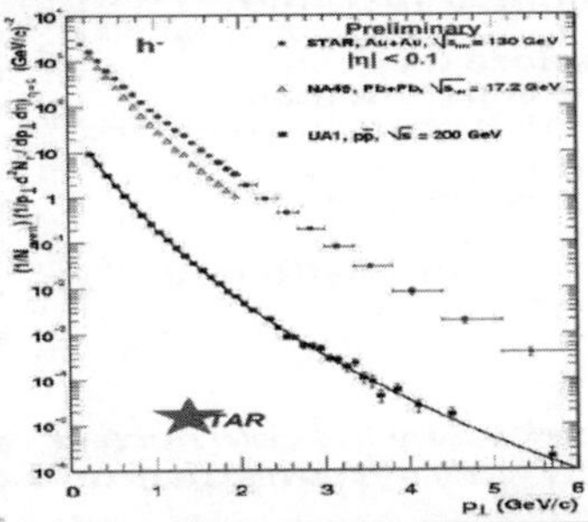

A*B scaling at RHIC?

- **NO!**
- **Both STAR and PHENIX data see deficit of particles at high transverse momenta,*relative to expected A*B scaling:***

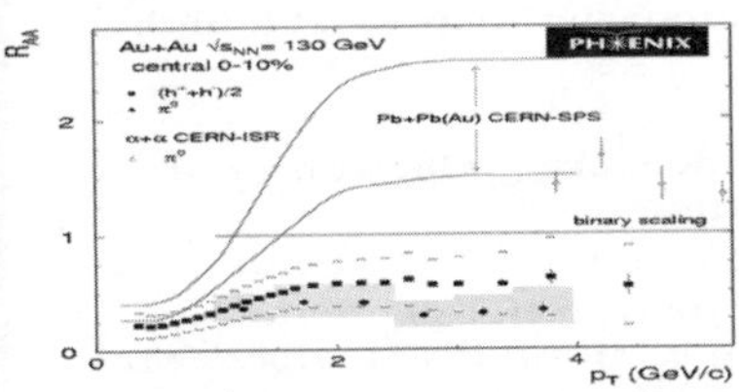

Deficit opposite sign of enhancements previously seen in nuclear collisions (Cronin effect)

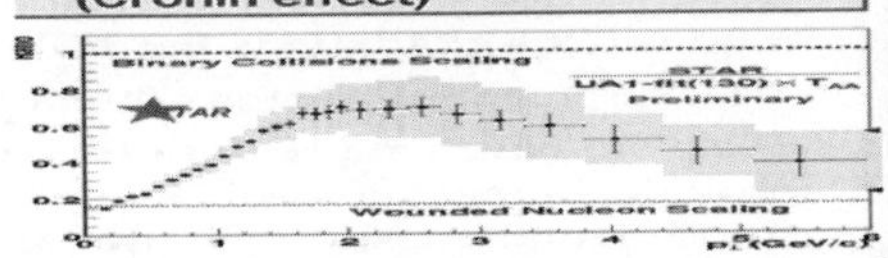

Additional Evidence

- **PHENIX sees an equal or larger effect in the π^0 spectrum**
- **Consistent with crude estimates of additional energy loss in a *deconfined medium***

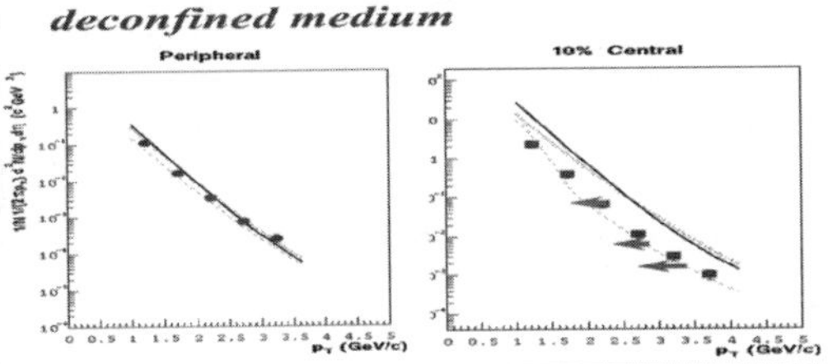

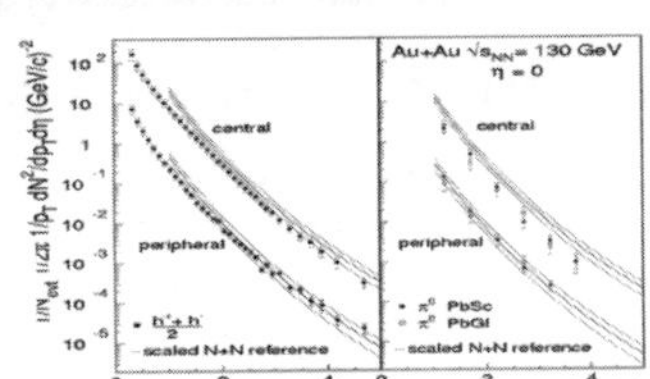

Discovery of Jet Quenching? **NOT YET**

- **Some theorists claimed just that**
- **before this can be experimentally established, we have to:**
  - □ **Improve p_T range (Run-2)**
  - □ **Measure p-p spectrum (Run-2)**

arXiv:nucl-th/0104035 10 Apr 2001

Discovery of Jet Quenching at RHIC and the Opacity of the Produced Gluon Plasma

P. Lévai[a], G. Papp[b], G. Fai[c],

[a] IIAS KFKI Research Institu
Budapest 1525, Hungary
[b] IIAS Research Group for T
Budapest 1117, Hungary
[c] Physics Department, Kent S
[d] Physics Department, Colum
10027, USA

The predicted quenching o
STAR and PHENIX in prelim
theory [1] of QCD radiative en
produced in $Au + Au$ collision
two Cronin enhancement of π
of $p_T \sim 2-4$ GeV π^0 report
static plasma opacity $L/\lambda_g \approx$

1. Energy Loss of High E

The moderately high $p_T <$
π^0 production [2], reported at

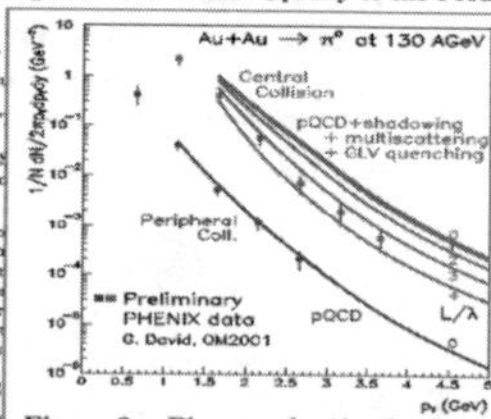

Figure 3. Pion production in Au+Au collision including jet quenching with opacity $L/\lambda = 1, 2, 3, 4$. Data are from Ref. [2].

☛ ***Study in "cold nuclear matter" with p-A collisions* (Run-3?)**
□ **Vary system size (Run-3 or 4??)**

RHIC Run 2 ended in Nov 2001

• Increase of collision energy from 65 GeV/A to 100 GeV/A.
• We collected 20 times more data than in the first run.

Some new analyses that will be possible:

- Increased statistics on all previous measurements - at a higher energy.
- **More data on jets.**
- Rare particles that can be seen by measuring electrons and muons. One example is the J/Ψ particle.
- Measurement of photons directly from the collision center.
- Also we have data with **polarized proton beams** which will provide a great "control" sample (in addition it will answer the fundamental question: where the proton gets its spin)

Summary

- **RHIC has had a very successful start!**
  - **Higher energy densities**
  - **Central region almost baryon free**
  - **Clear evidence of an expanding hadronic system**
  - **Indication of parton energy loss in a dense matter**
- **Analysis of high transverse momentum jets may be providing a glimpse of the QGP BUT, NO CONCLUSIVE EVIDENCE**

The Quark-Gluon Plasma is still at large, so...

The Quest continues.

More information at www.bnl.gov/RHIC/

RHIC Publications (I)

+ **BRAHMS**

"Pseudorapidity distributions of charged particles from Au+Au collisions at the maximum RHIC energy "
Submitted to Phys. Rev. Lett. (Dec. 2001), nucl-ex/0112001.

"Charged particle densities from Au+Au collisions at sqrt(s_NN)=130 GeV"
Phys. Lett. B 523 (2001) 227, nucl-ex/0108016.

"Rapidity dependence of antiproton to proton ratios in Au+Au collisions at sqrt(s_NN) = 130 GeV"
Phys. Rev. Lett. Volume 87, (2001) 112305

+ **STAR**

Measurement of inclusive antiprotons from Au+Au collisions at sqrt(snn) = 130 GeV
Phys. Rev. Lett. 87, 262302-1 (2001).

Antideuteron and Antihelium production in Au+Au collisions at sqrt(snn) = 130 GeV
Phys. Rev. Lett. 87, 262301-1 (2001).

Mid-rapidity phi production in Au+Au collisions at sqrt(snn) = 130 GeV
to be submitted to Phys. Rev. Lett.

Identified Particle Elliptic Flow in Au+Au Collisions at sqrt(snn) = 130 GeV
Phys. Rev. Lett. 87, 182301 (2001).

Multiplicity distribution and spectra of negatively charged hadrons in Au+Au collisions at sqrt(snn) = 130 GeV
Phys. Rev. Lett. 87, 112303 (2001).

Pion Interferometry of sqrt(snn) = 130 GeV Au+Au collisions at RHIC
Phys. Rev. Lett. 87 , 082301 (2001).

Midrapidity Antiproton-to-Proton Ratio from Au+Au sqrt(snn) = 130 GeV
Phys. Rev. Lett. 86 pp. 4778-4782 (2001).

Elliptic Flow in Au+Au Collisions at sqrt(snn) = 130 GeV
Phys. Rev. Lett. 86 pp. 402-407 (2001).

RHIC Publications (II)

+ **PHOBOS**

Centrality Dependence of the Charged Particle Multiplicity near Mid-Rapidity in Au+Au Collisions at sqrt(s_NN) = 130 and 200 GeV
submitted to Phys. Rev. C - Rapid Communications, nucl-ex/0201005

Energy dependence of particle multiplicities near mid-rapidity in central Au+Au collisions
Phys. Rev. Lett. 88, 22302 (2002)

Centrality Dependence of Charged Particle Multiplicity at η=0 in Au+Au Collisions at sqrt(sNN)=130 GeV
Accepted: Phys. Rev. C (Dec 2001)

Charged-particle pseudorapidity density distributions from Au+Au collisions at sqrt(sNN)=130 GeV
Phys. Rev. Lett. 87, 102303 (2001)

Ratios of charged particles to antiparticles near mid-rapidity in Au+Au collisions at sqrt(sNN)=130 GeV
Phys. Rev. Lett. 87, 102301 (2001)

Charged particle multiplicity near mid-rapidity in central Au+Au collisions sqrt(s) = 56 and 130 AGeV
Phys. Rev. Lett. 85, 3100 (2000)

+ **PHENIX**

"Centrality Dependence of Charged Particle Multiplicity in Au-Au Collisions at sqrt(s) = 130 GeV".
PRL 86 (2001) 3500, preprint nucl-ex/0012008.

"Measurement of the Midrapidity Transverse Energy Distribution from sqrt(s) = 130 GeV Au-Au Collisions at RHIC",
PRL 87 (2001) 052301, preprint nucl-ex/0104015,

"Suppression of Hadrons with Large Transverse Momentum in Central Au+Au Collisions at sqrt(s) = 130 GeV",
Phys. Rev. Lett. 88, 022301 (2001), preprint nucl-ex/0109003

"Centrality dependence of pi+-, K+-, p and pbar production from sqrt(s)=130 GeV Au + Au collisions at RHIC",
Submitted to PRL, preprint nucl-ex/0112006

"Transverse momentum dependence of the two-pion correlation for Au+Au collisions at sqrt(s_NN)=130 GeV",
Submitted to PRL, preprint nucl-ex/0201008

LIST OF PARTICIPANTS

J. W. GOETHE UNIVERSITÄT — GERMANY

Horst Stöcker stoecker@th.physik.uni-frankfurt.de

FRIEDRICH ALEXANDER UNIVERSITÄT — GERMANY

Frieder Lenz flenz@theorie3.physik.uni-erlangen.de

Manfred Dillig mdillig@theorie3.physik.uni-erlangen.de

CORNELL UNIVERSITY — USA

Peter Lepage gpl@mail.lns.cornell.edu

UNIVERSITY OF NOTRE DAME — USA

Fridolin Weber fweber@sciences.sdsu.edu

Shantilal G. Goradia shantilallg@aol.com

BROOKHAVEN NATIONAL LABORATORY — USA

Helio Takai takai@bnl.gov

IOWA STATE UNIVERSITY — USA

Marzia Rosati mrosati@iastate.edu

UNIVERSITÉ LIBRE DE BRUXELLES — BELGIUM

Laurent Favart lfavart@ulb.ac.be

IZMIRAN — RUSSIA

Lev Leinson leinson@izmiran.rssi.ru

UNIVERSIDADE LIVRE DE COIMBRA — PORTUGAL

Manuel Fiolhais tmanuel@teor.fis.uc.pt

UNIVERSIDAD NACIONAL DE LA PLATA — ARGENTINA

Carlos Alberto Garcia Canal garcia@fisica.unlp.edu.ar
Daniel Gomez Dumm dumm@fisica.unlp.edu.ar

BRAZIL

UNIVERSIDADE FEDERAL DA PARAÍBA

Maria Luiza Cescato mcescato@fisica.ufpb.br

UNIVERSIDADE FEDERAL DE PERNAMBUCO

Helio Teixeira Coelho heliotcoelho@terra.com.br

UNIVERSIDADE FEDERAL DE SERGIPE

Mario Everaldo de Souza msouza@fisica.ufs.br

ESCOLA FEDERAL DE ENGENHARIA DE ITAJUBÁ

Carlos Javier Solano Salinas javier@cbpf.br

CENTRO BRASILEIRO DE PESQUISAS FÍSICAS

Anna Maria Freire Endler endler@cbpf.br
Fernando Raimundo Aranha Simão simao@cbpfsu1.cat.cbpf.br
Herman Julio Mosquera Cuesta hermanjc@cbpfsu1.cat.cbpf.br
Ignacio Hickman Bediaga bediaga@cbpf.br
Javier Magnin jmagnin@cbpf.br
Jussara Marques de Miranda jussara@cbpf.br
Sergio José Barbosa Duarte sbd@cbpf.br

UNIVERSIDADE DO ESTADO DO RIO DE JANEIRO

Alberto Lozéa Feijó Soares	albertoloze@bol.com.br
Jorge Eduardo Cieza Montalvo	cieza@uerj.br
José de Sá Borges F^o	saborges@uerj.br
Marcelo Chiapparini	chiappa@uerj.br
Mirian Enriqueta Bracco	bracco@uerj.br

OBSERVATÓRIO NACIONAL

Flávio Irineu Mendes Pereira	flavio@.on.br

UNIVERSIDADE FEDERAL DO RIO DE JANEIRO

Ana Gabriela Grunfeld	gabriela@if.ufrj.br
Bernardo Mattos Tavares	bernardo@if.ufrj.br
Carlos Eduardo Aguiar	carlos@if.ufrj.br
Clóvis José Wotzasek	clovis@if.ufrj.br
Erasmo Ferreira	erasmo@if.ufrj.br
Fernando Nicácio	nicacio@if.ufrj.br
Fernando Marroquim L. de Almeida Jr.	marroqui@if.ufrj.br
Graziela Barreiro Grise	graziel@if.ufrj.br
Hans-Thomas Elze	thomas@if.ufrj.br
Licinio Lima Silva Portugal	licinio@if.ufrj.br
Luiz Felipe Canto	canto@if.ufrj.br
Marcelo Pereira de Souza Pinheiro	pinheiro@if.ufrj.br
Maria Aline Barros do Vale	aline@if.ufrj.br
Paolo Lotti	lotti@if.ufrj.br
Rui Alberto dos Santos Nazareth	rui@if.ufrj.br
Stenio Wulck Alves de Melo	steniow@if.ufrj.br
Takeshi Kodama	tkodama@if.ufrj.br
Wagner Luiz Ferreira Marcolino	wlfm@if.ufrj.br
Yara do Amaral Coutinho	yara@if.ufrj.br

UNIVERSIDADE FEDERAL FLUMINENSE

João Batista da Silva	jbsilva@if.uff.br
Manuel M. Bastos Malheiro de Oliveira	mane@if.uff.br

UNIVERSIDADE FEDERAL DE MATO GROSSO DO SUL

Isabela Porto Cavalcante ipc@dfi@ufms.br

INSTITUTO TECNOLÓGICO DA AERONÁUTICA

Tobias Frederico tobias@fis.ita.br
Wilson R. Barbosa de Araújo wbaraujo@fis.ita.cta.br

UNIVERSIDADE ESTADUAL DE CAMPINAS

Álvaro Favinha Martini martini@ifi.unicamp.br
Emerson G. de Souza Luna luna@ifi.unicamp.br
Márcio José Menon menon@ifi.unicamp.br
Regina Fonseca Ávila rfa@ifi.unicamp.br

UNIVERSIDADE ESTADUAL PAULISTA, CAMPUS DE GUARATINGUETÁ

Álvaro de Souza Dutra dutra@feg.unesp.br
Antonio Soares de Castro castro@feg.unesp.br
Marcelo Batista Hott hott@feg.unesp.br

UNIVERSIDADE ESTADUAL PAULISTA, CAMPUS DE RIO CLARO

Luiz Antonio Barreiro barreiro@rc.unesp.br

UNIVERSIDADE ESTADUAL PAULISTA, INSTITUTO DE FÍSICA TEÓRICA

Alexandre Alves aalves@ift.unesp.br
Antonio Mihara mihara@ift.unesp.br
Bruto Max Pimentel Escobar pimentel@ift.unesp.br
Gastão Inácio Krein gkrein@ift.unesp.br
João Pacheco B. C. de Melo pacheco@ift.unesp.br
Joel César Pupin pupin@femanet.com.br
Jorge Henrique de Oliveira Sales henrique@ift.unesp.br

Lauro Tomio	tomio@ift.unesp.br
Leandro Batista Holanda	holanda@ift.unesp.br
Patrícia Soares Morales	patricia@ift.unesp.br
Prafulla Kumar Panda	panda@ift.unesp.br
Raquel Santos M. de Carvalho	raquel@ift.unesp.br
Sérgio Marcolino Antunes	antunes@ift.unesp.br
Victor Enrique Vizcarra Ruiz	indhy@ift.unesp.br
Wanderson Gonçalves Wanzeller	wanzeler@ift.unesp.br

UNIVERSIDADE DE SÃO PAULO, CAMPUS DE SÃO CARLOS

Attilio Cucchieri	attilio@if.sc.usp.br
Tereza Mendes	mendes@if.sc.usp.br

UNIVERSIDADE DE SÃO PAULO

Fábio Braghin	braghin@if.usp.br
Fernando Navarra	navarra@axpfep1.if.usp.br
Maria Simone Kugeratski Souza	msks@axpfep1.if.usp.br
Marina Nielsen	mnielsen@axpfep1.if.usp.br
Ricardo D'Elia Matheus	ricardo-matheus@hotmail.com
Rômulo Rodrigues da Silva	romulo@if.usp.br

UNIVERSIDADE DE TAUBATÉ

Everton Murilo C. de Abreu	everton@feg.unesp.br

UNIVERSIDADE ESTADUAL DE PONTA GROSSA

Luis Augusto Trevisan	latrevis@uepg.br

UNIVERSIDADE ESTADUAL DE LONDRINA

Paulo Laerte Natti	plnatti@uel.br

UNIVERSIDADE NORTE DO PARANÁ

Érica Regina Takano Natti	erica.takano@uol.com.br
Marcelo Donizeti de Oliveira	marcelod.oliveira@prof.unopar.br

UNIVERSIDADE FEDERAL DE SANTA CATARINA

Débora Peres Menezes	debora@server.fsc.ufsc.br
Eduardo Veitenheimer Costa	veitenhe@if.ufrgs.br
Sidney dos Santos Avancini	fsc1sav@fsc.ufsc.br

UNIVERSIDADE FEDERAL DE SANTA MARIA

Orimar Antonio Battistel	orimar@ccne.ufsm.br

UNIVERSIDADE DO VALE DO RIO DOS SINOS

Bardo Ernst Josef Bodmann	bardo@exatas.unisinos.br
Sergio Mittmann dos Santos	mittmann@exatas.unisinos.br

UNIVERSIDADE FEDERAL DO RIO GRANDE DO SUL

Alberto Sperotto dos Santos Rocha	sperotto@if.ufrgs.br
Alexandre Mesquita	falcon@if.ufrgs.br
André Ribeiro Taurines	taurines@if.ufrgs.br
César Augusto Zen Vasconcellos	cesarzen@if.ufrgs.br
Cristiano Brenner Mariotto	mariotto@if.ufrgs.br
Eliane Angela Veit	eav@if.ufrgs.br
Fernando Gonçalves Pilotto	pilotto@if.ufrgs.br
Guilherme Frederico Marranghello	gfm@if.ufrgs.br
Horacio Alberto Dottori	dottori@if.ufrgs.br
Luiz Fernando Mackedanz	thunder@if.ufrgs.br
Magno Valério Trindade Machado	magnus@if.ufrgs.br
Marcos André Betemps	mandrebe@if.ufrgs.br
Maria Beatriz de Leone Gay	gay@if.ufrgs.br
Moisés Razeira	razeira@if.ufrgs.br
Pedro Castro M. X. de Mello e Silva	pcmxms@bol.com.br
Rafael Ban Jacobsen	jacobsen@mailbr.com.br
Sergio Souto Rocha	ssrocha@if.ufrgs.br
Theodor August Johannes Maris	maris@if.ufrgs.br
Victoria Elnecave Herscovitz	victoria@if.ufrgs.br

UNIVERSIDADE FEDERAL DE PELOTAS

Álvaro Leonardi Ayala F^o	ayala@ufpel.tche.br
Daniel Tavares da Silva	neodanielsil@bol.com.br
Diego Rossi Gratieri	gratieri@ufpel.tche.br
Dimiter Hadjimichef	dimiter@ufpel.tche.br
Karen Silva Torres	karentorres@bol.com.br
Mario Luiz Lopes da Silva	mariolls@ufpel.tche.br
Victor Paulo Barros Gonçalves	barros@ufpel.tche.br

FUNDAÇÃO UNIVERSIDADE DE RIO GRANDE

Fernando Kokubun	kokubun@einstein.sf.dfis.furg.br